Sustainable Food Fortification

Gaurav Sanghvi
Ashok Kumar Bishoyi
Sanket J. Joshi • Hesham A. El Enshasy
Editors

Sustainable Food Fortification

Biobased Approaches and Strategies

Springer

Editors
Gaurav Sanghvi
Department of Microbiology
Marwadi University
Rajkot, Gujarat, India

Ashok Kumar Bishoyi
Department of Microbiology
Marwadi University
Rajkot, Gujarat, India

Sanket J. Joshi
Amity Institute of Microbial Technology
Amity University Rajasthan
Jaipur, Rajasthan, India

Hesham A. El Enshasy
Chemical Engineering and Energy
Universiti Teknologi Malaysia
Johor Bahru, Malaysia

ISBN 978-981-95-1216-4 ISBN 978-981-95-1217-1 (eBook)
https://doi.org/10.1007/978-981-95-1217-1

This Springer imprint is published by the registered company Springer Nature Singapore Pte Ltd.
The registered company address is: 152 Beach Road, #21-01/04 Gateway East, Singapore 189721, Singapore

Dedicated to all Health conscious individuals and food researchers.

Preface

In recent years, the global food system has faced challenges associated with population growth, environmental degradation, and widespread nutritional deficiencies. Traditional food fortification methods, while beneficial, often fall short in terms of sustainability, accessibility, and long-term efficacy. With an increasing global emphasis on sustainable development and health equity, there is a pressing need for innovative, biologically grounded solutions that can address these shortcomings and help ensure a healthier future for all.

This book, *Sustainable Food Fortification: Biobased Approaches and Strategies*, aims to target that need. It brings together a collection of insightful and interdisciplinary chapters written by experts in Microbiology, Food Science, Biotechnology, Agriculture, and Public Health. These contributions highlight the role of eco-friendly, biobased approaches in enhancing the nutritional profile of food products. From microbial and plant-derived bioactive compounds to nano-enabled delivery systems and functional bakery innovations, the book provides a comprehensive exploration of novel techniques for fortifying food while minimizing environmental impact.

Our intent is not only to present the latest scientific advancements but also to contextualize their real-world applications for food producers, researchers, policy makers, and students. By bridging the gap between scientific innovation and societal benefit, this volume aspires to serve as both a reference and a roadmap for advancing sustainable food fortification practices. We hope this work will inspire further dialogue, research, and action in the quest for nutrition security across the globe.

We extend our deepest gratitude to all contributors and collaborators who helped shape this interdisciplinary endeavor. It is our hope that this book will support new innovations and novel information that positively impact food systems and human health for years to come.

Rajkot, Gujarat, India — Gaurav Sanghvi
Rajkot, Gujarat, India — Ashok Kumar Bishoyi
Jaipur, Rajasthan, India — Sanket J. Joshi
Johor Bahru, Malaysia — Hesham A. El Enshasy

Acknowledgments

We sincerely wish to thank Swati Sharma and Nandhini Viswanathan from Springer for their excellent guidance, support, and coordination of this fascinating project. We would like to thank all the contributors for sharing their knowledge and experiences in the form of contributed chapters. We greatly appreciate the support from all the reviewers for spending their valuable time reviewing the chapters of this book. We are also thankful to our family members, friends, and other academic colleagues who showed patience when we deprived them of our much-needed attention during the course of the book assignment. We would like to acknowledge the support received from the management and competent authority of Marwadi University (India), Amity University (India), and University Technology Malaysia (UTM). Last but not least, we are also thankful to Almighty God who has awarded us the wisdom to explore such novel thought and complete this book. We believe that despite our dedication, the first version always comes with some errors that may have crept into this compilation. We are happy to receive constructive feedback from readers to improve our future volumes.

Abbreviations

SDG	Sustainable Development Goals
GHO	Global Health Observatory
WHO	World Health Organization
FAO	Food and Drug Administration
FPP	Ferric pyrophosphate
FBG	Ferrous bis-glycinate
NaFeEDTA	Sodium iron ethylendiamine-tetraacetic acid
HDPE	High-density polyethylene
HPMC	Hydroxypropyl methylcellulose
MC	Methylcellulose
CMC	Carboxypropyl methylcellulose
AA	Ascorbic acid
DFC	Dietary fiber concentrate
FI	Functional food ingredient
CFU	Colony-forming units
EC	Edible coating
RDI	Recommended Daily Intake
GRDs	Gluten-related disorders
GF	Gluten-free
GFBs	GF breads
MNDs	Micronutrient deficiencies
DF	Dietary fiber
SV	Specific volume
GI	Glycemic index
RS	Resistant starch
DV	Daily value
MFA	Multiple factor analysis
LAB	Lactic acid bacteria
GFB-M	Millet
GFB-RB	Rice bran
GFB-T	Teff
GFB-B	Buckwheat
GFB-L	Lupin

GFB-Q	Quinoa
GFB-S	Sorghum
GFB-BR	Brown rice
GFB-C	Carob

Contents

About the Editors and Contributors

About the Editors

Gaurav Sanghvi Dr. Gaurav Sanghvi is presently working as Associate Professor at Department of Microbiology, Marwadi University. He has completed his PhD from The Maharaja Sayajirao University of Baroda, Vadodara. He also had experience as post-doctoral researcher for 3 years from Max Planck Institute of Developmental Biology, Tubingen, Germany. He has experience of more than 11 years in the fields of Microbiology, Enzymology, and Food Microbiology. As of now, he has published 40 peer-reviewed papers with an h-index of 8 and i10 index of 20. He also served as reviewer for more than 10 journals of Elsevier, Springer, Wiley, and Hindawi Publications. He was honored as the young scientist in the conference held at Nagaland, India, which was sponsored by CSIR, DRDO, and DST, Govt. of India. He has also presented his research work at international platforms (Germany, United Kingdom, and France).

Ashok Kumar Bishoyi Dr. Ashok Kumar Bishoyi is presently working as Associate Professor (Microbiology) at Marwadi University, Rajkot, Gujarat, India. Prior to academic carrier, he worked as scientist B at Gujarat Biotechnology Research Centre, Department of Science and Technology, Govt. of Gujarat. During his scientific carrier, he has executed five different research projects related to transcriptomics, genomics, plant tissue culture, animal tissue culture, virology, molecular marker development, etc. He has completed a PhD at ICAR—Directorate of Medicinal and Aromatic Plants Research & CHARUSAT, with doctorate in Biotechnology in the year 2017, and a master degree in Biotechnology at Berhampur University, Odisha. He has been selected/awarded ICAR-SRF 2009, ICAR-SRF 2013, ICAR-NET 2014, and ICAR-NET 2015. He has published more than 30 research publications/books/book chapters in different reputed peer-reviewed international/national journals for his research credit. He is an active reviewer of many Springer, Elsevier, and Wiley research journals.

Sanket J. Joshi Dr. Sanket Joshi is currently serving as a Professor and Deputy Director, at Amity Institute of Microbial Technology, Amity University Rajasthan,

Jaipur, India. He served as Deputy Director, Oil & Gas Research Center (2018–2023), and an Application Specialist, at Central Analytical and Applied Research Unit, Sultan Qaboos University, Oman (2013–2023). He holds BSc and MSc degrees from Sardar Patel University, India, and a PhD degree from M. S. University of Baroda, India—all in Microbiology. He has 19 years of academic teaching and research experience, and 4 years of industrial R&D experience, in India and Oman. While working in Indian pharma companies, he undertook several turnkey projects (both upstream and downstream) on antibiotics (β-lactams, macrolides, glycopeptides), antidiabetic drug (for type II diabetes), and co-enzyme Q10. His current research interests encompass: energy (in-situ/ex-situ microbial enhanced light/heavy oil recovery; chemical enhanced oil recovery; biofuels); microbial products (biosurfactants, biopolymers, R&D and scale-up); and environmental bioremediation (crude oil pollution; analysis, mitigation, and control of souring by sulfate-reducing bacteria; HPAM contaminated sites). Those research projects are multidisciplinary, involving both biotechnology and engineering disciplines. Since he joined SQU, his research team has received over 5.44 million USD in research grants and service contracts and established bioproducts for petroleum and environmental applications. He has 140 scientific publications as papers in international journals (100), book chapters (24), international books (15), Australian Innovation Patent (1), and conference abstracts (70), to his credit. He serves as an Academic/Associate Editor for some of the highly reputed journals: *Frontiers in Microbiology*, *PeerJ*, *BMC Biotechnology*, *Ecotoxicology*, *Petroleum Science and Technology*, *Heliyon*, and *Open Biotechnology Journal*. He is book series editor of Advances in Biotechnology and Bioengineering, Elsevier INC.; guest editor for *Frontiers in Microbiology*, *Energies*, *Sustainability*, *Scientifica*, *Biocatalysis and Agricultural Biotechnology*, *Applied Biochemistry and Biotechnology*, *Environmental Quality Management*, and *Open Biotechnology Journal*; and a recognized reviewer for more than 51 international journals (Elsevier, Springer Nature, T&F, Wiley/AOCS, RSC, ACS, Hindawi, DE Gruyter, MDPI, and others). He served as an intermediate guide for 30 undergraduate/graduate students and 3 PhD students, in India and in Oman, and an active member of MEOR research team at SQU since 2009; he established a state-of-the-art research and service laboratory at SQU. He was awarded as "NRI Senior Scientist Award," from Microbiologists Society, India, during academic year, 2019–2020. He has 5000 citations, *h-index* of 33, and *i10* index of 74 (Google Scholar), and was listed among the 2% of the "Stanford University – Elsevier" list of the highly cited scientists in the world for 2022 and 2023, for excellence in scientific research in "Energy" and "Biotechnology" disciplines.

Hesham A. El Enshasy Prof. Hesham Ali El Enshasy is Professor of bioprocess engineering, Faculty of Chemical Engineering and Energy, and Director for Innovation and Products Development, Institute of Bioproducts Development, University Technology Malaysia (UTM). He is also the current director for Food, Pharmaceutical& Bio-Engineering Division, Division 15A (Food Director), American Institute of Chemical Engineering (AIChE). He is active as editorial board member in many scientific journals, and he is currently the Editor-in-Chief for *International Journal of Biotechnology for Wellness Industries* (IJBWI). He is also the co-editor for the book entitled *Bioprocessing Technologies in Biorefinery for Sustainable Production of Fuels, Chemicals, and Polymers*, published by John Wiley and Sons, 2013. He has two technology patents and more than 150 publications in peer-reviewed international journals, book chapters, and books. He was also invited as keynote, plenary, and guest speaker in more than 50 international conferences in the field of industrial biotechnology. He is also active as advisory board member, scientific and bio-business consultant for biotechnology companies in Egypt, Belgium, Greece, the USA, China, and Malaysia for the important three Ts (Technology Transfer, Technology Implementation and Technology Development) as well as process and platform designer for Biotechnology industries.

Contributors

Anuska Adhikari National Food Research Centre, Nepal Agricultural Research Council, Khumaltar, Nepal

Sarra Akermi Laboratory of Microbial Biotechnology, Enzymatic and Biomolecules, Centre of Biotechnology of Sfax, Sfax, Tunisia

Jerold C. Alcantra Kasiska Division of Health Sciences, Department of Medical Laboratory Science, College of Health, Idaho State University, Meridian, ID, USA

Fakhreddine Ben Amara Laboratory of Microbial Biotechnology, Enzymatic and Biomolecules, Centre of Biotechnology of Sfax, Sfax, Tunisia

Syed Amir Ashraf Department of Clinical Nutrition, College of Applied Medical Sciences, University of Ha'il, Ha'il, Saudi Arabia

Grishma Baidhya National Food Research Centre, Nepal Agricultural Research Council, Khumaltar, Nepal

Yashvi Baldhiya Faculty of Science, Department of Microbiology, Marwadi University, Rajkot, Gujarat, India

Humera Banu Department of Clinical Nutrition, College of Applied Medical Sciences, University of Ha'il, Ha'il, Saudi Arabia

Arusha Bashyal National Food Research Centre, Nepal Agricultural Research Council, Khumaltar, Nepal

Sudiksha Bhattarai National Food Research Centre, Nepal Agricultural Research Council, Khumaltar, Nepal

Vaibhav D. Bhatt School of Applied Sciences and Technology, Gujarat Technological University, Ahmedabad, Gujarat, India

Ashok Kumar Bishoyi Department of Microbiology, Marwadi University, Rajkot, Gujarat, India

Camila Carmona Facultad de Bromatología, UNER, Gualeguaychú, Entre Ríos, Argentina

Adriana P. Castellanos-Fuentes Universidad de Buenos Aires (UBA), Facultad de Ciencias Exactas y Naturales (FCEN), Departamento de Industrias, Ciudad Autónoma de Buenos Aires, Argentina

CONICET – Universidad de Buenos Aires, Instituto de Tecnología de Alimentos y Procesos Químicos (ITAPROQ), Ciudad Autónoma de Buenos Aires, Argentina

Esparza-González Sandra Cecilia School of Chemistry, Autonomous University of Coahuila, Saltillo, Coahuila, Mexico

Moufida Chaari Laboratory of Microbial Biotechnology, Enzymatic and Biomolecules, Centre of Biotechnology of Sfax, Sfax, Tunisia

Sara Chadorshabi Student Research Committee, Faculty of Nutrition and Food Science, Department of Food Science and Technology, Tabriz University of Medical Sciences, Tabriz, Iran

Nentia Chib Division of Plant Pathology, FoA, Sher-E-Kashmir University of Agriculture Science and Technology, Jammu, Jammu and Kashmir, India

Gift Chisapo Department of Agriculture and Food Systems, Natural Resources College, Lilongwe University of Agriculture and Natural Resources, Lilongwe, Malawi

Nery-Flores Sendar Daniel School of Chemistry, Autonomous University of Coahuila, Saltillo, Coahuila, Mexico

Soriano-Melgar Lluvia de Abril Alexandra School of Chemistry, Autonomous University of Coahuila, Saltillo, Coahuila, Mexico

Marina F. de Escalada Pla Facultad de Ciencias Exactas y Naturales (FCEN), Departamento de Industrias, Universidad de Buenos Aires (UBA), Buenos Aires, Argentina

Instituto de Tecnología de Alimentos y Procesos Químicos (ITAPROQ), CONICET – Universidad de Buenos Aires, Buenos Aires, Argentina

Sabhya Dhakal National Food Research Centre, Nepal Agricultural Research Council, Khumaltar, Nepal

Sandip P. Dholakia Department of Pharmaceutical Chemistry and Quality Assurance, L. M. College of Pharmacy, Ahmedabad, Gujarat, India

Salazar-Villa Edith School of Biology, Autonomous University of Sinaloa, Culiacán, Sinaloa, Mexico

Khaoula Elhadef Laboratory of Microbial Biotechnology, Enzymatic and Biomolecules, Centre of Biotechnology of Sfax, Sfax, Tunisia

Nancy N. Espósito Instituto de Ciencia y Tecnología de los Alimentos de Entre Ríos (ICTAER), CONICET-Universidad Nacional de Entre Ríos (UNER), Gualeguaychú, Entre Ríos, Argentina

Facultad de Bromatología, UNER, Gualeguaychú, Entre Ríos, Argentina

Consejo Nacional de Investigaciones Científicas y Técnicas (CONICET), Ciudad Autónoma de Buenos Aires, Argentina

Silvia K. Flores Universidad de Buenos Aires (UBA), Facultad de Ciencias Exactas y Naturales (FCEN), Departamento de Industrias, Ciudad Autónoma de Buenos Aires, Argentina

CONICET – Universidad de Buenos Aires, Instituto de Tecnología de Alimentos y Procesos Químicos (ITAPROQ), Ciudad Autónoma de Buenos Aires, Argentina

Carolina E. Genevois Instituto de Ciencia y Tecnología de los Alimentos de Entre Ríos (ICTAER), CONICET-UNER, Gualeguaychú, Entre Ríos, Argentina

Facultad de Bromatología, UNER, Gualeguaychú, Entre Ríos, Argentina

Consejo Nacional de Investigaciones Científicas y Técnicas (CONICET), Ciudad Autónoma de Buenos Aires, Argentina

Mitaliba Gohil Faculty of Science, Department of Microbiology, Marwadi University, Rajkot, Gujarat, India

Nikitasinh Gohil Faculty of Science, Department of Microbiology, Marwadi University, Rajkot, Gujarat, India

Campos-Múzquiz Lizeth Guadalupe School of Chemistry, Autonomous University of Coahuila, Saltillo, Coahuila, Mexico

Boutheina Ben Hadj Hmida Laboratory of Biochemistry and Enzymatic Engineering of Lipases, National Engineering School of Sfax (ENIS), Sfax, Tunisia

Nosyba A. Ibrahim Department of Public Health, College of Public Health & Health Informatics, University of Ha'il, Ha'il, Saudi Arabia

Karen F. Irigoytia Instituto de Ciencia y Tecnología de los Alimentos de Entre Ríos (ICTAER), CONICET-Universidad Nacional de Entre Ríos (UNER), Gualeguaychú, Entre Ríos, Argentina

Facultad de Bromatología, UNER, Gualeguaychú, Entre Ríos, Argentina

Consejo Nacional de Investigaciones Científicas y Técnicas (CONICET), Ciudad Autónoma de Buenos Aires, Argentina

Nyadani Jali Department of Agriculture and Food Systems, Natural Resources College, Lilongwe University of Agriculture and Natural Resources, Lilongwe, Malawi

Simran Jha National Food Research Centre, Nepal Agricultural Research Council, Khumaltar, Nepal

Sunayana Jonchhe National Food Research Centre, Nepal Agricultural Research Council, Khumaltar, Nepal

Kondwani Kammwamba Department of Agriculture and Food Systems, Natural Resources College, Lilongwe University of Agriculture and Natural Resources, Lilongwe, Malawi

Kabina Karki National Food Research Centre, Nepal Agricultural Research Council, Khumaltar, Nepal

Roman Karki National Food Research Centre, Nepal Agricultural Research Council, Khumaltar, Nepal

Mukesh N. Kher Department of Pharmaceutical Chemistry and Quality Assurance, L. M. College of Pharmacy, Ahmedabad, Gujarat, India

Prajakta Kulkarni Department of Botany, PDEA'S Waghire College, Pune, Maharashtra, India

VijayaDurga V. V. Lekkala Department of Genetics and Genomics, Yogi Vemana University, Kadapa, Andhra Pradesh, India

Palomo-Ligas Lissethe School of Chemistry, Autonomous University of Coahuila, Saltillo, Coahuila, Mexico

Dakshayani Lomada Department of Genetics and Genomics, Yogi Vemana University, Kadapa, Andhra Pradesh, India

Maryam Mahmoudzadeh Faculty of Nutrition and Food Science, Department of Food Science and Technology, Tabriz University of Medical Sciences, Tabriz, Iran

Sofia Marchioni Universidad de Buenos Aires (UBA), Facultad de Ciencias Exactas y Naturales (FCEN), Departamento de Industrias, Ciudad Autónoma de Buenos Aires, Argentina

CONICET – Universidad de Buenos Aires, Instituto de Tecnología de Alimentos y Procesos Químicos (ITAPROQ), Ciudad Autónoma de Buenos Aires, Argentina

Alicia Matyatz Universidad de Buenos Aires (UBA), Facultad de Ciencias Exactas y Naturales (FCEN), Departamento de Industrias, Ciudad Autónoma de Buenos Aires, Argentina

Mansi Mekhiya Faculty of Science, Department of Microbiology, Marwadi University, Rajkot, Gujarat, India

Lotfi Mellouli Laboratory of Microbial Biotechnology, Enzymatic and Biomolecules, Centre of Biotechnology of Sfax, Sfax, Tunisia

Demian A. Molina Consejo Nacional de Investigaciones Científicas y Técnicas (CONICET), Argentina

Facultad de Ciencias Exactas y Naturales (FCEN), Departamento de Industrias, Universidad de Buenos Aires (UBA), Buenos Aires, Argentina

Instituto de Tecnología de Alimentos y Procesos Químicos (ITAPROQ), CONICET – Universidad de Buenos Aires, Buenos Aires, Argentina

Martha Monzerrath Orozco-Sifuentes School of Chemistry, Autonomous University of Coahuila, Saltillo, Coahuila, Mexico

Divya Mori Faculty of Science, Department of Microbiology, Marwadi University, Rajkot, Gujarat, India

Justice Munthali Alliance of Bioversity International & CIAT, Malawi, Agricultural Research Station, Lilongwe, Malawi

Sydney Namaumbo Department of Land and Water Resources, Natural Resources College, Lilongwe University of Agriculture and Natural Resources, Lilongwe, Malawi

Patrick Ndovie Department of Human Nutrition and Health, Lilongwe University of Agriculture and Natural Resources, Lilongwe, Malawi

Department of Agriculture and Food Systems, Natural Resources College, Lilongwe University of Agriculture and Natural Resources, Lilongwe, Malawi

Smith G. Nkhata Department of Agriculture and Food Systems, Natural Resources College, Lilongwe University of Agriculture and Natural Resources, Lilongwe, Malawi

Pravin Ojha National Food Research Centre, Nepal Agricultural Research Council, Khumaltar, Nepal

College of Food Science and Technology, Hebei Agricultural University, Baoding, China

Ranju Parajuli National Food Research Centre, Nepal Agricultural Research Council, Khumaltar, Nepal

Prachi Patel Department of Botany, The Maharaja Sayajirao University of Baroda, Vadodara, Gujarat, India

Vidya Patil-Patankar Department of Botany, PDEA'S Waghire College, Pune, Maharashtra, India

Mayuri Pawar Department of Botany, PDEA'S Waghire College, Pune, Maharashtra, India

Dewaki Rai National Food Research Centre, Nepal Agricultural Research Council, Khumaltar, Nepal

T. Sri Ranjani Department of Zoology, D.K. Government College for Women(A), Nellore, Andhra Pradesh, India

Rodríguez-Herrera Raúl School of Chemistry, Autonomous University of Coahuila, Saltillo, Coahuila, Mexico

L. V. Ravishankar Division of Plant Pathology, FoA, Sher-E-Kashmir University of Agriculture Science and Technology, Jammu, Jammu and Kashmir, India

Madhava C. Reddy Department of Biotechnology and Bioinformatics, Yogi Vemana University, Kadapa, Andhra Pradesh, India

Aadisha Saini Department of Food Science and Technology, National Institute of Food Technology Entrepreneurship and Management, Sonipat, Haryana, India

Lizzie Saka Department of Agriculture and Food Systems, Natural Resources College, Lilongwe University of Agriculture and Natural Resources, Lilongwe, Malawi

Jesús Andrés Salas-Tovar School of Chemistry, Autonomous University of Coahuila, Saltillo, Coahuila, Mexico

Gaurav Sanghvi Department of Microbiology, Marwadi University, Rajkot, Gujarat, India

Marion Sanuka Department of Agriculture and Food Systems, Natural Resources College, Lilongwe University of Agriculture and Natural Resources, Lilongwe, Malawi

Ritabrata Sarkar Department of Genetics and Plant Breeding, ICAR-Indian Institute of Agricultural Biotechnology, Ranchi, Jharkhand, India

Vidhi V. Shah Department of Pharmaceutics, L. M. College of Pharmacy, Ahmedabad, Gujarat, India

Noelia E. Silva Universidad de Buenos Aires (UBA), Facultad de Ciencias Exactas y Naturales (FCEN), Departamento de Industrias, Ciudad Autónoma de Buenos Aires, Argentina

CONICET – Universidad de Buenos Aires, Instituto de Tecnología de Alimentos y Procesos Químicos (ITAPROQ), Ciudad Autónoma de Buenos Aires, Argentina

Slim Smaoui Laboratory of Microbial Biotechnology, Enzymatic and Biomolecules, Centre of Biotechnology of Sfax, Sfax, Tunisia

Ashish Warghane School of Applied Sciences and Technology, Gujarat Technological University, Ahmedabad, Gujarat, India

Dhara Zaveri Department of Botany, The Maharaja Sayajirao University of Baroda, Vadodara, Gujarat, India

Biobased Fortification: Current Challenges and Perspectives

1

Roman Karki, Kabina Karki, Arusha Bashyal, Ranju Parajuli, Dewaki Rai, Anuska Adhikari, Simran Jha, Sudiksha Bhattarai, Sabhya Dhakal, Sunayana Jonchhe, Grishma Baidhya, and Pravin Ojha

Abstract

Micronutrient malnutrition remains a widespread global health concern, especially in low- and middle-income countries. Biofortification, defined as the enhancement of micronutrient content and bioavailability in food through conventional breeding, agronomic practices, or modern biotechnological approaches, offers a sustainable strategy to combat hidden hunger. This chapter critically evaluates recent progress in the biofortification of crops. Significant advancements have been achieved in enriching staple crops such as rice, wheat, maize, pearl millet, legumes, sweet potatoes, cassava, and banana with essential micronutrients like iron, zinc, and provitamin A through traditional breeding and genetic engineering. Newer techniques, including genome editing, multi-nutrient transgenic events, and CRISPR-Cas9, are expanding the scope of crop biofortification. Strategies such as dietary modification, genetic selection, and nano-enabled interventions are gaining interest for enhancing micronutrient levels. This chapter also discusses key challenges, including inconsistent nutrient bioavailability, antinutritional compounds, nutrient stability during processing, and consumer acceptance. Regulatory barriers and public perception still influence the wider adoption and scaling of biofortified products. Biofortification is not a panacea but a component of a broader strategy to improve nutritional outcomes. To maximize its effectiveness, future research must prioritize improving

R. Karki (✉) · K. Karki · A. Bashyal · R. Parajuli · D. Rai · A. Adhikari · S. Jha · S. Bhattarai · S. Dhakal · S. Jonchhe · G. Baidhya
National Food Research Centre, Nepal Agricultural Research Council, Khumaltar, Nepal
e-mail: rmn_karki@narc.gov; parajuliranju84@gail.com

P. Ojha
National Food Research Centre, Nepal Agricultural Research Council, Khumaltar, Nepal

College of Food Science and Technology, Hebei Agricultural University, Baoding, China

G. Sanghvi et al. (eds.), *Sustainable Food Fortification*,
https://doi.org/10.1007/978-981-95-1217-1_1

bioavailability, confirming long-term health benefits, and integrating biofortification into national nutrition and agricultural frameworks. A multidisciplinary effort involving breeders, food scientists, nutritionists, agronomists, biotechnologists, and policymakers is vital to fully harness biofortification's potential in addressing global micronutrient deficiencies and achieving Sustainable Development Goals.

Keywords

Biofortification · Micronutrient deficiency · Agronomic practices · Conventional breeding · Nanotechnology in agriculture · CRISPR-Cas9 gene editing

1.1 Introduction

The human body requires minute quantities of vitamins and minerals, known as micronutrients, for growth, brain development, and immunity. They must be supplied through diet or an external source, as the human body cannot synthesize them. Inadequate intake or absorption of these micronutrients, also known as "hidden hunger," results in learning disabilities, impaired work capacity, illness, and even mortality. This "hidden hunger" affects over 2 billion people globally and remains a critical public health challenge (Nestel et al., 2006; Sharma et al., 2017; Garcia-Casal et al., 2017), with zinc, iron, and vitamin A insufficiencies disproportionately impacting low- and middle-income countries (LMICs) (Cakmak & Kutman, 2018). Years of program experience have identified several solutions to hidden hunger that are not mutually exclusive but complement one another. Solutions include ingestion of oral supplements (tablets, capsules, and syrups), public health measures, food fortification, and other food-based approaches, such as biofortification. Supplementation with high doses of micronutrients is a short-term measure to address severe deficiency, and its use is envisaged to decrease as longer-term measures are established. Food fortification is a medium- to long-term solution for alleviating specific nutrient deficiencies in a population. Another key long-term method in biobased fortification is biofortification, which aims to enhance the nutrient levels in crops during their growth rather than through postharvest processing. This method can be achieved through agronomic practices, conventional plant breeding, or modern biotechnology (Sharma et al., 2017; Dhaliwal et al., 2022b). Unlike conventional fortification, which adds nutrients during food processing, biofortification aims to enhance nutrient levels during the growth of the plant itself. This approach is particularly beneficial for populations where supplementation and conventional fortification might be challenging to implement (Talsma & Pachón, 2017).

Plants can be biofortified through several techniques, including agronomic and microbial methods, selective breeding, gene editing, genetic engineering, and fermentation (Sharma et al., 2017; Dhaliwal et al., 2022b). Additionally, controlled environment agriculture and the reformulation of existing food products are also employed to increase nutrient levels. Biofortification not only focuses on enhancing the levels of essential nutrients but also on improving their bioavailability, making them easier for the body to absorb and use. Moreover, zinc-fertilized forage crops improve livestock nutrition, indirectly benefiting human health (Dhaliwal et al., 2022b).

The concept of biofortification gained traction during the Green Revolution when it became evident that higher productivity of staple crops did not necessarily translate to improved nutritional intake, especially in developing countries (Talsma & Pachón, 2017). Initially, biofortification efforts focused on enriching crops with a limited range of vitamins and minerals such as pro-vitamin A carotenoids, B1, B6, iron, zinc, and selenium. Over time, the scope has expanded to include essential amino acids, health-promoting fatty acids, beneficial phytonutrients, and a broader spectrum of vitamins and minerals (Talsma & Pachón, 2017). Biofortification projects have targeted a variety of crops, enhancing them with those micronutrients. For instance, efforts have been made to biofortify rice, beans, sweet potato, cassava, wheat, maize, and sorghum (Andersson et al., 2017; HarvestPlus, n.d.). Instances of successful biofortification initiatives include the augmentation of iron content in crops such as rice, beans, and sweet potatoes (Andersson et al., 2017).

These initiatives aim to provide a cost-effective strategy to improve public health by enabling consumers to increase their mineral and vitamin intake without additional expense (Global Food and Nutrition Security, 2024). Moreover, biofortification has the advantage of reaching populations where traditional fortification methods may be challenging to implement, making it a vital strategy for addressing micronutrient deficiencies in developing regions (Global Food and Nutrition Security, 2024). In 2008, biofortification was ranked fifth among the most cost-effective solutions to address global challenges, including hidden hunger, by a panel of economists that included five Nobel Laureates (Andersson et al., 2017). The 2016 World Food Prize further highlighted the importance of biofortification.

After two decades of investment, biofortified crops are now grown in over 40 countries and consumed by hundreds of millions of people (Global Food and Nutrition Security, 2024). The efficacy of these nutrient-rich crops has been validated through numerous published nutrition trials, positioning biofortification as a recognized strategy within the international nutrition community to combat micronutrient deficiencies (Global Food and Nutrition Security, 2024). While conventional interventions like supplementation and industrial fortification are effective, they often fail to reach rural and resource-constrained populations (Bouis & Welch, 2010). Biofortification presents a cost-effective and sustainable alternative that aligns with existing agricultural systems and dietary traditions. This approach offers several advantages, as it necessitates only a one-time investment in crop development, thereby eliminating the recurring expenses associated with industrial fortification. For instance, provitamin A-enriched maize can fulfill 50% of the daily vitamin A requirements without necessitating additional infrastructure, effectively reducing malnutrition at half the cost of supplementation programs (Nestel et al., 2006; Hulshof et al., 2007). For example, biofortified maize varieties in Zambia have reached 500,000 farming households, reducing vitamin A deficiency by 30% (Finkelstein et al., 2019). Biofortified orange-fleshed sweet potatoes reduced vitamin A deficiency by 14–20% in Ugandan children (Hotz et al., 2012). Genetically engineered Golden Rice with β-carotene addresses the vitamin A deficiency in Asia; trials in the Philippines demonstrated a 60% increase in serum retinol levels (Steiger et al., 2014). Iron-fortified pearl millet reversed anaemia in 80% of Indian adolescents within 6 months (Shubham et al., 2020). Zinc-biofortified wheat has been scaled to 50,000 farming households in Pakistan, increasing zinc intake by 40% (Tablante et al., 2019).

Moreover, biofortified crops are well adopted by farmers and also preferred by consumers; high-iron pearl millet was adopted by 60% of Indian farmers within 5 years due to comparable yields and market premiums (Finkelstein et al., 2019). In Uganda, orange-fleshed sweet potatoes gained acceptance through community-led campaigns, doubling vitamin A intake (Hotz et al., 2012). Similarly, Biofortification reduces reliance on synthetic supplements and industrial fortification, lowering carbon footprints (Lowe, 2025). For example, iron-biofortified beans in Rwanda reduced fertilizer use by 25% while improving soil health (Díaz-Gómez et al., 2017). Despite advancements, biofortification programs face implementation hurdles, including scientific, socioeconomic, and policy-related barriers that limit scalability and impact and policy gaps (Geny et al., 2023). Antinutrients like phytic acid reduce mineral bioavailability, necessitating complementary processing strategies (Díaz-Gómez et al., 2017).

However, regulatory delays and public skepticism hinder the adoption of genetically modified (GM) biofortified crops (Dubock, 2017), while soil health variability limits agronomic efficiency. Furthermore, socioeconomic barriers, including weak market linkages and cultural preferences for crop traits, affect farmer adoption (Siwela et al., 2020). Thus, there is a need to critically examine the current biofortification strategies, innovations, their efficacy, challenges, and propose actionable strategies to advance biofortification's future directions, emphasizing innovations that bridge laboratory research and real-world application. This chapter synthesizes evidence from several peer-reviewed studies to evaluate the efficacy, challenges, and prospects of biofortifications unresolved issues, including bioavailability constraints and scalability gaps, and proposes solutions such as Clustered Regularly Interspaced Short Palindromic Repeats (CRISPR)-driven multi-nutrient biofortification (a gene-editing technology that functions like a precise genetic scalpel. It's a simplified version of a bacterial immune system, using a guide RNA to direct a DNA-cutting enzyme, most commonly Cas9, to a specific location in DNA to make a targeted edit), and climate-resilient crop varieties. By integrating interdisciplinary insights, this work aims to guide policymakers, breeders, and nutritionists in optimizing biofortification's impact on global food systems.

1.2 Foundational Studies and Global Initiatives

The conceptual and operational foundations of biofortification were established in the early 2000s through pioneering work by Bouis (2002, 2003); Bouis and Welch (2010) and Bouis et al. (2011) and Pfeiffer and Mcclafferty (2007), who positioned biofortification as a cost-effective, scalable intervention to address micronutrient deficiencies in resource-limited settings. Bouis (2003) argued that biofortification uniquely targets the rural population, groups often excluded from urban-centric fortification and supplementation programs, by integrating nutrient enrichment directly into staple crops grown and consumed by smallholder farmers. This approach capitalizes on the self-replicating nature of seeds, ensuring sustained benefits across harvest cycles without recurring costs (Bouis et al., 2011). Early studies focused on three micronutrients with the highest deficiency burdens: iron (Fe), zinc (Zn), and vitamin A, prioritizing crops such as rice, wheat, maize, and beans due to their dietary dominance in LMICs (Pfeiffer & Mcclafferty, 2007).

The launch of HarvestPlus in 2003, under the CGIAR system, marked a watershed moment in translating theoretical frameworks into actionable programs. HarvestPlus adopted a three-phase model: (a) nutrient discovery and breeding, (b) efficacy trials, and (c) dissemination (Andersson et al., 2017). By 2017, the program had released 150 biofortified varieties across 26 countries, reaching 50 million people. Notable successes include iron-biofortified beans in Rwanda (30% higher Fe content) and zinc-enriched rice in Bangladesh (40% Zn increase), which reduced anemia and improved immune function in target populations (Andersson et al., 2017; HarvestPlus, n.d.). However, the reliance on conventional breeding limited micronutrient gains compared to transgenic methods. For example, provitamin A levels in conventionally bred maize reached 8–10 μg/g, whereas transgenic approaches achieved 15–20 μg/g (Nuss & Tanumihardjo, 2010).

The evolution of HarvestPlus also revealed gaps in long-term impact assessments. Early studies focused on short-term metrics like crop yield and micronutrient content, neglecting longitudinal data on health outcomes (e.g., stunting reduction) and sustainability (e.g., farmer retention rates) (Bouis & Welch, 2010). Furthermore, the program's initial exclusion of genetically modified (GM) crops, driven by regulatory and public acceptance concerns, delayed the adoption of high-impact solutions like Golden Rice, which could have accelerated progress in vitamin A-deficient regions (Dubock, 2017).

By the 2010s, biofortification gained traction in global policy agendas. The 2014 Global Hunger Index highlighted biofortification as a key tool for achieving the UN Sustainable Development Goals (SDGs), prompting partnerships with governments and NGOs (von Klaus et al., 2014). For example, Nigeria integrated biofortified cassava into its National Agricultural Seed Policy, while Brazil mandated Zn-biofortified wheat in school feeding programs (Garcia-Casal et al., 2017). These initiatives, however, faced uneven implementation due to fragmented funding and a lack of infrastructure for seed distribution, particularly in conflict-affected regions.

While foundational studies emphasized technical feasibility and cost-effectiveness, they often underestimated sociocultural and logistical barriers. For instance, orange-fleshed sweet potatoes (OFSP), biofortified with β-carotene, faced low adoption in Uganda and Mozambique due to consumer preferences for white-fleshed varieties and limited awareness of nutritional benefits (Siwela et al., 2020). Similarly, zinc-biofortified wheat in Pakistan encountered resistance from farmers accustomed to traditional varieties, despite proven agronomic resilience (Andersson et al., 2017). These examples underscore the necessity of embedding participatory approaches, engaging farmers and consumers in crop selection and trait prioritization, into biofortification programs. However, the foundational phase established biofortification as a viable strategy but exposed a disconnect between research and real-world complexity. Future programs must balance technical innovation with sociocultural adaptability, ensuring biofortified crops align with local dietary habits, agroecological conditions, and market dynamics.

1.3 Strategies for Leveraging International Research Partnerships for Biofortification

One of the primary strategies to leverage international research partnerships for biofortification involves fostering collaborative frameworks between global research institutions and local agricultural programs. A significant example of this is the Biofortification Challenge Program, initiated in 2003 under the management of the International Food Policy Research Institute (IFPRI) and the International Center for Tropical Agriculture or *Centro Internacional de Agricultura Tropical* (CIAT), which was later renamed HarvestPlus. This program successfully bridged the gap between agriculture and nutrition through biofortification, a process of breeding added nutritional value into staple crops like rice and maize, which are commonly consumed in developing countries (CGIAR, 2003).

These partnerships facilitate the exchange of knowledge and technology, enabling the adaptation of biofortification strategies to local contexts. For instance, the International Maize and Wheat Improvement Center (CIMMYT) in Mexico and HarvestPlus in Washington, D.C., have played pivotal roles in advancing preharvest and postharvest biofortification technologies (Listman et al., 2019). By working closely with local institutions, these organizations ensure that biofortified crops are not only developed but also successfully disseminated and adopted by smallholder farming communities.

Furthermore, addressing micronutrient deficiency, a significant impediment to achieving Sustainable Development Goal 2 (zero hunger), necessitates the continuous evaluation and adaptation of biofortification practices. This includes exploring new technological options such as crop genome editing and nanotechnology to enhance the efficacy of biofortification. International partnerships must also adopt a responsible research and innovation approach, considering local consumer preferences and needs. Policy recommendations, such as those proposed in the G20 brief, emphasize the importance of committing to international initiatives that support the localization of biofortified crops to ensure they are both effective and culturally accepted (Global Food and Nutrition Security, 2023).

Local agricultural programs have also seen significant advancements through the integration of biofortification practices. One notable example of successful biofortification initiatives can be observed in Nigeria and Tanzania. Through a multi-stakeholder partnership led by the International Potato Center (CIP), biofortification has been prioritized in 11 national and regional policy and strategy documents. This partnership has mobilized approximately $6.5 million, influencing seven programs and building the capacities of over 40 agencies. As a result, four learning toolkits were developed, training 11,433 change agents, including 5976 females, who now lead gender-sensitive projects using a nutritious food-basket approach. In these regions, almost 1 million households are growing seven fast-tracked biofortified crop varieties (Paredes, 2019).

Moreover, training programs for farmers in LMICs have been pivotal for the success of biofortification. These programs not only educate farmers on the benefits and techniques of cultivating biofortified crops but also establish robust local support systems. Such systems include extension services and demonstration plots that facilitate the dissemination of knowledge and practical skills (Satish, 2023). For instance, the Indian Council of Agricultural Research (ICAR) has developed 142 biofortified varieties of both field and horticultural crops since 2014, ensuring these varieties meet traditional water and nutrient requirements without adversely affecting soil and water conditions (Satish, 2023).

Furthermore, biofortification efforts in a variety of crops are in the works, including iron-biofortified beans, cowpea, and pearl millet, zinc-biofortified maize, rice, and wheat, and provitamin A carotenoid-biofortified cassava, maize, rice, and sweet potato (Talsma & Pachón, 2017). These initiatives are crucial in addressing micronutrient deficiencies and enhancing food security in LMICs, showcasing the socioeconomic impact and community acceptance of biofortified crops.

Further emphasizing the global nature of these efforts, researchers from the Quadram Institute in the United Kingdom and the United States have engaged in collaborative efforts to enhance the nutrition of plant-based foods. For instance, in March 2023, scientists from the Quadram Institute visited the United States to share knowledge and build collaborations with nutrition experts at UC Davis, focusing on biofortification and its applications in food biotechnology (Zhang, 2023). This collaboration underscores the significance of international cooperation in improving health outcomes through enhanced nutrition.

In addition to these initiatives, advancements in artificial intelligence (AI) are being explored to revolutionize food and nutrition research. The USDA ARS Western Health and Nutrition Research Center, in collaboration with UC Davis, is developing the "Periodic Table of Food," a global initiative to catalog the biomolecular composition of the food supply. This comprehensive database aims to transform agriculture, food, and health industries by providing essential tools, data, and training (Zhang, 2023). These efforts highlight the pivotal role of international collaborations in leveraging technology to advance biofortification and its broader applications.

1.4 Methods of Biobased Fortification

Biobased fortification involves various innovative techniques aimed at improving the nutritional quality of food crops. Each of these methods, agronomic practices, conventional breeding, biotechnological innovations, and mutagenesis, contributes uniquely to the field of biofortification, collectively addressing the complex and multifaceted issue of micronutrient malnutrition. By integrating these approaches, researchers and practitioners can develop comprehensive strategies that leverage the strengths of each method, ensuring the production of nutrient-rich crops that satisfy the dietary needs of global populations. The journey of biofortification, spearheaded by researchers like Dr. Howarth Bouis, has evolved from a conceptual idea to a practical solution for "hidden hunger." By breeding vitamins and minerals into staple crops, biofortification aims to prevent deficiencies in essential nutrients like iron, zinc, and vitamin A, which can lead to severe health issues, particularly in children (CGIAR, 2018).

1.4.1 Agronomic Biofortification

Agronomic biofortification represents a promising strategy for addressing global micronutrient deficiencies by improving the nutrient composition of staple crops. This approach involves the application of mineral fertilizers, foliar sprays, soil amendments, nanofertilizers, and precision agriculture to enhance the bioavailability and uptake of essential nutrients (White & Broadley, 2009). The use of micronutrient-enriched fertilizers, including zinc sulphate ($ZnSO_4$), ferrous sulphate ($FeSO_4$), and potassium iodate (KI), has been extensively studied for improving crop nutrient density(Cakmak, 2007). Zinc-enriched soils improve nutrient uptake, though efficacy depends on soil pH and organic matter (Cakmak, 2007; Cakmak et al., 2010; Cakmak & Kutman, 2018). Granular $ZnSO_4$ (10–25 kg/ha) applied pre-sowing improves zinc availability in cereal rhizospheres, increasing wheat grain zinc by 15–20% (Cakmak & Kutman, 2018). In selenium-deficient regions, soil application of sodium selenate (5–10 g/ha) boosts selenium levels in rice grains by 40–50%, addressing deficiencies linked to cardiovascular diseases (Steiger et al., 2014). The efficacy of these fertilizers varies with application method, soil characteristics, and plant species (Singh et al., 2017).

1.4.2 Foliar Application

Foliar application serves as an effective alternative to soil fertilization, particularly in nutrient-deficient soil. Foliar spraying of $ZnSO_4$ and $FeSO_4$, applied during flowering, increases grain zinc by 20–30% in chickpea and wheat (Rathod et al., 2020; Adetola et al., 2022). Zinc deficiency, as one of the significant factors of malnutrition, impairs the immune system, growth, and intellectual development of humans, particularly in cereal-dependent populations(Wei et al., 2012; Lu et al., 2023b). Foliar Zn application, which avoids soil fixation problems, provides a direct route for nutrients into the plant tissue to enhance Zn concentration in food to tackle Zn deficiency(Xue et al., 2023) (Sánchez-Palacios et al., 2023). Foliar-applied Zn is absorbed through leaf cuticles and stomata, with subsequent translocation to grains mediated by phloem transport. Key factors influencing efficacy include formulation, adjuvants, and timing of application. Conventional $ZnSO_4$ outperforms nano-Zn (e.g., ZnO nanoparticles) in maize, increasing grain Zn by 37–40% (Xue et al., 2023). Urea, as an adjuvant, enhances Zn penetration by increasing leaf permeability, while amino acids (e.g., Zn-glycine) improve translocation efficiency (Wei et al., 2012; Pal et al., 2021). Applications at the flowering and grain-filling stages optimize Zn remobilization. For example, split applications of wheat increased grain Zn by 28–34% (Jalal et al., 2022).

Foliar spraying biofortification method showed the efficacy across several crops. Wheat and rice, the staple foods of over half the global population, have been studied extensively for foliar applications. In wheat, foliar $ZnSO_4$ (0.5–1.0%) increased grain Zn by 22–58%, with simultaneous reductions in cadmium (Cd) accumulation (44.5%) and phytic acid (PA)/Zn ratios (28.9%), enhancing bioavailability(Ning et al., 2021; Lu et al., 2023b). Similarly, Zn-amino acid and $ZnSO_4$ foliar sprays increased polished rice's Zn bioavailability by 68.37% and 64.43%, respectively, without a yield penalty (Wei et al., 2012). Similarly, a Zn-Fe-Se-N cocktail solution elevated maize grain Zn (60%), Fe (28%), and Se (1830%) without yield penalties (Xue et al., 2023). In chickpea, tank-mix foliar $ZnSO_4$ (0.5%) + $FeSO_4$ (0.5%) + urea (2%) doubled grain Zn (44 mg/kg) and Fe (74 mg/kg) concentrations (Pal et al., 2021). And, also in broccoli, foliar $ZnSO_4$ (0.25%w/v) increased floret Zn by 23 mg/kg, providing 36% of the recommended daily intake (RDI) per 100 g boiled florets (Rivera-Martín et al., 2020).

1.4.3 Soil Amendments and Use of Microbiota

Soil amendments such as biochar, organic compost, and mycorrhizal fungi enhance nutrient solubility, nutrient retention, root accessibility and colonization, and stress resilience (Dias et al., 2018; Zulfiqar et al., 2020). But the efficacy varies with soil pH, temperature, and microbial strain compatibility. Beneficial soil microbes, including rhizobacteria and mycorrhizal fungi, facilitate the solubilization, bioavailability, and uptake of nutrients through symbiotic interactions, offering a sustainable approach to biofortification (Ku et al., 2019; Dhiman et al., 2023). These strategies are cost-effective and environmentally sustainable, particularly in resource-limited settings. Plant growth-promoting rhizobacteria (PGPR) such as zinc-solubilizing *Pseudomonas* and *Bacillus* species solubilize soil zinc via organic acid production, increasing bioavailability for legumes like chickpea (Ramesh et al., 2014; Upadhayay et al., 2022). *Pseudomonas fluorescens* inoculation boosts grain zinc content by 35–40% in zinc-deficient soils (Kaur et al., 2020). Similarly, arbuscular mycorrhizal fungi (AMF) improve zinc uptake in nutrient-poor soils (Gomes et al., 2021a). AMF hyphae extend root absorption zones, improving phosphorus and zinc uptake in nutrient-poor soils. For instance, *Rhizophagus irregularis* increases zinc content in chickpea grains by 25% (Nachimuthu et al., 2014; Gomes et al., 2021a). AMF also mitigates drought stress in legumes, indirectly enhancing nutrient accumulation (Dhaliwal, Sharma, Shukla, Shivay, et al., 2022a). Inoculating biofortified maize with AMF in sub-Saharan Africa improved zinc retention by 18%

compared to non-inoculated controls (Pfeiffer & Mcclafferty, 2007; Suganya et al., 2021). In addition, tank mix application, combining micronutrients with macronutrient fertilizers (e.g., urea, NPK), enhances cost-effectiveness and nutrient synergy. For instance, tank-mixing $ZnSO_4$ (1.5 kg/ha) with urea (50 kg/ha) in chickpea cultivation increases protein content by 12% and zinc uptake by 18%, compared to standalone applications (Pal et al., 2021). This method also reduces labour costs, as farmers apply multiple nutrients in a single operation.

However, several factors affect the efficacy of agronomic biofortification and micronutrient uptake and translocation. Despite its effectiveness, agronomic biofortification faces challenges related to soil variability, nutrient interactions, and environmental sustainability (Cakmak, 2007). Drought stress in wheat increased grain Zn by 58% but reduced yields by 52%, highlighting yield-nutrition trade-offs (Ning et al., 2021). Co-application with phosphorus (KH_2PO_4) reduced grain Zn by 20% due to Zn-P antagonism(Ning et al., 2021). Moreover, the efficiency of foliar-applied nutrients depends on factors such as leaf morphology, cuticle permeability, and environmental conditions (Smoleń, 2012). Soil composition significantly influences the bioavailability of essential micronutrients such as zinc (Zn), iron (Fe), selenium (Se), and iodine (I) (Garousi, 2017; Ahmad, 2020; Ahmad et al., 2023). Factors such as soil pH, organic matter content, and microbial activity play a crucial role in determining nutrient solubility and root absorption efficiency (Fageria et al., 2008; Fageria et al., 2010). The alkaline soils (pH >8) reduce zinc solubility, necessitating chelated forms (e.g., Zn-EDTA) or organic amendments (e.g., compost) to improve bioavailability (Pal et al., 2021). Phytic acid in cereals reduces Zn absorption by 50%, but foliar Zn combined with low-P fertilizers lowers PA/Zn ratios by 27–45% (Lu et al., 2023a, b). Finally, there are also several socioeconomic barriers; farmer adoption depends on cost, with nano-Zn formulations (e.g., oxozinc nanofibers) requiring 90% less Zn than conventional $ZnSO_4$ but facing scalability challenges (Ahmad et al., 2023). Concerns regarding fertilizer runoff, environmental contamination, and cost-effectiveness limit the widespread adoption of agronomic biofortification (Singh et al., 2021). Sustainable practices, including integrated nutrient management and resource-efficient fertilization, are critical for long-term implementation (Zhao & Mcgrath, 2009).

1.4.4 Use of Nanotechnology

But recent advances in nanotechnology, chelation strategies, and precision agriculture have improved the uptake and translocation of foliar-applied micronutrients (Dimkpa & Bindraban, 2018). Nanoparticle-based fertilizers and nano-chelated micronutrients have gained attention due to their enhanced solubility, controlled release properties, and higher bioavailability (Rico et al., 2011). Nano-delivery systems, such as zinc oxide nanofibers, ZnONF, and ZnCNPs boosted translocation to grains, and seed priming increased Zn absorption efficiency. Multiple nutritional enhancements through Zn, Fe, Se, and urea delivery using cocktail sprays accomplished biofortification without any disadvantages (Naeem et al., 2021; Xue et al.,

2023). Research has demonstrated that nano-fertilizers can significantly improve plant nutrient content with minimal environmental impact (Dimkpa & Bindraban, 2018). Zinc oxide nanoparticles (20–50 nm) applied at 2–5 ppm increase wheat grain zinc by 35% while reducing soil toxicity (Younas et al., 2023). Nano-Zn formulations, such as ZnONF, have higher uptake efficiency (34% in wheat) than soil application of $ZnSO_4$ (Ahmad et al., 2023). Cocktail sprays of Zn, Fe, Se, and urea have also been found to enhance nutrient content. It was demonstrated that a cocktail solution enhanced maize grain Zn (61.5%), Fe (28.6%), and Se (186%) content with synergies (Xue et al., 2023). Drones equipped with hyperspectral sensors identify zinc-deficient crop zones, enabling targeted foliar sprays and reducing fertilizer use by 20% (Rathod et al., 2020).

1.4.5 Conventional Plant Breeding

A key component of biofortification is still conventional plant breeding, which uses the genetic variety that exists naturally in crop species to increase the nutrients they contain. This method involves selecting and crossbreeding plants with desirable traits, such as higher concentrations of essential vitamins and minerals, to produce offspring with improved nutritional profiles (Yadav et al., 2020). For instance, the development of high-zinc wheat varieties through selective breeding has significantly contributed to combating zinc deficiency in South Asia, where this mineral is crucial for immune function and growth (Yadav et al., 2020). Similarly, the creation of orange-fleshed sweet potatoes, rich in provitamin A, illustrates the effectiveness of traditional breeding in addressing vitamin A deficiencies (Bhardwaj et al., 2022). High-zinc lentil varieties also retain nutrients across diverse environments (Nachimuthu et al., 2014). These successes demonstrate the adaptability of traditional breeding techniques to local agricultural systems and their compatibility with existing farming practices, facilitating widespread adoption and sustained impact. The success of conventional breeding for biofortification depends on the presence of sufficient natural variation in nutrient content within a given crop species and the heritability of traits that govern this variation. The natural variation in nutrient content is assessed by screening germplasm that represents the genetic diversity available for a particular species. Germplasm with higher nutrient concentrations is then selected and enters a multistage breeding process.

In addition to genetic variation, environmental conditions during crop development also influence the nutrient content of staple foods. As such, quantifying the relative contributions of genetics, environment, and gene-environment interactions is a crucial component of the breeding process. When stable genetic lines with increased nutrient content are identified, these lines are back-crossed with existing varieties that possess desirable agronomic and consumer-preferred traits. However, if environmental factors significantly impact nutrient content or if the natural genetic variation for the nutrient is low, alternative biofortification methods may be considered. Some of the biofortification efforts through conventional and biotechnological innovations conducted on the different crops are given in Table 1.1.

Table 1.1 Some of the biofortified crops developed in the world

Crop	Nutrient	Concentration/impact	Countries (key regions)	References
Orange-fleshed sweet potato (OFSP)	Provitamin A (β-carotene)	100–1600 μg RAE/100 g fresh weight; improves serum retinol in children	Mozambique, Uganda, Kenya, Malawi	Low et al. (2007), Talsma and Pachón (2017)
Golden rice	Vitamin A (β-carotene)	1.6–2.0 μg/g dry weight; addresses vitamin A deficiency	Philippines, Bangladesh	Bouis et al. (2011), Dubock (2017)
Iron-biofortified beans	Iron (Fe)	50–100 mg/kg (vs. 30–50 mg/kg); improves iron stores in women	Rwanda, India, Haiti	Bouis et al. (2011), Garg et al. (2018)
Zinc-biofortified wheat	Zinc (Zn)	40–50 mg/kg (vs. 25–30 mg/kg); enhances zinc absorption	India, Pakistan, Ethiopia	Garg et al. (2018)
Vitamin a cassava	Provitamin A	7–12 μg/g fresh weight; increases serum β-carotene levels	Nigeria, Kenya, Zambia	Talsma et al. (2016), Talsma and Pachón (2017)
Iron pearl millet	Iron (Fe)	70–80 mg/kg (vs. 40–50 mg/kg); reverses iron deficiency in children	India, Nepal	Bouis et al. (2011)
Zinc rice	Zinc (Zn)	28–30 mg/kg (vs. 15–20 mg/kg); improves zinc status in adults	Bangladesh, Philippines	Garg et al. (2018)
Quality protein maize (QPM)	Lysine and tryptophan	0.8–1.0% lysine (vs. 0.2%); supports child growth	Sub-Saharan Africa, Mexico	Bouis et al. (2011)
Selenium-biofortified rice	Selenium (Se)	0.3–0.5 mg/kg (vs. 0.05 mg/kg); enhances antioxidant capacity	China, India	Srivastav et al. (2022)
Vitamin a maize	Provitamin A	6–15 μg/g dry weight; reduces childhood blindness	Zambia, Nigeria	HarvestPlus (n.d.)
Iron-biofortified lentils	Iron (Fe)	8–10 mg/100 g cooked; reduces anemia in women	Nepal, Bangladesh	Saltzman et al. (2017)
Zinc sorghum	Zinc (Zn)	35–40 mg/kg (vs. 20–25 mg/kg); combats zinc deficiency	Sudan, Ethiopia	Saltzman et al. (2017), HarvestPlus (n.d.)

1.4.5.1 Rice

Rice serves as a primary source of energy for populations in many Asian countries, providing up to 80% of the energy intake for the poor. Efforts to develop high-zinc rice varieties have been led by the International Rice Research Institute (IRRI) and the Bangladesh Rice Research Institute (BRRI). The initial breeding target for high-zinc rice was set at 24 ppm, an increase of 8 ppm over the baseline zinc concentration found in commercially available rice. Currently, high-yielding varieties with more than 75% of the target are undergoing official registration trials in Bangladesh and India. In Brazil, a high-zinc rice variety was identified and registered in 2012, while in China, a high-iron rice variety was released in 2011, with ongoing research to incorporate the high-zinc trait into this variety.

Processing also impacts the retention of these nutrients; studies indicate that parboiling and milling have less impact on zinc retention in polished rice compared to iron, as zinc is more evenly distributed throughout the brown rice grain (Liang et al., 2008; Patindol et al., 2017). A study conducted by BRRI in Bangladesh found that approximately 10% of the zinc was lost during the washing of milled rice before cooking (Taleon et al., 2022). Furthermore, boiling rice in excess water, which is discarded before serving, resulted in an additional zinc loss of 10–14% (Dipti, 2012). Demonstrating the efficacy of such interventions to improve zinc intake is challenging, as serum zinc concentration is not highly sensitive to modest increases in zinc intake. More sensitive biomarkers of zinc status are needed to evaluate the impact of biofortified rice on human health.

1.4.5.2 Wheat

The CIMMYT has spearheaded the development of high-zinc wheat varieties for India, Pakistan, and Nepal in collaboration with respective governmental agricultural research councils such as the ICAR, Pakistan Agricultural Research Council (PARC), and Nepal Agricultural Research Council (NARC). The initial breeding target for whole wheat was set at 33 ppm zinc, an increase of 8 ppm above the baseline concentration. In addition to improving zinc content, CIMMYT's breeding efforts have focused on enhancing wheat resistance to emerging strains of yellow and stem rust. In 2011, China released a wheat variety with a zinc concentration of 44 ppm, well above the target level (Saltzman et al., 2015).

Wheat mineral losses during milling are directly proportional to the duration and intensity of milling, although bioavailability may improve due to a reduction in phytate content. Studies conducted by the Punjab Agricultural University have assessed the iron and zinc losses associated with traditional milling and cooking methods. A study in Mexico among women showed that total zinc absorption was significantly greater from biofortified wheat compared to non-biofortified wheat (Rosado et al., 2009).

1.4.5.3 Maize

The breeding of provitamin A maize is being led by CIMMYT, the International Institute of Tropical Agriculture (IITA), and the National Agriculture Research and Extension System (NARES) in Southern Africa. Germplasm screening revealed genetic variation for provitamin A carotenoids in temperate maize, which was then bred into tropical varieties. Marker-assisted selection technologies have expedited the identification of genes associated with provitamin A content. In 2012, maize varieties providing 25% of the Estimated Average Requirement (EAR) for adult women and preschool children were released in Zambia and Nigeria, with similar varieties also registered in Brazil (Campos & Caligari, 2017; Simpungwe, 2017; Andersson et al., 2017; Prasanna et al., 2020; Yadesa & Diro, 2023). Varieties capable of providing 50% of the EAR are currently in development.

Studies on African food processing and cooking methods indicate that provitamin A losses are typically below 25% (Li et al., 2010). However, storage stability studies on the 2012 varieties are ongoing, with prior studies indicating that provitamin A decay can range from 25% to 60% after drying and 4 months of dark storage at 25 °C (Burt et al., 2010). Bioavailability studies have revised the assumed bioconversion rate of beta-carotene to retinol, showing more efficient bioconversion rates of 3:1 and 6.5:1 Li et al., 2010; Muzhingi et al., 2011).

1.4.5.4 Pearl Millet

Pearl millet is an important staple in some regions of India, where its biofortification efforts have focused on increasing iron and zinc content (Singhal et al., 2022). The breeding target for iron content was set at 77 ppm, an increase of 30 ppm over the baseline (Govindaraj et al., 2024) . The International Crops Research Institute for the Semi-Arid Tropics (ICRISAT) has led these efforts in collaboration with NARES and the private sector. The improved variety ICTP 8203-Fe was officially released in 2012, containing the full target amount of iron (Karandikar et al., 2017). Hybrid varieties that achieve similar iron content are under development (Govindaraj et al., 2024). While the bioavailability of iron from pearl millet is low, studies have demonstrated improved iron absorption from the biofortified variety (Cercamondi et al., 2013). A study among children in India found a significant difference in total iron absorbed between the biofortified and control varieties, with about a 7% increase in absorption when consumed as porridge (Kodkany et al., 2013; Saltzman et al., 2013; Huey et al., 2017).

1.4.5.5 Beans

High-iron beans have been developed for Rwanda and the Democratic Republic of Congo (DRC), with the breeding process led by CIAT and the Rwandan Agricultural Board (RAB) (Beebe & Andersson, 2014). The initial breeding target for iron was set at 94 ppm, an increase of 44 ppm over the baseline. Several varieties with more than 60% of the target iron content have been delivered (Beebe & Andersson, 2014; Mulambu, 2017). In 2012, five biofortified bean varieties were released in Rwanda, with more varieties in development for South America, particularly in Brazil.

Retention studies suggest that beans will retain nearly all their iron content after cooking, as the cooking water is not discarded (Carvalho et al., 2012; Huertas et al., 2022). While iron bioavailability from a single meal is typically low (3–5%), longer-term studies suggest that absorption may improve as the body adapts to higher levels of dietary phytates (Petry et al., 2020).

1.4.5.6 Orange Sweet Potato (OSP)

Biofortified orange sweet potato (OSP) varieties, developed by CIP and NARES scientists, have been shown to retain more than 80% of provitamin A after boiling or steaming, and at least 75% after solar drying (Haskell et al., 2004; Van Jaarsveld et al., 2006). The varieties are highly bioavailable, significantly improving vitamin A stores in the body (Low et al., 2007). Mozambique and Uganda released these varieties in 2002 and 2007, respectively, and biofortified OSP is now being promoted across Africa, South America, and China (Saltzman et al., 2013). The Sweet Potato for Profit and Health Initiative (SPHI) aims to reach 10 million households in Africa by 2020, focusing on empowering women farmers and enhancing the breeding pipeline for OSP varieties (Bouis et al., 2020).

1.4.5.7 Cassava

Biofortified cassava is under development for Nigeria and the DRC, with an initial breeding target of 15 ppm provitamin A (Ilona et al., 2017; Okwuonu et al., 2021). Varieties with sufficient provitamin A to provide 25% of the EAR for women and preschool children were released in Nigeria in 2011 (Ilona et al., 2017; Okwuonu et al., 2021; Kumar et al., 2024). Screening efforts have also identified similar varieties for the DRC. Despite the promising genetic potential, processing losses for cassava are significant, with up to 80% of provitamin A lost during traditional processing methods like gari production (Chávez et al., 2007; Thakkar et al., 2009). Further research is needed to assess retention and bioavailability, particularly with newer varieties.

1.4.5.8 Additional Crops

Several other crops are being conventionally bred for higher micronutrient levels, though target levels for these crops have not yet been determined. The International Centre for Agricultural Research in the Dry Areas (ICARDA) leads research on lentils, with multilocation testing beginning in 2009 across several countries, including Bangladesh, Ethiopia, India, Nepal, and Syria (Gupta et al., 2019; Sen Gupta et al., 2023). Similarly, cowpea research in India, Brazil, and other regions has led to the development of high-iron and zinc varieties, which are now being disseminated to farmers.

In addition to these crops, biofortified varieties of bananas, sorghum, potatoes, and sweet potatoes are under development. The development of high-provitamin A bananas, for example, is focused on varieties that provide up to 20 ppm of provitamin A, with efforts to expand their availability in multiple African and South American countries (Saltzman et al., 2013).

1.4.6 Biotechnological Innovations

Biotechnological approaches represent the cutting edge of biofortification, enabling precise genetic modifications to enhance crop nutrient content (Banerjee et al., 2023). One of the primary methods used in biofortification is genetic engineering, which involves manipulating an organism's genetic material to introduce desired traits. When natural genetic variation in nutrient content is insufficient for a target food crop, transgenic approaches present a viable pathway to achieve biofortification (Nawaz et al., 2025). This involves introducing novel genes from external organisms or upregulating existing genes in specific tissues to enhance the accumulation of nutrients (Banerjee et al., 2023; Nawaz et al., 2025). While much of this research remains in early developmental stages, proof-of-concept has been established for multiple nutrients across diverse crops, underscoring the potential of genetic engineering to address global malnutrition (Tahakik et al., 2024).

This technique has led to the development of genetically modified crops that are resistant to pests, diseases, and environmental stresses, ultimately increasing agricultural productivity and food security (Kavhiza et al., 2022; Khokharvoytas et al., 2023). Genetic engineering has proven to be more efficient and effective in producing nutrient-dense crops compared to traditional breeding techniques due to its precision and ability to introduce specific traits rapidly(Sun et al., 2024). Genetic engineering has facilitated the development of biofortified crops such as Golden Rice, which is engineered to produce beta-carotene, a precursor of vitamin A, addressing critical deficiencies in developing countries (Dubock, 2017). Joshi et al. (2023) identified precise meta-QTLs and haplotypes for efficient Zn biofortification of rice, aiding in the development of Zn-enriched rice varieties. Some of the genetically biofortified crops are given below.

1.4.6.1 Golden and High-Iron Rice

Golden rice, initially developed at the Swiss Federal Institute of Technology and advanced by Syngenta, represents a landmark achievement in provitamin A biofortification. The GR2 transgenic events achieved provitamin A levels of up to 37 ppm in a US rice variety (Al-Babili & Beyer, 2005; Amna et al., 2020; Mallikarjuna Swamy et al., 2021). Following Syngenta's decision to discontinue commercial development, these events were donated to the Golden Rice Network for humanitarian use. Since 2006, GR2 has been backcrossed into locally adapted rice varieties in the Philippines, Indonesia, India, and Bangladesh (Beyer, 2010; Amna et al., 2020; Mallikarjuna Swamy et al., 2021), with field trials ongoing. Human bioavailability studies confirm that Golden Rice effectively addresses vitamin A deficiency, demonstrating a beta-carotene-to-retinol conversion ratio of 3.8:1 (Tang et al., 2009). A critical component of deploying genetically engineered crops involves comprehensive safety evaluations, including molecular characterization of inserted DNA and assessments of allergenicity and toxicity for novel proteins. A multi-tiered safety framework study conducted by Oliva et al. (2020) underscores the rigorous protocols employed to ensure provitamin A biofortified rice (GR2E) meets global safety standards before commercialization. Molecular analyses confirmed the insertion of a single, intact transfer-DNA copy into the genome, inherited stably as a single locus without unintended nucleotide modifications. The newly expressed proteins, phytoene synthase and carotene desaturase, were evaluated for potential risks. In silico analyses revealed no sequence homology to known allergens or toxins, while in vitro studies demonstrated rapid digestion in simulated gastric fluid and inactivation of enzymatic activity under heat treatment. Acute oral toxicity testing in mice at biologically relevant doses showed no adverse effects. These findings, combined with compositional equivalence studies, substantiated the absence of identifiable hazards, corroborating conclusions from regulatory agencies such as the US Food and Drug Administration (FDA), Health Canada, and Food Standards Australia New Zealand.

In parallel, high-iron rice developed by the University of Melbourne and IRRI targets iron deficiency, a pervasive issue in regions reliant on rice-based diets. This transgenic variety accumulates 14 ppm iron in polished grains, with iron strategically localized in the endosperm to minimize binding by phytic acid, thereby enhancing bioavailability (Kumar & Dhaka, 2020).

1.4.6.2 BioCassava Plus

The BioCassava Plus (BC+) initiative employs genetic engineering to enhance cassava, a staple crop critical for food security in sub-Saharan Africa (Sayre et al., 2011). Key traits under development include elevated iron and provitamin A levels, extended shelf life, reduced cyanogenic glycosides, and improved disease resistance (Sayre et al., 2011; Kumar et al., 2024). Initial field trials for provitamin A-enriched cassava began in 2009, followed by high-iron cassava trials. Retention and bioavailability studies of transgenic cassava align with findings from conventional biofortification efforts led by HarvestPlus, confirming comparable nutrient stability (Failla et al., 2012).

1.4.6.3 Banana Biofortification

Collaborative efforts between Queensland University of Technology and Uganda's National Agricultural Research Organization were focused on provitamin A and iron-enriched bananas tailored for Ugandan diets. Transgenic bananas with provitamin A levels reaching 20 ppm have been developed, with field trials initiated in Uganda(Namanya, 2011; Paul et al., 2017).

1.4.6.4 African Biofortified Sorghum

A consortium led by Africa Harvest and Pioneer Hi-Bred is engineering transgenic sorghum to address multiple micronutrient deficiencies. The modified sorghum exhibits elevated provitamin A (21 ppm), reduced phytate content (35–80%), and an optimized protein profile (Botha & Viljoen, 2008). Transgenic sorghum lines targeting vitamin A deficiency in sub-Saharan Africa Biofortified sorghum (*Sorghum bicolor* (L.) Moench) lines targeting vitamin A deficiency in sub-Saharan Africa demonstrated elevated β-carotene levels in transgenic germplasms (3.3–14.0 μg/g DW) versus non-transgenic lines (1.0–1.5 μg/g DW), with porridge formulations delivering up to 250 μg/100 g FW. Micellarization efficiency of all-trans-β-carotene was lower ($p < 0.05$) in transgenic lines (1–5%) compared to null/non-transgenic (6–11%), but increased lipid content (5% to 10% w/w) improved bioaccessibility ($p < 0.05$). Event *Homo188-A* yielded four- to eightfold higher bioaccessible β-carotene than controls, supporting sorghum biofortification's potential to enhance provitamin A delivery, though in vivo validation of bioavailability remains critical (Lipkie et al., 2013).

1.4.6.5 Novel Biotechnical Tools

Future research should prioritize multi-nutrient biofortification and climate-resilient traits to maximize impact. Integrating advances in gene editing (e.g., CRISPR-Cas9) with traditional methods could further streamline development, offering scalable solutions to global malnutrition. Recent advancements in CRISPR-Cas technology have further revolutionized biobased fortification by allowing for multiplex genome editing, which enables the simultaneous modification of multiple genes related to nutritional traits in crops (Bhat et al., 2022). This technology's integration with artificial intelligence (AI) has enhanced the precision of gene editing by predicting potential off-target effects and optimizing gene targets for better nutritional outcomes (Bhat et al., 2022). AI-driven biofortification can also tailor nutrient-enriched crops to address specific malnutrition profiles prevalent in low- and middle-income countries, thereby maximizing the health benefits for vulnerable populations (Groff, 2024).

Advances in genome editing technologies, such as CRISPR-Cas9, further expand the possibilities for biofortification by allowing targeted modifications to crop genomes to enhance nutrient profiles and improve resilience to environmental stressors (Kong et al., 2022). They stated that editing genes like FRD3) through CRISPR/Cas9 technology improves iron/zinc translocation to edible parts in rice grains by upregulating metal transporters (Kong et al., 2022). Modifying OsNAS genes increases nicotianamine synthase activity, boosting zinc and iron accumulation in polished rice (Majumder et al., 2019). CRISPR-edited wheat with reduced phytic acid (via TaABCC1 gene disruption) improves zinc bioavailability by 60% (Gomes et al., 2021b). These tools also enable selenium enrichment, as seen in rice engineered for higher selenocysteine methyltransferase activity (Kong et al., 2022).

CRISPR's precision reduces off-target effects compared to traditional breeding, accelerating the release of climate-resilient, high-yielding varieties. Combining zinc, iron, and vitamin A in a single crop addresses multiple deficiencies simultaneously. For instance, biofortified sorghum developed through transgenic stacking of IRT1 (iron transporter) and PSY1 (carotenoid biosynthesis) genes shows 40% higher iron and 30% higher β-carotene than non-biofortified varieties (Dwivedi et al., 2023). CRISPR-edited cassava with enhanced provitamin A and iron content is under trial in Nigeria, demonstrating stability across diverse agroclimatic zones (Gao et al., 2022; Ogbu et al., 2024). However, balancing multi-nutrient traits with agronomic performance remains challenging. Genetic trade-offs, such as yield reductions in early iron-biofortified pearl millet (Sperotto et al., 2012), necessitate iterative testing and trait stacking to optimize both nutrition and productivity.

These biotechnological innovations offer precision and efficiency in nutrient enhancement, potentially leading to crops with superior nutritional qualities and increased yield stability. However, the adoption of genetically modified organisms (GMOs) continues to be a topic of debate, with regulatory frameworks and public perception posing significant challenges to their implementation.

1.4.7 Mutagenesis

Mutagenesis, the induction of genetic mutations via chemicals or radiation, serves as a complementary strategy to transgenic approaches (Oladosu et al., 2016). While natural mutations underpin genetic diversity, induced mutagenesis accelerates trait development, followed by conventional breeding to stabilize desired characteristics (Keerthana et al., 2025). This method has yielded low-phytic-acid variants in rice, wheat, maize, barley, sorghum, and soybean, with phytic acid reductions ranging from 50% to 90% (Raboy, 2020; Shahzad et al., 2021; Asif et al., 2022). However, excessively low phytic acid levels correlate with diminished agronomic performance, necessitating a balance between nutrient enhancement and crop viability.

1.5 Combining Omics with Biofortification Methods

Integrating multi-omics technology and data, including transcriptomics, proteomics, and metabolomics, is pivotal in advancing biobased fortification strategies by providing a comprehensive understanding of molecular mechanisms that enhance nutrient bioavailability (Pandohee et al., 2023; Nasar et al., 2024). A single tool, such as the 3Omics Web-based system, can facilitate the visualization and integration of these datasets, allowing researchers to perform correlation networking, co-expression analysis, phenotyping, pathway enrichment, and GO (Gene Ontology) enrichment simultaneously (Nasar et al., 2024).

Gene co-expression analysis, integrated with metabolomics data, and gene-metabolite network approaches are commonly used methods to achieve this integration. Similarity network fusion (SNF) and enzyme-metabolite-based networks further enhance the ability to analyze complex interactions between different omics layers (Global Food and Nutrition Security, 2024). Pathway enrichment analysis, using differentially expressed genes and metabolites, and integrating genome-scale models with metabolomics and transcriptomics data, are key strategies to elucidate the biochemical pathways involved in nutrient fortification (Global Food and Nutrition Security, 2024).

Recent advancements in transcriptomics methodologies, particularly single-cell RNA sequencing (scRNA-seq), have revolutionized our understanding of gene expression at an unprecedented resolution (Yan et al., 2024). scRNA-seq allows for the dissection of complex tissues and identification of cell-type-specific gene expression patterns, which is critical for elucidating the regulatory networks involved in biofortification. This technology has enabled researchers to pinpoint specific genes and their expression profiles, providing deeper insights into biosynthetic pathways and identifying potential targets for genetic manipulation to enhance nutrient content in crops (Yan et al., 2024).

By employing scRNA-seq, scientists are now capable of overcoming previous limitations in transcriptomic analyses that averaged gene expression data across heterogeneous cell populations. This advancement not only enhances the precision of our gene expression measurements but also opens up new possibilities for targeted bio-fortification strategies. For instance, understanding the cellular heterogeneity within plant tissues can reveal differential gene expression related to nutrient accumulation, stress responses, and metabolic pathways, thus offering new avenues for improving crop nutritional quality through bioengineering and selective breeding (Yan et al., 2024).

The integration of scRNA-seq data with other omics approaches, such as proteomics and metabolomics, further amplifies our ability to construct comprehensive models of metabolic pathways. These models are instrumental in identifying key regulatory nodes and metabolic bottlenecks that can be targeted to enhance the efficiency of nutrient biosynthesis and accumulation. Consequently, the use of scRNA-seq is paving the way for more effective and precise biofortification techniques, ultimately contributing to the development of crops with improved nutritional profiles to combat micronutrient deficiencies globally.

Moreover, combining proteomic and transcriptomic data through differential expression analysis and interactome analysis allows for the observation of temporal delays between different omics data, providing deeper insights into the dynamic processes of nutrient absorption and metabolism (Global Food and Nutrition Security, 2024). Machine learning applications in omics data also contribute to this integrative approach by uncovering patterns and predictive models that can inform fortification strategies (Yan et al., 2024).

Another key area is using metabolomics to improve crop nutrition. Metabolomics has emerged as a crucial tool in understanding and enhancing crop nutrition. Recent studies have demonstrated its capacity to identify key metabolic pathways that can be leveraged to improve the nutritional content of crops. For instance, a comprehensive analysis combining transcriptomics, physiological assessments, and metabolomics has been conducted to examine the response of potato tuber formation to nitrogen, highlighting the intricate metabolic networks involved in nutrient uptake and assimilation (Ding et al., 2024). Researchers from various institutions, including the University of Lisbon and the University of Melbourne, have contributed to the field by exploring current and emerging methodologies in metabolomics, aiming to refine crop biofortification strategies (Sousa Silva et al., 2023). These advances underscore the potential of metabolomics to pinpoint critical metabolic pathways that influence nutrient density and availability in crops.

Additionally, significant contributions from Chinese research institutions, such as the Hebei Key Laboratory and the Hainan Key Laboratory of Tropical Oil Crops Biology, have focused on the metabolic pathways that enhance crop quality and safety (Liu et al., 2023; Lu et al., 2023a). These studies have provided valuable insights into the biochemical mechanisms that can be targeted to improve the nutritional profile of staple crops.

Moreover, the role of genetic variability among crop varieties has been highlighted as a key factor influencing the outcomes of biofortification programs. The interplay between genetic factors and metabolomic profiles determines the effectiveness of nutrient enhancement strategies (Vinoth & Ravindhran, 2017). This consideration is essential for the successful application of metabolomics in field conditions, ensuring that biofortification efforts are tailored to the specific genetic makeup of different crop varieties (Tiwari et al., 2024). Experts in metabolomics emphasize the importance of understanding the metabolic changes induced by biofortification, which can lead to more targeted and effective strategies. Integrating metabolomic data with transcriptomic and proteomic information helps identify key metabolic pathways and potential side effects, ensuring that nutrient enhancements are both beneficial and safe (Shahzad et al., 2021; Kong et al., 2022; Srivastav et al., 2022). By leveraging these advanced integrative techniques, researchers can develop more precise and effective biobased fortification strategies, potentially leading to significant improvements in global nutritional health (Srivastav et al., 2022).

1.6 Nutritional Impact, Bioavailability, and Processing Stability

The nutritional efficacy of biofortified crops hinges not only on their micronutrient content but also on bioavailability, the proportion of nutrients absorbed and utilized by the human body. While biofortification enhances micronutrient density, intrinsic plant compounds (e.g., phytates, polyphenols) and processing methods can significantly modulate bioavailability, often limiting real-world health impacts (La Frano et al., 2014). Some of the nutritional benefits, bioavailability, and impact of antinutrient factors are discussed here.

1.6.1 Zinc-Biofortified Wheat

Zinc-biofortified wheat has demonstrated significant health benefits, particularly in reducing morbidity and improving micronutrient status. Clinical trials show that consumption of zinc-enriched wheat reduces pneumonia episodes by 17% and vomiting by 40% in children, highlighting its role in strengthening immune function (Sazawal et al., 2018). These outcomes are linked to increased dietary zinc intake, which ranges from 20% to 80% higher compared to non-biofortified varieties, addressing critical gaps in populations reliant on staple crops (Bouis, 2002; Sazawal et al., 2018).

Biochemical evidence further supports these findings, with studies indicating that zinc-biofortified wheat improves serum zinc levels by 10–12% in adolescents and enhances growth metrics in malnourished children (Shahzad et al., 2024). Agronomic interventions, such as foliar zinc application, amplify these effects by increasing grain zinc content by 16–30%, ensuring consistent nutrient delivery even in zinc-deficient soils (Jan et al., 2020; Hasnain et al., 2023).

1.6.2 Iron-Biofortified Crops

While the focus here is on zinc, complementary evidence from iron-biofortified beans and pearl millet underscores broader biofortification impacts. These crops increase serum ferritin levels by 15–20%, reducing iron-deficiency anemia (Boy et al., 2017). Similarly, iron nanoparticle-fortified wheat achieves 55–66% higher grain iron content, improving bioaccessibility (Hassan et al., 2023).

1.6.3 Vitamin A-Enriched Crops

Provitamin A maize retains 80% of carotenoids after extrusion cooking, maintaining nutritional value during processing (Nuss & Tanumihardjo, 2010). However, vitamin A bioavailability remains constrained by matrix interactions, necessitating complementary strategies like lipid addition during meal preparation (La Frano et al., 2014).

1.6.4 Iron-Biofortified Beans

Iron-biofortified beans, developed through conventional breeding and genetic selection, significantly improve iron status in populations reliant on legumes. Clinical trials demonstrate that regular consumption increases serum ferritin levels by 15–20%, effectively reducing iron-deficiency anemia (Boy et al., 2017). These beans are particularly impactful in regions like Rwanda and India, where dietary iron intake is often inadequate. Agronomic practices, such as intercropping with zinc-fertilized cereals, further enhance iron bioavailability by reducing phytic acid interference (Díaz-Gómez et al., 2017).

1.6.5 Iron Nanoparticle Rice

Iron nanoparticle technology represents a breakthrough in enhancing grain iron content. Studies show that applying phytogenic iron nanoparticles to wheat crops increases grain iron by 55–66%, surpassing traditional fortification methods (Hassan et al., 2023). While Hassan et al. (2023) focused on wheat, similar nanobiofortification strategies in rice improve iron stability during cooking and digestion. However, challenges persist, including potential nanoparticle toxicity at high concentrations and the need for cost-effective scaling (Younas et al., 2023).

1.6.6 Bioavailability of Nutrients

In vitro models and in vivo models are used to assess bioavailability. The in vitro Caco-2 cell line, simulating human intestinal epithelium, predicts Fe bioavailability with 85% accuracy relative to human trials (Tako et al., 2016). However, these models oversimplify digestive dynamics, such as microbiota interactions and gastric pH variations. In vivo Models such as the Gallus gallus (chicken) model, which shares intestinal Fe transporters with humans, bridge the lab-to-field gaps. For instance, Zn-biofortified rice tested in Gallus gallus showed 20% higher serum Zn levels, corroborating findings in human efficacy trials (Tako et al., 2016).

1.6.7 Antinutrients and Absorption Barriers

A major inhibitor of Fe and Zn absorption is phytic acid, which forms insoluble complexes with these minerals in the gastrointestinal tract. Phytic acid is a natural antinutrient in cereals and legumes that binds with those minerals, reducing their bioavailability by 20–40% (Díaz-Gómez et al., 2017). For example, Fe-biofortified beans contain 50–70 mg/100 g phytic acid, reducing Fe absorption by 30–50% compared to $FeSO_4$ supplements (La Frano et al., 2014; Boy et al., 2017). Unprocessed chickpeas retain only 15% of their intrinsic iron due to phytate interactions (Sebastiá et al., 2001). Moreover, polyphenols, especially tannins in cereals like sorghum and millet, chelate Fe, further exacerbating deficiencies in populations reliant on these staples (Brnić et al., 2016). Nondigestible fibers in whole grains also physically entrap minerals, reducing accessibility to digestive enzymes (Baye et al., 2017). Several complementary processing strategies mitigate this; among them some have been practiced traditionally in Asia and Africa for millennia, such as:

- Soaking: Reduces phytic acid by 30–50% in legumes, enhancing mineral solubility (Zhang et al., 2022).
- Fermentation: Lactic acid bacteria degrade phytates, improving iron and zinc bioaccessibility by 50% in fermented legumes (Sebastiá et al., 2001), and enhance Zn absorption by 25% in pearl millet (Krishnan & Meera, 2018).
- Germination: Activates phytase enzymes, reducing phytic acid in cereals by 60% (Kumar & Anand, 2021), and also reduces the tannin concentration in millets (Karki et al., 2025).
- Nixtamalization: Alkaline cooking of maize degrades phytic acid by 50–80%, increasing Fe bioavailability by 40% (Nuss & Tanumihardjo, 2010).
- Milling: Polishing rice reduces fiber and polyphenol content but sacrifices micronutrient-rich bran layers, necessitating trade-offs (Brnić et al., 2016).
- Dietary synergists: Pairing biofortified crops with vitamin C-rich foods (e.g., citrus) boosts nonheme Fe absorption by 300% (La Frano et al., 2014).
- Extrusion cooking: One of the methods for producing fortified maize products; it retains 80% of carotenoids, preserving nutritional value despite high-temperature exposure (Nuss & Tanumihardjo, 2010). This stability ensures that processed foods like porridge or snacks derived from biofortified maize deliver consistent vitamin A benefits.

These methods are particularly critical in low-income regions, where traditional processing aligns with cultural practices and infrastructure limitations.

1.7 Agronomic and Economic Benefits

1.7.1 Yield Stability

Zinc-enriched wheat and rice exhibit 10–15% higher yields in micronutrient-deficient soils compared to non-biofortified varieties (Cakmak, 2007; Cakmak & Kutman, 2018). This yield advantage arises from improved plant resilience to abiotic stresses, such as drought and salinity, which are exacerbated in nutrient-poor soils. For example, zinc biofortification enhances root development and enzymatic activity, enabling crops to take up water and nutrients more efficiently (Dhaliwal et al., 2022a). Similarly, iron-biofortified pearl millet shows 20% higher biomass in alkaline soils, demonstrating broader agronomic benefits beyond nutrient enhancement (Valença et al., 2017).

1.7.2 Climate Resilience

Emerging evidence highlights biofortified crops' tolerance to climate extremes. For instance, zinc-enriched rice maintains yields under drought conditions, addressing both nutritional and food security goals (Blancquaert et al., 2014).

1.7.3 Cost-Effectiveness and Economic Viability

Biofortification is economically advantageous compared to industrial fortification or supplementation. For example, iron fortification of rice in India yields a benefit-cost ratio of 8.2:1 due to improved productivity and reduced healthcare expenditures, meaning every dollar invested yields $8.20 in economic gains through enhanced agricultural productivity and reduced healthcare expenditures (Qureshi et al., 2020, 2021). Moreover, biofortification provides comparative cost-effectiveness in contrast to large-scale food fortification (LSFF), while LSFF programs avert 3 million annual cases of vitamin A deficiency(Keats et al., 2019). They depend on centralized processing facilities and stable supply chains; those infrastructures are often absent in rural areas. In contrast, biofortification bypasses these requirements, making it 50% cheaper per beneficiary in remote regions (Yadav et al., 2020). For example, HarvestPlus' zinc-biofortified wheat in Pakistan reduced anemia-related healthcare costs by $12 million annually within 5 years of adoption (Andersson et al., 2017). Vitamin A supplementation costs $0.50–$1.00 per dose and achieves only 64% coverage in sub-Saharan Africa due to logistical challenges (Fitzpatrick et al., 2017). Biofortified crops, however, passively deliver nutrients through daily diets, eliminating distribution costs and reaching populations inaccessible to supplementation campaigns.

Bouis et al. (2011) also substantiated the cost-effectiveness of biofortification; they stated that biofortification delivers a $1:10 return on investment, making it one of the most economically viable strategies for addressing malnutrition. Improved micronutrient intake lowers expenditures on deficiency-related illnesses, such as diarrheal diseases and anemia (Sazawal et al., 2018). While cost-effective, biofortification requires upfront investments in agricultural R&D, limiting scalability in resource-poor regions (Nestel et al., 2006).

1.7.4 Sustainability and SDG Goals

Biofortification is increasingly being recognized as an affordable, sustainable, and climate-smart intervention to combat global malnutrition by enhancing the nutrient density of staple crops (Groff, 2024). The principle of sustainability, which involves development that meets the present needs without compromising the future, aligns well with the goals of biofortification. This approach ensures that essential nutrients are replenished in crops without depleting resources, thus supporting long-term agricultural productivity (Lohmann & Graf, 2021).

In the context of the United Nations' Sustainable Development Goals (SDGs), biofortification directly contributes to SDG 2, which aims to end hunger, achieve food security and improved nutrition, and promote sustainable agriculture by 2030 (Davey, 2024). By improving the nutritional quality of crops, biofortification addresses micronutrient deficiencies that are prevalent in developing countries, helping to alleviate malnutrition and improve public health (Global Food and Nutrition Security, 2023). This is particularly relevant in regions where poor nutritional quality and micronutrient deficiencies are significant barriers to achieving better health outcomes (Global Food and Nutrition Security, 2023). Moreover, biofortification can be integrated with advanced technologies such as genome editing and nanotechnology to enhance its effectiveness and adaptability to local conditions (Global Food and Nutrition Security, 2023). These innovations promise to make biofortification a more dynamic and robust solution, ensuring it can meet the evolving needs of food security and sustainability (Global Food and Nutrition Security, 2023).

Efforts to promote biofortified crops have been supported by research demonstrating their production and nutrition benefits, which facilitates their adoption by farmers. Awareness of these benefits compared to traditional crop varieties is crucial for wider acceptance and integration into farming practices (Groff, 2024). Policy interventions and international collaborations are recommended to support the continued development and dissemination of biofortified crops, ensuring they align with local needs and consumer preferences (Global Food and Nutrition Security, 2023).

1.7.5 Long-Term Savings

The initial costs of breeding and developing biofortified crops are outweighed by long-term advantages. These benefits include improved labor efficiency due to reduced anemia and lower medical expenses for treating micronutrient deficiencies.

Once biofortified varieties are adopted, they require no additional inputs beyond standard farming practices, which ensures their sustainability (Tripathi et al., 2018). This sets them apart from the ongoing costs associated with other agricultural practices. Of supplementation (e.g., vitamin A capsules) or industrial fortification (e.g., iodized salt production), biofortification requires a one-time investment in breeding or agronomic R&D. Once biofortified seeds are disseminated, farmers can replant them for multiple seasons, ensuring sustained benefits.

1.7.6 Farmer Income Gains

Biofortified crops often command premium prices in markets. In Nigeria, vitamin A-enriched cassava is sold at a 15–20% price premium, directly boosting smallholder incomes (HarvestPlus, n.d.). Biofortified crops like zinc-rich wheat command higher prices in niche markets, incentivizing farmer adoption (Gómez-Galera et al., 2010). Additionally, improved crop resilience (e.g., drought-tolerant zinc rice) enhances yield stability, further safeguarding farmer livelihoods.

1.7.7 Scalability and Equity

Biofortification leverages existing agricultural systems to reach remote populations, with 80% of smallholder farmers in sub-Saharan Africa and South Asia adopting zinc- or iron-biofortified varieties within 5 years of release incomes (HarvestPlus, n.d.). This scalability is critical for regions where fortification programs and supplementation are logistically challenging (Siwela et al., 2020).

1.8 Key Challenges in Biobased Fortification

1.8.1 Technical Challenges

Biofortification faces several technical challenges, particularly in LMICs. A significant challenge in biofortification is ensuring that the enhanced micronutrients are bioavailable and stable throughout the food supply chain. Anti-nutritional compounds such as phytates and polyphenols, naturally present in crops like cereals and legumes, can inhibit the absorption of iron and zinc in the human body (Blanco-Rojo & Vaquero, 2019). For instance, even iron-biofortified crops may fail to alleviate deficiencies if their mineral content is rendered inaccessible by these compounds. The biochemical interactions and stability of fortified nutrients during food processing, such as milling, present another challenge. Additionally, micronutrients such as vitamin A and zinc can degrade during postharvest processing, storage, and cooking, diminishing their nutritional value (Shubham et al., 2020). For instance, the milling of zinc-biofortified wheat affects its nutritional components, including vitamin B, mineral content, polyphenols, and phytates, impacting the bioaccessibility of zinc (Jiang et al., 2023). Innovations like nanoemulsions and encapsulation technologies have shown potential to stabilize nutrients, but their scalability and cost-effectiveness in low-resource settings remain unproven (Gharibzahedi & Jafari, 2017). This necessitates a comprehensive understanding of how different processing methods influence nutrient retention and efficacy.

Another significant challenge is ensuring seed availability and affordability. Farmers need access to high-quality biofortified seeds, which requires efficient distribution networks and affordability mechanisms to be in place (Walton, 2021; Global Food and Nutrition Security, 2024). The cost-effectiveness of biofortification is a crucial aspect, as it piggybacks on agronomically superior varieties developed at agricultural research centers, promoting adoption by farmers due to the high yield and profitability of these crops (Global Food and Nutrition Security, 2024).

Furthermore, the scalability of biofortified crops is essential for addressing micronutrient deficiencies in LMICs. Although HarvestPlus-led initiatives have seen success, with an estimated 9.7 million smallholder households in 19 countries growing biofortified crops, benefiting approximately 48.5 million people; expanding this impact requires overcoming obstacles related to seed distribution, cost, and local climatic conditions, which can affect crop performance and farmer adoption rates.

Moreover, there are knowledge gaps in the biosynthesis and regulation of key nutrients, such as carotenoids in sorghum, which limit the effectiveness of breeding strategies for pro-vitamin A biofortification (Cruet-Burgos & Rhodes, 2023). Addressing these gaps through advanced transcriptomic studies and other biotechnological approaches can improve the precision and efficiency of biofortification efforts. From a biotechnological perspective, advancements such as CRISPR technology offer promising solutions. These tools can enhance nutrient consistency without compromising crop yield and quality, providing a robust approach to biofortification (Kumar et al., 2022). However, these technologies also come with their own set of challenges, including regulatory hurdles, public acceptance, and the need for technical expertise.

1.8.2 Policy, Regulatory, and Ethical Challenges

Biofortification, the process of enhancing the nutritional quality of food crops through agronomic practices, conventional plant breeding, or modern biotechnology, faces numerous regulatory challenges both at the national and international levels. These challenges can hinder the adoption and widespread use of biofortified crops, thus impacting efforts to combat malnutrition and improve public health.

One of the primary regulatory challenges at the national level is the establishment and enforcement of biofortification policies. Governments must develop comprehensive frameworks that include the safety evaluation, environmental impact assessment, and nutritional efficacy of biofortified crops. These frameworks must be harmonized with existing agricultural and food safety regulations to ensure a coherent policy environment. For instance, countries need to ensure that biofortified crops do not adversely affect soil and water conditions, as has been demonstrated with the biofortified varieties developed under the ICAR (Satish, 2023).

One of the primary regulatory hurdles is the complex and varying standards across different countries, which can lead to significant discrepancies in the approval and acceptance of biofortified crops. For example, national regulations for biofortified crops differ widely, making it difficult to achieve a consistent regulatory framework globally (Nestel et al., 2006; Satish, 2023). This lack of harmonization can result in situations where biofortified products deemed safe in one country are not recognized as such in another, leading to potential confiscation and destruction of these products (Food Safety Magazine Editorial Team, n.d.). Such regulatory inconsistencies can delay the market entry of biofortified crops and increase costs for developers, particularly affecting small- and medium-sized enterprises (SMEs) that may lack the resources to navigate complex regulatory landscapes (Ahonobadha, 2024; Food Safety Magazine Editorial Team, n.d.).

International harmonization of safety standards for biobased fortified products could significantly streamline the approval process, reduce costs, and ensure wider acceptance across different regions (Eruaga, 2024; Ahonobadha, 2024). Different countries have varying requirements for the approval and cultivation of genetically modified organisms (GMOs), including biofortified crops. This variation can create barriers to trade and limit the global distribution of biofortified foods. International bodies such as the World Health Organization (WHO) and the Food and Agriculture Organization (FAO) play a crucial role in setting guidelines that can help standardize biofortification practices and facilitate their global adoption (GAIN et al., 2025). By developing global scientific consensus documents, regulatory agencies and industry stakeholders could work towards aligning the most critical regulations, thus reducing the risk of noncompliance and facilitating the smoother movement of biofortified products across borders (Food Safety Magazine Editorial Team, n.d.). The efforts in East Africa to harmonize food and nutrition standards are a step in the right direction, aiming to improve food safety and nutrition outcomes through regional collaboration (Ahonobadha, 2024).

In addition to these regulatory complexities, the rapid advancements in biotechnology pose another set of challenges. Biofortification often involves genetic modification, which can trigger stringent regulatory scrutiny and public resistance (Satish, 2023). The introduction of genetically modified organisms (GMOs) into the food supply chain requires rigorous safety assessments and compliance with regulatory frameworks set by bodies like the US Food and Drug Administration (FDA) and European Medicines Agency (EMA), which have become more intricate over time (BioBoston Consulting, 2024). The compliance burden can be overwhelming, making it essential for regulatory agencies to stay ahead of changing regulations and for companies to adapt quickly to new requirements (Olivera, 2024; BioBoston Consulting, 2024). For example, genetically modified (GM) crops, such as Golden Rice and vitamin E-enhanced Indica rice, face protracted regulatory approval, often exceeding 10 years, due to stringent biosafety laws and public skepticism (Dubock, 2017; Sundararajan et al., 2021). In contrast, non-GM biofortified crops like Zn-rice in Bangladesh benefited from streamlined national policies but still struggled with inconsistent funding and political prioritization (Andersson et al., 2017).

Moreover, the lack of ethical policies across regions stifles biofortification's scalability. Ethical concerns, such as corporate control over seed patents and intellectual property rights, further complicate adoption in LMICs. Transparent risk assessments, equitable licensing agreements, and inclusive policy frameworks are essential to building trust and ensuring access to biofortified crops for smallholder farmers. Additionally, public awareness campaigns and educational programs can help build consumer acceptance and trust in biofortified foods, further supporting their integration into national and international food systems (Talsma & Pachón, 2017).

1.8.3 Societal Concerns and Cultural Barriers to Adoption

Societal concerns surrounding biofortified crops, particularly those involving genetically modified organisms (GMOs), encompass various ethical, sociocultural, and health-related issues. One major concern is the potential impact on self-determination, liberties, and food justice, which underscores the necessity for community involvement in biofortification initiatives to minimize societal burdens and non-maleficence (Fanzo & Glass, 2018). For example, orange-fleshed sweet potatoes (OFSP), rich in β-carotene, initially faced resistance in sub-Saharan Africa due to their unfamiliar color and taste compared to traditional white-fleshed varieties.

The public perception of GMOs is fraught with misconceptions, primarily due to fears about health risks and environmental damage (Blancke, 2015; Thayer, 2016)]. Despite overwhelming scientific evidence supporting the safety and environmental benefits of GMOs, public opposition remains high. This discrepancy arises from deeply ingrained beliefs that GMOs are harmful or poisonous, which are not entirely unfounded given concerns like herbicide resistance and the role of multinational corporations (Blancke, 2015). However, it is crucial to note that these concerns are not exclusive to GMOs and often reflect broader issues within agricultural practices. Finally, fair access to seeds and foods, especially for farmers and consumers, is another critical issue linked to distributive justice (Fanzo & Glass, 2018).

1.8.4 Equity and Access

Biofortification's promise of targeting rural, low-income populations is often hampered by systemic inequities in seed distribution and agricultural extension services. For example, marginalized communities, particularly women and smallholder farmers in sub-Saharan Africa and South Asia, frequently lack access to biofortified seeds due to cost barriers, limited market infrastructure, and gendered disparities in resource ownership (Garcia-Casal et al., 2017). Women, who constitute 43% of the agricultural labor force in LMICs, are often excluded from biofortification programs due to limited land ownership and decision-making power. In Pakistan, only

8% of women farmers adopted Zn-wheat despite bearing primary responsibility for household nutrition (Andersson et al., 2017). In India, only 12% of farmers cultivating Zn-biofortified wheat were women, reflecting entrenched gender biases in agricultural decision-making (Andersson et al., 2017). Gender-inclusive programs, such as India's "*Mahila Kisan Sashaktikaran Pariyojana*" (Women Farmer Empowerment Project), increased women's participation in OFSP cultivation by 25% through targeted training and microloans (Siwela et al., 2020). In Nigeria, despite government mandates to integrate biofortified cassava into school feeding programs, rural farmers reported limited access to planting materials due to poor road networks and monopolization by urban agro-dealers (HarvestPlus, n.d.).

1.8.5 Consumer Acceptance

Sensory preferences and cultural perceptions significantly influence the adoption of biofortified crops. Provitamin A maize, with its distinctive yellow pigmentation, faced rejection in South Africa and Zambia, where consumers associated white maize with higher quality and tradition (Siwela et al., 2020). Similarly, iron-biofortified pearl millet in India was perceived as having a "bitter" taste, reducing its appeal despite proven health benefits (Kumar et al., 2019). Nutrition education campaigns, such as recipe demonstrations for orange-fleshed sweet potatoes (OFSP) in Uganda, improved acceptance by 40%, highlighting the need for culturally tailored communication strategies (HarvestPlus, n.d.).

1.8.6 Economic Viability

Agronomic biofortification methods, such as foliar Zn application, incurrent costs (e.g., $20–30/hectare for fertilizers), which are prohibitive for smallholders earning less than $2/day (Jan et al., 2020). Conversely, genetic biofortification requires high upfront R&D investments ($5–10 million per crop variety), deterring private-sector participation in low-profit markets (Kumar et al., 2019). Furthermore, biofortified seeds are often priced 15–30% higher than conventional varieties, deterring adoption among subsistence farmers (De Steur et al., 2017). Public-private partnerships, like HarvestPlus' collaboration with Bayer AG for vitamin A cassava in Nigeria, demonstrate potential but remain rare in LMICs (HarvestPlus, n.d.).

1.8.7 Challenges of Marginalized Populations in LMIC

The deployment of Community Health Workers (CHWs) is a widely endorsed strategy to mitigate health disparities in low- and middle-income countries (LMICs). However, there remains a scarcity of robust evidence on the effectiveness and mechanisms by which CHW programs can achieve these objectives (Ahmed et al., 2022). Similarly, the provision of mental health services in LMICs faces significant challenges due to resource constraints and the complexity of designing interventions that cater to marginalized populations, including older adults (Giebel et al., 2024).

One critical avenue for addressing health inequities in LMICs is the implementation of health insurance schemes aimed at achieving universal health coverage (UHC). Nonetheless, there is a notable gap in empirical evidence assessing the impact of these schemes beyond typical parameters like income level. Understanding how social inclusion influences access to equitable health financing arrangements remains crucial (van Hees et al., 2019). Furthermore, the expansion of public health insurance is intended to enhance healthcare access, reduce catastrophic healthcare expenditures, and improve health outcomes. However, the real-world efficacy of these goals in reducing health inequities continues to warrant extensive investigation (Erlangga et al., 2019).

Community-based health insurance (CBHI) schemes offer another potential solution. These schemes have been systematically evaluated for their impact on healthcare utilization and financial risk protection, yet their effectiveness in substantially reducing health inequities in LMICs is still under scrutiny (Eze et al., 2023). Food fortification and biofortification have been recognized as critical strategies to combat micronutrient deficiencies in vulnerable populations. The success of these programs hinges not only on the biological efficacy of fortified foods but also on their sustainable implementation, which necessitates ongoing monitoring, quality assurance, and high compliance rates (Osendarp et al., 2018).

Addressing the challenges faced by marginalized populations in LMICs requires an integrative approach that combines healthcare interventions, social insurance schemes, and nutritional programs to create inclusive policies and pathways that mitigate health and economic inequities.

1.8.8 Policy Formulation

Effective policy formulation and robust infrastructure development are essential to ensure the sustainability and scalability of biofortification programs in low- and middle-income countries (LMICs). Key aspects of policy formulation include continuous funding, quality control, and fostering public-private sector collaboration (Global Food and Nutrition Security, 2024). These elements are crucial to embedding biofortification in food systems and expanding its impact. By integrating biofortification into agricultural research centers, the program can leverage agronomically superior varieties, ensuring that farmers are motivated to adopt high-yielding, profitable crops. This high productivity subsequently leads to lower food prices, making nutritious food more accessible (Global Food and Nutrition Security, 2024).

1.8.9 Conflict and Infrastructure Deficits

Conflict zones, such as Yemen and South Sudan, face acute challenges in disseminating biofortified crops due to disrupted supply chains and humanitarian priorities. Mobile seed distribution networks and partnerships with NGOs (e.g., World Food Programme) have shown promise in crisis settings but require stable funding and political will.

1.9 Using Artificial Intelligence in Biofortification

Artificial intelligence (AI) offers a promising solution to some of these regulatory challenges. By streamlining regulatory processes, improving decision-making, and enhancing data analysis, AI can help companies maintain compliance more efficiently (BostonBiotech Advisors, 2024; Pavan, n.d.). AI-driven biofortification can provide transparent and data-driven insights, assisting policymakers in understanding and supporting biofortification initiatives (Olivera, 2024). Additionally, AI can tailor nutrient-enriched crops to specific malnutrition profiles prevalent in low- and middle-income countries, thereby maximizing the health benefits for vulnerable populations (Team Hoodin, 2024; BioBoston Consulting, 2024). This approach not only addresses regulatory compliance but also ensures that biofortified crops are optimized for addressing malnutrition in targeted regions.

As AI continues to evolve and integrate into various fields, the life sciences industry has begun leveraging this transformational technology to revolutionize regulatory affairs. Given the complexity and volume of regulatory requirements that companies must comply with, AI offers a promising solution to streamline processes and enhance accuracy.

In 2024, AI applications have significantly proliferated, focusing on assisting life sciences organizations to navigate the stringent regulatory landscape more efficiently. These applications are designed to enhance the management of regulatory data, making it more transparent and providing data-driven insights. This is particularly important for biofortification initiatives, where clear and accurate regulatory data is crucial for gaining the support of policymakers.

AI has shown considerable potential in the pharmaceutical industry, which is heavily regulated and faces substantial challenges in meeting regulatory requirements while accelerating the market entry of new drugs. By streamlining regulatory processes, improving decision-making, and enhancing data analysis, AI helps pharmaceutical companies overcome these challenges (Pavan, n.d.). The capabilities of AI in this context include providing insights that aid in regulatory compliance and optimizing the development and approval processes for new pharmaceutical products (Pavan, n.d.).

In regulatory compliance, staying ahead of changing regulations and maintaining adherence to them can be an overwhelming task, especially for the financial industry. Generative AI (gen AI) is revolutionizing how organizations access knowledge, enabling faster, more accurate access to critical information that drives compliance strategies and operational decisions (Olivera, 2024). According to IDC's Worldwide Generative Artificial Intelligence 2024 Predictions, "by 2025, two-thirds of businesses will leverage a combination of gen AI and retrieval-augmented generation (RAG) to power domain-specific self-service knowledge discovery, enhancing decision efficacy by 50%" (Olivera, 2024). Particularly through its ability to interact directly with documents, gen AI is transforming compliance by offering intuitive conversational access to up-to-date regulatory texts, empowering compliance teams to make faster, more informed decisions in an increasingly complex regulatory environment (Olivera, 2024).

A silent revolution is taking place in the life sciences industry, where AI is reshaping the landscape of regulatory compliance (BioBoston Consulting, 2024). As regulatory demands grow increasingly complex, AI is helping life sciences companies streamline their processes, ensuring compliance with greater efficiency and precision. Regulatory compliance has always been a cornerstone of the life sciences sector, ensuring the safety, efficacy, and quality of products that directly impact human health (BioBoston Consulting, 2024). The regulatory environment, governed by bodies like the US Food and Drug Administration (FDA) and European Medicines Agency (EMA), has become more intricate over time, leading to increased pressure on pharmaceutical, biotechnology, and medical device companies to meet stringent standards (BioBoston Consulting, 2024).

The integration of AI into compliance signifies a major shift in how businesses approach regulatory requirements (Suppress, 2024). Gone are the days when compliance was solely about manual reviews and checklists; today, AI introduces a new paradigm, transforming compliance into a dynamic, efficient, and more accurate process (Suppress, 2024). AI is reshaping the compliance landscape by automating routine tasks, analysing complex datasets, and identifying potential compliance issues before they escalate (Suppress, 2024). This technological advancement offers numerous benefits, including increased accuracy in compliance reporting, enhanced ability to predict regulatory risks, and improved efficiency in compliance-related tasks (Suppress, 2024).

In the ever-evolving landscape of biotechnology and pharmaceuticals, maintaining regulatory compliance stands as a paramount concern for industry stakeholders (BostonBiotech Advisors, 2024). Boston Biotech Advisors, renowned for its innovative solutions and forward-thinking strategies, has pioneered an AI-driven approach to streamline regulatory processes, FDA submissions, and compliance protocols (BostonBiotech Advisors, 2024). By harnessing the power of artificial intelligence, this consultation firm has revolutionized the paradigm of regulatory compliance, propelling clients towards unparalleled efficiency and success (BostonBiotech Advisors, 2024).

Regulatory affairs in sectors like pharmaceuticals, medical devices, biotechnology, and MedTech are facing unprecedented challenges (Team Hoodin, 2024). The increasingly complex regulations, such as the Medical Device Regulation (MDR) in Europe, combined with the growing demands of international compliance standards, are causing companies to struggle to keep up (Team Hoodin, 2024). According to a MedTech Europe Survey, over 85% of medical devices previously certified under older regulations have yet to transition to the new Medical Device Regulation (MDR), creating a backlog that has delayed market entry for many life-saving devices (Team Hoodin, 2024). Small- and medium-sized enterprises (SMEs) are hit the hardest, with many lacking access to certification bodies, forcing some to deprioritize key markets (Team Hoodin, 2024).

1.10 Future Directions and Perspectives

Despite significant advancements, biofortification faces critical gaps that hinder its scalability and impact. Addressing these gaps requires interdisciplinary research, policy innovation, and community engagement. Bridging these gaps demands a paradigm shift from isolated interventions to systemic, inclusive approaches. Prioritizing longitudinal studies, climate-adaptive crops, and equitable policies will position biofortification as a cornerstone of global food security. By addressing technical, social, and structural barriers, the scientific community can transform biofortification from a promising concept into a universally accessible solution for hidden hunger. Some of the key directions and their perspectives are discussed here.

1.10.1 Enhancing Public Acceptance Involvement

Efforts to enhance public acceptance of biofortified crops must address these misconceptions through education and transparent communication. Policymakers and scientists need to work collaboratively to demystify GMOs, highlighting their rigorous testing and proven safety (Thayer, 2016). Enhancing public awareness of the health benefits and sustainability of biofortified crops can foster better acceptance and support (Groff, 2024). Community engagement is vital in ensuring the successful adoption of biofortified crops, particularly in LCICs. Involving local communities in the planning and implementation stages can help tailor interventions to local needs and preferences, thereby improving their effectiveness and sustainability (Admin, 2019; Satish, 2023; Greenfield, 2024). Programs that educate and empower local farmers, like those led by HarvestPlus, have successfully integrated biofortified crops into rural communities, benefiting millions of households (Walton, 2021; HarvestPlus, n.d.). Additionally, integrating biofortified crops into community-based nutrition programs can significantly improve nutritional outcomes in vulnerable urban slum areas (Walton, 2021).

Training programs for farmers, supported by local extension services and demonstration plots, are also essential for the successful adoption and sustained use of biofortified crops (Okello et al., 2019). By educating farmers on the benefits and techniques of biofortification, these programs can establish robust local support systems that facilitate the long-term success of biofortification initiatives (Walton, 2021).

1.10.2 Improving Public Awareness and Education

Improving public awareness and education about the health benefits of biofortified crops is crucial in enhancing their acceptance and support. A series of research papers and a free online data dashboard have sought to boost the use of biofortification by increasing the concentrations of essential nutrients in staple crops, which offers an affordable, sustainable, and climate-smart solution to global malnutrition (Groff, 2024). The most recent paper, published in Nutrition Research Reviews, highlighted that better awareness of the crop production and nutrition benefits of biofortified crops facilitated their adoption, particularly when farmers were informed about the relative advantages compared to traditional varieties (Groff, 2024). Enhanced public awareness and education can thus lead to better acceptance and support for biofortified crops, which in turn can amplify their economic impacts.

1.10.3 Mitigating Societal Misconceptions and Concerns

Biofortified crops, particularly those involving genetically modified organisms (GMOs), have faced significant societal concerns and misconceptions. One primary concern is the belief that GMOs are inherently harmful or pose a risk to human health. This misconception persists despite extensive scientific evidence indicating that genetically modified foods are among the most thoroughly tested, with no evidence of harm to human health (Thayer, 2016). The rigorous testing process for GMOs, which takes an average of 13 years and $136 million per seed, ensures their safety (Thayer, 2016). Another common misconception is that GMOs are ubiquitous. Despite this, public opposition remains strong, fueled by fears that GMOs are poisonous and environmentally damaging. However, scientific research consistently shows that GMOs do not pose a health risk and can make agriculture more sustainable by reducing the need for insecticides (Blancke, 2015).

Ethical and sociocultural concerns also play a significant role in shaping public opinion. These concerns include issues related to self-determination, food justice, and equitable access to biofortified crops (Fanzo & Glass, 2018). To address these issues, it is crucial to involve communities early in the research and implementation of biofortification programs, ensuring that the benefits are fairly distributed and that the interventions are culturally acceptable (Fanzo & Glass, 2018). To improve public acceptance and trust in biofortified crops, policymakers and scientists need to work collaboratively. This can be achieved by transparently addressing ethical considerations, educating the public about the safety and benefits of GMOs, and ensuring that biofortification strategies do not disproportionately impact vulnerable populations. By tackling these misconceptions and societal concerns, biofortified crops can more effectively contribute to public health goals and sustainable agriculture (Blancke, 2015).

1.10.4 Developing Community Engagement and Adoption Strategies

Community engagement and education are essential for ensuring the adoption and equitable access to biofortified crops, particularly in LMICs. Malnutrition remains a pervasive issue globally, affecting millions, especially in sub-Saharan Africa (SSA), where children and women are the most impacted (Omari et al., 2019). Biofortification, which enhances the nutritional content of food crops through methods like conventional breeding and genetic engineering, has emerged as a viable solution to combat micronutrient deficiencies (Satish, 2023).

The success of biofortification initiatives significantly depends on the involvement and education of local communities. Engaging local communities allows for the tapping into valuable indigenous knowledge and expertise that can facilitate the adaptation and sustainability of biofortified crops (Walton, 2021). Community-based nutrition programs integrated with biofortification efforts have shown promise in improving nutritional outcomes, especially in urban slum areas within LMICs (Walton, 2021). Such programs can address both malnutrition and food insecurity among the most vulnerable populations by ensuring that biofortified crops are not only available but also culturally accepted and appropriately utilized.

A critical aspect of successful biofortification initiatives involves training programs for farmers. These programs should encompass education on the benefits and techniques of growing biofortified crops while establishing robust local support systems, including extension services and demonstration plots (Walton, 2021). For instance, the HarvestPlus initiative has significantly advanced the adoption of biofortified crops by engaging millions of smallholder households and ensuring they are well-informed about the advantages of these nutrient-rich crops (Walton, 2021).

The role of women and youth in biofortification efforts cannot be overstated. Women, being the primary caregivers and responsible for food preparation and child feeding, are pivotal in promoting household nutrition and food security (Malapit et al., 2015). Initiatives that engage women, such as the Lead Mother initiative in Uganda, demonstrate how empowering women with knowledge about biofortified crops can lead to improved health and nutrition outcomes for entire communities (Malapit et al., 2015). Moreover, addressing gender-specific barriers and enhancing women's involvement in biofortification initiatives can significantly improve adoption rates and leverage their roles in household food security and nutrition education (Birol & Saltzman, 2014; Malapit et al., 2015; Groff, 2024). Engaging youth in biofortification efforts is equally important as they represent the future of agricultural innovation and sustainability (Omari et al., 2019; Lowe, 2025). Programs that focus on training young farmers and involving them in the cultivation and advocacy of biofortified crops can ensure the long-term success and scalability of these initiatives.

To effectively engage communities in biofortification projects, it is crucial to understand and respect local cultural practices (MacDonald, 2017; Huey et al., 2024). Integrating traditional agricultural knowledge with modern genetic modification techniques can help in creating biofortified crops that are not only nutritionally superior but also culturally accepted and sustainable (MacDonald, 2017; Green, 2024). This approach ensures that the adoption of biofortified crops does not disrupt existing farming practices and respects the cultural values of the communities involved.

Finally, we outline key priority areas for future work, grounded in evidence from existing studies and aligned with global health and sustainability goals, which are:

- Conduct decade-long cohort studies to assess biofortification's role in breaking cycles of malnutrition.
- Integrate biofortified crops into national health surveillance systems to track population-level outcomes.
- Breed climate-resilient biofortified varieties using phenotyping and genomic selection (e.g., drought-tolerant Zn-rice).
- Evaluate nutrient retention in biofortified crops under projected climate scenarios (e.g., IPCC RCP 4.5/8.5).
- Establish regional regulatory harmonization frameworks (e.g., African Union's biosafety guidelines) to accelerate GM crop adoption.
- Incentivize private-sector investment through tax breaks or public-private partnerships (e.g., HarvestPlus-Bayer collaborations).
- Deploy culturally tailored marketing (e.g., mobile apps with localized recipes for biofortified crops).
- Engage influencers (e.g., community leaders, celebrities) to shift perceptions of biofortified foods.
- Develop low-phytate biofortified varieties using CRISPR-Cas9 gene editing.
- Engineer synergistic traits (e.g., high vitamin C in staples to enhance Fe absorption).
- Implement subsidized seed voucher programs targeting women farmers.
- Leverage digital platforms (e.g., blockchain) for transparent seed distribution.
- Create cross-sector consortia (e.g., nutritionists, climate scientists, economists) to design holistic biofortification strategies.
- Integrate biofortification into national nutrition programs and farmer Incentives.
- Integrate biofortification into university curricula to build capacity in LMICs.

1.11 Conclusion

Biofortification stands as a transformative intervention in addressing micronutrient malnutrition, directly contributing to the United Nations Sustainable Development Goals (SDGs), notably SDG 2 (Zero Hunger) and SDG 3 (Good Health and Well-being). Over two decades, initiatives like HarvestPlus have validated the efficacy of biofortified crops, such as iron-enriched beans, zinc-fortified rice, and provitamin A maize, in reducing anaemia prevalence by 20–30%, improving immune function, and mitigating nutrient gaps in vulnerable populations. However, translating experimental success into widespread adoption requires addressing persistent technical, socioeconomic, and regulatory challenges.

Scalability of biofortified crops is hindered by bioavailability constraints, nutrient degradation during processing, and climate-induced instability. For example, phytic acid in iron-biofortified beans reduces iron absorption by 40–50%, while provitamin A maize experiences up to 60% carotenoid loss under high temperatures. Innovations such as CRISPR-Cas9-mediated development of low-phytate crops and nanotechnology-based nutrient encapsulation offer promising solutions to enhance nutrient retention. Concurrently, breeding programs must prioritize climate-resilient traits, like drought tolerance in sorghum and heat-stable carotenoids in maize, to ensure agronomic viability under changing climatic conditions.

Adoption of biofortified crops is impeded by systemic inequities, including limited seed access for smallholder farmers and gendered disparities in agricultural decision-making. For instance, women farmers in sub-Saharan Africa face 30–50% higher barriers to adopting biofortified varieties due to cost and distribution gaps. Cultural resistance, such as consumer aversion to the yellow hue of provitamin A maize, necessitates community-centric strategies. Successful models include India's *Mahila Kisan Sashaktikaran Pariyojana*, which integrates nutrition education with participatory breeding, increasing adoption rates by 25% in pilot regions.

Fragmented regulatory frameworks and underfunded programs delay deployment, exemplified by the 20-year approval timeline for Golden Rice. Harmonizing biosafety regulations, as seen in the African Union's regional guidelines, and fostering partnerships like HarvestPlus-Bayer AG collaborations can accelerate progress. Nigeria's mandate for biofortified cassava, which reduced vitamin A deficiency by 15% in target populations, underscores the impact of policy integration.

The future of biofortification demands interdisciplinary synergy. Longitudinal studies tracking health impacts over generations, digital tools for equitable seed distribution, and climate-nutrition consortia are critical. For example, mobile platforms in Kenya improved smallholder access to biofortified seeds by 40%. Policymakers, researchers, and communities must co-design solutions that balance innovation with cultural relevance, ensuring biofortified crops reach those who are most in need.

Finally, biofortification is not merely a scientific pursuit but a moral imperative. By resolving technical bottlenecks, dismantling socioeconomic inequities, and fortifying policy frameworks, this strategy can transition from niche intervention to a cornerstone of global food systems. As climate change and population growth exacerbate food insecurity, biofortification must ensure equitable access to nourishing, resilient, and culturally acceptable food. The time for decisive, collaborative action is now.

References

Adetola, O., Duodu, K., & Taylor, J. (2022). Food-to-food fortification of ready-to-eat cereal-based porridges: A potential sustainable food-based strategy to improve bioavailable iron and zinc in plant-based diets. *Current Developments in Nutrition, 6*. https://doi.org/10.1093/cdn/nzac077.001

Admin. (2019). *Integrating indigenous knowledge and traditional practices for biodiversity conservation in a modern world*. Environmental Reports. https://doi.org/10.51470/ER.2019.1.2.04.

Ahmad, S. (2020). *Geochemical dynamics and bioavailability of iodine and selenium*. https://eprints.nottingham.ac.uk/60480/. Accessed 5 Apr 2025.

Ahmad, W., Zou, Z., Awais, M., Munsif, F., Khan, A., Nepal, J., Ahmad, M., Akbar, S., Ahmad, I., Khan, M. S., Qamar, Z., & Khan, H. (2023). Seed-primed and foliar oxozinc nanofiber application increased wheat production and zn biofortification in calcareous-alkaline soil. *Agronomy, 13*. https://doi.org/10.3390/agronomy13020400

Ahmed, S., Chase, L. E., Wagnild, J., Akhter, N., Sturridge, S., Clarke, A., Chowdhary, P., Mukami, D., Kasim, A., & Hampshire, K. (2022). Community health workers and health equity in low- and middle-income countries: systematic review and recommendations for policy and practice. *International Journal for Equity in Health, 21*, 49. https://doi.org/10.1186/s12939-021-01615-y

Ahonobadha, E. (2024). Harmonization of East African food and nutrition standards. In: Millers for Nutrition. https://millersfornutrition.com/2024/12/harmonization-of-east-african-food-and-nutrition-standards/. Accessed 30 Mar 2025.

Al-Babili, S., & Beyer, P. (2005). Golden rice – Five years on the road – Five years to go? *Trends in Plant Science, 10*, 565–573. https://doi.org/10.1016/j.tplants.2005.10.006

Amna, Q. S., Tantray, A. Y., Bashir, S. S., Zaid, A., & Wani, S. H. (2020). Golden rice: Genetic engineering, promises, present status and future prospects. In *Rice research for quality improvement: genomics and genetic engineering* (pp. 581–604). Springer Singapore.

Andersson, M., Saltzman, A., Virk, P., & Pfeiffer, W. (2017). Progress update: Crop development of biofortified staple food crops under HarvestPlus. *African Journal of Food, Agriculture, Nutrition and Development, 17*, 11905–11935. https://doi.org/10.18697/AJFAND.78.HARVESTPLUS05

Asif, M., Qureshi, I., Bangroo, S., Mahdi, S. S., Sheikh, F. A., Bhat, M. A., Alie, B. A., Khan, M. H., Dar, N. A., Dar, Z. A., Jha, U. C., & Lone, A. A. (2022). Reduction of phytic acid and enhancement of bioavailable micronutrients in common beans (Phaseolus vulgaris L.) in changing climatic scenario. In *Developing climate resilient grain and forage legumes* (pp. 59–76). Springer Nature Singapore.

Banerjee, S., Roy, P., Nandi, S., & Roy, S. (2023). Advanced biotechnological strategies towards the development of crops with enhanced micronutrient content. *Plant Growth Regulation, 100*, 355–371. https://doi.org/10.1007/s10725-023-00968-4

Baye, K., Guyot, J.-P., & Mouquet-Rivier, C. (2017). The unresolved role of dietary fibers on mineral absorption. *Critical Reviews in Food Science and Nutrition, 57*, 949–957. https://doi.org/10.1080/10408398.2014.953030

Beebe. S., & Andersson, M. (2014). Iron Common Bean (Phaseolus vulgaris).

Beyer, P. (2010). Golden Rice and 'Golden' crops for human nutrition. *New Biotechnology, 27*, 478–481. https://doi.org/10.1016/j.nbt.2010.05.010

Bhardwaj, A., Chejara, S., Malik, K., Kumar, R., Kumar, A., & Yadav, R. (2022). Agronomic biofortification of food crops: An emerging opportunity for global food and nutritional security. *Frontiers in Plant Science, 13*. https://doi.org/10.3389/fpls.2022.1055278

Bhat, A. A., Nisar, S., Mukherjee, S., Saha, N., Yarravarapu, N., Lone, S. N., Masoodi, T., Chauhan, R., Maacha, S., Bagga, P., Dhawan, P., Akil, A. A.-S., El-Rifai, W., Uddin, S., Reddy, R., Singh, M., Macha, M. A., & Haris, M. (2022). Integration of CRISPR/Cas9 with artificial intelligence for improved cancer therapeutics. *Journal of Translational Medicine, 20*. https://doi.org/10.1186/s12967-022-03765-1

BioBostonConsulting. (2024). *AI's quiet revolution: Transforming regulatory compliance*. https://biobostonconsulting.com/ais-revolution-transforming-regulatory-compliance-in-life-sciences/, https://biobostonconsulting.com/ais-revolution-transforming-regulatory-compliance-in-life-sciences/, https://www.example.com/ai-revolution-in-regulatory-compliance. Accessed 30 Mar 2025.

Biofortificationhub. (2025). What is biofortification? In: Biofortification Hub. https://biofortificationhub.org/about/what-is-biofortification/. Accessed 30 Mar 2025.

Birol, E., & Saltzman, A. (2014). *Promising opportunities: The role and importance of gender in biofortification research*. https://a4nh.cgiar.org/2014/11/17/the-role-and-importance-of-gender-in-biofortification-research/. Accessed 30 Mar 2025.

Blancke, S. (2015). Why people oppose GMOs even though science says they are safe | Scientific American. In: SCIAM. https://www.scientificamerican.com/article/why-people-oppose-gmos-even-though-science-says-they-are-safe/. Accessed 30 Mar 2025

Blanco-Rojo, R., & Vaquero, P. (2019). Iron bioavailability from food fortification to precision nutrition. A review. *Innovative Food Science & Emerging Technologies, 51*. https://doi.org/10.1016/J.IFSET.2018.04.015

Blancquaert, D., De Steur, H., Gellynck, X., & Van Der Straeten, D. (2014). Present and future of folate biofortification of crop plants. *Journal of Experimental Botany, 65*(4), 895–906. https://doi.org/10.1093/jxb/ert483

BostonBiotech Advisors. (2024). *Streamlining regulatory processes: Boston biotech advisors' approach to ai-driven compliance*. In: Boston Biotech Advis. https://www.bostonbiotechadvisors.com/post/streamlining-regulatory-processes-boston-biotech-advisors-approach-to-ai-driven-compliance. Accessed 30 Mar 2025.

Botha, G. M., & Viljoen, C. D. (2008). Can GM sorghum impact Africa? *Trends in Biotechnology, 26*, 64–69. https://doi.org/10.1016/j.tibtech.2007.10.008

Bouis, H. (2002). Plant breeding: A new tool for fighting micronutrient malnutrition. *The Journal of Nutrition, 132*(3), 491–494. https://doi.org/10.1093/JN/132.3.491S

Bouis, H. (2003). Micronutrient fortification of plants through plant breeding: Can it improve nutrition in man at low cost? *Proceedings of the Nutrition Society, 62*, 403–411. https://doi.org/10.1079/PNS2003262

Bouis, H., & Welch, R. (2010). Biofortification – A sustainable agricultural strategy for reducing micronutrient malnutrition in the global south. *Crop Science, 50*. https://doi.org/10.2135/CROPSCI2009.09.0531

Bouis, H., Hotz, C., Mcclafferty, B., Meenakshi, J., & Pfeiffer, W. (2011). Biofortification: A new tool to reduce micronutrient malnutrition. *Food and Nutrition Bulletin, 32*. https://doi.org/10.1177/15648265110321S105

Bouis, H., Birol, E., Boy, E., Gannon, B. M., Haas, J. D., Low, .J, Mehta, S., Michaux, K., Mudyahoto, B., & Pfeiffer, W. (2020). *Food biofortification: Reaping the benefits of science to overcome hidden hunger*. Council for Agricultural Science and Technology (CAST).

Boy, E., Haas, J., Petry, N., Cercamondi, C., Gahutu, J., Mehta, S., Finkelstein, J., & Hurrell, R. (2017). Efficacy of iron-biofortified crops. *African Journal of Food, Agriculture, Nutrition and Development, 17*, 11879–11892. https://doi.org/10.18697/AJFAND.78.HARVESTPLUS03

Brnić, M., Wegmüller, R., Melse-Boonstra, A., Stomph, T., Zeder, C., Tay, F., & Hurrell, R. (2016). Zinc absorption by adults is similar from intrinsically labeled zinc-biofortified rice and from rice fortified with labeled zinc sulfate. *The Journal of Nutrition, 146*(1), 76–80. https://doi.org/10.3945/jn.115.213421

Burt, A. J., Grainger, C. M., Young, J. C., Shelp, B. J., & Lee, E. A. (2010). Impact of postharvest handling on carotenoid concentration and composition in high-carotenoid maize (Zea mays L.) kernels. *Journal of Agricultural and Food Chemistry, 58*, 8286–8292. https://doi.org/10.1021/jf100161r

Cakmak, I. (2007). Enrichment of cereal grains with zinc: Agronomic or genetic biofortification? *Plant and Soil, 302*, 1–17. https://doi.org/10.1007/s11104-007-9466-3

Cakmak, I., & Kutman, U. (2018). Agronomic biofortification of cereals with zinc: A review. *European Journal of Soil Science, 69*. https://doi.org/10.1111/ejss.12437

Cakmak, I., Pfeiffer, W., & Mcclafferty, B. (2010). Biofortification of durum wheat with zinc and iron. *Cereal Chemistry, 87*, 10–20. https://doi.org/10.1094/CCHEM-87-1-0010

Campos, H., & Caligari, P. D. S. (2017). *Genetic improvement of tropical crops*. Springer International Publishing. https://doi.org/10.1007/978-3-319-59819-2

Carvalho, L. J., Corrêa, M., Pereira, E., Nutti, M., Carvalho, J. V., Ribeiro, E. G., & Freitas, S. (2012). Iron and zinc retention in common beans (Phaseolus vulgaris L.) after home cooking. *Food & Nutrition Research, 56*(1), 15618.

Cercamondi, C. I., Egli, I. M., Mitchikpe, E., Tossou, F., Zeder, C., Hounhouigan, J. D., & Hurrell, R. F. (2013). Total iron absorption by young women from iron-biofortified pearl millet composite meals is double that from regular millet meals but less than that from post-harvest iron-fortified millet meals. *The Journal of Nutrition, 143*, 1376–1382. https://doi.org/10.3945/jn.113.176826

CGIAR. (2003). *Innovation: Crops bred to improve nutrition*. https://www.cgiar.org/innovations/crops-bred-to-improve-nutrition/. Accessed 30 Mar 2025.

CGIAR. (2018). *How the science of biofortification grew from an idea to a food revolution*. https://www.cgiar.org/news-events/news/science-biofortification-grew-idea-food-revolution. Accessed 30 Mar 2025.

Chávez, A., Sánchez, T., Ceballos, H., Rodriguez-Amaya, D., Nestel, P., Tohme, J., & Ishitani, M. (2007). Retention of carotenoids in cassava roots submitted to different processing methods. *Journal of the Science of Food and Agriculture, 87*, 388–393. https://doi.org/10.1002/jsfa.2704

Cruet-Burgos, C., & Rhodes, D. H. (2023). Unraveling transcriptomics of sorghum grain carotenoids: A step forward for biofortification. *BMC Genomics, 24*. https://doi.org/10.1186/s12864-023-09323-3

Davey, R. (2024). *How biofortification strengthens food security*. In: AZO LIFESCIENCES. https://www.azolifesciences.com/article/How-Biofortification-Strengthens-Food-Security.aspx. Accessed 30 Mar 2025.

De Steur, H., Mehta, S., Gellynck, X., & Finkelstein, J. (2017). GM biofortified crops: Potential effects on targeting the micronutrient intake gap in human populations. *Current Opinion in Biotechnology, 44*, 181–188. https://doi.org/10.1016/j.copbio.2017.02.003

Dhaliwal, S., Sharma, V., Shukla, A., Shivay, Y., Hossain, A., Verma, V., Gill, M., Singh, J., Bhatti, S., Verma, G., Singh, J., & Singh, P. (2022a). Agronomic biofortification of forage crops with Zn and Cu for enhancing nutritive potential – A systematic review. *Journal of the science of food and agriculture*. https://doi.org/10.1002/jsfa.12353

Dhaliwal, S., Sharma, V., Shukla, A., Verma, V., Kaur, M., Shivay, Y., Nisar, S., Gaber, A., Brestič, M., Bárek, V., Skalický, M., Ondrisik, P., & Hossain, A. (2022b). Biofortification – A frontier novel approach to enrich micronutrients in field crops to encounter the nutritional security. *Molecules, 27*. https://doi.org/10.3390/molecules27041340

Dhiman, K., Sharma, D., Kumari, R., & Tomar, P. (2023). Biofortification of crops using microbes – A promising sustainable agriculture strategy. *Journal of Plant Nutrition, 46*, 2912–2935. https://doi.org/10.1080/01904167.2022.2160755

Dias, D. M., Costa, N., Nutti, M., Tako, E., & Martino, H. (2018). Advantages and limitations of in vitro and in vivo methods of iron and zinc bioavailability evaluation in the assessment of biofortification program effectiveness. *Critical Reviews in Food Science and Nutrition, 58*, 2136–2146. https://doi.org/10.1080/10408398.2017.1306484

Díaz-Gómez, J., Twyman, R., Zhu, C., Farré, G., Serrano, J., Portero-Otín, M., Muñoz, P., Sandmann, G., Capell, T., & Christou, P. (2017). Biofortification of crops with nutrients: factors affecting utilization and storage. *Current Opinion in Biotechnology, 44*, 115–123. https://doi.org/10.1016/j.copbio.2016.12.002

Dimkpa, C. O., & Bindraban, P. S. (2018). Nanofertilizers: New products for the industry? *Journal of Agricultural and Food Chemistry, 66*, 6462–6473. https://doi.org/10.1021/acs.jafc.7b02150

Ding, K., Shan, Y., Wang, L., Zhang, Y., & Tian, G. (2024). Transcriptomics combined with physiological analysis and metabolomics revealed the response of potato tuber formation to nitrogen. *BMC Plant Biology, 24*. https://doi.org/10.1186/s12870-024-05758-2

Dipti, S. (2012). *Bioavailability of selected minerals in different processing and cooking methods of rice (Oryza sativa L.)*. Human Nutrition Doctoral Thesis.

Dubock, A. (2017). An overview of agriculture, nutrition and fortification, supplementation and biofortification: Golden Rice as an example for enhancing micronutrient intake. *Agriculture & Food Security, 6*, 1–20. https://doi.org/10.1186/s40066-017-0135-3

Dwivedi, S., Garcia-Oliveira, A., Govindaraj, M., & Ortiz, R. (2023). Biofortification to avoid malnutrition in humans in a changing climate: Enhancing micronutrient bioavailability in seed, tuber, and storage roots. *Frontiers in Plant Science, 14*. https://doi.org/10.3389/fpls.2023.1119148

Erlangga, D., Suhrcke, M., Ali, S., & Bloor, K. (2019). The impact of public health insurance on health care utilisation, financial protection and health status in low- and middle-income countries: A systematic review. *PLoS One, 14*, e0219731. https://doi.org/10.1371/journal.pone.0219731

Eruaga, M. A. (2024). Enhancing global food safety standards through international collaboration and policy harmonization. *International Journal of Scholarly Research in Multidisciplinary Studies, 4*. https://doi.org/10.56781/ijsrms.2024.4.1.0027

Eze, P., Ilechukwu, S., & Lawani, L. O. (2023). Impact of community-based health insurance in low- and middle-income countries: A systematic review and meta-analysis. *PLoS One, 18*, e0287600. https://doi.org/10.1371/journal.pone.0287600

Fageria, N., Baligar, V., & Li, Y. (2008). The role of nutrient efficient plants in improving crop yields in the twenty first century. *Journal of Plant Nutrition, 31*, 1121–1157. https://doi.org/10.1080/01904160802116068

Fageria, N. K., Baligar, V. C., & Jones, C. A. (2010). *Growth and mineral nutrition of field crops*. CRC Press.

Failla, M. L., Chitchumroonchokchai, C., Siritunga, D., De Moura, F. F., Fregene, M., Manary, M. J., & Sayre, R. T. (2012). Retention during processing and bioaccessibility of β-carotene in high β-carotene transgenic cassava root. *Journal of Agricultural and Food Chemistry, 60*, 3861–3866. https://doi.org/10.1021/jf204958w

Fanzo, J., & Glass, S. (2018). Ethical and sociocultural considerations of biofortified crops: Ensuring value and sustainability for public health. In *Advances in food security and sustainability* (pp. 93–133). Elsevier.

Finkelstein, J. L., Fothergill, A., Hackl, L. S., Haas, J. D., & Mehta, S. (2019). Iron biofortification interventions to improve iron status and functional outcomes. *Proceedings of the Nutrition Society, 78*, 197–207. https://doi.org/10.1017/s0029665118002847

Fitzpatrick, C., Nwankwo, U., Lenk, E., de Vlas, S. J., & Bundy, D. A. P. (2017). An investment case for ending neglected tropical diseases.

Food Safety Magazine Editorial Team. (n.d.). International harmonization of food safety standards. In: *Food Safety Magazine*. https://www.food-safety.com/articles/6364-international-harmonization-of-food-safety-standards. Accessed 30 Mar 2025.

Futureharvest. (2025). Beginner farmer training program. In: *Future Harvest*. https://futureharvest.org/programs/beginner-farmer-training-program/. Accessed 6 Apr 2025.

GAIN. (2025). Thde Columbia Climate School. Cornell University College of Agriculture and Life Sciences. National biofortification policies adopted and in-process – Map – Food Systems Dashboard. In: Food Systems Dashboard. https://www.foodsystemsdashboard.org/indicators/food-supply-chains/production-systems-and-input-supply/national-biofortification-policies-and-programs-policy-adopted-and-in-process/map. Accessed 30 Mar 2025.

Gao, C., Kikulwe, E., Kuzma, J., Lema, M., Lidder, P., Robinson, J., Wessler, J., & Zhao, K. (2022). *Gene editing and agrifood systems*. FAO.

Garcia-Casal, M., Peña-Rosas, J., & Giyose, B. (2017). Staple crops biofortified with increased vitamins and minerals: considerations for a public health strategy. *Annals of the New York Academy of Sciences, 1390*. https://doi.org/10.1111/nyas.13293

Garg, M., Sharma, N., Sharma, S., Kapoor, P., Kumar, A., Chunduri, V., & Arora, P. (2018). Biofortified crops generated by breeding, agronomy, and transgenic approaches are improving lives of millions of people around the world. *Frontiers in Nutrition, 5*. https://doi.org/10.3389/fnut.2018.00012

Garousi, F. (2017). Selenium in soil–plant–food systems. *Acta Universitatis Sapientiae, Alimentaria, 10*, 91–106. https://doi.org/10.1515/ausal-2017-0006

Geny, A., Petitjean, M., Van Wymelbeke-Delannoy, V., & Sulmont-Rossé, C. (2023). Impact of food-based fortification on nutritional outcomes and acceptability in older adults: systematic literature review. *Frontiers in Nutrition, 10*. https://doi.org/10.3389/fnut.2023.1232502

Gharibzahedi, S. M. T., & Jafari, S. M. (2017). The importance of minerals in human nutrition: Bioavailability, food fortification, processing effects and nanoencapsulation. *Trends in Food Science & Technology, 62*, 119–132. https://doi.org/10.1016/j.tifs.2017.02.017

Giebel, C., Gabbay, M., Shrestha, N., Saldarriaga, G., Reilly, S., White, R., Liu, G., Allen, D., & Zuluaga, M. I. (2024). Community-based mental health interventions in low- and middle-income countries: a qualitative study with international experts. *International Journal for Equity in Health, 23*, 19. https://doi.org/10.1186/s12939-024-02106-6

Global Food and Nutrition Security. (2023). Biofortification: A responsible research and innovation strategy of the G20 | Knowledge for policy. In: Knowledge for policy. https://knowledge4policy.ec.europa.eu/publication/biofortification-responsible-research-innovation-strategy-g20_en. Accessed 30 Mar 2025.

Global Food and Nutrition Security. (2024). Biofortification: future challenges for a newly emerging technology to improve nutrition security sustainably. In: Knowledge for policy. https://knowledge4policy.ec.europa.eu/publication/biofortification-future-challenges-newly-emerging-technology-improve-nutrition-security_en. Accessed 30 Mar 2025.

Gomes, M., Martino, H., & Tako, E. (2021a). Zinc-biofortified staple food crops to improve zinc status in humans: a systematic review. *Critical Reviews in Food Science and Nutrition, 63*, 4966–4978. https://doi.org/10.1080/10408398.2021.2010032

Gomes, M. J. C., Martino, H. S. D., & Tako, E. (2021b). Effects of iron and zinc biofortified foods on gut microbiota in vivo (Gallus gallus): A systematic review. *Nutrients, 13*. https://doi.org/10.3390/nu13010189

Gómez-Galera, S., Rojas, E., Sudhakar, D., Zhu, C., Pelacho, A., Capell, T., & Christou, P. (2010). Critical evaluation of strategies for mineral fortification of staple food crops. *Transgenic Research, 19*, 165–180. https://doi.org/10.1007/s11248-009-9311-y

Govindaraj, M., Pujar, M., Srivastava, R., Gupta, S. K., & Pfeiffer, W. H. (2024). Genetic biofortification of pearl millet: Trait priority, breeding and genomic progress. In *Pearl millet in the 21st century* (pp. 221–246). Springer Nature Singapore.

Green, W. (2024). Traditional farming practices and their contribution to crop diversity – HusFarm. https://husfarm.com/article/traditional-farming-practices-and-their-contribution-to-crop-diversity. Accessed 30 Mar 2025.

Greenfield, E. (2024). Inclusive conservation: How engaging local communities is key to successful environmental protection. In: Sigma Earth. https://sigmaearth.com/inclusive-conservation-how-engaging-local-communities-is-key-to-successful-environmental-protection/. Accessed 6 Apr 2025.

Groff, E. (2024). Research boosts potential of biofortification on nutrition policy, intervention. In: Cornell Chronicle. https://news.cornell.edu/stories/2024/11/research-boosts-potential-biofortification-nutrition-policy-intervention. Accessed 30 Mar 2025.

Gupta, D., Dadu, R. H. R., Sambasivam, P., Bar, I., Singh, M., Beera, N., & Biju, S. (2019). Toward climate-resilient lentils: Challenges and opportunities. In *Genomic designing of climate-smart pulse crops* (pp. 165–234). Springer International Publishing.

HarvestPlus. (n.d.). Enriching 100 million lives. In: HarvestPlus. https://www.harvestplus.org/. Accessed 5 Apr 2025.

Haskell, M. J., Jamil, K. M., Hassan, F., Peerson, J. M., Hossain, M. I., Fuchs, G. J., & Brown, K. H. (2004). Daily consumption of Indian spinach (Basella alba) or sweet potatoes has a positive effect on total-body vitamin A stores in Bangladeshi men. *The American Journal of Clinical Nutrition, 80*, 705–714. https://doi.org/10.1093/ajcn/80.3.705

Hasnain, M., Nazir, Q., Saleem, I., Shah, S., Ali, M. A., & Ain, N. (2023). Effectiveness of bio-activated zinc coated urea (Engro Zabardast Urea) for growth, yield, and biochemical parameters of rice crop (Oryza sativa L.). *Pakistan Journal of Agricultural Research, 36*. https://doi.org/10.17582/journal.pjar/2023/36.1.80.88

Hassan, U. U., Khan, M., Younas, Z., Raja, N., & Mashwani, Z. (2023). Effect of phytogenic iron nanoparticles on the bio-fortification of wheat varieties. *Green Processing and Synthesis, 12*. https://doi.org/10.1515/gps-2023-8002

Hotz, C., Loechl, C., Lubowa, A., Tumwine, J. K., Ndeezi, G., Nandutu Masawi, A., Baingana, R., Carriquiry, A., De Brauw, A., Meenakshi, J. V., & Gilligan, D. O. (2012). Introduction of β-carotene–rich orange sweet potato in rural Uganda resulted in increased vitamin a intakes among children and women and improved vitamin a status among children3. *The Journal of Nutrition, 142*, 1871–1880. https://doi.org/10.3945/jn.111.151829

Huertas, R., William Allwood, J., Hancock, R. D., & Stewart, D. (2022). Iron and zinc bioavailability in common bean (Phaseolus vulgaris) is dependent on chemical composition and cooking method. *Food Chemistry, 387*, 132900. https://doi.org/10.1016/j.foodchem.2022.132900

Huey, S., Venkatramanan, S., Udipi, S., Finkelstein, J., Ghugre, P., Haas, J., Thakker, V., Thorat, A., Salvi, A., Kurpad, A., & Mehta, S. (2017). Acceptability of iron- and zinc-biofortified pearl millet (ICTP-8203)-based complementary foods among children in an urban slum of Mumbai, India. *Frontiers in Nutrition, 4*. https://doi.org/10.3389/fnut.2017.00039

Huey, S. L., Islam, S., Mehta, N. H., Konieczynski, E. M., Friesen, V. M., Krisher, J. T., Mbuya, M. N. N., Monterrosa, E. C., Nyangaresi, A. M., & Mehta, S. (2024). Review of the facilitators and barriers to adoption of biofortified foods and food products. *Nutrition Research Reviews, 1–22*. https://doi.org/10.1017/S0954422424000258

Hulshof, P., Kosmeijer-Schuil, T., West, C., & Hollman, P. (2007). Quick screening of maize kernels for provitamin A content. *Journal of Food Composition and Analysis, 20*, 655–661. https://doi.org/10.1016/J.JFCA.2006.04.014

Ilona, P., Bouis, H. E., Palenberg, M., Moursi, M., & Oparinde, A. (2017). Vitamin A cassava in Nigeria: Crop development and delivery. *African Journal of Food, Agriculture, Nutrition and Development, 17*, 12000–12025.

Jalal, A., Galindo, F., De Freitas, L. A., Oliveira, C., Lima, B., Pereira, Í., Ferraz, G. F., De Souza, J. S., Da Costa, K. N., Nogueira, T., & Filho, M. T. (2022). Yield, zinc efficiencies and biofortification of wheat with zinc sulfate application in soil and foliar nanozinc fertilisation. *Crop and Pasture Science, 73*, 749–759. https://doi.org/10.1071/CP21458

Jan, B., Bhat, T., Sheikh, T., Wani, O., Bhat, M., Nazir, A., Fayaz, S., Mushtaq, T., Farooq, A., Wani, S., & Rashid, A. (2020). Agronomic bio-fortification of rice and maize with iron and zinc: A review. *International Research Journal of Pure and Applied Chemistry, 28–37*. https://doi.org/10.9734/IRJPAC/2020/V21I1630257

Jiang, Z., Zhou, S., Peng, Y.-Y., Wen, X., Ni, Y., & Li, M. (2023). Effect of milling on nutritional components in common and zinc-biofortified wheat. *Nutrients, 15*. https://doi.org/10.3390/nu15040833

Joshi, G., Soe, Y. P., Palanog, A., Hore, T., Nha, C. T., Calayugan, M. I., Inabangan-Asilo, M. A., Amparado, A., Pandey, I. D., Cruz, P. S., Hernandez, J., & Swamy, B. M. (2023). Meta-QTL s and haplotypes for efficient zinc biofortification of rice. *The Plant Genome, 16*. https://doi.org/10.1002/tpg2.20315

Karandikar, B., Smale, M., Birol, E., & Tedla-Diressie, M. (2017). India's Pearl Millet Seed Industry: Prospects for high-iron hybrids.

Karki, R., Ojha, P., Maharjan, S., Manandhar, U., & Maharjan, S. (2025). Optimization of the germination time of proso and foxtail millets to enhance the bioactive properties, antioxidant activity, and enzymatic power and reduce antinutritional factor. *Current Research in Food Science, 10*, 100987.

Kaur, T., Rana, K. L., Kour, D., Sheikh, I., Yadav, N., Kumar, V., Yadav, A. N., Dhaliwal, H. S., & Saxena, A. K. (2020). Microbe-mediated biofortification for micronutrients: present status and future challenges. In *New and future developments in microbial biotechnology and bioengineering* (pp. 1–17). Elsevier.

Kavhiza, N. J., Zargar, M., Prikhodko, S. I., Pakina, E. N., Murtazova, K. M.-S., & Nakhaev, M. R. (2022). Improving crop productivity and ensuring food security through the adoption of genetically modified crops in Sub-Saharan Africa. *Agronomy, 12*, 439. https://doi.org/10.3390/agronomy12020439

Keats, E. C., Neufeld, L. M., Garrett, G. S., Mbuya, M. N. N., & Bhutta, Z. A. (2019). Improved micronutrient status and health outcomes in low- and middle-income countries following large-scale fortification: Evidence from a systematic review and meta-analysis. *The American Journal of Clinical Nutrition, 109*, 1696–1708. https://doi.org/10.1093/ajcn/nqz023

Keerthana, R., Bhat, D., Sivamani, Y., Ganesan, S., & Elayaperumal, S. (2025). Genetic modification in vitamin-rich crop biofortification. In *Methods and protocols in food science* (pp. 189–212). Springer.

Khokharvoytas, A., Shahbaz, M., Maqsood, M. F., Zulfiqar, U., Naz, N., Iqbal, U. Z., Sara, M., Aqeel, M., Khalid, N., Noman, A., Zulfiqar, F., Al Syaad, K. M., & Alshaqhaa, M. A. (2023). Genetic modification strategies for enhancing plant resilience to abiotic stresses in the context of climate change. *Functional & Integrative Genomics, 23*. https://doi.org/10.1007/s10142-023-01202-0

Kodkany, B. S., Bellad, R. M., Mahantshetti, N. S., Westcott, J. E., Krebs, N. F., Kemp, J. F., & Hambidge, K. M. (2013). Biofortification of pearl millet with iron and zinc in a randomized controlled trial increases absorption of these minerals above physiologic requirements in young children. *The Journal of Nutrition, 143*, 1489–1493. https://doi.org/10.3945/jn.113.176677

Kong, D., Khan, S., Wu, H., Liu, Y., & Ling, H. (2022). Biofortification of iron and zinc in rice and wheat. *Journal of Integrative Plant Biology, 64*. https://doi.org/10.1111/jipb.13262

Krishnan, R., & Meera, M. S. (2018). Pearl millet minerals: Effect of processing on bioaccessibility. *Journal of Food Science and Technology, 55*, 3362–3372. https://doi.org/10.1007/s13197-018-3305-9

Ku, Y.-S., Rehman, H., & Lam, H. (2019). Possible roles of rhizospheric and endophytic microbes to provide a safe and affordable means of crop biofortification. *Agronomy, 9*. https://doi.org/10.3390/agronomy9110764

Kumar, S., & Anand, R. (2021). Effect of germination and temperature on phytic acid content of cereals. *The International Journal of Agricultural Science, 8*, 24–35.

Kumar, N., & Dhaka, A. (2020). The conventional and transgenic breeding approaches for biofortification of rice grains with high fe & zn contents and enhancing their bioavailability to combat micronutrient malnutrition-a review. *Progressive Agriculture, 20*, 170. https://doi.org/10.5958/0976-4615.2020.00026.5

Kumar, S., Palve, A., Joshi, C., & Srivastava, R. (2019). Crop biofortification for iron (Fe), zinc (Zn) and vitamin A with transgenic approaches. *Heliyon, 5*. https://doi.org/10.1016/j.heliyon.2019.e01914

Kumar, D., Yadav, A., Ahmad, R., Dwivedi, U. N., & Yadav, K. (2022). CRISPR-based genome editing for nutrient enrichment in crops: A promising approach toward global food security. *Frontiers in Genetics, 13*. https://doi.org/10.3389/fgene.2022.932859

Kumar, N., Sinha, S., & Tiwari, S. (2024). Biofortification of cassava: Addressing micronutrient deficiencies in subsistence farming. In *Harnessing crop biofortification for sustainable agriculture* (pp. 203–223). Springer Nature Singapore.

La Frano, M., De Moura, F., Boy, E., Lönnerdal, B., & Burri, B. (2014). Bioavailability of iron, zinc, and provitamin A carotenoids in biofortified staple crops. *Nutrition Reviews, 72*(5), 289–307. https://doi.org/10.1111/nure.12108

Li, S., Nugroho, A., Rocheford, T., & White, W. S. (2010). Vitamin A equivalence of the β-carotene in β-carotene–biofortified maize porridge consumed by women. *The American Journal of Clinical Nutrition, 92*, 1105–1112. https://doi.org/10.3945/ajcn.2010.29802

Liang, J., Li, Z., Tsuji, K., Nakano, K., Nout, M. J. R., & Hamer, R. J. (2008). Milling characteristics and distribution of phytic acid and zinc in long-, medium- and short-grain rice. *Journal of Cereal Science, 48*, 83–91. https://doi.org/10.1016/j.jcs.2007.08.003

Lipkie, T. E., De Moura, F. F., Zhao, Z.-Y., Albertsen, M. C., Che, P., Glassman, K., & Ferruzzi, M. G. (2013). Bioaccessibility of carotenoids from transgenic Provitamin A biofortified sorghum. *Journal of Agricultural and Food Chemistry, 61*, 5764–5771. https://doi.org/10.1021/jf305361s

Listman, G. M., Guzmán, C., Palacios-Rojas, N., Pfeiffer, W. H., Vicente, F. S., & Govindan, V. (2019). Improving nutrition through biofortification: Preharvest and postharvest technologies. In: Cereals & Grains Association. https://www.cerealsgrains.org/Pages/default.aspx. Accessed 6 Apr 2025.

Liu, S., Zenda, T., Tian, Z., & Huang, Z. (2023). Metabolic pathways engineering for drought or/and heat tolerance in cereals. *Frontiers in Plant Science, 14*. https://doi.org/10.3389/fpls.2023.1111875

Lohmann, B., & Graf, P. (2021). Bioeconomy and the UN sustainable development goals I Bioökonomie.de. In: Biooekonomie.De. https://biooekonomie.de/en/topics/in-depth-reports/bioeconomy-and-un-sustainable-development-goals. Accessed 30 Mar 2025.

Low, J. W., Arimond, M., Osman, N., Cunguara, B., Zano, F., & Tschirley, D. (2007). A food-based approach introducing orange-fleshed sweet potatoes increased vitamin a intake and serum retinol concentrations in young children in rural mozambique, 3. *The Journal of Nutrition, 137*, 1320–1327. https://doi.org/10.1093/jn/137.5.1320

Lowe, N. M. (2025). Fortification or biofortification: complimentary strategies or duplication of effort? *Proceedings of the Nutrition Society, 84*, 65–74. https://doi.org/10.1017/S0029665124000041

Lu, L., Chen, S., Yang, W., Wu, Y., Liu, Y., Yin, X., Yang, Y., & Yang, Y. (2023a). Integrated transcriptomic and metabolomic analyses reveal key metabolic pathways in response to potassium deficiency in coconut (Cocos nucifera L.) seedlings. *Frontiers Plant Science, 14*. https://doi.org/10.3389/fpls.2023.1112264

Lu, M., Yuan, C., Liu, Y., Feng, Y., Qi, B., He, Z., & Yang, X. (2023b). Cadmium (Cd) Minimization and Zinc (Zn) Biofortification in Wheat (Triticum aestivum L.) Grains by spraying with the Foliar Zn fertilizer in Cd-contaminated fields. *Agronomy, 14*. https://doi.org/10.3390/agronomy14010018

MacDonald, C. (2017). Integrating biofortified crops into community development programs. *African Journal of Food, Agriculture, Nutrition and Development, 17*, 12063.

Majumder, S., Datta, K., & Datta, S. (2019). Rice biofortification: High Iron, Zinc, and Vitamin-A to Fight against "Hidden Hunger.". *Agronomy, 9*. https://doi.org/10.3390/agronomy9120803

Malapit, H. J. L., Kadiyala, S., Quisumbing, A. R., Cunningham, K., & Tyagi, P. (2015). Women's empowerment mitigates the negative effects of low production diversity on maternal and child nutrition in Nepal. *The Journal of Development Studies, 51*, 1097–1123. https://doi.org/10.1080/00220388.2015.1018904

Mallikarjuna Swamy, B. P., Marundan, S., Samia, M., Ordonio, R. L., Rebong, D. B., Miranda, R., Alibuyog, A., Rebong, A. T., MaA, T., Suralta, R. R., Alfonso, A. A., Biswas, P. S., Kader, M. A., Reinke, R. F., Boncodin, R., & Mackenzie, D. J. (2021). Development and characterization of GR2E Golden rice introgression lines. *Scientific Reports, 11*. https://doi.org/10.1038/s41598-021-82001-0

Mulambu, J. (2017). Iron beans in Rwanda: Crop development and delivery experience. *African Journal of Food, Agriculture, Nutrition and Development, 17*, 12026–12050. https://doi.org/10.18697/ajfand.78.harvestplus10

Muzhingi, T., Gadaga, T. H., Siwela, A. H., Grusak, M. A., Russell, R. M., & Tang, G. (2011). Yellow maize with high β-carotene is an effective source of vitamin A in healthy Zimbabwean men. *The American Journal of Clinical Nutrition, 94*, 510–519. https://doi.org/10.3945/ajcn.110.006486

Nachimuthu, V. V., Robin, S., Sudhakar, D., Rajeswari, S., Raveendran, M., Subramanian, K., Tannidi, S., & Pandian, B. (2014). Genotypic variation for micronutrient content in traditional and improved rice lines and its role in biofortification programme. *Indian Journal of Science and Technology, 7*, 1414–1425. https://doi.org/10.17485/IJST/2014/V7I9.15

Naeem, A., Aslam, M., Ahmad, M., Asif, M., Yazici, M., Cakmak, I., & Rashid, A. (2021). Biofortification of diverse basmati rice cultivars with Iodine, Selenium, and Zinc by individual and cocktail spray of micronutrients. *Agronomy, 12*. https://doi.org/10.3390/agronomy12010049

Namanya, P. (2011). *Towards the biofortification of banana fruit for enhanced micronutrient content*. PhD, Queensland University of Technology.

Nasar, S., Muhammad, M., Siddiqi, E. H., Majeed, M., Ameen, N., Ullah, S., Kousar, S., & Abdullah, M. (2024). OMICS-based knowledge for achieving food and nutritional security. In *OMICs-based techniques for global food security* (pp. 67–90). Wiley.

Nawaz, T., Nelson, D., Saleem, A., Bibi, M., Joshi, N., Fahad, S., Saud, S., Adnan, M., Rasheed Khan, M. N., & Hassan, S. (2025). Genetic engineering for crop biofortification. In *Crop biofortification: biotechnological approaches for achieving nutritional security under changing climate* (pp. 263–294).

Nestel, P., Bouis, H. E., Meenakshi, J., & Pfeiffer, W. (2006). Biofortification of staple food crops. *The Journal of Nutrition, 136*, 1064–1067. https://doi.org/10.1093/jn/136.4.1064

Ning, P., Zhang, X., Wu, T., Li, Y., Wang, S., Fei, P., Dong, J., Shi, J.-L., & Tian, X. (2021). Biofortification of wheat with zinc as affected by foliar applications of zinc, pesticides, phosphorus and biostimulants. *Crop and Pasture Science, 73*, 3–12. https://doi.org/10.1071/CP20455

Nuss, E., & Tanumihardjo, S. (2010). Maize: A paramount staple crop in the context of global nutrition. *Comprehensive Reviews in Food Science and Food Safety, 9*(4), 417–436. https://doi.org/10.1111/J.1541-4337.2010.00117.X

Ogbu, A. C., Ume, O. A., Nwogiji, O. C., Akeredolu, S. O., Eze, O. G., Onwe, C. V., Oduguwa, O. I., & Chigozie, U. V. (2024). Advances in the use of genome editing tools in Africa: A review. *Asian Journal of Biochemistry, Genetics and Molecular Biology, 16*, 48–67.

Okello, J., Kwikiriza, N., Muoki, P., Wambaya, J., & Heck, S. (2019). Effect of Intensive Agriculture-Nutrition Education and Extension Program Adoption and Diffusion of Biofortified Crops. Journal of Agricultural & Food Information, 20:254–276. https://doi.org/10.1080/10496505.2018.1515632

Okwuonu, I. C., Narayanan, N. N., Egesi, C. N., & Taylor, N. J. (2021). Opportunities and challenges for biofortification of cassava to address iron and zinc deficiency in Nigeria. *Global Food Security, 28*, 100478. https://doi.org/10.1016/j.gfs.2020.100478

Oladosu, Y., Rafii, M. Y., Abdullah, N., Hussin, G., Ramli, A., Rahim, H. A., Miah, G., & Usman, M. (2016). Principle and application of plant mutagenesis in crop improvement: a review. *Biotechnology & Biotechnological Equipment, 30*, 1–16. https://doi.org/10.1080/13102818.2015.1087333

Oliva, N., Florida Cueto-Reaño, M., Trijatmiko, K. R., Samia, M., Welsch, R., Schaub, P., Beyer, P., Mackenzie, D., Boncodin, R., Reinke, R., Slamet-Loedin, I., & Mallikarjuna Swamy, B. P. (2020). Molecular characterization and safety assessment of biofortified provitamin A rice. *Scientific Reports, 10*. https://doi.org/10.1038/s41598-020-57669-5

Olivera, J. M. (2024). Enhancing regulatory compliance in the AI age by grounding documents with generative AI | IBM. https://www.ibm.com/think/insights/enhancing-regulatory-compliance-ai-age. Accessed 30 Mar 2025.

Omari, R., Zotor, F., Tagwireyi, J., & Lokosang, L. (2019). Advocacy for scaling up biofortified crops for improved micronutrient status in Africa: Approaches, achievements, challenges and lessons. In *Proceedings of the nutrition society* (Vol. 78, pp. 567–575). Cambridge University Press.

Osendarp, S., Martínez, H., Garrett, G., Neufeld, L., De-Regil, L., Vossenaar, M., & Darnton-Hill, I. (2018). Large-scale food fortification and biofortification in low- and middle-income countries: a review of programs, trends, challenges, and evidence gaps. *Food and Nutrition Bulletin, 39*, 315–331. https://doi.org/10.1177/0379572118774229

Pal, V., Singh, G., & Dhaliwal, S. (2021). A new approach in agronomic biofortification for improving zinc and iron content in Chickpea (Cicer arietinum L.) grain with simultaneous foliar application of zinc sulphate, ferrous sulphate and urea. *Journal of Soil Science and Plant Nutrition, 21*, 883–896. https://doi.org/10.1007/s42729-021-00408-0

Pandohee, J., Basu, R., Dasgupta, S., Sundarrajan, P., Shaikh, N., Patel, N., & Noor, A. (2023). Applications of multi-omics approaches for food and nutritional security. In *Sustainable agriculture in the Era of the OMICs revolution* (pp. 103–118). Springer International Publishing.

Paredes, D. (2019). Successful scaling up of biofortified crops to fight against hidden hunger in Nigeria, Tanzania and surrounding region through strong advocacy efforts and supportive policy environment. In: CIP International Potato Center. https://mel.cgiar.org/projects/409/110/successful-scaling-up-of-biofortified-crops-to-fight-against-hidden-hunger-in-nigeria-tanzania-and-surrounding-region-through-strong-advocacy-efforts-and-supportive-policy-environment. Accessed 30 Mar 2025.

Patindol, J., Fragallo, L., Wang, Y., & Durand-Morat, A. (2017). Impact of feedstock, parboiling condition, and nutrient concentration on simultaneous fortification of two U.S. long-grain rice cultivars with iron and zinc. *Cereal Chemistry, 94*, 984–990. https://doi.org/10.1094/cchem-05-17-0115-r

Paul, J.-Y., Khanna, H., Kleidon, J., Hoang, P., Geijskes, J., Daniells, J., Zaplin, E., Rosenberg, Y., James, A., Mlalazi, B., Deo, P., Arinaitwe, G., Namanya, P., Becker, D., Tindamanyire, J., Tushemereirwe, W., Harding, R., & Dale, J. (2017). Golden bananas in the field: Elevated fruit pro-vitamin A from the expression of a single banana transgene. *Plant Biotechnology Journal, 15*, 520–532. https://doi.org/10.1111/pbi.12650

Pavan. (n.d.). Artificial intelligence in pharmaceutical industry regulatory affairs. In: acubit blog. https://blog.accubits.com/artificial-intelligence-in-pharmaceutical-industry-regulatory-affairs/. Accessed 30 Mar 2025.

Petry, N., Nizamov, F., Woodruff, B., Ishmakova, R., Komilov, J., Wegmüller, R., Wirth, J., Arifdjanova, D., Guo, S.-F., & Rohner, F. (2020). Risk factors for anemia and micronutrient deficiencies among women of reproductive age – The impact of the wheat flour fortification program in Uzbekistan. *Nutrients, 12*. https://doi.org/10.3390/nu12030714

Pfeiffer, W., & Mcclafferty, B. (2007). HarvestPlus: Breeding crops for better nutrition. *Crop Science, 47*. https://doi.org/10.2135/CROPSCI2007.09.0020IPBS

Prasanna, B. M., Palacios-Rojas, N., Hossain, F., Muthusamy, V., Menkir, A., Dhliwayo, T., Ndhlela, T., San Vicente, F., Nair, S. K., Vivek, B. S., Zhang, X., Olsen, M., & Fan, X. (2020). Molecular breeding for nutritionally enriched maize: Status and prospects. *Frontiers in Genetics, 10*. https://doi.org/10.3389/fgene.2019.01392

Qureshi, M. T., Iqbal, N., Noorka, I., & Waheed, H. (2020). Bio-fortification of iron and zinc improves the biomass, uptake, distribution and yield of different maize varieties. *Journal of Plant Nutrition, 44*, 120–129. https://doi.org/10.1080/01904167.2020.1806309

Qureshi, M. T., Ahmad, M. F., Iqbal, N., Waheed, H., Hussain, S., Brestič, M., Anjum, A., & Noorka, I. (2021). Agronomic bio-fortification of iron, zinc and selenium enhance growth, quality and uptake of different sorghum accessions. *Plant, Soil and Environment, 67*. https://doi.org/10.17221/137/2021-pse

Raboy, V. (2020). Low phytic acid crops: Observations based on four decades of research. *Plants, 9*, 140. https://doi.org/10.3390/plants9020140

Ramesh, A., Sharma, S. K., Sharma, M. P., Yadav, N., & Joshi, O. P. (2014). Inoculation of zinc solubilizing Bacillus aryabhattai strains for improved growth, mobilization and biofortification of zinc in soybean and wheat cultivated in Vertisols of central India. *Applied Soil Ecology, 73*, 87–96. https://doi.org/10.1016/j.apsoil.2013.08.009

Rathod, P., Patil, D., Bellad, S., & Haveri, R. (2020). Biofortification of Zn and Fe in Chickpea through agronomic intervention in medium black soils of Karnataka. *LEGUME Research –An International Journal*. https://doi.org/10.18805/LR-4341

Rico, C. M., Majumdar, S., Duarte-Gardea, M., Peralta-Videa, J. R., & Gardea-Torresdey, J. L. (2011). Interaction of nanoparticles with edible plants and their possible implications in the food chain. *Journal of Agricultural and Food Chemistry, 59*, 3485–3498. https://doi.org/10.1021/jf104517j

Rivera-Martín, A., Broadley, M., & Poblaciones, M. (2020). Soil and foliar zinc application to biofortify broccoli (Brassica oleracea var. italica L.): effects on the zinc concentration and bioavailability. *Plant, Soil and Environment, 66*. https://doi.org/10.17221/14/2020-pse

Rosado, J., Hambidge, K., Miller, L., Garcia, O., Westcott, J., González, K., Conde, J., Hotz, C., Pfeiffer, W., Ortiz-Monasterio, I., & Krebs, N. (2009). The quantity of zinc absorbed from wheat in adult women is enhanced by biofortification. *The Journal of Nutrition, 139*(10), 1920–1925. https://doi.org/10.3945/jn.109.107755

Saltzman, A., Birol, E., Bouis, H. E., Boy, E., De Moura, F. F., Islam, Y., & Pfeiffer, W. H. (2013). Biofortification: Progress toward a more nourishing future. *Global Food Security, 2*, 9–17. https://doi.org/10.1016/j.gfs.2012.12.003

Saltzman, A., Andersson, M., Asare-Marfo, D., Lividini, K., Moura, F., Moursi, M., Oparinde, A., & Taleon, V. (2015). Biofortification techniques to improve. *Food Security.* https://doi.org/10.1016/B978-0-08-100596-5.03078-X

Saltzman, A., Birol, E., Oparinde, A., Andersson, M. S., Asare-Marfo, D., Diressie, M. T., Gonzalez, C., Lividini, K., Moursi, M., & Zeller, M. (2017). Availability, production, and consumption of crops biofortified by plant breeding: current evidence and future potential. *Annals of the New York Academy of Sciences, 1390*, 104–114. https://doi.org/10.1111/nyas.13314

Sánchez-Palacios, J., Henry, D., Penrose, B., & Bell, R. (2023). Formulation of zinc foliar sprays for wheat grain biofortification: a review of current applications and future perspectives. *Frontiers in Plant Science, 14*. https://doi.org/10.3389/fpls.2023.1247600

Satish, S. (2023). Biofortification – ClearIAS. https://www.clearias.com/biofortification/. Accessed 30 Mar 2025.

Sayre, R., Beeching, J. R., Cahoon, E. B., Egesi, C., Fauquet, C., Fellman, J., Fregene, M., Gruissem, W., Mallowa, S., Manary, M., Maziya-Dixon, B., Mbanaso, A., Schachtman, D. P., Siritunga, D., Taylor, N., Vanderschuren, H., & Zhang, P. (2011). The biocassava plus program: Biofortification of cassava for Sub-Saharan Africa. *Annual Review of Plant Biology, 62*, 251–272. https://doi.org/10.1146/annurev-arplant-042110-103751

Sazawal, S., Dhingra, U., Dhingra, P., Dutta, A., Deb, S., Kumar, J., Devi, P., & Prakash, A. (2018). Efficacy of high zinc biofortified wheat in improvement of micronutrient status, and prevention of morbidity among preschool children and women – A double masked, randomized, controlled trial. *Nutrition Journal, 17*. https://doi.org/10.1186/s12937-018-0391-5

Sebastiá, V., Barberá, R., Farré, R., & Lagarda, M. (2001). Effects of legume processing on calcium, iron and zinc contents and dialysabilities. *Journal of the Science of Food and Agriculture, 81*, 1180–1185. https://doi.org/10.1002/JSFA.927

Sen Gupta, D., Kumar, J., Barpate, S., Parihar, A. K., Chandra, A., Roy, A., & Djalovic, I. (2023). Lentils (Lens culinaris Medik): Nutritional profile and biofortification prospects. In *Compendium of crop genome designing for nutraceuticals* (pp. 1–27). Springer Nature Singapore.

Shahzad, R., Jamil, S., Ahmad, S., Nisar, A., Khan, S., Amina, Z., Kanwal, S., Aslam, H. M. U., Gill, R. A., & Zhou, W. (2021). Biofortification of cereals and pulses using new breeding techniques: Current and future perspectives. *Frontiers in Nutrition, 8*. https://doi.org/10.3389/fnut.2021.721728

Shahzad, B., Holt, R. R., Gupta, S., Zaman, M., Shahzad, M., Lowe, N. M., & Hall, A. G. (2024). Effects of zinc-biofortified wheat intake on plasma markers of fatty acid metabolism and oxidative stress among adolescents. *Nutrients, 16*, 4265. https://doi.org/10.3390/nu16244265

Sharma, P., Aggarwal, P., & Kaur, A. (2017). Biofortification: A new approach to eradicate hidden hunger. *Food Reviews International, 33*, 1–21. https://doi.org/10.1080/87559129.2015.1137309

Shubham, K., Anukiruthika, T., Dutta, S., Kashyap, A., Moses, J., & Anandharamakrishnan, C. (2020). Iron deficiency anemia: A comprehensive review on iron absorption, bioavailability and emerging food fortification approaches. *Trends in Food Science & Technology, 99*. https://doi.org/10.1016/j.tifs.2020.02.021

Simpungwe, E. (2017). Orange maize in Zambia: Crop development and delivery experience. *African Journal of Food, Agriculture, Nutrition and Development, 17*, 11973–11999. https://doi.org/10.18697/ajfand.78.harvestplus08

Singh, P., Thakur, A., Dogra, S., Garg, P., Srivastav, L., & Kler, N. (2017). Comparison of osmolality of human milk after fortification with three different fortifiers. *Current Medicine Research and Practice, 7*, 81–83. https://doi.org/10.1016/J.CMRP.2017.02.003

Singh, U., Malviya, D., Singh, S., Singh, P., Ghatak, A., Imran, M., Rai, J., Singh, R., Manna, M., Sharma, A., & Saxena, A. (2021). Salt-tolerant compatible microbial inoculants modulate physio-biochemical responses enhance plant growth, Zn biofortification and yield of wheat grown in saline-sodic soil. *International Journal of Environmental Research and Public Health, 18*. https://doi.org/10.3390/ijerph18189936

Singhal, T., Tara Satyavathi, C., Singh, S. P., Mallik, M., Anuradha, N., Sankar, S. M., Bharadwaj, C., & Singh, N. (2022). Achieving nutritional security in India through iron and zinc biofortification in pearl millet (Pennisetum glaucum (L.) R. Br.). *Physiology and Molecular Biology of Plants, 28*, 849–869. https://doi.org/10.1007/s12298-022-01144-0

Siwela, M., Pillay, K., Govender, L., Lottering, S., Mudau, F., Modi, A., & Mabhaudhi, T. (2020). Biofortified crops for combating Hidden Hunger in South Africa: Availability, acceptability, micronutrient retention and bioavailability. *Foods, 9*. https://doi.org/10.3390/foods9060815

Smoleń, S. (2012). Foliar nutrition: Current state of knowledge and opportunities. In *Advances in citrus nutrition* (pp. 41–58). Springer.

Sousa Silva, M., Roessner, U., & Cordeiro, C. (2023). Editorial: Metabolomics in crop research – Current and emerging methodologies, volume II. *Frontiers in Plant Science, 14*. https://doi.org/10.3389/fpls.2023.1292878

Sperotto, R., Ricachenevsky, F., Waldow, V., & Fett, J. (2012). Iron biofortification in rice: It's a long way to the top. *Plant Science, 190*, 24–39. https://doi.org/10.1016/j.plantsci.2012.03.004

Srivastav, P., Vutukuru, M., Ravindran, G., & Awad, M. (2022). Biofortification – Present scenario, possibilities and challenges: A scientometric approach. *Sustainability, 14*. https://doi.org/10.3390/su141811632

Steiger, G., Müller-Fischer, N., Cori, H., & Conde-Petit, B. (2014). Fortification of rice: Technologies and nutrients. *Annals of the New York Academy of Sciences, 1324*, 29–39. https://doi.org/10.1111/nyas.12418

Suganya, A., Saravanan, A., Baskar, M., Pandiyarajan, P., & Kavimani, R. (2021). Agronomic biofortification of maize (Zea mays L.) with zinc by using of graded levels of zinc in combination with zinc solubilizing bacteria and Arbuscular mycorrhizal fungi. *Journal of Plant Nutrition, 44*, 988–994. https://doi.org/10.1080/01904167.2020.1845383

Sun, L., Lai, M., Ghouri, F., Nawaz, M. A., Ali, F., Baloch, F. S., Nadeem, M. A., Aasim, M., & Shahid, M. Q. (2024). Modern plant breeding techniques in crop improvement and genetic diversity: From molecular markers and gene editing to artificial intelligence – A critical review. *Plants, 13*, 2676. https://doi.org/10.3390/plants13192676

Sundararajan, S., Rajendran, V., Sivakumar, H., Nayeem, S., Chandra, H., Sharma, A., & Ramalingam, S. (2021). Enhanced vitamin E content in an Indica rice cultivar harbouring two transgenes from Arabidopsis thaliana involved in tocopherol biosynthesis pathway. *Plant Biotechnology Journal, 19*, 1083–1085. https://doi.org/10.1111/pbi.13578

Suppress. (2024). *AI in compliance: Streamlining regulatory compliance with generative AI*. https://gradient-ascent.com/ai-in-compliance/. Accessed 30 Mar 2025.

Tablante, E. C., Pachón, H., Guetterman, H., & Finkelstein, J. (2019). Fortification of wheat and maize flour with folic acid for population health outcomes. *The Cochrane Database of Systematic Reviews, 7*. https://doi.org/10.1002/14651858.CD012150.pub2

Tahakik, R. R., Deshmukh, A. G., Moharil, M. P., Jadhav, P. V., Kogade, V. T., More, K. D., & Shinde, V. P. (2024). Transitioning from the green revolution to the gene revolution: strengthening nutritional security using climate resilient traditional crops. *Bulletin of the National Research Centre, 48*. https://doi.org/10.1186/s42269-024-01281-4

Tako, E., Bar, H., & Glahn, R. (2016). The combined application of the Caco-2 cell bioassay coupled with in vivo (Gallus gallus) feeding trial represents an effective approach to predicting Fe bioavailability in humans. *Nutrients, 8*. https://doi.org/10.3390/nu8110732

Taleon, V., Hasan, M. Z., Jongstra, R., Wegmüller, R., & Bashar, M. K. (2022). Effect of parboiling conditions on zinc and iron retention in biofortified and non-biofortified milled rice. *Journal of the Science of Food and Agriculture, 102*, 514–522. https://doi.org/10.1002/jsfa.11379

Talsma, E. F., & Pachón, H. (2017). *Biofortification of crops with minerals and vitamins*. https://www.who.int/tools/elena/bbc/biofortification. Accessed 30 Mar 2025.

Talsma, E. F., Brouwer, I. D., Verhoef, H., Mbera, G. N., Mwangi, A. M., Demir, A. Y., Maziya-Dixon, B., Boy, E., Zimmermann, M. B., & Melse-Boonstra, A. (2016). Biofortified yellow cassava and vitamin A status of Kenyan children: A randomized controlled trial. *The American Journal of Clinical Nutrition, 103*, 258–267. https://doi.org/10.3945/ajcn.114.100164

Tang, G., Qin, J., Dolnikowski, G. G., Russell, R. M., & Grusak, M. A. (2009). Golden Rice is an effective source of vitamin A. *The American Journal of Clinical Nutrition, 89*, 1776–1783. https://doi.org/10.3945/ajcn.2008.27119

Team Hoodin. (2024). *AI and automation in regulatory affairs: Where we are now and what to expect in the future*. https://www.hoodin.com/post/ai-and-automation-in-regulatory-affairs-where-we-are-now-and-what-to-expect-in-the-future. Accessed 30 Mar 2025.

Thakkar, S. K., Huo, T., Maziya-Dixon, B., & Failla, M. L. (2009). Impact of style of processing on retention and bioaccessibility of β-Carotene in Cassava (Manihot esculanta, Crantz). *Journal of Agricultural and Food Chemistry, 57*, 1344–1348. https://doi.org/10.1021/jf803053d

Thayer, B. (2016). 6 Myths About GMOs | Henry Ford Health – Detroit, MI. In: Henry Ford Health. https://www.henryford.com/blog/2016/09/6-myths-about-gmos. Accessed 30 Mar 2025.

Tiwari, A., Chaudhary, E., Giri, L., Sharma, A., & Garg, M. (2024). Omics in biofortification: Unlocking nutritional potential. In *Biofortification for nutrient-rich crops* (pp. 272–291). CRC Press.

Tripathi, A., Choudhary, J., & Prakash, N. R. (2018). Bio fortification-breeding for nutritional security. *Journal of Pharmacognosy and Phytochemistry, 7*, 1266–1270.

Upadhayay, V. K., Singh, A. V., Khan, A., & Sharma, A. (2022). Contemplating the role of zinc-solubilizing bacteria in crop biofortification: An approach for sustainable bioeconomy. *Frontiers in Agronomy, 4*, 903321.

Valença, A., Bake, A., Brouwer, I., & Giller, K. (2017). Agronomic biofortification of crops to fight hidden hunger in sub-Saharan Africa. *Global Food Security, 12*, 8–14. https://doi.org/10.1016/J.GFS.2016.12.001

van Hees, S. G. M., O'Fallon, T., Hofker, M., Dekker, M., Polack, S., Banks, L. M., & Spaan, E. J. A. M. (2019). Leaving no one behind? Social inclusion of health insurance in low- and middle-income countries: a systematic review. *International Journal for Equity in Health, 18*, 134. https://doi.org/10.1186/s12939-019-1040-0

Van Jaarsveld, P. J., Marais, D. W., Harmse, E., Nestel, P., & Rodriguez-Amaya, D. B. (2006). Retention of β-carotene in boiled, mashed orange-fleshed sweet potato. *Journal of Food Composition and Analysis, 19*, 321–329. https://doi.org/10.1016/j.jfca.2004.10.007

Vinoth, A., & Ravindhran, R. (2017). Biofortification in millets: A sustainable approach for nutritional security. *Frontiers in Plant Science, 8*. https://doi.org/10.3389/fpls.2017.00029

von Klaus, G., Amy, S., Ekin, B., Doris, W., Nilam, P., Sandra, Y., Yisehac, Y., Purnima, M., Jennifer, T., & Andrea, S. (2014). *Synopsis: 2014 global hunger index: The challenge of hidden hunger*. International Food Policy Research Institute.

Walton, J. (2021). Digitalizing biofortified food supply chains to facilitate scale. In: CGAIR: Research program on agriculture for nutrition and health. https://a4nh.cgiar.org/2021/12/14/digitalizing-biofortified-food-supply-chains-to-facilitate-scale/. Accessed 30 Mar 2025.

Wei, Y.-L., Shohag, M., & Yang, X. (2012). Biofortification and bioavailability of rice grain zinc as affected by different forms of foliar zinc fertilization. *PLoS One, 7*. https://doi.org/10.1371/journal.pone.0045428

White, P., & Broadley, M. (2009). Biofortification of crops with seven mineral elements often lacking in human diets – Iron, zinc, copper, calcium, magnesium, selenium and iodine. *The New Phytologist, 182*(1), 49–84. https://doi.org/10.1111/j.1469-8137.2008.02738.x

Xue, Y.-F., Li, X.-J., Yan, W., Miao, Q., Zhang, C.-Y., Huang, M., Sun, J.-B., Qi, S.-J., Ding, Z.-H., & Cui, Z.-L. (2023). Biofortification of different maize cultivars with zinc, iron and selenium by foliar fertilizer applications. *Frontiers in Plant Science, 14*. https://doi.org/10.3389/fpls.2023.1144514

Yadav, D. N., Bansal, S., Tushir, S., Kaur, J., & Sharma, K. (2020). Advantage of biofortification over fortification technologies. In *Wheat and barley grain biofortification* (pp. 257–273). Woodhead Publishing.

Yadesa, L., & Diro, D. (2023). Nutritional and specialty maize production, consumption, and promising impact on Ethiopia's food and nutrition security: A review. *EAS J Nutr Food Sci, 5*, 142–157. https://doi.org/10.36349/easjnfs.2023.v05i05.003

Yan, Y., Zhu, S., Jia, M., Chen, X., Qi, W., Gu, F., Valencak, T. G., Liu, J.-X., & Sun, H.-Z. (2024). Advances in single-cell transcriptomics in animal research. *Journal of Animal Science and Biotechnology, 15*. https://doi.org/10.1186/s40104-024-01063-y

Younas, Z., Mashwani, Z., Ahmad, I., Khan, M., Zaman, S., & Sawati, L. (2023). Mechanistic approaches to the application of nano-zinc in the poultry and biomedical industries: A comprehensive review of future perspectives and challenges. *Molecules, 28*. https://doi.org/10.3390/molecules28031064

Zhang, L. (2023). Bringing UK and US biofortification bioscience together. In: Quadram Institute. https://quadram.ac.uk/blogs/bringing-uk-and-us-biofortification-bioscience-together/. Accessed 30 Mar 2025.

Zhang, Y., Stockmann, R., Ng, K., & Ajlouni, S. (2022). Bioprocessing of pea protein can enhance fortified Fe but reduce Zn in vitro bioaccessibility. *Journal of Agricultural and Food Chemistry, 70*. https://doi.org/10.1021/acs.jafc.1c05796

Zhao, F.-J., & Mcgrath, S. P. (2009). Biofortification and phytoremediation. *Current Opinion in Plant Biology, 12*, 373–380. https://doi.org/10.1016/j.pbi.2009.04.005

Zulfiqar, U., Maqsood, M., Hussain, S., & Anwar-ul-Haq, M. (2020). Iron nutrition improves productivity, profitability, and biofortification of bread wheat under conventional and conservation tillage systems. *Journal of Soil Science and Plant Nutrition, 20*, 1298–1310. https://doi.org/10.1007/s42729-020-00213-1

Sustainability and Economic Aspects of Food Fortification

2

Noelia E. Silva, Sofia Marchioni, Adriana P. Castellanos-Fuentes, Alicia Matyatz, Carolina E. Genevois, Silvia K. Flores, and Marina de Escalada Pla

Abstract

Circular economy models and sustainable production practices are some of the actions required to achieve the Sustainable Development Goals. Driving innovation at every stage of production to reduce waste generation should be promoted for responsible consumption. The prevalence of anaemia is one of the concerns for achieving food security and improving nutrition. Anaemia due to iron deficiency could be mitigated mainly by food fortification. Nevertheless, iron reactivity presents some technological difficulties to be solved. Some challenges are associated with sensory characteristics, while other difficulties concerning mineral bioaccessibility can appear according to the fortified food matrix. In this chapter, a pumpkin-based iron-fortified food was selected as an example to explain the fortification process. Active coating proposals for improving the

N. E. Silva · S. Marchioni · A. P. Castellanos-Fuentes · S. K. Flores · M. de Escalada Pla (✉)
Universidad de Buenos Aires (UBA), Facultad de Ciencias Exactas y Naturales (FCEN), Departamento de Industrias, Ciudad Autónoma de Buenos Aires, Argentina

CONICET – Universidad de Buenos Aires, Instituto de Tecnología de Alimentos y Procesos Químicos (ITAPROQ), Ciudad Autónoma de Buenos Aires, Argentina
e-mail: nsilva@di.fcen.uba.ar; sflores@di.fcen.uba.ar; marina@di.fcen.uba.ar

A. Matyatz
Universidad de Buenos Aires (UBA), Facultad de Ciencias Exactas y Naturales (FCEN), Departamento de Industrias, Ciudad Autónoma de Buenos Aires, Argentina

C. E. Genevois
Instituto de Ciencia y Tecnología de los Alimentos de Entre Ríos (ICTAER), CONICET-UNER, Gualeguaychú, Entre Ríos, Argentina

Facultad de Bromatología, UNER, Gualeguaychú, Entre Ríos, Argentina

Consejo Nacional de Investigaciones Científicas y Técnicas (CONICET), Ciudad Autónoma de Buenos Aires, Argentina
e-mail: carolina.genevois@uner.edu.ar

G. Sanghvi et al. (eds.), *Sustainable Food Fortification*,
https://doi.org/10.1007/978-981-95-1217-1_2

product shelf-life and/or iron bioaccessibility will be discussed. In order to contribute to the sustainable development 12-goal, profiting from the main by-products and currents originated in the process will also be considered. Understanding the effects of the variables of this process on the properties of the developed fortified food contributes to thinking about novel strategies for adapting to new challenges and future consumer trends. Avoiding or reducing waste generation should be a key characteristic when selecting the future process for producing fortified foods.

Keywords

Iron deficiency · Iron fortification · Vegetable matrices · Pumpkin · Impregnation · Probiotics

2.1 Introduction

The 2030 Agenda for Sustainable Development is an action plan for people's prosperity and the planet that integrates three dimensions, economic, social, and the environmental, established by the United Nations in 2015. However, a general lack of progress was registered in the Sustainable Development Goals (SDG) reports for 2023 and 2024. Climate crisis, wars in different regions producing a fragile global economy, can explain, in part, why the progress towards achieving the Goals has been impeded. Most of these obstacles were suffered mainly by the world's poorest and most vulnerable populations (E/2023/64, 2023; E/2024/54, 2024).

Food systems are intricate networks that encompass diverse activities, processes, and individuals collaborating to produce, process, distribute, and consume food (von Braun et al., 2021). These systems feature value chains that span all stages, from agricultural inputs to consumption, including waste management and creating economic opportunities for those involved. The SDGs act as a guiding framework for desired outcomes in food systems, connecting them deeply to human health, environmental stewardship, and economic growth. Notably, food system activities contribute directly to achieving nine of the SDGs.

Relating to Social SDGs, food processing contributes to 2, 3, and 12 SDGs, since this activity is highly connected to reducing hunger, improving health and well-being, and aligning food production to responsible resource consumption. These approaches can take advantage of fossil fuel opportunities and use, drive innovation, conserve energy, and mitigate emissions (E/2024/54, 2024). Food processing provides alternatives to achieve access to safe and nutritional foods, reducing waste and optimising the resources and nutrients. In this sense, the prevalence of anaemia is one of the indices followed by Global Health Observatory (GHO) and World Health Organization (WHO) to analyse the progress of SD2-G. It will be further discussed due to the special interest in this chapter.

Efficient practices along the value chain process should be applied to maximise resource profit in food systems. The use of the raw-material and drinking water should be optimised, taking care of depletion of non-renewable energy and waste generation, emissions to air, water, and ground. With this behaviour, the food processors can contribute to the environmental SDGs 6, 13, 14 and 15 related to water, climate, and land, respectively.

The economic SDGs related to food involve all the individuals and organisations from agricultural raw materials to the creation of food products (Food and Drug Administration (FAO), 2014). Once optimised energy and materials, labour involved in food production is wide, also considering transport, storage and distribution steps. Therefore, food systems contribute to equitable livelihoods.

This chapter aims to provide a review of the challenges to overcome when projecting iron fortified vegetable foods, considering all aspects related to the SDGs involved. Pumpkin fortified snack development experience is shared as an example to be discussed.

2.2 Deficiencies of Essential Nutrients and Prevalence of Anaemia

According to recent global publications, 372 and 1200 million pre-school children and reproductive-age women, respectively, are estimated to suffer from vitamin and mineral deficiencies. Poor diet is mainly responsible for the lack of essential nutrients, which constitute a major public health problem worldwide (Darwar et al., 2023). Micronutrient deficiencies cause serious problems in children's growth and cognitive development, weakening immunity and increasing susceptibility to diseases. It also has significant economic and health consequences. An annual loss of gross domestic product of between 2% and 5% with direct costs around US$ 20–30 billion each year due to micronutrient deficiencies has been estimated for low and middle-income countries.

Anaemia is a global public health concern, impacting around 1620 million individuals—equivalent to nearly one-quarter of the world's population—across both developed and developing nations (Chaudhary et al., 2024). It has been estimated that anaemia reduces productivity in heavy manual labour, up to 17%, and produces a 2.5% loss of income due to lower cognitive skills (Osendarp et al., 2018). Globally, the frequency progress since 2015 has remained stagnant from 28.8% to 29.9% in 2019, in women aged 15–49 years, and a similar scenario occurs for all regions (E/2024/54, 2024). Nowadays, India registers more than half of women affected by anaemia, showing an increase in prevalence (Chaudhary et al., 2024; Menon et al., 2014). According to the WHO, anaemia is defined as the prevalence of haemoglobin concentrations lower than normal (110 $g.L^{-1}$ for pregnant women and 120 $g.L^{-1}$ for non-pregnant women), due to the lack of one or more important nutrients. Not enough dietary iron intake has been considered responsible for poor erythropoiesis and haemoglobin production.

Iron in the diet is categorised into haem-iron, which is found in meat, fish, and poultry, and non-haem-iron, present in vegetable foods in ferric form (Fe^{3+}) as well as in elemental iron used for supplementation and food fortification. Iron absorption predominantly occurs in the duodenum through distinct mechanisms based on these iron types. Haem-iron is absorbed more efficiently due to specific haem transporters that facilitate its movement across cell membranes and into the bloodstream. In contrast, non-haem-iron absorption relies heavily on the interaction between inhibitors and enhancers of absorption within the food matrix. This is a key factor for individuals with limited access to animal-based foods. Enhancers, such as ascorbic acid, carotenes, organic acids, and certain amino acids, play a key role in converting ferric iron to the more bioavailable ferrous iron (Fe^{2+}). Conversely, inhibitors like tannins, phytates, polyphenols, calcium, and casein either bind to iron or compete for its absorption. A diet lacking diversity or rich in foods with high levels of antinutrient factors can impair iron absorption (WHO, 2017). Among the four recommended strategies for combating iron deficiency—supplementation, fortification, nutrition education, and dietary diversity—food fortification has been recognised as the most effective and cost-efficient approach (Kaur et al., 2022).

2.3 Strategies for the Sustainable Implementation of Fortification Programmes

Food fortification boosts the levels of essential micronutrients in foods, enhancing their nutritional value and health benefits, taking care to minimise health risks. Depending on the population reached, "mass fortification" targets a broad range while "targeted fortification" focusses on a specific population. Mass fortification programmes are one means of addressing nutritional deficiencies by adding micronutrients, during food processing, to usually consumed foods, such as wheat and corn flour, cooking oil, salt, and rice. (Darwar et al., 2023). Such programmes have been implemented in industrialised countries since the early twentieth century, and more recently in middle and low-income countries. Targeted fortification is a systematic and methodological approach to designing and planning a food fortification programme, which includes nutritional objectives, communicating to a specific group of people to help them decide to consume fortified foods, and monitoring and evaluating programmes (Darwar et al., 2023; WHO, 2006).

"Market-driven (or commercial) fortification" is applied by the food processing industry, and "household fortification" involves distribution of supplements at the household or community level (Teshome et al., 2017). In addition, microbial fortification improves the release of micronutrients mainly through fermentation, and biofortification is associated with breeding or biotechnological strategies (Liyanage & Hettiarachchi, 2011; Ohanenye et al., 2021; WHO, 2006).

Given its widespread deficiency, food fortification has emerged as a key strategy to counter this problem. This process involves adding iron to regularly consumed foods, ensuring a constant and readily available source of this essential nutrient, without the impediments caused by seasonal food availability, and achieving better nutritional quality of foods with health benefits. Currently, more than 63 countries have implemented mass programmes, targeting foods such as cereals (rice, wheat, corn), cooking fats and oils (fat-soluble vitamins such as A and D), salt, sugar, spices, sauces, and processed dairy products.

Many iron compounds have been considered for food fortification. Nineteen iron compounds have been reported (WHO, 2006). The most relevant iron compounds are ferrous sulfate ($FeSO_4$), iron sodium ethylendiamin-tetraacetic acid (NaFeEDTA), ferrous salt of gluconate, fumarate and bis-glycinate (FBG), ferric pyrophosphate (FPP), and elemental iron powders. In addition, iron-rich foods such as moringa oleifera and some grains like amaranth, sesame and finger millet, are also being considered as potential fortifiers (Kaur et al., 2022; Orsango et al., 2020).

Research and development activities have led to the fortification of iron in a wide range of food, including dairy products like Cheddar cheese, powdered milk, yogurt drinks (Gupta et al., 2015; Santillán-Urquiza et al., 2017; Siddique & Park, 2019); orange juice (Blanco-Rojo et al., 2011), fish oil (Choi et al., 2009), soy-based infant cereals (Hurrell et al., 1989), infant formulas and dry drink mixes (Shubham et al., 2020), potato chips (Tiwari et al., 2022), bouillon cubes (Cercamondi et al., 2016), powdered fruit drink (Mehansho, 2006), roasted ground coffee (Soares et al., 2019), tempeh (Sudargo et al., 2013), and tea (Dueik et al., 2017). While some of these products are commercially available, others are still in the research or testing phase. Table 2.1 summarises, partially fortified or supplemented foods offered in different countries.

As can be observed, wheat flour, sugar and rice are the main matrices to be fortified, selected by different governments as vehicles for fortification programmes. Contrarily, dairy and probiotics formulations are provided by the industry as commercial options for improving the nutritional profile of the formulated food product.

2.4 Considering Sustainable Development Goals when Selecting Plant-Food Process Fortification: Technological and Nutritional Challenges to Achieve

Biofortification improves micronutrient content in crops by indirectly performing its addition during plant-breeding or by means of agronomic practices or modern biotechnology. Iron, zinc, carotenoids, and biofortification can be mentioned as examples applied (Kaur et al., 2022; WHO, 2017). Nevertheless, globally, the consumption of whole grains is scarce. In general, it is usually consumed around 20% and 30% of total grain (Boyle et al., 2024; Milani et al., 2024). The milling process removes bran and germ, which represent the main nutrient parts of the grain. The refined carbohydrate-rich and low-fibre part (endosperm) is selected by consumers. In this way, precious natural resources like dietary fibre and bioactive compounds are wasted (Milani et al., 2024). Therefore, the consumption profile should be considered to develop a suitable combination of nutrients and food vehicles.

Contrary to crops, fruit and vegetables contain higher water content (>50%) and therefore, their processing requires "hurdle technology". Water activity (a_w) and pH reduction, along with a temperature pre-treatment, are examples of the main barriers adopted to assure microbiological stability and food safety (de Escalada Pla et al., 2020). In this context, the vegetable tissue impregnation has been widely applied as a main step of the hurdle technology. Impregnation is a non-thermal

Table 2.1 Enriched or fortified foods available in different countries' markets

Fortified food	Origin	Fortified/enriched	
 Sugar	Honduras	Fortified with vitamin A	Mass fortification programme Agreement No. 4667-2016[a]
 Salt	Argentina	Fortified with iodine	Mass fortification programme Law 17.259[b]
 Salt	Colombia	Fortified with iodine and fluoride	Mass fortification programme Decree 0547 of 1996[c]
 Wheat flour	Argentina	Enriched with iron, folic acid, thiamine, riboflavin, and niacin	Mass fortification programme Law 25.630[d]
 Vegetable oil	Bolivia	Fortified with vitamin A	Mass fortification programme Supreme Decree No. 28094[e]

(continued)

Table 2.1 (continued)

Fortified food	Origin	Fortified/enriched	
 Rice	India	Fortified with iron, folic acid, and vitamin B12	Mass fortification programme FSSAI 2018[f]
 Yogurt	Argentina	Enriched with probiotic strain *Lactobacillus rhamnosus* CRL1505	Targeted fortification
 Mixed vegetable strain	Switzerland	Fortified with zinc, iron, and folic acid	Commercial fortification
 Yogurt	Spain	Enriched in iron and vitamin C	Commercial fortification
 Mashed potatoes	Mexico	Fortified with folic acid and vitamins A, B6, B12, C, E, and B3	Commercial fortification

(continued)

Table 2.1 (continued)

Fortified food	Origin	Fortified/enriched	
 Milk	Argentina	Fortified with iron and vitamins A, C, and D	Commercial fortification
 Cereals	Mexico, Bolivia	Fortified with vitamins and minerals: calcium carbonate, iron, zinc, and vitamins B3, B5, B1, B6, B2, and B9	Commercial fortification

[a]https://arsateca.arsa.hn/wp-content/uploads/2022/02/Acuerdo-No.-4667-2016-Reglamento-de-Fortificacion-de-azucar-.pdf
[b]https://www.argentina.gob.ar/normativa/nacional/ley-17259-197392/texto
[c]https://www.minsalud.gov.co/Normatividad_Nuevo/Decreto%200547%20DE%201996.pdf
[d]https://www.argentina.gob.ar/normativa/nacional/ley-25630-77088/texto
[e]http://gacetaoficialdebolivia.gob.bo/normas/buscar/28094
[f]https://www.fssai.gov.in/upload/uploadfiles/files/Gazette_Notification_Food_Fortification_10_08_2018.pdf

treatment to partially remove water and is generally used as pre-treatment before drying or freezing. During the mass transfer processes of water leaving the vegetable tissue, physiologically active compounds can be introduced to the food matrix (de Escalada Pla et al., 2020; Miano & Rojas, 2023; Vinod et al., 2024). Compounds lowering aw and pH, as well as other composites as texture enhancers and micronutrients, can be transferred inside fruit and vegetable tissues, and lose water simultaneously, modifying the composition of food material (Vinod et al., 2024).

Impregnation may be performed by submerging vegetable pieces in solutions prepared at different osmotic pressures that can contain the compounds to be incorporated. Hypertonic solutions are necessary in the case of osmotic dehydration since the exit of water is promoted by the chemical potential difference. However, when performing vacuum impregnation or ultrasound-assisted impregnation, the mass transfer is promoted by pressure difference and acoustic cavitation effects, respectively. In all cases, the solutions must be non-toxic and without undesirable sensory effects (Miano & Rojas, 2023).

Particularly, dry infusion is a sustainable process for vegetable matrix impregnation, since the fruit pieces are sprinkled with the dry powdered ingredients. The solution begins to be formed when water from the vegetable tissue dissolves the surrounding solid ingredients. Therefore, drinking water should not be used for impregnation solution preparation (de Escalada Pla et al., 2020).

All these impregnation technologies have been applied in developing fortified foods, extending shelf life, with improved sensory attributes. Table 2.2 summarises different technological alternatives for fortifying fruit and vegetable tissues.

In recent years, considering the iron fortification of vegetables after breeding, research has mainly focused on impregnation processes assisted by ultrasound or vacuum pressure. Although fruit and vegetable fortification by impregnation is an efficient and feasible method, there are several challenging tasks. Considering the aim previously discussed, when a process is proposed, each step must be revised in order to contribute to SDGs, particularly SD-12G. Firstly, the fruit and vegetable processing to obtain homogeneous pieces generates by-products such as peels, seeds, and pulp pieces of heterogeneous size, in large quantities. These by-products contain a high concentration of valuable bioactive components as well as nutrients that should be exploited (Mohd Basri et al., 2021). On the other hand, fortification with iron may present several issues, such as undesirable changes in colour, texture or sensory properties (de Escalada Pla et al., 2020). These effects are related to the oxidation of fatty acids, vitamins and amino acids, due to the reactivity nature of iron. Besides altering sensory characteristics, these reactions also decrease the nutritional value and shelf life of the food. These oxidative reactions are a function of temperature as well as the presence of air (Kumari & Chauhan, 2022). Increasing the shelf life of the formulated foods reduces losses and, therefore, contributes to SD-12G. Meanwhile, selecting an adequate food matrix and fortifier compound may enhance absorption of iron in the body, contributing to nutritional aims, SD-2G (Hurrell, 2018; Kaur et al., 2022).

Therefore, to reduce losses and wastes, two main challenges must be addressed: (1) improving product stability and shelf life; (2) reduction of by-products.

2.4.1 Evaluation of Stability and Improving Shelf Life of Fortified Foods

Once formulated and manufactured, fortified food products must be conveniently managed to ensure microbiological, nutritional, physicochemical and sensorial quality during storage. One of the critical factors in assessing the success of any fortification programme is the stability of the micronutrients in the food supply to avoid undesirable changes over time. The study of appropriate packaging material, as well as the convenience of additional technologies to extend the shelf-life of fortified foods, has promoted much research. Among the factors affecting micronutrients are heat, moisture, air, light and acid or alkaline environments (Akhtar et al., 2011). In addition, the study of the food formulation is mandatory to avoid negative interactions among components. Particularly, some fat content could be oxidised by added mineral salts, causing off-flavours. Hemery et al. (2018) studied the vitamin A stability in wheat flour over 6 months of storage under the combined influence of different temperatures (25 °C or 40 °C), relative humidity (65% or 85% R.H.),

Table 2.2 Different technologies for iron fortification of vegetable matrix

Vegetable matrix	Iron salt	Iron content (mg.100 g^{-1})	Technology	References
Lettuce (*Lactuca sativa L.*)	Iron sulfate	107	Biofortification foliar	Preciado-Rangel et al. (2022)
Verdin bean (*Phaseolus vulgaris L.*)	Iron sulfate	9	Biofortification foliar and edaphic	Félix et al. (2024)
Sweet potato (*Ipomoea batatas*)	Ferrous sulfate	106	Ultrasound + vacuum impregnation	Purizaca-Santisteban et al. (2023)
Pineapple chips (*Comosus cv. Perola*)	Ferrous sulfate	25	Microencapsulation, ethanol + ultrasound + convective drying	Carvalho et al. (2021)
Potato tubers (*Toyoshiro*)	Ferric pyrophosphate	4.1	Vacuum impregnation	Erihemu et al. (2014)
Potato chips (*S. tuberosum L. cv. Sante*)	Ferrous sulfate	16.5	Vacuum impregnation	Tiwari et al. (2022)
Carioca kidney beans (*Phaseolus v.*)	Ferrous sulfate	60	Soaking hydrating process + ultrasound	Miano et al. (2018)
Pumpkin (*Cucurbita moschata*)	Ferrous sulfate microencapsulated	33	Ultrasound + ethanol + microencapsulated iron salt	Rojas et al. (2019)
Pumpkin (*Cucurbita moschata*)	Ferrous sulfate	35	Dry impregnation + tapioca starch coating with iron	Genevois et al. (2016)
Pumpkin (*Cucurbita moschata*)	Ferrous sulfate	55	Dry impregnation + HPMC coating containing *L. casei*	Genevois et al. (2017)
Pumpkin (*Cucurbita moschata*)	Ferrous gluconate	20	Vacuum impregnation	Lencina et al. (2022)
Chickpea (*Cicer arietinum L.*)	Ferrous sulfate heptahydrate	23	Spray and dry method	Bonto et al. (2020)
Rice (*var. IR64*)	Ferrous sulfate	32	Ultrasound	Jahan et al. (2019)

packaging with high (paper bags) and low (polyethylene terephthalate/aluminium bags) oxygen-permeability and premix formulation using or not ferrous sulphate. The authors observed a fast vitamin A loss while iron and zinc concentrations were stable, concluding that temperature and packaging materials were significant in vitamin reduction, but not iron presence. Similar results were reported by Kuong et al. (2016) for fortified rice storage during a year. However, Alam et al. (2007) reported for wheat flour enriched with ferrous sulphate and stored for 2 months at 28–39 °C, a complete retention of total iron, whereas ferrous cation decreased due to the conversion to ferric. Regarding other kinds of food, a fermented cassava product popular in Africa, Gari, was fortified with iron sulfate, iron fumarate and sodium iron ethylendiamin-tetraacetic acid (NaFeEDTA) (Sanni, 2011). The packaging selected for storage at ambient conditions (30 °C, 5 months) was high-density polyethylene (HDPE) zip-lock bags and plastic jars for 5 months. The HDPE showed a better protective barrier, reducing iron oxidation, probably associated with the reduced moisture absorption from the environment.

New approaches have been studied to extend the stability and shelf-life of fortified foods regarding sustainable and eco-friendly technologies and materials. Among others, biodegradable or edible films and coatings are extensively studied to minimise the negative impact of petroleum-derived and non-sustainable plastics (Díaz-Montes & Castro-Muñoz, 2021). Films are flexible and stand-alone matrices obtained by wet process (solvent casting at laboratory scale) or dry processes (extrusion, thermoforming, injection moulding at pilot or industrial scale) techniques, used as packaging material to wrap, separate or contain food products, while coatings are thin covers uniformly formed on food surfaces after deposition and drying (Kupervaser et al., 2023). Films and coatings can be constituted by edible lipids (waxes, oils), biopolymers (starches, derivative cellulose, chitosan, gums, caseins, gelatine, etc.), or their mixes, and added with plasticisers to make them stretchable (Chen et al., 2021). One of the main characteristics of biopolymer-based films is their low oxygen permeability and high barrier to UV radiation; therefore, they could provide convenient conditions to avoid undesirable oxidation reactions in mineral fortified foods. In a pioneering work, Park and Zhao (2004) stated the relevant effect of the high levels of a mixture of calcium gluconate and lactate or zinc lactate on the properties of chitosan edible films. More research is necessary to test the performance of biopolymer-based packaging when are applied to preserve fortified foodstuff.

Edible coatings (EC) are interesting because of their advantages as good gas barriers, micronutrient carriers, attractive appearance, etc. For coating application, several strategies can be used: dipping, spraying, spreading (brushing, rolling) or panning followed by a drying step (Aayush et al., 2022; Suhag et al., 2020). Many of such coating methodologies could be implemented at the industrial level with relative simplicity and low investment (Berti et al., 2020).

Vitamin or mineral fortification of foods through EC application has been proposed to enhance calcium content in gelatine-based coating on apple slices (Amiri et al., 2018), to supplement with iron functionalised cut pineapples coated with chitosan allowing, in addition, an extension of whole quality (Ghosh et al., 2021), to extend the shelf-life and to increase the selenium content of strawberries by applying Se-nanoparticles added in a sodium alginate EC (Tran et al., 2023) and to battle the deficiencies of chromium, magnesium and vitamin D by consuming parboiled rice coated with hydroxypropyl methylcellulose (HPMC), methylcellulose (MC), carboxypropyl methylcellulose (CMC), gum arabic and rice starch (Yulianto et al., 2017).

In an early study, Iahtisham-Ul-Haq et al. (2015) explored the application of sodium caseinate-based ECs as a carrier of zinc (ZnO at 20 and 40 ppm levels) over apricots for reducing the zinc deficiency in the Pakistani population. In addition, an extension of the shelf life of the fruit was observed for coated samples associated with the water retention and selective gas transference. Significantly higher zinc levels were found to be nutritionally available for absorption in the body. In another research, whey protein concentrate-based EC was fortified using four different iron salts to enhance the nutritional quality of *Paneer*, a common dairy product (soft cheese) of India (Jotarkar et al., 2017). As best results, minimum effects on sensory attributes and maximum iron content (93.5 ppm) in the *Paneer* were observed for ferrous sulfate heptahydrate. Regarding ferrous sulfate-fortified rice premix followed by coating with HPMC, MC, zein or their combinations, Mridula and Pooja (2013) observed that steaming and washing removed part of the iron, which was improved when coating was applied.

In another study, De'Nobili et al. (2017) developed an iron fortified pumpkin-based food product by compartmentalisation of ascorbic acid (AA) in an alginate EC while iron was incorporated in the pumpkin mesocarp by dry impregnation. It was observed that during storage (25 °C for 23 days), AA loss and colour changes were reduced when vitamin was added to the edible coating, enhancing this retention and improving global quality of iron fortified pumpkin.

Similarly, Genevois et al. (2017) proposed the simultaneous fortification with iron and probiotics (*Lacticaseibacillus casei*) of pumpkin tissue, as an alternative to dairy products. The mineral was incorporated by dry infusion while *L. casei* was supported in EC. This strategy improved the iron bioaccessibility (50–58%) by promoting the solubility of iron in a simulated digestion assay. Interestingly, lactic bacteria count was higher than 10^7 CFU.g^{-1} for 2 weeks, and the sensory test revealed that the fortified pumpkin was accepted for consumption by the untrained panel.

All reports demonstrate that EC technology is a sustainable strategy with high potential to develop fortified foods with improved nutritional and sensory quality while, at the same time, extending the shelf-life.

2.4.2 Reduction of By-products

Fruit and vegetable processing produces significant waste, mainly including pomace, peels, rinds and seeds, representing ≈25–30% of the total product. The generation of waste is creating serious economic and environmental problems. On the other hand, the most common fruit and vegetable wastes are an important source of a wide range of compounds with interesting bioactivity. For example, avocados peel, lemons and grapes pomaces, jackfruit seeds and mangoes contain 15% higher phenolic concentrations when compared with the pulp of the corresponding fruits (Kumar et al., 2020). Therefore, the utilisation of fruit and vegetable waste as raw material in new industrial products production contributes to reducing the environmental impact and promotes the circular economy within the agricultural and industrial sectors.

Peels can be used as a renewable energy source by means of processes such as combustion or anaerobic digestion—in biofuels, offering a sustainable alternative to fossil fuels—to produce organic fertilisers that improve soil quality. Another peel and agro-industrial by-products application is the development of biodegradable packaging materials (Berti et al., 2021; Ollé Resa et al., 2021). There are food and pharmaceutical applications, extracting bioactive compounds, and obtaining biodegradable polymers that reduce the use of conventional plastics. Also, in functional foods, to enrich food products with dietary fibre and other beneficial nutrients (Kumar et al., 2020). Bhardwaj et al. (2022) mention some applications and the potential of functional foods that incorporate fruit and vegetable peels, emphasising their nutritional benefits as well as contributing to healthy well-being. Some applications include (1) development of new food products: incorporating peels into the formulation of products such as cookies, bread, smoothies and snacks, increasing their nutritional value; (2) dietary supplements: producing supplements containing peel extracts to take advantage of their beneficial properties, such as antioxidants and anti-inflammatories; (3) improving digestive health: using fibre-rich peels to create foods that promote intestinal health and improve digestion, as fibre-rich foods and some compounds present in peels act as prebiotics; (4) strengthening immune systems: taking advantage of the bioactive properties of peels to develop ingredients and foods that support the immune system. In addition, bioactive compounds as well as dietary fibre may perform different functions in the food formulation. For instance, dietary fibre concentrate (DFC) was obtained from pumpkin peel and was applied as a bread improver. Lower crumb firmness was determined 24 h after baking breads containing 10 g DFC.kg^{-1} flour; remaining significantly softer breads along 9 days of storage, delaying starch retrogradation and enhancing shelf life (de Escalada Pla et al., 2013). Nieto-Calvache et al. (2020) produced DFC from papaya by-products and assayed its application as stabilisers in oil:water emulsions. In that opportunity, DFCs were obtained from the peel or pulp of papaya by-products. DFC from pulp resulted better emulsion stabiliser to creaming (18 weeks) than DFC from peel (6 weeks). Along with storage, peel DFC reduced lipid peroxidation of oil:water emulsion, presenting up to 30% reduction when compared with the reference system without DFC incorporation. DFCs from papaya or blueberry pomace were also used in muffin formulations for improving functionality (Nieto-Calvache et al., 2023; Irigoytia et al., 2022). Generally, the incorporation of these DFCs reduced crumb hardness in muffins when they were used up to 15%. For muffins containing 7.7% DFC from papaya peel, lower hardness values were observed when compared with the control formulation (without DFC) during the 22 days storage, tending to a lower rate of crumb hardening and preserving anti-radical and reducing capacity despite the processes applied (Nieto-Calvache et al., 2023). Colourful and softer muffins were obtained when wheat flour was replaced by dried blueberry pomaces. In this case, hardness was reduced ~29% when compared with the reference system. The overall acceptability of these muffins received 6 points in the 9-scored hedonic scale, being classified as "like slightly" (Irigoytia et al., 2022).

Another application explored is the production of essential enzymes in the food industry, such as phytase, pectinase, amylase, cellulase and proteases, as well as organic acids such as acetic and lactic acid by means of a biotechnology process on fruit by-products (Zahid et al., 2024). In this context, pumpkin by-products, soy expeller or okara generated after dry or wet milling, respectively, were assayed as substrate and support of *Lacticaseibacillus casei.* From the corresponding optimised process, functional food ingredients (FIs) were obtained (Castellanos-Fuentes et al., 2020; Genevois et al., 2016). On the other hand, the supernatant represented an effluent. This supernatant showed a pH of 4.64 ± 0.01 and a titratable acidity of 0.455 ± 0.01 mEq of lactic acid.g^{-1} of dehydrated pumpkin, being lactic acid the main responsible for the observed acidity (Genevois et al., 2019). This compound has various applications in the food industry as an acidulant, flavouring or preservative, and also in the manufacture of biodegradable polymers. Therefore, the process carried out generates effluents that could be used, which would contribute to reducing the generation of new by-products and would represent an additional advantage in terms of sustainability.

The main final products, FIs, were characterised by the probiotic cells count supported in a food matrix enriched in cell wall material (lignin, cellulose, no cellulosic polysaccharides) and protein. According to the matrix support, different improvements were recorded on the probiotic survival to the process applied, storage and simulated gastrointestinal digestion, keeping other probiotic characteristics. For instance, the cell count of *L. casei* 10^7 colony forming units (CFU).g^{-1} pumpkin FI after 28 days storage at 22 °C was observed (Genevois et al., 2019). Meanwhile, probiotic cell counts above 10^6 CFU.g^{-1} were recorded for FIs prepared from okara or soy expeller, at 63 and 42 days of storage at 25 °C, respectively (Castellanos-Fuentes et al., 2020). FI based on pumpkin was evaluated as a dietary supplement in two beverages: one made of chocolate milk, and another made of soy and apple juice. Approximately 80% of the cells, considering log count, survived the simulated gastrointestinal conditions. Both supplemented beverages were well received by consumers. When chocolate milk containing the FI was heated, probiotic cells survived almost 4 log cycles more than free cells after 90 min at 63 °C (Genevois et al., 2019). For nutritional purposes, Castellanos-Fuentes et al. (2022) proposed to replace 50% of oat in a porridge with the FI based on soy expeller, increasing protein and lipid content. In addition, ≈4.8 Log CFU.g^{-1} porridge of probiotic cells could survive for 2 min. microwave heating (Castellanos-Fuentes et al., 2022). Phytase activity is one of the probiotic properties of *L. casei.* This property remained in the FI based on pumpkin, despite the matrix support and the processes applied (Genevois et al., 2018). This result contributes to the aim of this chapter for reviewing the progress of different alternatives or proposals to enhance iron intake and/or bioaccessibility in diet, since phytic acids or its salts generate complexes with the minerals as previously discussed. For instance, when FIs based on soy expellers were used at 6% in a gluten-free bread formulation, a slight but significantly better iron bioaccessibility could be measured (Castellanos-Fuentes et al., 2023).

2.5 Pumpkin Fortified Snack Development through a Sustainable Process

Pumpkins pertain to the species from the cucurbits genre. Cucurbits are present in all continents, being an important source of food products and dietary fibre (Bisognin, 2002). *Cucurbita moschata* was spread from Mexico to South America and nowadays is one of the most important vegetable crops in Africa, having adapted to a wide range of climatic and soil conditions (Bisognin, 2002; Gwanama et al., 2000). Therefore, it represents fresh raw materials available almost year-round with relatively stable and affordable prices. In addition, pumpkin is one of the first foods that pediatricians suggest incorporating into babies' diets after the sixth month, and it is also recommended by gerontologists for older adults. As can be seen, it is a food widely accepted and adopted by a large part of the population. Figure 2.1 shows the scheme of the process applied for obtaining iron fortified food based on pumpkin through a dry impregnation process performed as part of research at the food pilot plant of the Department of Industries, School of Science, University of Buenos Aires. Briefly, the production process is carried out in six separate stages. In the first stage, the pumpkins are received and washed. In the second stage, the peeling, cutting and selection of the pieces are carried out. The third stage consists of cooking and impregnation with the addition of saccharose, glucose and the rest of the additives. In the fourth stage, the syrup is drained, obtaining the impregnated pumpkin cubes and syrup as a by-product. The fifth stage consists of coating with an EC formulated with HPMC containing *L. casei*, as previously explained, and subsequent drying. Finally, the sixth stage includes packaging in plastic bags and storage at 8 °C for 55 days. Pictures of different steps of the process are provided in Fig. 2.2 and can also be visualised in the following video (https://www.conicet.gov.ar/desarrollan-calabaza-fortificada-con-hierro-y-probioticos-para-disminuir-la-incidencia-de-la-anemia/), which helps in understanding the process.

As a final product, a ready-to-eat food was obtained. The pumpkin-based product fortified with iron and probiotics was presented in 10 mm cubes (Fig. 2.2, panel 5) with the following characteristics: a_w 0.913; 41.34% moisture content; pH 4.53; 37.21°Brix; 45 mg Fe.100 g^{-1} db and *L. casei* cell viability of 8.54 log CFU.g^{-1} db. A hygienic process, along with technological barriers incorporated into the product formulation (reduced pH and the addition of potassium sorbate) and storage at 8 °C contributed to the product's microbiological stability, giving a count for total coliforms (CFU.g^{-1}) < 10 and for yeast and fungi (CFU.g^{-1}) < 10. Additionally, the probiotic count remained stable, around an average value of 8.2 log (CFU.g^{-1} of pumpkin db.) over 45 storage days. Nevertheless, a significant slight reduction in a_w values was recorded from the seventh storage day, being 0.897 ± 0.002 at 45 days of storage. It demonstrated moisture transfer from the product to the environmental,

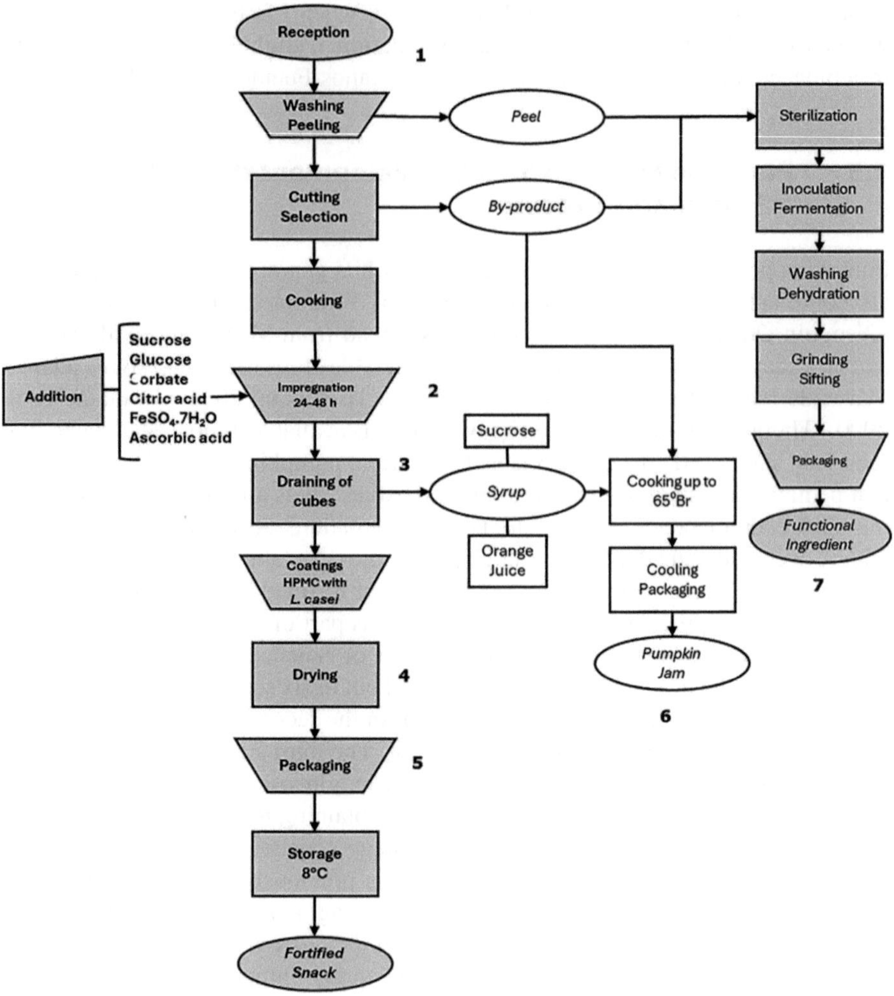

Fig. 2.1 Flow sheet for pumpkin snack fortified with iron and for alternative processes for the by-products generated

due to the non-hermetic packaging selected. Meanwhile slight and significant reduction was recorded for pH, reaching 4.34 ± 0.01 values, mainly due to the metabolic activity of the *L. casei* viable cells. Non-significant changes were recorded for iron content and viable probiotic cells, resulting in a pumpkin fortified product (Fig. 2.2, panel 5) with 45 ± 6 mg Fe .100 g^{-1} db and 8.2 log (CFU *L. casei*.g^{-1} db) at the end of the storage. In this context, it can be estimated that a 25 g serving of the product would meet 39% of the Recommended Daily Intake (RDI) for women between 19 and 50 years and 42% of the RDI for pregnant women. On the other hand, probiotic cells should survive the conditions of the different gastrointestinal tracts (Genevois et al., 2017). In this regard, the probiotic's survival rate under in vitro

Fig. 2.2 Picture related to different steps of the scheme shown in Fig. 2.1

gastrointestinal conditions on day 1 was 64%, while at 45 days of storage, the survival rate was 40.8%, when comparing log CFU, before and after digestion. Regarding iron bioaccessibility, it is defined as the iron fraction released from the food matrix in the gastrointestinal condition and remaining available for intestinal absorption (Genevois et al., 2017). It was observed that approximately 63 ± 9% of the Fe incorporated into the pumpkin tissue was bioaccessible, a value that decreased at the end of storage, reaching ≈30 ± 5% at 45 days, possibly as a result of the reduction in probiotic survival. These values were similar to those reported by the research group for laboratory-scale processes (Genevois et al., 2017).

Following the line of sustainable thinking, it must pay attention to the currents that appear as by-products and or effluents. In Fig. 2.1, it can be observed that by-products generated during the cutting and selection step: peel and pumpkin trimmings that did not meet the geometry specifications for the impregnation step. On the other hand, a liquid current originating from the syrup was produced as a result of the tissue impregnation. This syrup was an effluent with high added value, as it contains dissolved sugars (glucose and sucrose), potassium sorbate, citric acid, AA, and ferrous sulphate. It was determined to have a residual iron content of 8 mg of iron.100 g^{-1} of syrup after 24 h of impregnation. The preparation of an iron-fortified pumpkin and orange jam was proposed. It was prepared using as ingredients pumpkin trimmings, syrup effluent and orange juice, in the proportions 45%, 36% and 19%, respectively (Figs. 2.1 and 2.2, panel 6). The jam presented an iron content of 33 ± 4 mg Fe.100 g^{-1} db, with high bioaccessibility (≈91%). On the other hand, peel and pumpkin trimmings can also be applied as substrate for obtaining FIs supporting probiotics, as described in the previous section (Genevois et al., 2018, 2019) (Figs. 2.1 and 2.2, panel 7).

2.6 Conclusions

An overview of the food matrices and processes applied today for fortified food development was presented. The stability and shelf-life improvement of fortified foods by means of different coatings and packaging, as well as alternatives to reduce wastes during processing, were also deeply revised. Pumpkin is a raw material available almost year-round with relatively stable and affordable prices, accepted and adopted by a large part of the population, to be considered as a matrix for fortified food development. A process for producing a pumpkin-based product fortified with iron and probiotics was successfully scaled up. The product obtained was stable under the conditions tested for 45 days of refrigerated storage. Alternative processes were also discussed to avoid nutrient losses in the waste, by-products and effluent current generated. Pumpkin jam, and a food ingredient containing probiotics supported in a concentrate of dietary fibre from pumpkin, were proposed. It was possible to develop strategies for food processing and fortified food formulation, revaluing the by-products and contributing to sustainable production practices as some of the actions required to achieve Sustainable Development Goals.

Acknowledgements This work was supported by the Universidad de Buenos Aires (UBACYT-20020170100092BA), the National Scientific and Technical Research Council of Argentina (CONICET), PIP 2022-2024 GI 11220210100712CO, and ARCOR Foundation.

References

Aayush, K., McClements, D. J., Sharma, S., Sharma, R., Singh, G. P., Sharma, K., & Oberoi, K. (2022). Innovations in the development and application of edible coatings for fresh and minimally processed apple. *Food Control, 141*, 109188. https://doi.org/10.1016/j.foodcont.2022.109188

Akhtar, S., Anjum, F. M., & Anjum, M. A. (2011). Micronutrient fortification of wheat flour: Recent development and strategies. *Food Research International, 44*, 652–659. https://doi.org/10.1016/j.foodres.2010.12.033

Alam, S., Shah, H. U., Saleemullah, & Riaz, A. (2007). Comparative studies on storage stability of ferrous iron in whole wheat flour and flat bread (naan). *International Journal of Food Sciences and Nutrition, 58*, 5462. https://doi.org/10.1080/09637480601121979

Amiri, S., Akhavan, H. R., Zare, N., & Radi, M. (2018). Effect of gelatin-based edible coatings incorporated with aloe veraand greentea extracts on the shelf-life of fresh-cut apples. *Italian Journal of Food Science, 30*, 61.

Berti, S., Flores, S. K., & Jagus, R. J. (2020). Improvement of the microbiological quality of Argentinian Port Salut cheese by applying starch-based films and coatings reinforced with rice bran and containing natural antimicrobials. *Journal of Food Processing & Preservation, 44*(10), e14827. https://doi.org/10.1111/jfpp.14827

Berti, S., Jagus, R. J., & Flores, S. K. (2021). Effect of rice bran addition on physical properties of antimicrobial biocomposite films based on starch. *Food and Bioprocess Technology, 14*, 1700–1711. https://doi.org/10.1007/s11947-021-02669-0

Bhardwaj, K., Najda, A., Sharma, R., Nurzyńska-Wierdak, R., Dhanjal, D., Sharma, R., Manickam, S., Kabra, A., Kuča, K., & Bhardwaj, P. (2022). Fruit and vegetable peel-enriched functional foods: potential avenues and health perspectives. *Evidence-Based Complementary and Alternative Medicine, 1*, 8543881. https://doi.org/10.1155/2022/8543881

Bisognin, D. (2002). Origin and evolution of cultivated cucurbits. *Ciência Rural, 32*(5), 715–723.

Blanco-Rojo, R., Pérez-Granados, A., Toxqui, L., González-Vizcayno, C., Delgado, M., & Vaquero, M. (2011). Efficacy of a microencapsulated iron pyrophosphate-fortified fruit juice: A randomised, double-blind, placebo-controlled study in Spanish iron-deficient women. *British Journal of Nutrition, 105*(11), 1652–1659. https://doi.org/10.1017/S0007114510005490

Bonto, A., Jearanaikoon, N., Sreenivasulu, N., & Camacho, D. (2020). High uptake and inward diffusion of iron fortificant in ultrasonicated milled rice. *LWT - Food Science and Technology, 128*, 109459. https://doi.org/10.1016/j.lwt.2020.109459

Boyle, N., Adolphus, K., Caton, S., Croden, F., Dye, L., Glass, A., Halliwell, K., Hansen, G., Holm, L., Jackson, P., Makinwa, F., Stærk, B., & Wilkinson, N. (2024). Increasing fibre intake in the UK: Lessons from the Danish whole grain partnership. *The British Journal of Nutrition, 131*(4), 672–685. https://doi.org/10.1017/S0007114523002106

Carvalho, G. R., Massarioli, A. P., Alvim, I. D., & Augusto, P. E. D. (2021). Iron-fortified pineapple chips produced using microencapsulation, ethanol, ultrasound and convective drying. *Food Engineering Reviews, 13*, 726–739. https://doi.org/10.1007/s12393-020-09259-4

Castellanos-Fuentes, A., Genevois, C., Flores, S., & de Escalada Pla, M. (2020). Valorisation of soy by-products as substrate for food ingredients containing L. casei through solid state fermentation. *LWT - Food Science and Technology, 132*, 109779. https://doi.org/10.1016/j.lwt.2020.109779

Castellanos-Fuentes, A., Cortés, N., Genevois, C., Flores, S., & de Escalada Pla, M. (2022). Soybean extruded by-products as substrate for obtaining food ingredients containing probiotics. *International Journal of Food Science & Technology, 57*, 4825–4831. https://doi.org/10.1111/ijfs.15690

Castellanos-Fuentes, A., Genevois, C., Flores, S., & de Escalada Pla, M. (2023). Novel application of a food ingredient based on soybean extruded-expelled meal containing probiotics for improving gluten free bread quality. *International Journal of Food Science & Technology, 58*, 6855–6861.

Cercamondi, C. I., Duchateau, G. S., Harika, R. K., van den Berg, R., Murray, P., Koppenol, W. P., Zeder, C., Zimmermann, M. B., & Moretti, D. (2016). Sodium pyrophosphate enhances iron bioavailability from bouillon cubes fortified with ferric pyrophosphate. *The British Journal of Nutrition, 116*, 496–503. https://doi.org/10.1017/S0007114516002191

Chaudhary, A., Kumari, M., & Vyas, S. (2024). Strategies to combat iron deficiency anemia among lactating women in India: A review. *Food and Humanity, 2*, 100253.

Chen, W., Ma, S., Wang, Q., McClements, D. J., Liu, X., Ngai, T., & Liu, F. (2021). Fortification of edible films with bioactive agents: A review of their formation, properties, and application in food preservation. *Critical Reviews in Food Science and Nutrition*. https://doi.org/10.1080/10408398.2021.1881435

Choi, S. J., Decker, E. A., & McClements, D. J. (2009). Impact of iron encapsulation within the interior aqueous phase of water-in-oil-in-water emulsions on lipid oxidation. *Food Chemistry, 116*, 271–276. https://doi.org/10.1016/j.foodchem.2009.02.045

Darwar, R., Rowe, L., Chadha, M., Rosas, A., & Arabi, M. (2023). A blueprint for fortification planning and programming: Lessons learned from an analytical review of existing fortification frameworks. *Maternal & Child Nutrition*, e13571. https://doi.org/10.1111/mcn.13571

de Escalada Pla, M., Flores, S., & Genevois, C. (2020). Innovative strategies and nutritional perspectives for fortifying pumpkin tissue and other vegetable matrices with iron. *Food Science and Human Wellness, 9*, 103–111. https://doi.org/10.1016/j.fshw.2020.02.005

De'Nobili, M. D., Genevois, C., Rojas, A. M., Flores, S., & de Escalada Pla, M. (2017). Ch. 5: Stabilization of L-(+)-ascorbic acid in an iron fortified vegetable product (Cucurbita moschata Duchesne ex Poiret) using an alginate coating. In E. Parsons (Ed.), *Ascorbic acid: Properties, synthesis and applications* (pp. 105–122). Nova Science Publishers. ISBN: 978-1-63485-886-1.

Díaz-Montes, E., & Castro-Muñoz, R. (2021). Edible films and coatings as food-quality preservers: An overview. *Foods, 10*, 249. https://doi.org/10.3390/foods10020249

Dueik, V., Chen, B. K., & Diosady, L. L. (2017). Iron-polyphenol interaction reduces iron bioavailability in fortified tea: Competing complexation to ensure iron bioavailability. *Journal of Food Quality*. https://doi.org/10.1155/2017/1805047

Erihemu K., Hironaka Y., Oda H., Koaze (2014) Iron enrichment of whole potato tuber by vacuum impregnation LWT - Food Science and Technology 59(1):504–509. https://doi.org/10.1016/j.lwt.2014.04.043

E/2023/64, Annex. (2023). *Progress towards the Sustainable Development Goals: Towards a rescue plan for people and planet: report of the Secretary-General (special edition).* UN, Economic and Social Council. Available from: https://digitallibrary.un.org/record/4014344?v=pdf

E/2024/54, Annex. (2024). *Report of the Secretary-General. Progress towards the Sustainable Development Goals.* UN, Economic and Social Council. Available from: https://digitallibrary.un.org/record/4051646?v=pdf

FAO. (2014). *Developing sustainable food value chains. Guiding principles.* Available from: https://www.fao.org/sustainable-food-value-chains/library/details/en/c/265156/

Félix, J. W., Sánchez-Chávez, E., Tosquy-Valle, O., Preciado-Rangel, P., Márquez-Quiroz, C., & de la Cruz-Lázaro, E. (2024). Biofortificación de Frijol (Phaseolus vulgaris L.) Variedad Verdín con Quelato y Sulfato de Hierro. *Terra Latinoamericana, 42*, e1831. https://doi.org/10.28940/terra.v42i0.1831

Genevois, C., Flores, S., & de Escalada Pla, M. (2016). Byproduct from pumpkin (*Cucurbita moschata* Duchesne ex poiret) as a substrate and vegetable matrix to contain Lactobacillus casei. *Journal of Functional Foods, 23*, 210–219. https://doi.org/10.1016/j.jff.2016.02.030

Genevois, C., de Escalada Pla, M., & Flores, S. (2017). Novel strategies for fortifying vegetable matrices with iron and *Lactobacillus casei* simultaneously. *LWT - Food Science and Technology, 79*, 34–41. https://doi.org/10.1016/j.lwt.2017.01.019

Genevois, C. E., Castellanos Fuentes, A. P., Flores, S. K., & de Escalada Pla, M. F. (2018). The functional and organoleptic characterization of a dairy-free dessert containing a novel probiotic food ingredient. *Food and Function, 9*(11), 5697–5706. https://doi.org/10.1039/c8fo00805a

Genevois, C., Pieniazek, F., Messina, V., Flores, S., & de Escalada Pla, M. (2019). Bioconversion of pumpkin by-products in novel supplements supporting *Lactobacillus casei. LWT - Food Science and Technology, 105*, 23–29. https://doi.org/10.1016/j.lwt.2019.01.057

Ghosh, T., Nakano, K., Mulchandani, N., & Katiyar, V. (2021). Curcumin loaded iron functionalized biopolymeric nanofibre reinforced edible nanocoatings for improved shelf life of cut pineapples. *Food Packaging and Shelf Life, 28*, 100658. https://doi.org/10.1016/j.fpsl.2021.100658

Gupta, C., Chawla, E. P., Arora, S., Tomar, S., & Singh, A. (2015). Iron microencapsulation with a blend of gum arabic, maltodextrin and modified starch using modified solvent evaporation method—Milk fortification. *Food Hydrocolloids, 43*, 622–628. https://doi.org/10.1016/j.foodhyd.2014.07.021

Gwanama, C., Labuschagne, M., & Botha, A. (2000). An·lisis of genetic variation in Cucurbita moschata by random amplified polymorphic DNA (RAPD) markers. *Euphyrica, 113*, 19–24.

Hemery, Y. M., Laillou, A., Fontan, L., Jallier, V., Moench-Pfanner, R., Berger, J., & Avallone, S. (2018). Storage conditions and packaging greatly affects the stability of fortified wheat flour: Influence on vitamin A, iron, zinc, and oxidation. *Food Chemistry, 240*, 43–50. https://doi.org/10.1016/j.foodchem.2017.07.084

Hurrell, R. (2018). Chapter 20 – Efficacy and safety of iron fortification. In M. Venkatesh Mannar & R. Hurrell (Eds.), *Food fortification in a globalized world* (pp. 195–212). Academic Press. https://doi.org/10.1016/B978-0-12-802861-2.00020-1

Hurrell, R., Furniss, D., Burri, J., Whittaker, P., Lynch, S., & Cook, J. (1989). Iron fortification of infant cereals: A proposal for the use of ferrous fumarate or ferrous succinate. *The American Journal of Clinical Nutrition, 49*, 1274–1282. https://doi.org/10.1093/ajcn/49.6.1274

Iahtisham-Ul-Haq, Butt, M. S., Suleria, H. A. R., & Ahmed, W. (2015). Physicochemical behavior of zinc-fortified, sodium caseinate-based, edible-coated apricots during storage in controlled atmosphere. *Journal of Food Processing and Preservation, 39*, 2431–2441. https://doi.org/10.1111/jfpp.12493

Irigoytia, M., Irigoytia, K., Sosa, N., de Escalada Pla, M., & Genevois, C. (2022). Blueberry by-product as a novel food ingredient: Physicochemical characterization and study of its application in a bakery product. *Journal of the Science of Food and Agriculture, 102*(11), 4551–4560. https://doi.org/10.1002/jsfa.11812

Jahan, T. A., Vandenberg, A., Glahn, R. P., Tyler, R. T., Reaney, M. J. T., & Taran, B. (2019). Iron fortification and bioavailability of chickpea (Cicer arietinum L.) seeds and flour. *Nutrients, 11*(9), 2240. https://doi.org/10.3390/nu11092240

Jotarkar, P. S., Panjagari, N. R., Singh, A. K., & Arora, S. (2017). Effect of whey protein-iron based edible coating on the quality of paneer and process optimization. *International Journal of Dairy Technology, 70*. https://doi.org/10.1111/1471-0307.12478

Kaur, N., Agarwal, A., & Sabharwal, M. (2022). Food fortification strategies to deliver nutrients for the management of iron deficiency anaemia. *Current Research in Food Science, 5*, 2094–2107. https://doi.org/10.1016/j.crfs.2022.10.020

Kumar, E., Surekha, K., Rekha, K., & Sharma, S. (2020). Effect of various sources of zinc and iron on grain yield, nutrient uptake and quality parameters of finger millet (Eleusine coracana L.). *International Research Journal of Pure and Applied Chemistry, 21*(2), 46–55. https://doi.org/10.9734/irjpac/2020/v21i230154

Kumari, A., & Chauhan, A. K. (2022). Iron nanoparticles as a promising compound for food fortification in iron deficiency anemia: A review. *Journal of Food Science and Technology, 59*, 3319–3335. https://doi.org/10.1007/s13197-021-05184-4

Kuong, K., Laillou, A., Chea, C., Chamnan, C., Berger, J., & Wieringa, F. T. (2016). Stability of vitamin A, iron and zinc in fortified rice during storage and its impact on future national standards and programs—Case study in Cambodia. *Nutrients, 8*, 51. https://doi.org/10.3390/nu8010051

Kupervaser, M. G., Traffano-Schiffo, M. V., Dellamea, M. L., Flores, S. K., & Sosa, C. A. (2023). Trends in starch-based edible films and coatings enriched with tropical fruits extracts: A review. *Food Hydrocolloids for Health, 4*, 100138. https://doi.org/10.1016/j.fhfh.2023.100138

Lencina, M. S., Dos Santos, C., Archaina, D., G6mez, M. B., & Mazzobre, M. F. (2022). Stability and bioaccessibility of iron in pumpkin discs vacuum impregnated with ferrous gluconate, β-cyclodextrin and ascorbic acid. *LWT, 161*, 113342. https://doi.org/10.1016/j.lwt.2022.113342

Liyanage, C., & Hettiarachchi, M. (2011). Food fortification. Ceylon. *Medizinhistorisches Journal, 56*, 124–127. https://doi.org/10.4038/cmj.v56i3.3607

Marina, de Escalada Pla Ana María, Rojas Lía Noemí, Gerschenson (2013). Effect of Butternut (Cucurbita moschata Duchesne ex Poiret) Fibres on Bread Making Quality and Staling Food and Bioprocess Technology 6(3):828–838. https://doi.org/10.1007/s11947-011-0744-y

Mehansho, H. (2006). Iron fortification technology development: New approaches. *Journal of Nutrition, 136*(4), 1059–1063. https://doi.org/10.1093/jn/136.4.1059

Menon, K., Ferguson, E., Thomson, C., Gray, A., Zodpey, S., Saraf, A., Kumar, D. P., Pandav, C., & Skeaff, C. (2014). Iron status of pregnant Indian women from an area of active iron supplementation. *Nutrition, 30*, 291–296.

Miano, A., & Rojas, M. (2023). Engineering strategies for food fortification. *Current Opinion in Food Science, 51*, 101033. https://doi.org/10.1016/j.cofs.2023.101033

Miano, A., Esteves, P., & Augusto, D. (2018). The ultrasound assisted hydration as an opportunity to incorporate nutrients into grains. *Food Research International, 106*, 928–935. https://doi.org/10.1016/j.foodres.2018.02.006

Milani, P., Haddad, L., Steiner, R., Mkambula, P., Ehsani, M., Kamau, D., Ndung'u, D., & de Pee, S. (2024). Fortified whole grains and whole blends: A timely food systems shift. *Global Food Security, 42*, 100784. https://doi.org/10.1016/j.gfs.2024.100784

Mohd Basri, M. S., Abdul Karim Shah, N. N., Sulaiman, A., Mohamed Amin Tawakkal, I. S., Mohd Nor, M. Z., Ariffin, S. H., Abdul Ghani, N. H., & Mohd Salleh, F. S. (2021). Progress in the valorization of fruit and vegetable wastes: Active packaging, biocomposites, by-products, and innovative technologies used for bioactive compound extraction. *Polymers, 13*(20), 3503.

Mridula, D., & Pooja, J. (2013). Preparation of iron-fortified rice using edible coating materials. *International Journal of Food Science and Technology*. https://doi.org/10.1111/ijfs.12305

Nieto-Calvache, J., Gerschenson, L., & de Escalada Pla, M. (2020). Papaya by-products for providing stability and antioxidant activity to oil in water emulsions. *Journal of Food Science and Technology, 58*(5), 1693–1702. https://doi.org/10.1007/s13197-020-04679-w

Nieto-Calvache, J., Lorenzo, G., Califano, A., Gerschenson, L., & de Escalada Pla, M. (2023). Papaya dietary fibre concentrates for providing functionality to muffin formulations. *Journal of Science and Food Agriculture, 103*(3), 1326–1333. https://doi.org/10.1002/jsfa.12226

Ohanenye, I., Emenike, C., Mensi, A., Medina-Godoy, S., Jin, J., Ahmed, T., Sun, X., & Udenigwe, C. (2021). Food fortification technologies: Influence on iron, zinc and vitamin A bioavailability and potential implications on micronutrient deficiency in sub-Saharan Africa. *Scientific African, 11*, e00667. https://doi.org/10.1016/j.sciaf.2020.e00667

Ollé Resa, C. P., Jagus, R. J., & Gerschenson, L. N. (2021). Do fillers improve the physicochemical properties of antimicrobial tapioca starch edible films? *Journal of Food Safety, 41*, e12880. https://doi.org/10.1111/jfs.12880

Orsango, A. Z., Loha, E., Lindtjørn, B., & Engebretsen, I. M. S. (2020). Efficacy of processed amaranth-containing bread compared to maize bread on hemoglobin, anemia and iron deficiency anemia prevalence among two-to-five-year-old anemic children in Southern Ethiopia: A cluster randomized controlled trial. *PLoS One, 15*(9), e0239192. https://doi.org/10.1371/journal.pone.0239192

Osendarp, S., Martinez, S., Garrett, G., Neufeld, L., De-Regil, L., Vossenaar, M., & Darnton-Hill, I. (2018). Large-scale food fortification and biofortification in low- and middle-income countries: A review of programs, trends, challenges, and evidence gaps. *Food and Nutrition Bulletin, 39*(2), 315–331. https://doi.org/10.1177/0379572118774229

Park, S.-I., & Zhao, Y. (2004). Incorporation of a high concentration of mineral or vitamin into chitosan-based films. *Journal of Agricultural and Food Chemistry, 52*, 1933–1939. https://doi.org/10.1021/jf034612p

Preciado-Rangel, P., Valenzuela-García, A. A., Pérez-García, L. A., González-Salas, U., Ortiz-Díaz, S. A., Buendía-García, A., & Rueda-Puente, E. O. (2022). La biofortificación foliar con hierro mejora la calidad nutracéutica y la capacidad antioxidante en lechuga. *Revista Terra Latinoamericana, 40*. https://doi.org/10.28940/terra.v40i0.1060

Purizaca Santisteban, K., Ruiz-Flores, L., Sócola, Z., Chaves, E., & Espinoza, M. (2023). Ultrasound-assisted fortification of yellow sweet potato (Ipomoea batatas) with iron and ascorbic acid. *Frontiers in Sustainable Food Systems, 7*. https://doi.org/10.3389/fsufs.2023.1193496

Rojas, M. I., Dutra Alvim, P., & Esteves, D. A. (2019). Incorporation of microencapsulated hydrophilic and lipophilic nutrients into foods by using ultrasound as a pre-treatment for drying: A prospective study. *Ultrasonics Sonochemistry, 54*, 153–161. https://doi.org/10.1016/j.ultsonch.2019.02.004

Sanni, S. A. (2011). Stability of iron-fortified gari during storage in different packaging materials. *Journal of Food Processing and Preservation, 579*, 1–7. https://doi.org/10.1111/j.1745-4549.2011.00579.x

Santillán-Urquiza, E., Méndez-Rojas, M. Á., & Vélez-Ruiz, J. F. (2017). Fortification of yogurt with nano and micro-sized calcium, iron, and zinc, effect on the physicochemical and rheological properties. *LWT - Food Science and Technology, 80*, 462–469. https://doi.org/10.1016/j.lwt.2017.03.025

Shubham, T., Anukiruthika, S., Dutta, A. V., Kashyap, J. A., & Moses, C. A. (2020). Iron deficiency anemia: A comprehensive review on iron absorption, bioavailability, and emerging food fortification approaches. *Trends in Food Science & Technology, 99*, 58–75. https://doi.org/10.1016/j.tifs.2020.02.021

Siddique, A., & Park, Y. (2019). Effect of iron fortification on microstructural, textural, and sensory characteristics of caprine milk Cheddar cheeses under different storage treatments. *Journal of Dairy Science, 102*, 2890–2902. https://doi.org/10.3168/jds.2018-15427

Soares, A., Barros, N. M., Saint'Pierre, T. D., De Paula Lima, J., Calado, V., Donangelo, C. M., & Farah, A. (2019). Fortification of ground roasted coffees with iron, zinc, and calcium salts: Evaluation of minerals recovery in filtered and espresso brews. *Beverages, 5*, 4. https://doi.org/10.3390/beverages5010004

Sudargo, T., Nisa, F. Z., Helmiyati, S., Kusuma, R. J., Arjuna, T., & Septiana, R. D. (2013). Tempeh with iron fortification to overcome iron deficiency anemia. *Pakistan Journal of Nutrition, 12*, 815.

Suhag, R., Kumar, N., Petkoska, A. T., & Upadhyay, A. (2020). Film formation and deposition methods of edible coating on food products: A review. *Food Research International, 136*, 109582. https://doi.org/10.1016/j.foodres.2020.109582

Teshome, E., Otieno, W., Terwel, S., Osoti, V., Demir, A., Andango, P., Prentice, A., & Verhoef, H. (2017). Comparison of home fortification with two iron formulations among Kenyan children: Rationale and design of a placebo-controlled non-inferiority trial. *Contemporary Clinical Trials Communications, 7*, 1–10. https://doi.org/10.1016/j.conctc.2017.04.007

Tiwari, P., Thakur, M., Joshi, A., Raigond, P., & Arora, B. (2022). Development of iron fortified potato fries through vacuum assisted processing strategies. *Journal of Food Science and Technology, 59*, 4644–4652. https://doi.org/10.1007/s13197-022-05543-9

Tran, T. H., Le, X. C., Tran, T. N. M., Nguyen, N. T. T., Pham, B. N., & Vu, D. (2023). Nano selenium–alginate edible coating extends hydroponic strawberry shelf life and provides selenium fortification as a micro-nutrient. *Food Bioscience, 53*, 102597. https://doi.org/10.1016/j.fbio.2023.102597

Vinod, R., Asrey, R., Sethi, S., Menaka, M., Kumar Meena, N., & Shivaswamy, G. (2024). Recent advances in vacuum impregnation of fruits and vegetables processing: A concise review. *Heliyon, 10*, e28023. https://doi.org/10.1016/j.heliyon.2024.e28023

von Braun, J., Afsana, K., Fresco, L. O., Hassan, M., & Torero, M. (2021). Food system concepts and definitions for science and political action. *Nature Food, 2*, 748–750. https://doi.org/10.1038/s43016-021-00361-2

World Health Organization (WHO). (2017). *Nutritional anaemias: tools for effective prevention.* Available from: https://apps.who.int/iris/handle/10665/259425

World Health Organization, Food and Agricultural Organization of the United Nations. WHO & FAO. (2006). *Guidelines on food fortification with micronutrients.* Available from: https://www.who.int/publicationsdetail-redirect/9241594012

Yulianto, W. A., Susiati, A. M., & Adhini, H. A. N. (2017). Evaluation of resistant starch, glycemic index and fortificants content of premix rice coated with various concentrations and types of edible coating materials. International symposium on food and agro-biodiversity. *IOP Conference Series: Earth and Environmental Science, 102*, 012101. https://doi.org/10.1088/1755-1315/102/1/012101

Zahid, M., Khalid, S., Raana, S., Amin, S., Javaid, H., Arshad, R., Jahangeer, A., Ahmad, S., & Hassan, S. (2024). Unveiling the anti-oxidative potential of fruits and vegetables waste in prolonging the shelf stability of vegetable oils. *Future Foods, 10*, 100328. https://doi.org/10.1016/j.fufo.2024.100328

Plant-Based Solution for Food Fortification

3

Ritabrata Sarkar and Aadisha Saini

Abstract

With increasing consumer concern regarding the health effects of animal-based foods, ethical and environmental issues, a rise in the demand for plant-based food has been observed over the past few years. Plant-based foods are comparatively low in micronutrients as compared to animal-based foods, which has increased the risk of micronutrient deficiencies in a large part of the population. This shift in the food consumption trends, and to combat the problem of hidden hunger, researchers have come up with a new strategy—plant-based food to food fortification. This comprises of incorporation of plant-based fortifiers to food, resulting in plant-based fortified foods. This strategy would not only fortify the food but also diversify the diet. Plant-based food to food fortification would help in accomplishing the two strategies—dietary diversity and fortification out of the four strategies—fortification, dietary diversity, supplementation and nutrition education recommended by the World Health Organisation/Food Agriculture Organisation (2006) guidelines to overcome hidden hunger. This chapter will provide insights into the different plant-based fortifiers that have the potential to be used as fortifiers in food fortification.

Keywords

Plant-based food fortification · Micronutrient deficiencies · Agro-industry waste · Malnutrition · Bioavailability · Sustainable Development Goals (SDGs)

R. Sarkar
Department of Genetics and Plant Breeding, ICAR-Indian Institute of Agricultural Biotechnology, Ranchi, Jharkhand, India

A. Saini (✉)
Department of Food Science and Technology, National Institute of Food Technology Entrepreneurship and Management, Sonipat, Haryana, India

G. Sanghvi et al. (eds.), *Sustainable Food Fortification*,
https://doi.org/10.1007/978-981-95-1217-1_3

3.1 Introduction

Millions of individuals worldwide suffer from micronutrient deficiencies, which are especially prevalent in low- and middle-income nations (Olson et al., 2021). Iron, vitamin A, zinc, iodine and other essential vitamins and minerals are frequently deficient in diets, which can have serious health effects such as stunted growth, reduced cognitive function and increased susceptibility to infections (Black et al., 2013). Innovative and long-term strategies are needed to address these deficits and guarantee that all people receive enough nutrients (Muthayya et al., 2013). Food fortification, which entails enhancing the nutritional content of frequently consumed foods by adding necessary nutrients, is one of the most extensively used practices all over the world (Dary & Hurrell, 2006). Early efforts like the iodisation of salt in the 1920s and the addition of vitamins such as vitamin B to flour in the 1930s greatly decreased the incidence of associated deficiencies, suggesting years long history of this practice (Bishai & Nalubola, 2002). In order to fight malnutrition on a broad scale, fortification initiatives have grown throughout time to incorporate a range of staple foods, such as rice, milk and cooking oils (Stoltzfus, 2003). But there are drawbacks to traditional food fortification, including the possibility of nutrient deterioration, financial constraints and bioavailability problems (Hurrell, 2002a). Food-to-food fortification can be a promising substitute for traditional fortification. By using naturally occurring, whole-food constituents instead of separated micronutrients, this strategy aims to improve the nutrient profile of staple foods (Chadare et al., 2019). This chapter examines different methods of food fortification, with a focus on how plant-based and food-to-food fortification might enhance nutritional results. The potential of underutilised crops and food industry by-products to improve food security is also highlighted. By looking at the possible solutions, this chapter hopes to support further efforts to reduce malnutrition and provide more affordable, sustainable and easily accessible nutrition solutions for people who are at risk.

3.2 Fortification

In order to prevent and treat nutritional deficiencies, especially related to the micronutrients in the population, food fortification is a public health technique that involves adding necessary nutrients to food products. Although the idea of food fortification has been around for millennia, it only really came into focus in the twentieth century as a result of the prevalent vitamin inadequacies. In the 1920s, goitre and other iodine-related deficiencies were addressed by supplementing salt with iodine (Au & Markel, 1987). This programme was among the first extensive public health initiatives to use food in order to improve nutrition when the role of health supplements was followed. The initiative of food fortification was primarily done to prevent micronutrient deficiencies in children, which affect their physical as well as mental growth and development. Several nations began fortifying flour with vitamin B in the 1930s, specifically thiamine, riboflavin, niacin and later folic acid.

The prevalence of pellagra and other disorders related to nutrient deficiencies was a major factor in establishing food fortification as a technique to overcome this problem (Bishai & Nalubola, 2002). By introducing enriched flour in 1941, the United States was a leader in this movement and established a standard for comparable programmes around the world. Food fortification initiatives increased in the years following World War II (Stoltzfus, 2003). In order to combat malnutrition in underdeveloped nations, the World Health Organisation (WHO) and other international organisations started to promote the fortification of staple foods. This resulted in the introduction of fortified cooking oils with vitamin A and the fortification of other basic foods, such as salt, milk, flour and rice. Food fortification had gained widespread acceptance as a public health tactic by the end of the twentieth century. As a result of international fortification norms and regulations, fortified products like cereals, milk and sauces are also being developed (Dary & Hurrell, 2006). In order to tackle anaemia in the population, the emphasis moved to addressing specific dietary shortages that are common in different regions, which led to the development iron-fortified foods. Another initiative that has come up over the years is biofortification using genetic engineering to develop such crops that already have a predefined amount of otherwise deficient nutrients. This strategy seeks to address malnutrition in the long run, particularly in places where access to fortified meals may be restricted. Food fortification is a vital part of public health plans around the globe today, and more research and development need to be done to increase the effectiveness and accessibility of these interventions. The practice of fortifying foods is an attempt to provide nutritious food to all sections of society in order to improve their health and well-being (Bouis & Saltzman, 2017). However, taking into account environmental crises, human health and achieving SDGs in a short time span has led to a shift in dietary preference for plant-based foods, which has made researchers develop fortification methods using plant-based fortifiers.

3.3 Food to Food Fortification.

Conventional fortification approach such as breeding of varieties for higher nutrient content, agronomic practices which use higher nutrients in the field or enhancing the soil nutrient mobility in order to enhance the plant nutrient content, transgenic techniques which utilises biotechnological tools to enhance the synthesis of nutrients in the plant and another approach involves adding micronutrients to staple foods and promoting diet diversification in order to mitigate malnutrition in the population has different challenges. Producing varieties with higher nutrient content is a long-term approach that cannot keep pace with the growing population's needs for a nutrient-rich diet (Sharma et al., 2017). Moreover, processing of such foods leads to significant losses of these nutrients due to their labile nature, which destroys the purpose of developing such nutrient-rich varieties. Fortifying foods with elemental forms of nutrients is not feasible for all types of foods as they interact with other food components and negatively affect the food properties and is a nutrient-specific approach (Hurrell, 2002b). The availability of such foods in the market is less and the awareness among consumers regarding such foods is also limited. Diversification in diet depends upon the purchasing ability of a household; therefore, this approach to combat micronutrient deficiency also has to be supplemented with another approach (Nair et al., 2016). Such drawbacks in the conventional fortification techniques have led to the development of a new strategy known as food-to-food fortification as a complementary technique to conventional strategies in order to overcome nutrient deficiencies in the population. Food-to-food fortification can be described as a technique of preparing a food formulation at the household or industrial level by substituting antinutritional factors with micronutrient foods in order to enhance the bioavailability of micronutrients in the diet. Food-to-food fortification is different from all the above-mentioned conventional techniques as it aims not only to enhance the amount of micronutrients in the foods but also to make them bioavailable by substituting or adding another food rather than an individual micronutrient or a mixture. This technique can be used as a sole fortification method or in combination with other techniques (Kruger et al., 2020).

3.4 Plant-Based Fortificants

In order to develop a food-to-food fortified product with natural fortificants, the selection of the right type of fortificant is important. The fortificant selection can be dependent upon many factors some of which are (a) foods rich in particular type of micronutrient (b) other micronutrients present in the food (c) the amount of micronutrients present in the food (d) bioavailability of the micronutrients when incorporated in another food matrix (e) effect of the fortificant on the different properties of the carrier food (f) acceptability in the target population (Vishwakarma et al., 2022). Due to the drastic shift in consumer preference for vegan products, the fortificants have to be plant-based in order to meet the dietary requirements of a large proportion of the population. In adults, the major cause of food allergy is from plant-based foods as compared to animal foods, which has led to a concern among scientists to make a proper choice of food fortificants, which has undergone a proper clinical investigation. Due to the surge in plant-based foods, plant sources of protein in the form of concentrates, isolates from legumes (e.g. pea, fava beans, chickpea), tubers (e.g. potato), seaweed, etc., have gained popularity and do not have any allergen disclaimer associated with them as in the case of wheat and soy (Boukid et al., 2019). The cross-reactivity risk or allergenic potential of these crops should not be overlooked and all safety measures should be taken care of before preparing a fortified food which is meant for mass consumption. Besides the clinical concerns associated with plant-based fortificants, there is a wide scope for plant-origin foods as fortificants. Grain legumes are regarded as a "superfood" for their high nutritional value in terms of providing macronutrients for growth and development and micronutrients, which have multiple health benefits such as enhanced satiety and cholesterol-lowering effects (McCrory et al., 2010). As cereals form a staple diet for a major part of the population, it should be the primary objective of researchers to make staple foods complete in all nutritional aspects so that all the possible nutrients are available to all sections of society.

3.4.1 Type of Plant-Based Fortificants.

All kinds of food are rich in particular types of nutrients; however, the selection of the right type of fortificant with a suitable food vehicle needs to be determined. Using pulses, fruits and spices as food fortificants when consuming them in their real form would be a more preferable choice. Scientists have come up with the idea of incorporating agro-industry waste into food as food fortificants. It has been known that agro-industry waste, after being used for its intended purpose, still contains nutrients in amounts that can be used as fortificants in another food. Antonic et al. (2020) reviewed the potential of apple pomace as a food fortificant. This study shows that dried apple pomace has a sufficient amount of micronutrients such as sodium, phosphorus, calcium, potassium, magnesium, iron, copper and manganese. Table 3.1 shows the studies on the addition of agro-industry waste as a food fortificant, by-product from which industry, type of food vehicle and its effect on the carrier. However, the potential of whole crops as food fortificants cannot be undermined and studies have been conducted on incorporating different pulses such as gram, pea, lentil and tubers such as sweet potato, moringaand mushroom have been presented in Table 3.2.

3.4.1.1 Mushroom

Mushrooms have been used as an ingredient since ancient times in food cuisines due to their high nutritional value, organoleptic and medicinal properties. Different edible mushroom species such as *Hericium* spp., *Ganoderma* spp., shiitake and *Pleurotus* spp. have been known for their superior medicinal as well as nutritional properties. Kumar et al. (2022) reviewed the scope of incorporating mushrooms and its parts into different food products. Mushrooms are a good source of macronutrients such as proteins, fats, carbohydrates, especially glucans, mannans, xylan and micronutrients such as copper, selenium, carotenoids, tocopherols, ergosterols and many other secondary metabolites which hold pharmaceutical importance. Okafor et al. (2012) studied the quality characteristics of wheat bread enriched with oyster mushroom (*Pleurotus plumonarius*) (5–25%). The study concluded that 10% supplementation of mushroom powder in wheat bread enhanced the crude protein content, fibre and ash content. The 10% bread sample did not have any negative impact on the sensory properties of the bread in terms of crust colour, hardness, taste and overall acceptability.

3.4.1.2 Pulses

These nutritious seeds are potential low-cost, high-value plant-based food fortificants. They are a rich and affordable source of macro- and micronutrients and are regarded as a superfood. They are considered low-calorie products with abundant sources of dietary fibre, essential amino acids except methionine and minerals such as potassium (Crews & Peoples, 2004). Studies have been conducted on incorporating pulses in different types of food products as mentioned in Table 3.2. Balogun et al. (2012) performed a study in which they studied the quality parameters of tapioca meal fortified with defatted soybean flour. As tapioca meal lacks protein

Table 3.1 Studies conducted to evaluate the efficacy of agro-industry waste as fortifiers

Fortifier	Commodity	By-product from	Nutritional enhancement	References
Grape pomace powder	Wheat bread	Winemaking process	Increase in total phenolic compounds, dietary fiber content and antioxidant capacity	Tolve et al. (2021)
Olive mill waste water and olive paste	Bread and pasta	Oil industry	Improvement in phenolic contents and antioxidant activity	Cedola et al. (2020)
Apple pomace	Cake, biscuits, bread, cookies and muffins	Wine or juice industry	Increase in total dietary fiber, phenolic compounds and antioxidant activity	Alongi et al. (2019), Bchir et al. (2014), Jung et al. (2015), and Kırbaş et al. (2019)
Soy flour, carrot powder, mango peel powder and moringa leaves powder	Pasta	–	Increase in antioxidants, vitamins, carotenoids, dietary fiber, minerals, protein, flavonoids and phenolic compounds	Jalgaonkar et al. (2018)
Soy okara	Pasta	Soybean processing	Improved antioxidant profile	Kamble et al. (2019)
Okara	Noodle	Soymilk or tofu processing	Increase in total phenolics and flavonoids	Pan et al. (2018)
Onion skin powder	Wheat pasta	Onion bulb processing	Increase in total phenolic compounds, dietary fibre, ash, flavonoid content and antioxidant activity	Michalak-Majewska et al. (2020)
Coconut flour and coconut residue	Wheat pasta	Coconut industry	Increase in dietary fiber, protein and mineral content	Sykut-Domańska et al. (2020)
Watermelon rind powder	Noodle	–	Increase in ash, crude fat, crude fiber, carbohydrate and total phenolic content	Ho and Che Dahri (2016)
Papaya peel and seed	Bread	–	Increase in desirable fatty acids, carotenoids, flavonoids and phytosterols	da Silva and Jorge (2017)
Passion fruit peel and seed	Pork burger, fermented soymilk and preserved meat products	–	Increase in soluble fiber, pectin, carotenoids and flavonoids	López-Vargas et al. (2014)

(continued)

Table 3.1 (continued)

Fortifier	Commodity	By-product from	Nutritional enhancement	References
Pineapple peel and core	Cookies	Juice industry	Increase in vitamin A, C and E, carotenoids, fatty acid and hemicellulose content	de Toledo et al. (2017)
Mango peel and seed	Bread and edible film and coating	Juice industry	Increase in starch, proteins, amino acids, provitamin A, vitamin C and K, calcium, magnesium, flavonoids, aromatic compounds and fatty acids	Chen et al. (2019) and Rojas-Bravo et al. (2019)
Banana peel	Chicken or frankfurter sausage, gluten-free riss, cereal bar, bread and orange juice	–	Increase in protein content and amino acids, fatty acids, soluble fiber, carotenoids, dopamine, L-dopa and fructo oligosaccharides	Amini Khoozani et al. (2020) and Zaini et al. (2020)
Grape pomace extract	Yogurt, cheese, ice cream formulation	Winemaking process	Increase of total phenolic content and antioxidant activity	Demirkol and Tarakci (2018) and Marchiani et al. (2016)
Baobab (*Adansonia digitata*) fruit pulp powder	Yoghurt made from milk	Fruit industry	Enhancement in lipid, protein, fiber and mineral content	Abdullahi et al. (2014)
Stems of oyster mushroom	Chicken nuggets	Mushroom industry	Increase in moisture content; decrease in fat and protein content	Husain and Huda-Faujan (2020)
Stems of velvet shank mushroom	Goat meat nuggets	Mushroom industry	Improved cooking yield, enhances mineral and dietary fiber content and prolongs the shelf life of meat products by preventing lipid oxidation	Banerjee et al. (2020)
Stipes of shiitake mushroom	Fermented sausage; bread	Mushroom industry	Enhances lipid oxidation, boosts microbial stability, inhibits pathogen growth and increases the levels of total soluble sugars and free amino acids	Lin et al. (2008) and Van Ba et al. (2016)
Basal hyphae clumps of chestnut mushroom	Snacks	Mushroom industry	Modifies the glycaemic response of an individual	Brennan et al. (2012)

content, the researchers decided to fortify it with a protein-rich source i.e. soy. They substituted 100% cassava starch with defatted soy flour (0–20%) and conducted sensory and proximate studies of the samples. The cassava meal with 20% soy flour was the most acceptable sample in terms of nutrition and sensory profile. Boukid et al. (2019) reviewed the potential of pulses as fortificants in bread. Pulses have great potential in curing and preventing malnutrition and undernutrition in developing countries. Pulses are most suitable for integrating with other foods as they have a nutritionally complete profile, can be processed easily, economically and help in conserving the environment as its production leaves minimum carbon and water footprints.

3.4.1.3 Herbs and Spices

Foods processed from wheat are central to diets in much of the world, serving as the main source of carbohydrates some proteins and minimal amount of minerals and vitamins. In order to prevent micronutrient-related deficiencies in the population, it is required that we supplement the refined wheat flour with other food groups, such as herbs and spices. Herbs such as *Moringa oleifera* and sesameare rich in both macro- and micronutrients and thus can be used as food fortificants. Already, sesame, which is an excellent source of protein, calcium and phosphorus along with fat, is used in confectionary products all over the world, but its use as a food fortificant needs to be explored (Agrahar-Murugkar, 2020; Boz & Murat Karaoğlu, 2013; Vitali et al., 2008).

Moringa oleifera is known as a miracle tree due to the immense nutritional value present in its leaves, seeds and flowers, which makes it an important ingredient in most food preparations. It is known to contain the highest amount of vitamin A, C, E, folate and minerals such as iron. Its leaves are known to possess a wide range of phytochemicals, which are known for their great antioxidant potential. The leaves are also rich in essential amino acids, especially alpha linoleic acid. The leaves, tender pods and flowers also contain unsaturated fatty acids, which are required in the human diet. Therefore, *M. oleifera* has great potential to be utilised as a functional ingredient in foods (Oyeyinka & Oyeyinka, 2018; Saini et al., 2014a, b). Studies have been conducted on its application as a food fortificant, as shown in Table 3.2. Nadeem et al. (2012) conducted a study in which they incorporated *M. oleifera* leaves powder (1–3%) in butter milk to enhance the nutritional content of butter milk for preventing malnutrition in females and children. The study concluded that leaves did not have any negative effect on the physicochemical properties of butter milk, such as acidity and pH, but enhanced the nutritional value in terms of protein, vitamins and mineral content. The overall acceptability of fortified buttermilk was acceptable; hence, it can be used to improve the nutritional profile of the population.

3.4.1.4 Fruits and Vegetables

Fruits and vegetables are an integral part of the diet that contain almost all the essential nutrients required by the body. They are being used as ingredients in different kinds of foods, such as bakery items in the form of fruit pies and buns, but their role

Table 3.2 Studies conducted to determine the efficiency of plant-based fortifiers

Fortifier	Commodity	Outcomes	References
Cumin, moringa leaves, soy-butter malted finger millet and sesame	Breads and biscuits	Increase in protein, minerals, calcium, iron and zinc content	Agrahar-Murugkar (2020)
	Refined wheat flour utilised in making breads and biscuits	Enhancement in bio-accessible calcium, iron and zinc, protein, fat and phenolics content	
Germinated pulses flour (gram, pea, lentil and faba bean flour)	Bread	Enhancement in polyphenol and antioxidant contents; improvement in hydrolytic enzymatic activity and nutritional enrichment	Boukid et al. (2019)
Orange fleshed sweet potato tuber, cowpea, moringa leaves; onion and garlic bulb	Various cereal porridges	Increase in iron and zinc content	Gautam et al. (2010), Kruger et al. (2018), and Vilakati et al. (2016)
Sweet potato (*Ipomoea batatas*) tubers, cowpea leaves, carrot root	Different types of porridge (both fermented and thick porridge)	Increase in β-carotene (vitamin A) content	Kruger et al. (2018) and Ndiaye et al. (2020)
Moringa oleifera leaf powder	Labneh cheese, maize gruel powder	Increase in protein, fat, ash, crude fiber, potassium, calcium, magnesium, iron, zinc and copper content along with vitamins A, B1, B2 and E	Nadeem et al. (2012) and Salem et al. (2013)
Soybean flour or soybean residue	Gari made from cassava tubers (*Manihot esculenta*)	Enhancement in macro and micronutrient levels of the fortified gari	Kolapo and Sanni (2009)
Oyster mushroom (*Pleurotus plumonarius*) powder	Bread made from wheat flour	Increase in crude proteins, fiber and ash content	Okafor et al. (2012)
Pawpaw fruit slurry	Sorghum-maize gruel	Blends containing pawpaw effectively enhanced the nutritional value (protein, fat, ash, vitamins C and sugar) of maize gruel	Oluseyi et al. (2010)
Citrullus lanatus seeds flour	Bread from wheat flour	Increase in proteins, fat and ash content in flour	Meite et al. (2008)
Defatted soy flour	Tapioca meal	Enhancement in protein and ash content	Balogun et al. (2012)
Pulses such as beans, peas and lentils in the form of dehulled or hulled flour, hydrolysates, protein isolates, germinated flour or fermented flour	Bread made from wheat flour	Enhancement in protein, fat, dietary fiber and mineral content compared to bread made entirely from wheat flour	Indrani et al. (2011)

as fortifiers in staple foods has been known only to some extent. However, due to their high cost, they are not available to all sections of society. Due to their perishable nature, the nutrient content starts reducing as soon as they are harvested and until they reach the plates, most of the nutrition is lost due to poor handling, storage and processing conditions. Fruits such as apple, mango and coconut, which are abundant in terms of production and possess stable shelf life, can be explored as fortificants. Coconut, which is a source to both nutritious kernel and water, is a good source of energy, fibre, vitamins, especially vitamin B and minerals such as calcium. It is also a rich source of phytocompounds such as phenols, flavonoids, catechins and tannins, which have immense health benefits such as antioxidant activity against cancer-causing species, helping in reducing the loss of trabecular bone number and thickness (Hayatullina et al., 2012; Vysakh et al., 2014). Table 3.2 contains the studies that have incorporated fruits and vegetables such as pawpaw, sweet potato and carrot in different products. They also help in enhancing the bioavailability of micronutrients. Gautam et al. (2010) performed a study to enhance the bio-accessibility of zinc and iron from grains in the presence of sulfur-containing compounds from onion. Another study conducted by Kruger et al. (2018) showed that the availability of vitamin A, zinc and iron was enhanced by adding cowpea leaves and sweet potato relishes to cereal porridges. Therefore, different plant-based food groups can be used as potential fortifiers and help in reducing the growing malnutrition concerns in society. They not only act as a source of nutrients but also help in enhancing the bioavailability of nutrients present in the food.

3.5 Agro-industry Waste-Based Fortificants

Cedola et al. (2020) studied the effect of olive oil industry waste on bread and pasta as shown in Table 3.1. Bread and pasta were fortified with waste water and paste of the olive mill industry. The study showed that there was an enhancement in the nutritional content of the bread and pasta; however, the sensorial properties were compromised. The bitter and spicy character of olives was transferred to bread and pasta, which was not acceptable to the sensory panellists. Kamble et al. (2019) studied the effect of incorporating soymilk processing industry waste, known as okara, into pasta. As okara holds immense nutritional potential with hypoglycaemic effects, the study estimated the physicochemical attributes of okara incorporated (10–50%) durum wheat pasta along with in vitro protein digestibility, and glycaemic index lowering potential. It was found that the pasta fortified with okara possessed higher antioxidant potential with potentially lower glycaemic index as compared to the control pasta. Sykut-Domańska et al. (2020) performed a study in which they incorporated coconut industry waste, i.e. coconut flour and residue, into pasta as a source of dietary fibre, which is nearly absent in wheat pasta. The incorporation of coconut industry by-products led to a reduction in the output of the pasta extruder by 32%. However, the textural properties of pasta were affected by the incorporation of dietary fibre. The firmness of coconut fortified pasta was lower than that of the control. The incorporation of coconut industry residues did not negatively affect the sensory properties. The enhanced nutritional benefits attributed to the coconut fortified pasta were recorded in terms of protein, fibre and lipid content. The study concluded that coconut-industry by-products could be a potential source of dietary fibres. de Toledo et al. (2017) studied the phytochemical and sensory effect of incorporating melon, pineapple and apple industry by-products into cookies (5–15%). The details of the study are mentioned in Table 3.1. The utilisation of agro-industry by-products in the foods mostly has a positive influence on the nutritional aspects of the food, with certain changes in the textural and sensory attributes of the original product, which in some cases was acceptable and in few studies it was unacceptable. However, by-product powders can be a potential source of micronutrients and can be further explored as a potential fortificant.

3.6 Efficacy of Food Products Fortified with Plant-Based Fortificants

Jain et al. (2017) conducted a study to study the impact of garden cress-enriched biscuits on the nutritional status of malnourished school-going children in the age group of 7–9 years. The group under study consisted of 60 participants and was given 60 g of biscuits made with roasted garden cress seeds, while the control group was given biscuits devoid of garden cress seeds. The duration of the trial was 3 months. It was found that the students given garden cress supplemented biscuits; the height increased by 0.87% and the weight increased by 3.56% which was significant as compared to the control. Additionally, haemoglobin levels increased from 10.6 to 11 g/dL. Nine participants moved into the non-anaemic category, and minor improvements were noted in blood profile indicators such as albumin and protein. These results demonstrate how garden cress seed-fortified biscuits may help treat childhood malnutrition and anaemia at the early stages of growth. A study was conducted by Baruah et al. (2019) to determine how the haemoglobin levels of moderately anaemic teenage females (16–18 years old) were affected by ladoos supplemented with rice flakes, wheat, soybean, Bengal gram, garden cress seeds, gingelly seeds and jaggery. An intervention group and a control group were each assigned 60 participants. For 90 days, no treatment was given to the control group, but the intervention group ate two ladoos (80 g) every day. The intervention group's mean haemoglobin level was 9.15 g/dL at the start of the study and increased to 9.81 g/dL at the conclusion of the intervention, indicating a substantial improvement of 0.66 g/dL. According to this study, ladoos incorporated with iron rich ingredients can be helpful in preventing anemia among teenage girls. A study was conducted by Karkada et al. (2019) to assess the impact of ragi-fortified porridge on teenage girls' haemoglobin levels. A control group with haemoglobin levels between 10 and 11.9 g/dL and an intervention group with Hb levels between 12 and 12.5 g/dL were each assigned 60 subjects. For 90 days, the intervention group ate ragi porridge every day. Haemoglobin levels in the intervention group increased significantly after the intervention, going from 11.3 g% to 12.54 g% ($p < 0.0001$). The findings show that ragi-fortified porridge helps teenage girls' haemoglobin levels, although it might not have a major influence on other health indicators such as red cell distribution width, body mass index, scholastic performance etc. Boateng et al. (2019) studied the effect of supplemental foods supplemented with moringa on the haemoglobin levels of babies between the ages of 8 and 12 months. A cereal-legume blend containing moringa leaf powder (MLP) (MCL-35 g), a "sprinkle" version of moringa to be added to regular diets (MS-5 g), and a control group that received a supplemental food blend (CF-35 g) devoid of moringa were the three groups into which participants were split. After adjusting for baseline values, morbidity and wealth index, no significant differences between the study groups were found for 4 months after controlling for baseline values, morbidity and wealth index. However, the highest increase in haemoglobin levels was obtained in the CF-35 g group, followed by the MS-5 g group. The MCL-35 g group showed the least increase in haemoglobin levels. This study suggests that while moringa has potential as a food fortifier, its effectiveness may depend on the form in which it is administered. A study carried out in Malawi by Akomo et al. (2019) assessed how well a ready-to-use therapeutic food (RUTF) supplemented with sorghum, maize and soybeans treated anaemia and severe acute malnutrition (SAM) in children between the ages of 6 and 59 months. Three groups of participants were given different formulations: FSMS-RUTF, which had extruded soy, maize and sorghum but no milk; MSMS-RUTF, which contained 9% milk; and PM-RUTF, which was a normal peanut-based RUTF with 28% milk. The FSMS-RUTF group had the lowest anaemia prevalence after 21 days (12.0%), followed by the MSMS-RUTF group (18.2%) and the PM-RUTF group (24.5%). Similarly, the FSMS-RUTF group had the lowest prevalence of iron deficiency anaemia (IDA) at discharge (7.9%), followed by the MSMS-RUTF group (10.9%) and the PM-RUTF group (20.5%). These findings indicate that plant-based formulations, particularly those using soy, maize and sorghum, can be viable alternatives for treating anaemia in malnourished children. Another study conducted by Rizik 2020 assessed the impact of baobab-fortified mixed vegetables on the haemoglobin levels of adolescent girls with anaemia (Hb < 11.9 mg/dL). A control group was given vegetables without baobab powder, whereas the intervention group was given mixed veggies enhanced with 10 g of baobab powder every 225 g meal. Although the baseline mean Hb levels in the IG and control group were marginally lower (10.5 g/dL and 11.2 g/dL, respectively), the difference was not statistically significant. 45.2% of the teenage girls in the intervention group had normal haemoglobin levels (>11.9 g/dL) after 30 days. With a notable mean increase of 1.1 g/dL in the intervention group and 0.6 g/dL in the control group, the mean end haemoglobin levels were 11.6 g/dL in the intervention group and 11.8 g/dL in the control group.

3.7 Conclusion

Micronutrient deficiencies are a worldwide public health issue that necessitates quick, practical and affordable solutions. Even though there have been a number of food-based initiatives implemented both domestically and abroad, the number of cases with micronutrient deficiencies worldwide is still on the rise. One such food-based approach that has been embraced by most countries worldwide is food fortification. Micronutrient deficits could be affordably addressed on a large scale through conventional food fortification. Food scientists have discovered food-to-food fortification as a new food-based strategy to supplement conventional fortification because of the drawbacks of conventional fortification, such as inadequate internal and external evaluation of producers' compliance with regulations, worries about nutrient overdosage, the risk of fortified food cost inflation, costly fortification techniques, nutrient losses and exposure of fortified foods to sunlight during retail. To increase the micronutrient content of a staple food, this new tactic entails adding nutrient-dense food as a fortifier. Furthermore, food-to-food fortification makes use of crops that have been neglected because of the overproduction of staple crops and helps create local value-added food businesses that could support community economic growth in addition to offering nutritional and health advantages. Despite the availability of numerous food-based fortifiers with potential to reduce micronutrient deficiencies, there is still limited research dedicated to developing foods using this method and to evaluate the effectiveness of such developed foods in combating micronutrient deficiencies. Given the advantages and disadvantages of conventional fortification and food-to-food fortification, these two methods can be used in order to effectively address malnutrition and undernutrition in the population. Therefore, it is crucial to research this new food-based approach, use underutilised crops to create fortified meals, assess how well these products work to reduce micronutrient deficiencies, and promote them in the public and commercial sectors. The need and importance of crop domestication, crop diversification and the production of traditional and other crops useful in food-to-food fortification must be made clear to farmers with small landholdings who have stopped producing traditional crops that could be used in food-to-food fortification for productivity and financial reasons.

References

A comparative evaluation of the macronutrient and micronutrient profiles of soybean-fortified gari and tapioca. (2009).

Abdullahi, M., Zainab, F., Pedavoah, M., Bashir, U., Pedavoah, M., Sumayya, U., & Ibrahim, A. (2014). Evaluating the suitability of *Adansonia digitata* fruit pulp for the production of yoghurt. *International Journal of Biological and Chemical Sciences, 8*(2), 508. https://doi.org/10.4314/ijbcs.v8i2.10

Agrahar-Murugkar, D. (2020). Food to food fortification of breads and biscuits with herbs, spices, millets and oilseeds on bio-accessibility of calcium, iron and zinc and impact of proteins, fat and phenolics. *LWT, 130*. https://doi.org/10.1016/j.lwt.2020.109703

Akomo, P., Bahwere, P., Murakami, H., Banda, C., Maganga, E., Kathumba, S., Sadler, K., & Collins, S. (2019). Soya, maize and sorghum ready-to-use therapeutic foods are more effective in correcting anaemia and iron deficiency than the standard ready-to-use therapeutic food: Randomized controlled trial. *BMC Public Health, 19*(1). https://doi.org/10.1186/s12889-019-7170-x

Alongi, M., Melchior, S., & Anese, M. (2019). Reducing the glycemic index of short dough biscuits by using apple pomace as a functional ingredient. *LWT, 100*, 300–305. https://doi.org/10.1016/j.lwt.2018.10.068

Amini Khoozani, A., Kebede, B., & El-Din Ahmed Bekhit, A. (2020). Rheological, textural and structural changes in dough and bread partially substituted with whole green banana flour. *LWT, 126*. https://doi.org/10.1016/j.lwt.2020.109252

Antonic, B., Jancikova, S., Dordevic, D., & Tremlova, B. (2020). Apple pomace as food fortification ingredient: A systematic review and meta-analysis. *Journal of Food Science, 85*(10), 2977–2985. https://doi.org/10.1111/1750-3841.15449

Balogun, M. A., Karim, O. R., Kolawole, F. L., & Solarin, A. O. (2012). Quality attributes of tapioca meal fortified with defatted soy flour. *Agrosearch, 12*(1):61–68

Banerjee, D. K., Das, A. K., Banerjee, R., Pateiro, M., Nanda, P. K., Gadekar, Y. P., Biswas, S., McClements, D. J., & Lorenzo, J. M. (2020). Application of enoki mushroom (Flammulina Velutipes) stem wastes as functional ingredients in goat meat nuggets. *Foods, 9*(4). https://doi.org/10.3390/foods9040432

Baruah, U., Singh, V., & Vishakha Singh, C. (2019). Effect of multi nutrient supplementation on health status of malnourished and anaemic school age children. *International Journal of Chemical Studies, 6*, 107–109. https://doi.org/10.13140/RG.2.2.23722.03520

Bchir, B., Rabetafika, H. N., Paquot, M., & Blecker, C. (2014). Effect of pear, apple and date fibres from cooked fruit by-products on dough performance and bread quality. *Food and Bioprocess Technology, 7*(4), 1114–1127. https://doi.org/10.1007/s11947-013-1148-y

Bishai, D., & Nalubola, R. (2002). The history of food fortification in the United States: Its relevance for current fortification efforts in developing countries. *Economic Development and Cultural Change, 1*(51), 37–53.

Black, R. E., Victora, C. G., Walker, S. P., Bhutta, Z. A., & Christian, P. (2013). Maternal and child undernutrition and overweight in low-income and middle-income countries. *The Lancet, 382*(9890), 427–451.

Boateng, L., Quarpong, W., Ohemeng, A., Asante, M., & Steiner-Asiedu, M. (2019). Effect of complementary foods fortified with Moringa oleifera leaf powder on hemoglobin concentration and growth of infants in the eastern region of Ghana. *Food Science and Nutrition, 7*(1), 302–311. https://doi.org/10.1002/fsn3.890

Bouis, H. E., & Saltzman, A. (2017). Improving nutrition through biofortification: A review of evidence from HarvestPlus, 2003 through 2016. *Global Food Security, 12*, 49–58. https://doi.org/10.1016/j.gfs.2017.01.009

Boukid, F., Zannini, E., Carini, E., & Vittadini, E. (2019). Pulses for bread fortification: A necessity or a choice? *Trends in Food Science and Technology, 88*, 416–428. https://doi.org/10.1016/j.tifs.2019.04.007

Boz, H., & Murat Karaoğlu, M. (2013). Improving the quality of whole wheat bread by using various plant origin materials. *Czech Journal of Food Sciences, 31*(5), 457.

Brennan, M. A., Derbyshire, E., Tiwari, B. K., & Brennan, C. S. (2012). Enrichment of extruded snack products with coproducts from chestnut mushroom (Agrocybe aegerita) production: Interactions between dietary fiber, physicochemical characteristics, and glycemic load. *Journal of Agricultural and Food Chemistry, 60*(17), 4396–4401. https://doi.org/10.1021/jf3008635

Cedola, A., Cardinali, A., D'Antuono, I., Conte, A., & Del Nobile, M. A. (2020). Cereal foods fortified with by-products from the olive oil industry. *Food Bioscience, 33*. https://doi.org/10.1016/j.fbio.2019.100490

Chadare, F. J., Idohou, R., Nago, E., Affonfere, M., Agossadou, J., Fassinou, T. K., Kénou, C., Honfo, S., Azokpota, P., Linnemann, A. R., & Hounhouigan, D. J. (2019). Conventional and food-to-food fortification: An appraisal of past practices and lessons learned. *Food Science and Nutrition, 7*(9), 2781–2795. https://doi.org/10.1002/fsn3.1133

Chen, Y., Zhao, L., He, T., Ou, Z., Hu, Z., & Wang, K. (2019). Effects of mango peel powder on starch digestion and quality characteristics of bread. *International Journal of Biological Macromolecules, 140*, 647–652. https://doi.org/10.1016/j.ijbiomac.2019.08.188

Crews, T. E., & Peoples, M. B. (2004). Legume versus fertilizer sources of nitrogen: Ecological tradeoffs and human needs. *Agriculture, Ecosystems and Environment, 102*(3), 279–297. https://doi.org/10.1016/j.agee.2003.09.018

da Silva, A. C., & Jorge, N. (2017). Bioactive compounds of oils extracted from fruits seeds obtained from agroindustrial waste. *European Journal of Lipid Science and Technology, 119*(4). https://doi.org/10.1002/ejlt.201600024

Dary, O., & Hurrell, R. (2006). Guidelines on food fortification with micronutrients. World Health Organization, Food and Agricultural Organization of the United Nations: Geneva, Switzerland, 1–376

de Toledo, N. M. V., Nunes, L. P., da Silva, P. P. M., Spoto, M. H. F., & Canniatti-Brazaca, S. G. (2017). Influence of pineapple, apple and melon by-products on cookies: Physicochemical and sensory aspects. *International Journal of Food Science and Technology, 52*(5), 1185–1192. https://doi.org/10.1111/ijfs.13383

Demirkol, M., & Tarakci, Z. (2018). Effect of grape (Vitis labrusca L.) pomace dried by different methods on physicochemical, microbiological and bioactive properties of yoghurt. *LWT, 97*, 770–777. https://doi.org/10.1016/j.lwt.2018.07.058

Enhancing haemoglobin status of anaemic school girls using cooked vegetables enriched with baobab powder in Kilifi County-Kenya. Doctoral dissertation, University of Nairobi. (2020).

Gautam, S., Platel, K., & Srinivasan, K. (2010). Higher bioaccessibility of iron and zinc from food grains in the presence of garlic and onion. *Journal of Agricultural and Food Chemistry, 58*(14), 8426–8429. https://doi.org/10.1021/jf100716t

Hayatullina, Z., Muhammad, N., Mohamed, N., & Soelaiman, I. N. (2012). Virgin coconut oil supplementation prevents bone loss in osteoporosis rat model. *Evidence-based Complementary and Alternative Medicine, 2012*. https://doi.org/10.1155/2012/237236

Ho, L. H., & Che Dahri, N. (2016). Effect of watermelon rind powder on physicochemical, textural, and sensory properties of wet yellow noodles. *CYTA – Journal of Food, 14*(3), 465–472. https://doi.org/10.1080/19476337.2015.1134672

Hurrell, R. (2002a). Fortification: Overcoming technical and practical barriers. *The Journal of Nutrition, 132*(4), 806S–812S.

Hurrell, R. F. (2002b). Forging effective strategies to combat iron deficiency fortification: Overcoming technical and practical barriers. *The Journal of Nutrition, 132*, 806S–812S.

Husain, H., & Huda-Faujan, N. (2020). Potential application of grey oyster mushroom stems as halal meat replacer in imitation chicken nuggets. *Food Research, 4*, 179–186.

Indrani, D., Swetha, P., Soumya, C., Rajiv, J., & Venkateswara Rao, G. (2011). Effect of multigrains on rheological, microstructural and quality characteristics of north Indian parotta – An Indian flat bread. *LWT, 44*(3), 719–724. https://doi.org/10.1016/j.lwt.2010.11.017

Jain, T., Grover, K., & Gill, N. K. (2017). Impact of garden cress supplemented biscuits on nutritional profile of malnourished and anemic school children (seven–nine years). *Nutrition and Food Science, 47*(4), 553–566. https://doi.org/10.1108/NFS-09-2016-0144

Jalgaonkar, K., Jha, S. K., & Mahawar, M. K. (2018). Influence of incorporating defatted soy flour, carrot powder, mango peel powder, and moringa leaves powder on quality characteristics of wheat semolina-pearl millet pasta. *Journal of Food Processing and Preservation, 42*(4). https://doi.org/10.1111/jfpp.13575

Jung, J., Cavender, G., & Zhao, Y. (2015). Impingement drying for preparing dried apple pomace flour and its fortification in bakery and meat products. *Journal of Food Science and Technology, 52*(9), 5568–5578. https://doi.org/10.1007/s13197-014-1680-4

Kamble, D. B., Singh, R., Rani, S., & Pratap, D. (2019). Physicochemical properties, in vitro digestibility and structural attributes of okara-enriched functional pasta. *Journal of Food Processing and Preservation, 43*(12). https://doi.org/10.1111/jfpp.14232

Karkada, S., Upadhya, S., Upadhya, S., & Bhat, G. (2019). Beneficial effects of ragi (finger millet) on hematological parameters, body mass index, and scholastic performance among anemic adolescent high-school girls (AHSG). *Comprehensive Child and Adolescent Nursing, 42*(2), 141–150. https://doi.org/10.1080/24694193.2018.1440031

Kırbaş, Z., Kumcuoglu, S., & Tavman, S. (2019). Effects of apple, orange and carrot pomace powders on gluten-free batter rheology and cake properties. *Journal of Food Science and Technology, 56*(2), 914–926. https://doi.org/10.1007/s13197-018-03554-z

Kolapo, A. L., & Sanni, M. O. (2009). A comparative evaluation of the macronutrient and micronutrient profiles of soybean-fortified gari and tapioca. *Food and Nutrition Bulletin, 30*(1):90–94

Kruger, J., Breynaert, A., Pieters, L., & Hermans, N. (2018). Vegetable relishes, high in β-carotene, increase the iron, zinc and β-carotene nutritive values from cereal porridges. *International Journal of Food Sciences and Nutrition, 69*(3), 291–297. https://doi.org/10.1080/09637486.2017.1360259

Kruger, J., Taylor, J. R. N., Ferruzzi, M. G., & Debelo, H. (2020). What is food-to-food fortification? A working definition and framework for evaluation of efficiency and implementation of best practices. *Comprehensive Reviews in Food Science and Food Safety, 19*(6), 3618–3658. https://doi.org/10.1111/1541-4337.12624

Kumar, H., Bhardwaj, K., Kuča, K., Sharifi-Rad, J., Verma, R., Machado, M., Kumar, D., & Cruz-Martins, N. (2022). Edible mushrooms' enrichment in food and feed: A mini review. *International Journal of Food Science and Technology, 57*(3), 1386–1398. https://doi.org/10.1111/ijfs.15546

Lin, L. Y., Tseng, Y. H., Li, R. C., & Mau, J. L. (2008). Quality of shiitake stipe bread. *Journal of Food Processing and Preservation, 32*(6), 1002–1015. https://doi.org/10.1111/j.1745-4549.2008.00229.x

López-Vargas, J. H., Fernández-López, J., Pérez-Álvarez, J. Á., & Viuda-Martos, M. (2014). Quality characteristics of pork burger added with albedo-fiber powder obtained from yellow passion fruit (Passiflora edulis var. flavicarpa) co-products. *Meat Science, 97*(2), 270–276. https://doi.org/10.1016/j.meatsci.2014.02.010

Marchiani, R., Bertolino, M., Ghirardello, D., McSweeney, P. L. H., & Zeppa, G. (2016). Physicochemical and nutritional qualities of grape pomace powder-fortified semi-hard cheeses. *Journal of Food Science and Technology, 53*(3), 1585–1596. https://doi.org/10.1007/s13197-015-2105-8

Markel, H. (1987). When it rains it pours: endemic goiter, iodized salt, and David Murray Cowie, MD. *American Journal of Public Health, 77*(2):219–229

McCrory, M. A., Hamaker, B. R., Lovejoy, J. C., & Eichelsdoerfer, P. E. (2010). Pulse consumption, satiety, and weight management. *Advances in Nutrition, 1*(1), 17–30. https://doi.org/10.3945/an.110.1006

Meite, A., Kouame, K. G., Amani, N. G., Kati-Coulibaly, S., & Offoumou, A. (2008). Caractéristiques physico-chimiques et sensorielles des pains fortifiés avec les farines de graines de Citrullus lanatus. *Journal of Pharmaceutical and Biological Sciences, 9*, 32–43.

Michalak-Majewska, M., Teterycz, D., Muszyński, S., Radzki, W., & Sykut-Domańska, E. (2020). Influence of onion skin powder on nutritional and quality attributes of wheat pasta. *PLoS One, 15*(1). https://doi.org/10.1371/journal.pone.0227942

Muthayya, S., Rah, J. H., Sugimoto, J. D., Roos, F. F., Kraemer, K., & Black, R. E. (2013). The global hidden hunger indices and maps: An advocacy tool for action. *PLoS One, 8*(6). https://doi.org/10.1371/journal.pone.0067860

Nadeem, M., Javid, A., Abdullah, M., Arif, A. M., & Mahmood, T. (2012). Improving nutritional value of butter milk by blending with dry leaves of Moringa oleifera. *Pakistan Journal of Nutrition, 11*(9), 714–718. https://doi.org/10.3923/pjn.2012.812.816

Nair, M. K., Augustine, L. F., & Konapur, A. (2016). Food-based interventions to modify diet quality and diversity to address multiple micronutrient deficiency. *Frontiers in Public Health, 3*. https://doi.org/10.3389/fpubh.2015.00277

Ndiaye, C., Martinez, M. M., Hamaker, B. R., Campanella, O. H., & Ferruzzi, M. G. (2020). Effect of edible plant materials on provitamin A stability and bioaccessibility from extruded whole pearl millet (P. typhoides) composite blends. *LWT, 123*. https://doi.org/10.1016/j.lwt.2020.109109

Okafor, J. N. C., Okafor, G. I., Ozumba, A. U., & Elemo, G. N. (2012). Quality characteristics of bread made from wheat and nigerian oyster mushroom (Pleurotus plumonarius) powder. *Pakistan Journal of Nutrition, 11*(1), 5–10. https://doi.org/10.3923/pjn.2012.5.10

Olson, R., Gavin-Smith, B., Ferraboschi, C., & Kraemer, K. (2021). Food fortification: The advantages, disadvantages and lessons from sight and life programs. *Nutrients, 13*(4). https://doi.org/10.3390/nu13041118

Oluseyi, A. K., Olurotimi, O. K., Ajani, O. O., & James, O. O. (2010). *Improvement of nutritive value of sorghum-ogi fortified with pawpaw (Carica papaya L).* https://www.researchgate.net/publication/228496982

Oyeyinka, A. T., & Oyeyinka, S. A. (2018). Moringa oleifera as a food fortificant: Recent trends and prospects. *Journal of the Saudi Society of Agricultural Sciences, 17*(2), 127–136. https://doi.org/10.1016/j.jssas.2016.02.002

Pan, W. C., Liu, Y. M., & Shiau, S. Y. (2018). Effect of okara and vital gluten on physico-chemical properties of noodle. *Czech Journal of Food Sciences, 4*, 36.

Riziki, D. Y. (2020). Enhancing Haemoglobin Status of Anaemic School Girls Using Cooked Vegetables Enriched with Baobab Powder in Kilifi County Kenya (Doctoral dissertation, University of Nairobi)

Rojas-Bravo, M., Rojas-Zenteno, E. G., Hernández-Carranza, P., Ávila-Sosa, R., Aguilar-Sánchez, R., Ruiz-López, I. I., & Ochoa-Velasco, C. E. (2019). A potential application of mango (Mangifera indica L. cv Manila) peel powder to increase the total phenolic compounds and antioxidant capacity of edible films and coatings. *Food and Bioprocess Technology, 12*(9), 1584–1592. https://doi.org/10.1007/s11947-019-02317-8

Saini, R. K., Harish Prashanth, K. V., Shetty, N. P., & Giridhar, P. (2014a). Elicitors, SA and MJ enhance carotenoids and tocopherol biosynthesis and expression of antioxidant related genes in Moringa oleifera Lam. leaves. *Acta Physiologiae Plantarum, 36*(10), 2695–2704. https://doi.org/10.1007/s11738-014-1640-7

Saini, R. K., Shetty, N. P., Prakash, M., & Giridhar, P. (2014b). Effect of dehydration methods on retention of carotenoids, tocopherols, ascorbic acid and antioxidant activity in Moringa oleifera leaves and preparation of a RTE product. *Journal of Food Science and Technology, 51*(9), 2176–2182. https://doi.org/10.1007/s13197-014-1264-3

Salem, A. S., Salama, W. M., Hassanein, A. M., & El-Ghandour, H. M. A. (2013). Enhancement of nutritional and biological values of labneh by adding dry leaves of moringa oleifera as innovative dairy products. *World Applied Sciences Journal, 22*(11), 1194–1602. https://doi.org/10.5829/idosi.wasj.2013.22.11.13024

Sharma, P., Aggarwal, P., & Kaur, A. (2017). Biofortification: A new approach to eradicate hidden hunger. *Food Reviews International, 33*(1), 1–21. https://doi.org/10.1080/87559129.2015.1137309

Stoltzfus, R. J. (2003). Iron deficiency: Global prevalence and consequences. *Food and Nutrition Bulletin, 24*, S99–S103.

Sykut-Domańska, E., Zarzycki, P., Sobota, A., Teterycz, D., Wirkijowska, A., Blicharz-Kania, A., Andrejko, D., & Mazurkiewicz, J. (2020). The potential use of by-products from coconut industry for production of pasta. *Journal of Food Processing and Preservation, 44*(7). https://doi.org/10.1111/jfpp.14490

Tolve, R., Simonato, B., Rainero, G., Bianchi, F., Rizzi, C., Cervini, M., & Giuberti, G. (2021). Wheat bread fortification by grape pomace powder: Nutritional, technological, antioxidant, and sensory properties. *Foods, 10*(1). https://doi.org/10.3390/foods10010075

Van Ba, H., Seo, H. W., Cho, S. H., Kim, Y. S., Kim, J. H., Ham, J. S., Park, B. Y., & Pil Nam, S. (2016). Antioxidant and anti-foodborne bacteria activities of shiitake by-product extract in fermented sausages. *Food Control, 70*, 201–209. https://doi.org/10.1016/j.foodcont.2016.05.053

Vilakati, N., Taylor, J. R. N., MacIntyre, U., & Kruger, J. (2016). Effects of processing and addition of a cowpea leaf relish on the iron and zinc nutritive value of a ready-to-eat sorghum-cowpea porridge aimed at young children. *LWT, 73*, 467–472. https://doi.org/10.1016/j.lwt.2016.06.022

Vishwakarma, S., Genu Dalbhagat, C., Mandliya, S., & Niwas Mishra, H. (2022). Investigation of natural food fortificants for improving various properties of fortified foods: A review. *Food Research International, 156*. https://doi.org/10.1016/j.foodres.2022.111186

Vitali, D., Vedrina Dragojević, I., & Šebečić, B. (2008). Bioaccessibility of Ca, Mg, Mn and Cu from whole grain tea-biscuits: Impact of proteins, phytic acid and polyphenols. *Food Chemistry, 110*(1), 62–68. https://doi.org/10.1016/j.foodchem.2008.01.056

Vysakh, A., Ratheesh, M., Rajmohanan, T. P., Pramod, C., Premlal, S., Girish Kumar, B., & Sibi, P. I. (2014). Polyphenolics isolated from virgin coconut oil inhibits adjuvant induced arthritis in rats through antioxidant and anti-inflammatory action. *International Immunopharmacology, 20*(1), 124–130. https://doi.org/10.1016/j.intimp.2014.02.026

Zaini, H. B. M., Sintang, M. D. B., & Pindi, W. (2020). The roles of banana peel powders to alter technological functionality, sensory and nutritional quality of chicken sausage. *Food Science and Nutrition, 8*(10), 5497–5507. https://doi.org/10.1002/fsn3.1847

Micronutrients and Food Fortification

4

Syed Amir Ashraf, Humera Banu, Jerold C. Alcantra, and Nosyba A. Ibrahim

Abstract

Micronutrients play a crucial role in maintaining overall health and well-being, as they are essential for a wide range of physiological functions such as immune support, growth, development, and disease prevention. However, a lack of healthy food intake causes micronutrient deficiencies in many people around the world. By incorporating indispensable vitamins and minerals into frequently consumed meals, food fortification has become a vital strategy to address these deficiencies. Food fortification process aims to improve public health by providing a cost-effective and sustainable solution to nutritional gaps in populations, especially in low-income and developing regions. Common micronutrients used in food fortification include iodine, iron, folic acid, vitamin A, and zinc. Iodine fortification, typically in the form of iodized salt, helps prevent thyroid-related disorders. Iron fortification in staple foods such as flour and rice help to combat anemia, a widespread condition linked to iron deficiency. Vitamin A fortification is vital for preventing vision impairment and boosting immune function, especially in children and pregnant women. Folic acid, often added to grains, is critical in preventing neural tube defects in newborns. While food fortification has proven to be effective in reducing micronutrient deficiencies at the population level, its success depends on several factors, including the availability of appro-

S. A. Ashraf (✉) · H. Banu
Department of Clinical Nutrition, College of Applied Medical Sciences, University of Ha'il, Ha'il, Saudi Arabia

J. C. Alcantra
Kasiska Division of Health Sciences, Department of Medical Laboratory Science, College of Health, Idaho State University, Meridian, ID, USA

N. A. Ibrahim
Department of Public Health, College of Public Health & Health Informatics, University of Ha'il, Ha'il, Saudi Arabia

G. Sanghvi et al. (eds.), *Sustainable Food Fortification*,
https://doi.org/10.1007/978-981-95-1217-1_4

priate fortification technologies, regulatory frameworks, and public awareness. Moreover, fortification programs must be carefully designed to ensure that the right amounts of nutrients are added, avoiding both deficiency and toxicity. In combination with other nutritional interventions, food fortification has the potential to significantly improve global health outcomes, particularly in vulnerable communities. In this chapter, the current status and possible application of micronutrients for food fortification were discussed. In addition, the role of various micronutrients will also be addressed with its therapeutic potentials. Subsequently, this chapter will attract the attention of food scientists to improve the application of micronutrients for food supplementation purposes to aid and promote a healthy lifestyle and well-being.

Keywords

Micronutrients · Minerals · Iodine · Vitamins · Functional foods · Nutraceuticals

4.1 Introduction

Micronutrients are vital vitamins and minerals that play important roles in body processes and are needed in trace levels for optimum health. Minerals like iron, zinc, and calcium are essential for oxygen transport, immunological response, and bone strength, while vitamins A, C, and D promote immunity, bone health, and vision. Unlike macronutrients, they are required in smaller quantities but play a crucial role in maintaining overall health. However, micronutrient deficiencies remain a significant global public health concern, affecting an estimated 1.6 to 2 billion individuals, particularly due to deficiencies in iron, iodine, vitamin A, and zinc. Furthermore, a study conducted by Passarelli et al. (2024) highlights that over 5 billion people have inadequate intakes of iodine, vitamin E, and calcium, while more than 4 billion lack sufficient iron, riboflavin, folate, and vitamin C. Subsequently, deficiencies in vitamin and mineral status, particularly of folate, iron, vitamin A, and zinc, affect 50% of all preschool aged children and 67% of all women of reproductive age (WRA) worldwide. These deficiencies have severe health consequences, especially among vulnerable populations such as young girls, women, preschool children, and the elderly, leading to impaired cognitive development, weakened immunity, increased susceptibility to infectious diseases anemia, night blindness, neural tube defects, premature or low birth weight infants and higher mortality rates (WHO, 2023). Efforts to combat these deficiencies include food fortification, supplementation programs, and nutrition education aiming to improve the health and well-being of affected populations. Among different strategies to combat micronutrient deficiency, food fortification is been considered of very high significance. Since food fortification is a highly significant strategy to combat micronutrient deficiencies due to its cost-effectiveness, scalability, and ability to deliver essential nutrients like iron, iodine, vitamin A, and zinc to large populations without requiring dietary changes. Micronutrient food fortification is a deliberate process of enriching commonly consumed staple foods with essential vitamins and minerals, such as B-vitamins, vitamin D, calcium, iodine, iron, selenium, zinc, and omega-3 fatty acids. This strategy is designed to address widespread micronutrient deficiencies, which are often referred to as "hidden hunger," and significantly improve the overall nutritional status of populations. It plays a particularly critical role in supporting the health and well-being of vulnerable groups, including children, pregnant women, and lactating mothers, who are at a higher risk of nutrient deficiencies.

Food fortification has been an effective and economical approach to addressing widespread micronutrient deficiencies for more than a hundred years (Neufeld et al., 2016). The World Health Organization (WHO) identifies food fortification as one of the most efficient and advantageous public health strategies, alongside vaccinations, and considers it a safe, effective, and economical method to avert the adverse effects linked to serious micronutrient deficiencies (WHO, 2020). Globally, one in two preschool-aged children and two in three women of reproductive age suffer from at least one micronutrient deficiency (Stevens et al., 2022a, b). These deficiencies can have significant negative effects on an individual's survival, as well as their mental and physical development. Consequently, this also impacts the economies of countries as a whole (Horton et al., 2017). As a result, on January 31, 2023, the Executive Board of the World Health Organization made a recommendation for the World Health Assembly to pass a resolution aimed at speeding up initiatives to prevent micronutrient deficiencies and their related effects, such as spina bifida and other neural tube defects, via safe and effective food fortification (WHO, 2023). In 2019, 161 countries established either mandatory or voluntary laws regarding at least one category of fortified food (www.fortificationdata.org). Among these, salt emerged as the most frequently fortified item, with 145 countries having relevant legislation. Regionally, the Americas boasted the highest percentage of countries with fortification laws as all 35 countries in the region was fortifying wheat flour (www.fortificationdata.org). As of June 2021, nearly two-thirds of countries worldwide have enacted mandatory fortification laws for salt ($n = 125$). Regarding wheat flour, nearly half of all countries globally ($n = 91$), particularly in the Americas, West Africa, and East Africa, have implemented mandatory legislation for fortification. In the case of vegetable oil, only 32 countries have mandatory fortification initiatives, primarily located in West and East Africa, with a few in South America and Asia. For maize flour, just 19 countries across the Americas and Africa have established mandatory fortification regulations, while only 12 countries have mandatory fortification legislation for sugar, and 7 countries have it for rice (Rohner et al., 2023). These initiatives are essential for fostering healthier populations and ensuring that everyone has access to the necessary nutrition for their well-being. Furthermore, the safety of incorporating micronutrients into the food supply will continue to be a significant factor to consider. Therefore, national fortification initiatives should be properly managed to ensure health advantages while preventing excessive consumption, particularly when micronutrient supplementation occurs and fortified processed foods are accessible (Mannar et al., 2018).

This chapter provides a comprehensive overview of micronutrient deficiencies and food fortification. It defines micronutrients, explains their importance, and presents global deficiency data. In addition, It also highlights common deficiencies such as iron, iodine, calcium, vitamin A, vitamin D, and many more along with their health impacts and strategies for prevention. This chapter also emphasizes the significance of micronutrients and recommends future research and interventions to enhance fortification efforts.

4.2 Micronutrient Deficiencies and Their Impact on Human Health

Micronutrient deficiencies, involving vitamins and minerals like iron, vitamin A, vitamin B, vitamin C, vitamin D, vitamin E, iron, iodine, calcium, and zinc, profoundly impact human health. Vitamin A deficiency impairs vision and immunity, increasing infection risk. Iron deficiency causes anemia, reducing energy and cognitive function. Iodine deficiency leads to goiter and developmental delays, particularly in children. Zinc deficiency weakens immune responses and growth. These deficiencies disproportionately affect vulnerable groups, such as children and pregnant women, leading to higher morbidity, mortality, and impaired quality of life. Addressing them through strategies like food fortification is critical for public health (Bailey et al., 2015).

4.2.1 Vitamin A

Vitamin A deficiency (VAD) affects around 190 million children and 19 million pregnant women, mainly in low-income regions. The prevalence rate is been recorded highest in sub-Saharan Africa (48%), South Asia (45–50%), and indigenous communities in Latin America (10–20%) (Stevens et al., 2022a, b). Furthermore, children, pregnant/lactating women, populations with poor dietary diversity, and those with chronic illnesses are at high risk for VAD. Growth demands make children vulnerable, causing xerophthalmia in 5.2 million cases annually (Imdad et al., 2022). Pregnant women face increased risks of maternal mortality and preterm birth. Diets lacking animal-source foods and β-carotene-rich vegetables contribute to deficiency (Vishwakarma et al., 2022), while conditions like measles, diarrhea, and HIV further impair absorption (Behboudi et al., 2022). More importantly, Vitamin A, including retinoids and carotenoids, is essential for vision, immune function, cellular growth, and reproduction. It forms rhodopsin for low-light vision (Saari, 2016), regulates T cell differentiation and mucosal immunity, supports epithelial maintenance and fetal development (Ross & Harrison, 2007), and is crucial for spermatogenesis and placental health (Clagett-Dame & Knutson, 2011). Vitamin A deficiency interventions include supplementation, fortification, dietary diversification, biofortification, and public health education. High-dose capsules (200,000 IU) biannually lower child mortality by 24%, while postpartum dosing enhances breastmilk retinol (Gallo et al., 2020). Fortifying staple foods has reduced VAD by 30–50% in Nigeria and Guatemala (Mkambula et al., 2020). Promoting orange-fleshed sweet potatoes improved serum retinol by 9% in Uganda (Lowe et al., 2024), and biofortified Golden Rice offers potential solutions for rice-dependent populations (Moghissi et al., 2015). Commonly fortified staple foods, such as oils, flour, dairy, sugar, and processed foods, effectively combat micronutrient deficiencies due to their widespread consumption and cost-effectiveness. Fortification has significantly improved vitamin A levels, reduced deficiency-related diseases, prevented over 600,000 childhood deaths annually, and lowered night blindness cases by 70% in affected regions.

4.2.2 Vitamin C

Vitamin C is a crucial antioxidant involved in collagen synthesis, oxidative stress reduction, immune function, iron absorption, and neurotransmitter production. It stabilizes collagen fibrils, enhances immune cell activity, regenerates vitamin E, and improves non-heme iron bioavailability (Jayedi et al., 2021). Additionally, it supports dopamine and norepinephrine synthesis, while high-dose IV administration has shown a 30% reduction in sepsis mortality (Fowler III et al., 2019). Severe vitamin C deficiency (scurvy) is rare; marginal deficiency affects 10–20% of adults, particularly in low-income regions. Prevalence is highest in sub-Saharan Africa and South Asia (15–30%) due to limited fruit and vegetable intake, while 5–10% of cases occur in high-income countries, especially among the elderly and smokers with low vitamin C intake (Rowe & Carr, 2020). Individuals at high risk of vitamin C deficiency include those with low fruit and vegetable intake, smokers (30–50% have suboptimal levels) (Lykkesfeldt & Carr, 2023), and patients with chronic diseases like CKD and IBD, which increase metabolic demand (Alberts et al., 2025). Older adults face reduced absorption (Carr & Zawari, 2023), while refugees and displaced populations are vulnerable due to food insecurity. Interventions for vitamin C deficiency include dietary diversification, with school feeding programs to improve serum levels of vitamins C. Supplementation with 200–500 mg/day is effective, though doses >2 g/day may cause nephrolithiasis (Pullar et al., 2017). Fortifying maize flour in South Africa reduced deficiency in at-risk children from 15% to 5% (Olson et al., 2021). Public health campaigns like New Zealand's "5+ A Day" increased fruit and vegetable intake, lowering deficiency rates (Jones et al., 2022). Commonly fortified foods with vitamin C include fruit juices, breakfast cereals, plant-based milk alternatives, snack bars, and infant formulas, chosen for their high consumption and compatibility with water-soluble nutrients. However, challenges include vitamin C degradation from heat and storage, risks of overconsumption, and inconsistent labeling. Despite minimal production costs, proper regulation and stability measures are essential for maximizing fortification benefits.

4.2.3 Vitamin D

Vitamin D (calciferol) is a fat-soluble vitamin that acts as a hormone, crucial for regulating calcium-phosphate balance, bone mineralization, and neuromuscular function. It also enhances immune responses by boosting antimicrobial peptides and reducing inflammatory cytokines. Additionally, vitamin D improves cardiometabolic health by enhancing insulin sensitivity and endothelial function (Ahsan et al., 2023). Adequate vitamin D levels (>30 ng/mL serum 25(OH) D) are associated with a lower risk of autoimmune diseases, such as multiple sclerosis, and certain cancers (Manson et al., 2019). Vitamin D deficiency (serum 25(OH) D < 20 ng/mL) affects approximately 1 billion people globally, with higher prevalence in regions with limited sunlight, air pollution, or cultural practices that restrict skin exposure (Cui et al., 2023). In the Middle East and North Africa, deficiency rates range from 50% to 90% due to sun-avoidance practices (Al-Daghri et al., 2021), while South Asia sees 70–90% deficiency due to skin pigmentation and inadequate dietary intake (Ritu & Gupta, 2014). Northern Europe reports 25–40% deficiency despite fortification efforts (Lips et al., 2019). High-risk populations for vitamin D deficiency include the elderly (40–60% of adults >65 years), who experience reduced skin synthesis and dietary absorption (Amrein et al., 2020), and individuals with darker skin, such as African Americans, who have three times higher deficiency rates due to reduced UVB penetration. Pregnant women, especially in India, have a higher risk, with 60% of pregnancies showing insufficiency (Gallo et al., 2020). Urban populations also face higher deficiency rates due to limited sun exposure from indoor lifestyles and pollution, with 35% of urban Chinese adults affected (Ning et al., 2016). Additionally, chronic diseases like obesity, chronic kidney disease (CKD), and inflammatory bowel disease (IBD) impair vitamin D metabolism. Vitamin D deficiency is addressed through supplementation (600–2000 IU/day), food fortification, sunlight exposure, and national policies. Fortification of dairy and cereals has reduced deficiency by 30–50% (Dwyer et al., 2015), while biofortified eggs and UVB-exposed mushrooms have improved vitamin D intake in regions with limited sunlight (Neill et al., 2021). Sunlight exposure (10–30 min, two to three times/week) aids synthesis, and screening programs like those in Saudi Arabia, improved adherence (Al-Daghri et al., 2021). Fortified foods like dairy products, plant-based milks, cereals, orange juice, margarine, and infant formula help address vitamin D deficiency. These foods are fortified to ensure broad consumption and enhance nutrient absorption, especially when paired with fats, supporting public health efforts in deficiency-prone regions.

4.2.4 Folate

Folate (vitamin B9) is a water-soluble vitamin critical for numerous biological processes. Its primary roles include DNA synthesis and repair, methylation reactions, and amino acid metabolism, mediated through its active form, 5-methyltetrahydrofolate

(5-MTHF) (Bo et al., 2020). It is particularly vital during periods of rapid cell division, such as pregnancy and infancy, as it supports DNA replication, cell proliferation, and neural tube formation (Crider et al., 2022). It plays a key role in one-carbon metabolism, supporting epigenetic regulation and reducing neural tube defect (NTD) risk by 50–70% (Wang et al., 2015). Beyond development, folate is essential for cardiovascular and neurological health. It helps lower homocysteine levels, thereby reducing the risk of cardiovascular diseases such as stroke and coronary artery disease (Cosentino et al., 2021). Additionally, its role in neurotransmitter methylation links folate deficiency to cognitive decline and depression (Nalder et al., 2021). Folate also plays a complex role in cancer and immune regulation. While adequate folate intake helps maintain DNA integrity and prevents mutations, excessive intake may promote tumor growth (Pieroth et al., 2018). Moreover, folate supports immune function by enhancing T cell responses and antibody production. Deficiency in folate is associated with conditions such as megaloblastic anemia, neural tube defects, and elevated homocysteine levels, all of which contribute to significant health risks (Crider et al., 2022). Folate deficiency remains a significant public health concern, affecting an estimated 20–37% of the global population, with disparities driven by socioeconomic status, dietary patterns, and healthcare access (Kancherla, 2023). Regions with limited access to fortified foods, such as sub-Saharan Africa and South Asia, report the highest prevalence. Certain groups are at a higher risk of folate deficiency due to increased physiological demands, poor dietary intake, or impaired absorption. Pregnant women, especially in low-income countries, face a heightened risk, contributing to neural tube defects (NTDs) in approximately 1 in 1000 live births globally (Green et al., 2021). Infants and adolescents, undergoing rapid growth phases, are also vulnerable, with 32% of Indian adolescents exhibiting low serum folate levels due to inadequate intake (Jani et al., 2015). Older adults experience reduced folate bioavailability due to malabsorption, chronic diseases, and polypharmacy. Additionally, low-income populations reliant on unfortified staple foods often fail to meet folate requirements (Rogers et al., 2018). To address folate deficiency, various interventions have been implemented. Food fortification programs, such as mandatory folic acid fortification in over 90 countries, have reduced NTD prevalence by 30–50%, with South Africa's policy lowering rates from 1.4 to 0.3 per 1000 births (Kancherla, 2023). Prenatal folic acid supplementation (400–800 μg/day) is globally recommended but has low adherence, particularly in rural areas (<30%) (Caffrey et al., 2019). Biofortification of staple crops, like folate-enriched rice, has increased intake by 50–70% in Southeast Asia (Avnee et al., 2023). Additionally, community-based education programs in Ethiopia have improved the consumption of folate-rich foods by 40%, demonstrating the importance of awareness campaigns in combating deficiency (Anato & Reshid, 2025). Folate-fortified foods include breakfast cereals, breads, pasta, rice, snack bars, dairy products, and infant formulas, supporting growth and health. Supplements are also available for those needing additional intake.

4.2.5 Vitamin E

Vitamin E (α-tocopherol) is a vital fat-soluble antioxidant that protects cell membranes from oxidative damage, supports immune function, regulates antioxidant enzyme expression, and plays a crucial role in neurological and reproductive health (Lewis et al., 2019). Deficiency is linked to neurodegeneration and impaired fertility (Agarwal et al., 2021). Severe vitamin E deficiency is rare (<1%), but subclinical insufficiency (12–30 μmol/L) affects 10–20% of adults, especially in regions with poor dietary diversity. (Malik et al., 2021). Prevalence is 15–25% in South Asia and sub-Saharan Africa due to refined oil consumption (Jiang, 2020), while 5–10% of elderly and chronic disease patients in high-income countries are affected. Individuals with fat malabsorption disorders (e.g., cystic fibrosis, Crohn's) have impaired vitamin E absorption (Degrassi et al., 2023), while 30% of preterm infants are deficient due to poor placental transfer (Brion et al., 2003). Genetic conditions like a mutation in the α-TTP gene (e.g., Ataxia with Vitamin E Deficiency, AVED) cause severe deficiency. Elderly individuals are at risk of deficiency due to inadequate dietary intake and increased oxidative stress. Vitamin E deficiency can be addressed through supplementation, dietary diversification, fortification, and managing underlying conditions. High-dose α-tocopherol (800–1200 mg/day) reverses neurological symptoms in AVED patients, while enteral supplementation (10–25 IU/day) prevents deficiency in preterm infants (Brion et al., 2003).

4.2.6 Iron

Iron is a vital micronutrient essential for oxygen transport via hemoglobin and myoglobin, supporting cellular respiration (Camaschella, 2015). It acts as a cofactor in energy metabolism, aiding ATP production. Iron is crucial for immune function, facilitating immune cell proliferation and ROS generation (Cassat & Skaar, 2013). It supports DNA synthesis and repair (Charlebois & Pantopoulos, 2023) and influences neurodevelopment, neurotransmitter synthesis, and cognition (Mahmoud, 2024). However, imbalances in iron levels can impair immunity and cognition or cause oxidative stress-related damage. Iron deficiency (ID) affects 1.8 billion people globally, with disproportionately high rates in low-resource regions. Sub-Saharan Africa and South Asia bear the greatest burden, where 40–60% of women and 50–70% of children under five are affected, driven by inadequate diets, malaria, and helminth infections (Lin et al., 2025; Kedir et al., 2024). In contrast, high-income countries report lower prevalence (10–15% in women, 5–10% in children), though ID persists among vulnerable groups like the elderly and adolescents (Joosten, 2018). Key risk factors for ID include dietary reliance on iron-poor staples (e.g., rice, maize), chronic infections (e.g., malaria, HIV), and heavy menstrual bleeding (Sandalinas et al., 2023). ID during pregnancy and childhood elevates risks of maternal mortality, preterm birth, and impaired neurodevelopment (Da Silva Lopes et al., 2021). Interventions like food fortification, iron supplementation, and biofortified crops (e.g., iron-rich beans) show promise but face challenges such as poor adherence and logistical barriers. Addressing ID requires integrated strategies combining nutrition, healthcare, and infection control, particularly in LMICs. Common iron-fortified foods include staple cereals like wheat, maize flour, and rice, as well as breakfast cereals, infant formulas, plant-based milks, and condiments like fish sauce and bouillon cubes. These foods ensure broad daily consumption, are cost-effective, and enhance absorption when paired with vitamin C (Mahmoud, 2024).

4.2.7 Iodine

Iodine is a critical micronutrient essential for synthesizing thyroid hormones, triiodothyronine (T3) and thyroxine (T4), which regulate metabolic processes, energy production, and thermogenesis (Skeaff & Lonsdale-Cooper, 2013). These hormones are indispensable for normal growth, neurological development, and cellular function, particularly during pregnancy and early childhood. Maternal iodine deficiency can lead to irreversible cognitive impairments in offspring, including cretinism, characterized by severe developmental delays (Pearce et al., 2016). Beyond its role in thyroid function, iodine exhibits antioxidant properties and modulates immune responses by influencing thyroid-mediated activity of immune cells, enhancing the body's defense against infections (Sorrenti et al., 2021). Iodine deficiency disorders (IDDs), including goiter and hypothyroidism, affect approximately 2 billion people globally, with the highest risk in iodine-poor regions such as parts of Africa and Asia. Despite universal salt iodization (USI) reducing IDD prevalence by 72% since 1990, accessibility remains a challenge in remote and low-income communities (Andersson et al., 2020). According to Iodine Global Network (IGN, 2021), iodine deficiency persists in 21% of the global population, with disparities driven by regional inequities, particularly in sub-Saharan Africa, South Asia, and Eastern Europe. For example, 30% of rural Indians and 40% of Eastern European school-aged children have insufficient iodine intake (Chakrabarty et al., 2022; Andersson et al., 2022). Vulnerable groups, including pregnant women and children, face neurodevelopmental risks due to deficiency. Additionally, emerging threats like climate change, reducing soil iodine bioavailability and pandemic-related disruptions to iodized salt programs further exacerbate the issue. Iodine deficiency leads to various iodine deficiency disorders (IDDs). It causes goiter and hypothyroidism, impairing metabolism and energy regulation (Zimmermann, 2011). In pregnant women and infants, deficiency leads to cognitive impairment, reduced IQ, and severe conditions like cretinism (Andersson et al., 2020; Chakrabarty et al., 2022). It also weakens immunity, increases infection risk, and contributes to reproductive complications, including miscarriage and preterm birth (IGN, 2021). Despite progress through universal salt iodization (USI), iodine deficiency remains a challenge in vulnerable regions, necessitating strengthened fortification programs and targeted public health interventions. Enhancing salt iodization, implementing effective policies, and adopting climate-resilient strategies are crucial for eliminating iodine deficiency disorders (IDDs) globally. Additionally, ensuring adequate iodine intake through fortified foods, seafood, or supplements is essential for preventing IDDs and promoting lifelong health (Eveleigh et al., 2023). Iodine-fortified foods include iodized salt (mandatory in 130+ countries), bread, dairy products, processed foods, and infant formula. These are chosen for their universal consumption, cost-effectiveness, and iodine stability, ensuring widespread deficiency prevention.

4.2.8 Calcium

Calcium is a critical mineral for maintaining skeletal integrity, neuromuscular function, and cellular signaling. Over 99% of the body's calcium is stored in bones and teeth, where it provides structural strength and regulates bone remodeling processes (Vannucci et al., 2018). The remaining 1% circulates in blood, supporting muscle contraction, nerve impulse transmission, and blood clotting (Berchtold et al., 2020). Calcium also modulates enzyme activity, hormone secretion (e.g., insulin), and vascular tone, making it indispensable for cardiovascular and metabolic health (Betts et al., 2023). Despite its vital roles, calcium deficiency affects ~3.5 billion people globally, with the highest prevalence in South Asia and sub-Saharan Africa (Global Burden of Disease Study, 2020). This deficiency is primarily driven by low consumption of dairy products, limited access to calcium-rich foods such as leafy greens and fortified products, and poor dietary diversity. For instance, Harinarayan et al. (2021) reported that 70–90% of adults in India consume less than half the recommended daily calcium intake (500–600 mg vs. 1000–1300 mg). In Bangladesh and Nepal, over 60% of women of reproductive age have insufficient calcium intake (Bromage et al., 2016), while in sub-Saharan Africa, 50% of children in low-income regions exhibit inadequate calcium intake (Chanchlani et al., 2020). Certain populations are disproportionately vulnerable to calcium deficiency due to physiological changes and dietary limitations. Postmenopausal women, for instance, face heightened susceptibility to osteoporosis as declining estrogen levels hinder calcium absorption (Vannucci et al., 2018). Older adults are similarly at risk, as natural aging reduces vitamin D synthesis and kidney efficiency, further depleting calcium retention. Children and adolescents, undergoing rapid bone development, require elevated calcium intake during growth spurts to support skeletal mineralization (Golden et al., 2020). Pregnant individuals also need increased calcium to nourish the developing fetal skeleton, with deficiencies correlating to complications like preeclampsia. Additionally, those following vegan diets or managing lactose intolerance often struggle to meet calcium requirements, as their diets typically exclude dairy—a primary calcium source. Chronic insufficiency leads to hypocalcemia, osteoporosis, and fractures. In children, severe deficiency causes rickets, while adults face heightened risks of hypertension and colorectal cancer (Chanchlani et al., 2020). Intervention strategies include food fortification (e.g., calcium-fortified flour), supplementation, and promoting dietary diversity (Micha et al., 2020). Public health programs in LMICs, such as school-based milk initiatives, have reduced deficiency rates by 15–20% (Bolland et al., 2021). Common calcium-fortified foods include plant-based milks, fruit juices, breakfast cereals, tofu, snack bars, and bread. These options support dairy-free diets, ensure staple consumption, and enhance absorption, with calcium citrate malate in juices offering superior bioavailability.

4.2.9 Zinc

Zinc is an essential micronutrient involved in numerous physiological processes. It acts as a cofactor for over 300 enzymes, playing critical roles in DNA synthesis, protein production, cell division, immune function, cellular signaling, gene expression, and protein metabolism (Kiouri et al., 2025). Zinc is also vital for wound healing, taste perception, neurological functions, proper growth and development during pregnancy, childhood, and adolescence. Additionally, zinc has antioxidant properties that help protect cells from oxidative stress and inflammation (Kiouri et al., 2025). Zinc deficiency is a significant public health issue, particularly in low- and middle-income countries where dietary intake is inadequate. An estimated 17% of the global population is at risk of inadequate zinc intake, with higher prevalence rates in Asia (19%) and Africa (24%) (https://www.ifpri.org/blog/addressing-global-life-long-health-impacts-zinc-deficiency-call-action/). Certain populations are more vulnerable to zinc deficiency, including children, pregnant and lactating women, and individuals with chronic illnesses. Children are at high risk due to their increased nutritional needs for growth and development. Pregnant and lactating women require additional zinc to support fetal development and milk production. People with chronic illnesses, such as inflammatory bowel disease, may also have higher zinc requirements or impaired absorption. Zinc deficiency is associated with stunted growth in children, impaired immune function, and increased susceptibility to infections (Krebs et al., 2017). Chronic zinc deficiency can also contribute to neurological disorders, oxidative stress, and an increased risk of non-communicable diseases. Pregnant women with inadequate zinc levels may experience complications such as low birth weight and preterm delivery (Keats et al., 2019). Addressing this deficiency through dietary diversification, supplementation, food fortification, and biofortification remains a priority in global nutrition strategies (Bailey et al., 2015). Zinc supplementation has been shown to reduce child mortality and morbidity related to diarrhea and respiratory infections (Bhutta et al., 2018). Additionally, nutrition education programs promote awareness and encourage consumption of zinc-rich foods to prevent deficiency (Mayo-Wilson et al., 2014). Biofortification of staple crops, such as rice and wheat, has proven to be a sustainable strategy to improve zinc intake in vulnerable populations. Fortification of staple foods and zinc-enriched fertilizers also contributes to improving zinc status in deficient regions. Common zinc-fortified foods include breakfast cereals, snack bars, dairy products, breads, meat substitutes, infant formulas, and sports drinks. These help boost intake, especially for those with limited dietary sources, supporting growth and overall health (Andersson et al., 2017).

4.2.10 Selenium

Selenium is a crucial trace mineral involved in antioxidant defense, thyroid hormone regulation, immune function, cancer prevention, reproductive health, and neuroprotection. It protects against oxidative damage, supports thyroid hormone activation, enhances immune responses, and may reduce the risk of certain cancers (Drutel et al., 2013). Selenium also aids in spermatogenesis, reduces miscarriage risk, and supports cognitive health, with adequate levels linked to better cognitive function and lower dementia risk (Alahmar, 2019). Deficiency of selenium affects 500 million to 1 billion people worldwide, largely influenced by soil selenium content, diet, and socioeconomic factors (Jones et al., 2022). Regions with selenium-poor soils, such as parts of China, Central Europe, and sub-Saharan Africa, report the highest deficiency rates, with 72% of residents in selenium-depleted areas of Ethiopia showing suboptimal selenium status (Imran et al., 2023; Bai et al., 2025). Populations at high risk include those in low-selenium regions like the UK and Denmark, where 15–30% of adults consume insufficient selenium (Jones et al., 2017). Individuals with chronic diseases such as HIV/AIDS experience worsened immune dysfunction due to selenium deficiency (Muzembo et al., 2019). Pregnant women face increased risks of preterm birth and preeclampsia (McDougall et al., 2023), while vegetarians in low-selenium regions, such as India, often have inadequate intake. Additionally, older adults experience reduced selenium absorption and retention due to aging and inflammation (Rayman, 2020). Interventions to combat selenium deficiency include supplementation (50–200 μg/day), which improves status, and food fortification, such as Finland's selenium fertilization program, which eliminated deficiency (Ulfberg & Stehlik, 2021). Dietary diversification with selenium-rich foods like Brazil nuts (95 μg per nut) and agronomic biofortification, such as selenium-enriched fertilizers, has also proven effective. Excessive selenium intake (>400 μg/day) can lead to selenosis. Selenium-fortified foods include breakfast cereals, snack bars, breads, dairy products, infant formulas, and meat substitutes, providing essential support for growth and overall health. Supplements are also available for those needing additional intake (Cominetti et al., 2012) (Table 4.1).

4.3 Fortified Foods and Nutraceuticals: Combating Micronutrient Deficiencies

Food fortification and nutraceuticals have emerged as pivotal strategies to address global nutritional deficiencies and promote health in recent years. Food fortification involves the deliberate addition of essential micronutrients, such as vitamins and minerals, to staple foods to improve their nutritional quality and combat deficiencies that affect millions worldwide. Nutraceuticals, defined as food-derived products with health benefits beyond basic nutrition, encompass a broad range of items, including dietary supplements, functional foods, and bioactive compounds (Ashraf et al., 2020). Together, these approaches bridge the gap between food security and nutritional adequacy, tackling both undernutrition and the rising burden of

Table 4.1 Various micronutrient fortification strategies and its health benefits

Micronutrients	Types of food fortified	Health benefits	Study design	References
Vitamin A	Rice	Improved the vitamin A status of lactating women by increasing total body VA stores	RCT	Pinkaew et al. (2021)
	Cooking oil	Improved breast milk retinol levels in postpartum mothers Improved vitamin A status in children under 5 Reduced vitamin A deficiency in infants, young children, and women in Indonesia	Quasi-experiment design Cohort study	Salam et al. (2021), Sudikno et al. (2017), Sandjaja et al. (2015)
Vitamin D	Yogurt	Improved glycemic and anthropometric indices (waist-to-hip ratio WHR) and percentage of body fat (PBF)) in pre-diabetic individuals	RCT	Mostafai et al. (2018), Nikooyeh et al. (2011)
		Improved glycemic status in T2D patients	RCT	Hajimohammadi et al. (2017)
		Increased leptin and ghrelin levels but reduced leptin-to-ghrelin ratio in type 2 diabetes patients		
	Cheese	Improved vitamin D status (serum 25-OHD) in elderly	RCT	Johnson et al. (2005)
		Lowered TRAP 5b in postmenopausal women at moderate risk of osteoporosis fracture		Bonjour et al. (2012)
	Milk	Increased [25(OH)D] serum levels in children aged 10–14 years	RCT	Neyestani et al. (2014), Khadgawat et al. (2013)
	Milk and yogurt	Increased serum vitamin D levels and reduced the restless legs syndrome (RLS) severity in women with severe premenstrual syndrome (PMS)		Sharifan et al. (2024)
	Orange juice	Increased 25(OH)D concentration in young children within 12 weeks	RCT	Economos et al. (2014), Neyestani et al. (2014)
Vitamin E	Orange juice	Increased α-tocopherol concentrations in young children within 12 weeks	RCT	Economos et al. (2014)

(continued)

Table 4.1 (continued)

Micronutrients	Types of food fortified	Health benefits	Study design	References
Iron	Water	Reduced iron deficiency anemia	Experimental	Noubactep et al. (2023)
	Dairy products	Increased Hb concentrations among children 24 to 59 months old in northern Senegal	RCT	Le Port et al. (2017)
		Improved growth and iron status and reduced anemia in children 1–4 years old	RCT	Sazawal et al. (2010)
	Whole wheat flour Maize flour	Enhanced iron stores, prevented anemia, and improved overall health	RCT	Muthayya et al. (2012a, b)
	Biscuits	No differences in iron status, anemia, or hookworm. Significant increase in the number of enterobacteria, a decrease in lactobacilli, and a significant increase in fecal calprotectin, a marker of gut inflammation	Meta-analysis	Konikoff and Denson (2006)
	Micronutrient powders (MNPs) (complementary foods)	Reduced iron deficiency anemia (IDA) in children 6 to 23 months of age	Meta-analysis	De-Regil et al. (2013)
	Extruded rice	Increased iron stores and reduced the prevalence of iron deficiency in 6–13-y-old Indian school children	RCT	Moretti et al. (2006)
	Double-fortified salt (DFS) with iron and iodine	Increased hemoglobin concentrations and reduced the risk of anemia and iron deficiency anemia (IDA) in low- and middle-income countries (LMICs)	Systematic review and meta-analysis	Ramírez-Luzuriaga et al. (2018)
		Significantly reduced anemia prevalence and improved cognitive function in school-age children	RCT	Bathla and Grover (2016)
		Increased hemoglobin and ferritin levels and improved iron stores in women of reproductive age	RCT	Haas et al. (2014), Wenger et al. (2017)

Iodine	Salt	Improved both pregnant women and their infant's mental and cognitive development	RCT	Bougma et al. (2018), Mohammed et al. (2020)
	Bread	Improved iodine status in school children and adults		Skeaff and Lonsdale-Cooper (2013), Jones et al. (2016)
	Dairy products	Increased iodine status toddlers, women of childbearing age, pregnant women, and lactating mothers	RCT	Szymlek-Gay et al. (2020), O'Kane et al. (2018), Nazeri et al. (2017)
Calcium	Milk	Reduced age-related bone density loss and fracture risk in adults over 50 by restoring bone homeostasis through the modulation of the calcium-vitamin D-PTH axis, bone remodeling rate, and growth hormone/IGF-1 axis	RCT and meta-analysis	Daly et al. (2006), Hidayat Khemayanto et al. (2022)
	Water or flour	Improved calcium intake for women, adolescent girls, and young children in Bangladesh, Uganda, and Guatemala	Review	Knight et al. (2023)
	Fermented Milk products (yoghurt and sour milk)	Reduced the risk of hip fracture in postmenopausal women Increased circulating concentrations of SIRT1 and SIRT6 in T2D subjects	RCT	Nikooyeh et al. (2021)
	Cereal bars	Helped maintain and improve bone health of women	RCT	Lee et al. (2015)
Zinc	Water	Increased zinc intake and helped in preventing childhood infections, diarrhea and improved growth in pre-school children in rural Africa	RCT	Kujinga et al. (2018).
	Juice drink	Improved the zinc status of children Beneficial to the growth and development of children	RCT	Angeles-Agdeppa et al. (2011)
Folic acid/ folate	Wheat and maize flour	Reduced the risk of neural tube defects	RCT	Centeno Tablante et al. (2019), Sayed et al. (2008)

diet-related chronic diseases. Scientific advancements have refined these interventions, ensuring efficacy, safety, and accessibility, while addressing challenges like bioavailability, consumer acceptance, and regulatory frameworks. Food fortification has proven effective in reducing micronutrient deficiencies, particularly in low- and middle-income countries. For instance, fortifying staple foods like wheat flour with iron and folic acid has significantly decreased anemia and neural tube defects. A 2018 study highlighted that fortification of maize flour with vitamin A in sub-Saharan Africa improved serum retinol levels in children, reducing the risk of blindness and immune dysfunction (Bouis & Saltzman, 2017). Advances in biofortification, where crops are bred or genetically engineered to enhance nutrient content, have further expanded impact. Golden Rice, engineered to produce beta-carotene, has addressed vitamin A deficiency in Asia, with field trials showing a 20–30% increase in dietary vitamin A intake (Paine et al., 2015). These interventions demonstrate fortification's scalability, but challenges remain, including ensuring equitable access and minimizing nutrient losses during processing or cooking. Nutraceuticals, meanwhile, cater to both preventive and therapeutic health needs, driven by growing consumer awareness of diet-disease linkages. Nutraceuticals also include dietary supplements like vitamin D, critical in regions with limited sunlight exposure. A 2021 study linked vitamin D supplementation to a 15% reduction in respiratory infection risk, underscoring its public health potential (Martineau et al., 2021). However, the nutraceutical market faces scrutiny over exaggerated health claims and inconsistent regulation, necessitating robust clinical evidence to validate efficacy. Both fortification and nutraceuticals confront bioavailability challenges. Nutrients like iron and zinc often have low absorption rates, mitigated through innovations like microencapsulation, which protects compounds during digestion. A 2016 study demonstrated that microencapsulated iron in fortified rice improved hemoglobin levels by 10% in schoolchildren (Moretti et al., 2016). Similarly, nutraceuticals leverage delivery systems like liposomes to enhance the absorption of fat-soluble vitamins (Sharma et al., 2019). Consumer acceptance is another hurdle, as fortified foods must align with cultural preferences and sensory expectations. Nutraceuticals face skepticism due to cost and perceived complexity, particularly in low-resource settings. Future research must prioritize personalized nutrition, leveraging genomics to tailor interventions, and address sustainability in production. By integrating scientific rigor with equitable access, food fortification and nutraceuticals can sustainably enhance global health outcomes.

4.4 Guidelines for Food Fortification

Food fortification, the addition of essential nutrients to foods, is guided by international standards to address micronutrient deficiencies safely and effectively. The World Health Organization WHO and FAO provide frameworks emphasizing evidence-based nutrient selection, targeting vulnerable populations, and ensuring bioavailability. Guidelines recommend fortifying staple foods like flour with iodine, iron, and vitamin A, based on dietary patterns and deficiency prevalence. Safety is prioritized to avoid overconsumption, with monitoring systems advised. Codex Alimentarius standards ensure global consistency. Fortification programs have reduced anemia and neural tube defects significantly (https://www.who.int/publications/i/item/9241594012).

4.4.1 Iodine Fortification

Iodine comes in two chemical forms that can be used to fortify food: iodate and iodide. They are commonly added as potassium salt, but can also be calcium or sodium salt (WHO, 2006). Salt is the most commonly used food carrier for iodine fortification. WHO recommends universal salt iodization (USI) as a technique for controlling iodine deficiency. This involves iodizing all salt for human (food industry and household) and livestock consumption. USI is a mass fortification technique that can cover the iodine needs of the entire population, including vulnerable groups like pregnant and lactating women and infants (Sharma et al., 2019). Iodine fortification levels of 20–40 mg per kg salt are recommended, based on a typical 5–10 g of salt intake per day in adult populations, while those taking up to 25 g salt per day are deemed safe when dietary iodine intakes are insufficient (Andersson et al., 2012). Fortification levels should be tailored to the population's salt intake and modified on a regular basis. The WHO recommends an iodine intake of 150 μg per day for adults (https://data.unicef.org/topic/nutrition/iodine/). This can be achieved by consuming 5 g of salt iodized to 39 mg/kg or only 3 g of salt iodized to 60 mg/kg (Lammie et al., 2007). In certain countries, various staple foods or condiments, such as biscuits, milk, sauce, drinking water, yogurt, fruit beverages, seasoning powder, and bouillon cubes, are iodized voluntarily or mandatorily (https://iris.who.int/handle/10665/63322; https://iris.who.int/handle/10665/43781).

4.4.2 Iron Fortification

Iron is the most difficult micronutrient to incorporate into foods, primarily because the iron compounds that offer optimal bioavailability often react intensely with food components, leading to undesirable changes in flavor, color, and texture. When choosing an appropriate iron compound for food fortification, the primary goal is to identify one with the highest relative bioavailability (RBV) in comparison to ferrous sulfate, while also avoiding significant alterations to the sensory attributes of the food product. Additionally, cost is frequently a crucial factor to consider (WHO, 2006). Wheat and maize flours have well-defined international standards for iron fortification. The four recommended compounds for fortification are ferrous sulfate, ferrous fumarate, NaFeEDTA, and electrolytic iron powder (https://www.ffinetwork.org/plan-standards-and-legislation). The proposed fortification levels were determined based on sensory evaluations, effectiveness studies conducted with women and children in low- and middle-income countries (LMIC), the bioavailability of the iron compounds assessed through human isotopic studies, and the expected daily flour consumption (Hurrell, 2021). For instance, in a population where daily flour intake ranges from 150 to 300 grams, the WHO (2009) recommends fortifying wheat flour with 30 ppm iron in the form of ferrous sulfate or ferrous fumarate, 60 ppm iron as electrolytic iron powder, or 20 ppm iron as NaFeEDTA. The use of electrolytic iron powder is suggested solely for low-extraction flours consumed at levels of 150 g/d or more, and it is not advised for nixtamalized maize flour. NaFeEDTA is the only iron compound recommended by the WHO for wholegrain wheat and maize flours that are high in phytate. The absorption from wholegrain wheat bread rolls enriched with NaFeEDTA was four times greater than that from ferrous sulfate [17], and the use of NaFeEDTA-fortified wholegrain atta flour in chapattis over a period of 7 months in the school meals of Indian children greatly enhanced their iron levels (Muthayya et al., 2012a, b).

4.4.3 Vitamin A Fortification

The selection of a vitamin A fortification is critically influenced by the unique properties of the food vehicle, alongside a range of technological, regulatory, and religious factors. Given that preformed vitamin A (retinol) is inherently unstable, it is typically transformed into more stable esters through a process called esterification, often utilizing palmitic or acetic acid. As a result, the primary commercial forms of vitamin A available for food fortification include retinyl acetate, retinyl palmitate, and provitamin A (β-carotene). This strategic choice not only ensures stability in food products but also optimizes the nutritional benefits for consumers (WHO, 2006). Using oil as a base for vitamin A fortification offers a powerful solution for enhancing nutrition, given that vitamin A is fat-soluble. The processes for enriching vegetable oil with vitamin A are not only well-established but also remarkably straightforward and cost-effective (Dary & Mora, 2002). This makes it an accessible option for communities in need. Moreover, oil plays a crucial role in preserving retinol stability and preventing vitamin oxidation, ensuring that the essential nutrients remain intact (Dary & Mora, 2002). The recommended forms of vitamin A for this purpose are retinol acetate or retinol palmitate in an oil base, which are effective and reliable. Other than oil, margarine and other hydrogenated oil products are the most suitable vehicles for vitamin A. Several countries fortify margarine at levels ranging from 1 to 15 mg/kg as either a mandatory or a voluntary practice (Johnson, 1997). This provides 2–40% of the RDI, assuming a daily consumption of 15 g. Furthermore, fortifying cereal flours and meals with vitamin A is a technically feasible process, and the stability of nutrients in these products is typically high (Dary & Mora, 2002). In the Philippines, wheat flour used for making the popular bread pandesal is fortified with vitamin A at a level of 4.5 mg/kg, resulting in a pandesal that is fortified at 2.8 mg/kg (Solon, 1997). Daily consumption of 40 grams of this bread provides about 19% of the vitamin A RDI. Sugar has become the primary source of vitamin A even for infants and young children, contributing approximately 30% of the recommended daily intake (RDI) [8]. Preventive supplementation with vitamin A for children under 3 years old continues to be implemented to ensure sufficient intake, as noted in data from national reports. Sugar is enriched at the sugar mills with a theoretical amount of 15 mg/kg, although after losses that occur before consumption at the household level, the actual content is closer to 9 mg/kg (Dary & Mora, 2002).

4.5 Food Fortifying Agents: Market Size and Current Trends

The global food fortification market was valued at USD 96.86 billion in 2023 and is anticipated to reach USD 223 billion by 2032, with a compound annual growth rate (CAGR) of 9.8% throughout the forecast period from 2024 to 2032. The growing awareness among consumers about fortified foods and their health benefits is likely to serve as a driving force (https://www.grandviewresearch.com/industry-analysis/food-fortifying-agents-market-report; https://straitsresearch.com/report/food-fortification-market#:~:text=The%20global%20food%20fortification%20market,concerns%20about%20micro%2Dnutrient%20deficiency). In 2023, the food fortifying agents' market in North America represented a revenue share of 30.4%, driven by its developed food and beverage industry, heightened consumer awareness, and a favorable regulatory framework. The region boasts established guidelines and policies for food fortification, which facilitate the extensive application of fortifying agents in a variety of food products. The market for food fortifying agents in the United States is projected to experience a CAGR of 9.1% between 2024 and 2030. This market plays a crucial role in the North American market, influenced by high levels of consumer awareness, a wide range of product options, and strong regulatory support. The United States has a longstanding tradition of food fortification initiatives, such as enriching cereals with folic acid to reduce the incidence of neural tube defects (https://www.grandviewresearch.com/industry-analysis/food-fortifying-agents-market-report). In Europe, the market for food fortifying agents is projected to expand at a CAGR of 8.4% between 2024 and 2030, driven by rigorous food safety regulations, increasing consumer demand for healthy foods, and a strong emphasis on preventive healthcare. Key regulatory authorities in the region, including the European Food Safety Authority (EFSA), are instrumental in maintaining the quality and safety standards of fortified foods [12]. In the Asia Pacific region, the market for food fortifying agents in the Asia Pacific region is projected to expand at a CAGR of 9.7% between 2024 and 2030. Factors such as swift urbanization, growing disposable incomes, and heightened health consciousness among consumers are driving this growth. The region is confronting notable nutritional deficiencies, leading governments to initiate fortification initiatives to tackle public health issues (https://www.grandviewresearch.com/industry-analysis/food-fortifying-agents-market-report) (Fig. 4.1).

Fig. 4.1 Global food fortification market growth (2023–2032, billion USD) and regional market growth index (2024–2030) for the United States (CAGR 9.1%), Europe (CAGR 8.4%), and Asia-Pacific (CAGR 9.7%)

4.6 Safety and Risk Assessment

Vitamins and essential minerals, including trace elements, are micronutrients that are vital for the normal functioning of the human body and must be obtained through diet. However, like other chemical substances found in foods, consuming these micronutrients in excess can lead to negative health effects (EFSA NDA Panel, 2024). When discussing micronutrients, many are considered safe even when consumed in high amounts (European Commission 206), which gives us reassurance about our dietary decisions. Nevertheless, some micronutrients have established maximum safe intake limits known as tolerable upper intake levels (UL), set for our safety (WFP, 2022). A tolerable upper intake level (UL) serves as a scientifically based reference for policymakers and other stakeholders in addressing the risks associated with excessive nutrient consumption. The idea of a UL pertains to the highest daily intake from all dietary sources (including food and beverages, fortified products, and dietary supplements) that, if exceeded, could lead to adverse health outcomes(EFSA NDA Panel, 2024). These ULs are created with consumer protection in mind. They include a safety margin that helps to prevent daily chronic intakes that could lead to negative effects. The probability of surpassing the UL varies among different micronutrients. For certain population subgroups, high intake levels of vitamin A, iron, zinc, calcium, copper, and iodine can approach the UL (Flynn et al., 2003). Across the globe, particularly in low- and middle-income countries, reports indicate that many individuals consume micronutrients significantly below the estimated average requirement (EAR). The health risks stemming from the portion of the population with micronutrient intakes under the EAR are a far greater concern than those from intakes nearing the UL, which indicates a level likely considered safe (Flynn et al., 2003). Various models have been created to establish maximum safe limits for either fortified foods or dietary supplements (Kloosterman et al., 2007; Rasmussen et al., 2006). In 2000, the French Food Safety Agency (AFSSA) pioneered a groundbreaking probabilistic risk assessment approach to evaluate the safety of maximum permissible levels of vitamins and minerals. This innovative method specifically targets food fortification, ensuring that nutritional enhancements are not only effective but also safe for public consumption (http://www.afssa.fr/Documents/NUT2000sa0239Ra.pdf). Furthermore, the National dietary assessments indicate that the role of fortified foods in excessive micronutrient consumption is quite limited, even in affluent nations (Howlett & Binns, 2009). A recent investigation simulating the potential micronutrient contribution of fortified maize flour, oil, rice, salt, and wheat flour across 153 countries reveals a negligible risk of exceeding Tolerable Upper Intake Levels and harming populations due to excessive nutrient intake from fortifying staple foods (Pachón et al., 2021). This scenario is improbable with merely consuming fortified staple foods, since the fortification levels for micronutrients are established according to context-specific guidelines or regulations, and an exceedingly large amount of these staples would need to be eaten to exceed safe limits (WFP, 2022).

4.7 Conclusion

Food fortification demands country-specific regulations tailored to address distinct nutritional challenges and public health priorities. Effective programs rely on evidence-based policies that evaluate micronutrient deficiencies, target fortification efforts, and validate interventions with robust health outcome data. Choosing widely consumed staples as food vehicles and investing in biofortified crops can enhance sustainability. Ongoing monitoring is critical to track progress, address gaps, and adapt to shifting dietary trends, demographic changes, and emerging research. Future strategies should prioritize context-specific approaches, accounting for local dietary habits and deficiency rates. Smart fortification, using AI to optimize nutrient levels based on demographic data, alongside research on safe intake limits, can tackle both deficiencies and excesses. Additionally, fortification should incorporate climate-resilient solutions, such as heat-stable micronutrients, for regions facing extreme temperatures. Integrating fortification with complementary health initiatives, like deworming or hypertension education, can improve nutrient absorption and balance iodine sufficiency with sodium reduction. A multisectoral, data-driven approach is essential to scale fortification effectively, minimize unintended consequences, and deliver lasting public health benefits.

References

Agarwal, A., et al. (2021). Role of vitamin E in male infertility. *World Journal of Men's Health, 39*(1), 1–10. https://doi.org/10.5534/wjmh.210025

Ahsan, N., Imran, M., Mohammed, Y., Al Anouti, F., Khan, M. I., Banerjee, T., Adnan, M., Ashfaq, F., Kieliszek, M., Ashraf, S. A., & Haq, A. (2023). Mechanistic insight into the role of vitamin D and zinc in modulating immunity against COVID-19: A view from an immunological standpoint. *Biological Trace Element Research, 201*(12), 5546–5560. https://doi.org/10.1007/s12011-023-03620-4

Alahmar, A. T. (2019). Role of oxidative stress in male infertility: An updated review. *Journal of human reproductive sciences, 12*(1), 4–18. https://doi.org/10.4103/jhrs.JHRS_150_18

Alberts, A., Moldoveanu, E., et al. (2025). Vitamin C: A comprehensive review of its role in health, disease prevention, and therapeutic potential. *Molecules, 30*(3), 748. https://doi.org/10.3390/molecules30030748

Al-Daghri, N. M., et al. (2021). Vitamin D deficiency in the Middle East: A decade of progress. *BMC Public Health, 21*(1), 1–10.

Amrein, K., Scherkl, M., Hoffmann, M., Neuwersch-Sommeregger, S., Köstenberger, M., Tmava Berisha, A., Martucci, G., Pilz, S., & Malle, O. (2020). Vitamin D deficiency 2.0: An update on the current status worldwide. *European Journal of Clinical Nutrition, 74*(11), 1498–1513. https://doi.org/10.1038/s41430-020-0558-y

Anato, A., & Reshid, M. (2025). Effect of nutrition education and iron-folic acid supplementation on anemia among pregnant women in Ethiopia: A quasi-experimental study. *Scientific Reports, 15*(1), 3556. https://doi.org/10.1038/s41598-025-87957-x

Andersson, M., Karumbunathan, V., & Zimmermann, M. B. (2012). Global iodine status in 2011 and trends over the past decade. *Journal of Nutrition, 142*, 744–750.

Andersson, M. S., Saltzman, A., Virk, P. S., & Pfeiffer, W. H. (2017). Progress update: Crop development of biofortified staple food crops under HarvestPlus. *Frontiers in Plant Science, 8*, 1781. https://doi.org/10.3389/fpls.2017.01781

Andersson, M., Karumbunathan, V., & Zimmermann, M. B. (2020). Global iodine status in 2020 and trends over the past decade. *The Lancet Diabetes & Endocrinology, 8*(4), 300–310. https://doi.org/10.1016/S2213-8587(19)30407-5

Andersson, M., Hulthén, L., Filipsson Nyström, H., & Åberg, M. A. (2022). Iodine deficiency in eastern European school-aged children: A regional analysis. *European Journal of Nutrition, 61*(5), 2319–2329. https://doi.org/10.1007/s00394-022-02807-0

Angeles-Agdeppa, I., Magsadia, C. R., & Capanzana, M. V. (2011). Fortified juice drink improved iron and zinc status of schoolchildren. *Asia Pacific Journal of Clinical Nutrition, 20*(4), 535–543.

Ashraf, S. A., Adnan, M., Patel, M., Siddiqui, A. J., Sachidanandan, M., Snoussi, M., & Hadi, S. (2020). Fish-based bioactives as potent nutraceuticals: Exploring the therapeutic perspective of sustainable food from the sea. *Marine Drugs, 18*, 265. https://doi.org/10.3390/md18050265

Avnee, S., Chaudhary, D. R., Jhorar, P., & Rana, R. S. (2023). Biofortification: An approach to eradicate micronutrient deficiency. *Frontiers in Nutrition, 10*, 1233070. https://doi.org/10.3389/fnut.2023.1233070

Bai, S., Zhang, M., Tang, S., Li, M., Wu, R., Wan, S., Chen, L., Wei, X., & Feng, S. (2025). Effects and impact of selenium on human health, a review. *Molecules, 30*(1), 50. https://doi.org/10.3390/molecules30010050

Bailey, R. L., West, K. P., & Black, R. E. (2015). The epidemiology of global micronutrient deficiencies. *Annals of Nutrition and Metabolism, 66*(2), 22–33. https://doi.org/10.1159/000371618

Bastos Maia, S., Rolland Souza, A. S., et al. (2019). Vitamin A and pregnancy: A narrative review. *Nutrients, 11*(3), 681. https://doi.org/10.3390/nu11030681

Bathla, S. (2015). Efficacy of double fortified salt in combating the dual problem of iron and iodine deficiency. https://krishikosh.egranth.ac.in/handle/1/5810015975

Bathla, S., & Grover, K. (2016). Effect of supplementation of double fortified salt in mid day meal programme on the nutritional status of Anaemic children. *Indian Journal Of Agricultural Biochemistry, 29*(2), 189–194. https://doi.org/10.5958/0974-4479.2016.00030.7

Behboudi, E., Zeynali, P., & Hamidi Sofiani, V. (2022). Vitamin a and viral infection: A systematic review. *Jorjani Biomedicine Journal.* https://doi.org/10.52547/jorjanibiomedj.10.1.67

Berchtold, M. W., et al. (2020). Calcium signaling in muscle contraction and disease. *Cell Calcium, 89*, 102213. https://doi.org/10.1016/j.ceca.2020.102213

Betts, J. G., et al. (2023). Calcium's role in metabolic regulation. *Journal of Endocrinology, 256*(1), R1–R15. https://doi.org/10.1530/JOE-23-0015

Bhutta, Z. A., Salam, R. A., & Das, J. K. (2018). Meeting the challenges of micronutrient malnutrition in the developing world. *British Medical Bulletin, 125*(1), 1–14. https://doi.org/10.1093/bmb/ldy011

Bo, Y., Zhu, Y., Tao, Y., Li, X., Zhai, D., Bu, Y., Wan, Z., Wang, L., Wang, Y., & Yu, Z. (2020). Association between folate and health outcomes: An umbrella review of meta-analyses. *Frontiers in Public Health, 8*, 550753. https://doi.org/10.3389/fpubh.2020.550753

Bolland, M. J., et al. (2021). Calcium supplementation: Balancing benefits and risks. *BMJ, 372*, n180. https://doi.org/10.1136/bmj.n180

Bonjour, J. P., Benoit, V., Rousseau, B., & Souberbielle, J. C. (2012). Consumption of vitamin D-and calcium-fortified soft white cheese lowers the biochemical marker of bone resorption TRAP 5b in postmenopausal women at moderate risk of osteoporosis fracture. *The Journal of Nutrition, 142*(4), 698–703. https://doi.org/10.3945/jn.111.153692

Bougma, K., Aboud, F. E., Lemma, T. M., Frongillo, E. A., & Marquis, G. S. (2018). Introduction of iodised salt benefits infants' mental development in a community-based cluster-randomised effectiveness trial in Ethiopia. *The British Journal of Nutrition, 119*(7), 801–809. https://doi.org/10.1017/S0007114517003658

Bouis, H. E., & Saltzman, A. (2017). Improving nutrition through biofortification: A review of evidence from HarvestPlus, 2003 through 2016. *Global Food Security, 12*, 49–58. https://doi.org/10.1016/j.gfs.2017.01.009

Brion, L. P., Bell, E. F., & Raghuveer, T. S. (2003). Vitamin E supplementation for prevention of morbidity and mortality in preterm infants. *The Cochrane Database of Systematic Reviews, 4*, CD003665. https://doi.org/10.1002/14651858.CD003665

Bromage, S., Ahmed, T., & Fawzi, W. W. (2016). Calcium deficiency in Bangladesh: Burden and proposed solutions for the first 1000 days. *Food and Nutrition Bulletin, 37*(4), 475–493. https://doi.org/10.1177/0379572116652748

Caffrey, A., McNulty, H., Irwin, R. E., Walsh, C. P., & Pentieva, K. (2019). Maternal folate nutrition and offspring health: Evidence and current controversies. *Proceedings of the Nutrition Society., 78*(2), 208–220. https://doi.org/10.1017/S0029665118002689

Camaschella, C. (2015). Iron-deficiency anemia. *The New England Journal of Medicine, 372*, 1832–1843. https://doi.org/10.1056/NEJMra1401038

Carr, A. C., & Zawari, M. (2023). Does aging have an impact on vitamin C status and requirements? A scoping review of comparative studies of aging and institutionalisation. *Nutrients, 15*(4), 915. https://doi.org/10.3390/nu15040915

Cassat, J. E., & Skaar, E. P. (2013). Iron in infection and immunity. *Cell Host & Microbe, 13*(5), 509–519. https://doi.org/10.1016/j.chom.2013.04.010

Centeno Tablante, E., Pachón, H., Guetterman, H. M., & Finkelstein, J. L. (2019). Fortification of wheat and maize flour with folic acid for population health outcomes. *The Cochrane Database of Systematic Reviews, 7*(7), CD012150. https://doi.org/10.1002/14651858.CD012150.pub2

Chakrabarty, S., Sinha, R. K., Kapil, U., & Pandav, C. S. (2022). Iodine status in rural India: A cross-sectional study of urinary iodine concentration. *Indian Journal of Pediatrics, 89*(7), 675–681. https://doi.org/10.1007/s12098-022-04122-2

Chanchlani, R., Nemer, P., Sinha, R., Nemer, L., Krishnappa, V., Sochett, E., Safadi, F., & Raina, R. (2020). An overview of rickets in children. *Kidney International Reports, 5*(7), 980–990. https://doi.org/10.1016/j.ekir.2020.03.025

Charlebois, E., & Pantopoulos, K. (2023). Nutritional aspects of iron in health and disease. *Nutrients, 15*(11), 2441. https://doi.org/10.3390/nu15112441

Clagett-Dame, M., & Knutson, D. (2011). Vitamin A in reproduction and development. *Clinical Nutrition, 35*(4), 859–863. https://doi.org/10.1016/j.clnu.2016.05.009

Cominetti, C., de Bortoli, M. C., Garrido, A. B., Jr., & Cozzolino, S. M. (2012). Brazilian nut consumption improves selenium status and glutathione peroxidase activity and reduces atherogenic risk in obese women. *Nutrition Research, 32*(6), 403–407. https://doi.org/10.1016/j.nutres.2012.05.005

Cosentino, N., Campodonico, J., Milazzo, V., De Metrio, M., Brambilla, M., Camera, M., & Marenzi, G. (2021). Vitamin D and cardiovascular disease: Current evidence and future perspectives. *Nutrients, 13*(10), 3603. https://doi.org/10.3390/nu13103603

Crider, K. S., Qi, Y. P., Yeung, L. F., Mai, C. T., Head Zauche, L., Wang, A., Daniels, K., & Williams, J. L. (2022). Folic acid and the prevention of birth defects: 30 years of opportunity and controversies. *Annual Review of Nutrition, 42*, 423–452. https://doi.org/10.1146/annurev-nutr-043020-091647

Cui, A., Zhang, T., Xiao, P., Fan, Z., Wang, H., & Zhuang, Y. (2023). Global and regional prevalence of vitamin D deficiency in population-based studies from 2000 to 2022: A pooled analysis of 7.9 million participants. *Frontiers in Nutrition, 10*, 1070808. https://doi.org/10.3389/fnut.2023.1070808

Da Silva Lopes, K., Yamaji, N., Rahman, M. O., Suto, M., Takemoto, Y., Garcia-Casal, M. N., & Ota, E. (2021). Nutrition-specific interventions for preventing and controlling anaemia throughout the life cycle: An overview of systematic reviews. *The Cochrane Database of Systematic Reviews, 9*(9), CD013092. https://doi.org/10.1002/14651858.CD013092.pub2

Daly, R. M., Brown, M., Bass, S., Kukuljan, S., & Nowson, C. (2006). Calcium- and vitamin D3-fortified milk reduces bone loss at clinically relevant skeletal sites in older men: A 2-year randomized controlled trial. *Journal of Bone and Mineral Research, 21*(3), 397–405. https://doi.org/10.1359/JBMR.051206

Dary, O., & Mora, J. O. (2002). Food fortification to reduce vitamin a deficiency: International vitamin a consultative group recommendations. *The Journal of Nutrition, 132*(9), 2927S–2933S. https://doi.org/10.1093/jn/132.9.2927S

Degrassi, I., Leonardi, I., Di Profio, E., Montanari, C., Zuccotti, G., & Verduci, E. (2023). Fat-soluble vitamins deficiency in pediatric cholestasis: A scoping review. *Nutrients, 15*(11), 2491. https://doi.org/10.3390/nu15112491

De-Regil, L. M., Suchdev, P. S., Vist, G. E., Walleser, S., & Peña-Rosas, J. P. (2013). Home fortification of foods with multiple micronutrient powders for health and nutrition in children under two years of age. *Evidence-Based Child Health: A Cochrane Review Journal, 8*(1), 112–201. https://doi.org/10.1002/ebch.1895

Drutel, A., Archambeaud, F., & Caron, P. (2013). Selenium and the thyroid gland: More good news for clinicians. *Clinical Endocrinology, 78*(2), 155–164. https://doi.org/10.1111/cen.12066

Dwyer, J. T., Wiemer, K. L., Dary, O., Keen, C. L., King, J. C., Miller, K. B., Philbert, M. A., Tarasuk, V., Taylor, C. L., Gaine, P. C., Jarvis, A. B., & Bailey, R. L. (2015). Fortification and health: Challenges and opportunities. *Advances in Nutrition, 6*(1), 124–131. https://doi.org/10.3945/an.114.007443

Economos, C. D., Moore, C. E., Hyatt, R. R., Kuder, J., Chen, T., Meydani, S. N., Meydani, M., Klein, E., Biancuzzo, R. M., & Holick, M. F. (2014). Multinutrient-fortified juices improve vitamin D and vitamin E status in children: A randomized controlled trial. *Journal of the Academy of Nutrition and Dietetics, 114*(5), 709–717. https://doi.org/10.1016/j.jand.2013.07.027

EFSA NDA Panel, et al. (2024). Guidance for establishing and applying tolerable upper intake levels for vitamins and essential minerals. *EFSA Journal, 22*(11), e9052. https://doi.org/10.2903/j.efsa.2024.9052

European Commission. (2006). Discussion Paper on the setting of maximum and minimum amounts for itamins and minerals in foodstuffs.

Eveleigh, E. R., Coneyworth, L., & Welham, S. J. M. (2023). Systematic review and meta-analysis of iodine nutrition in modern vegan and vegetarian diets. *The British Journal of Nutrition, 130*(9), 1580–1594. https://doi.org/10.1017/S000711452300051X

Flynn, A., Moreiras, O., Stehle, P., Fletcher, R. J., Müller, D. J. G., & Rolland, V. (2003). Vitamins and minerals: A model for safe addition to foods. *Europian Journal of Nutrition, 42*, 118–130. https://doi.org/10.1007/s00394-003-0391-9

Fowler, A. A., III, Truwit, J. D., Hite, R. D., Morris, P. E., DeWilde, C., Priday, A., Fisher, B., Thacker, L. R., II, Natarajan, R., Brophy, D. F., Sculthorpe, R., Nanchal, R., Syed, A., Sturgill, J., Martin, G. S., Sevransky, J., Kashiouris, M., Hamman, S., Egan, K. F., Hastings, A., & Halquist, M. (2019). Effect of vitamin C infusion on organ failure and biomarkers of inflammation and vascular injury in patients with sepsis and severe acute respiratory failure: The CITRIS-ALI randomized clinical trial. *JAMA, 322*(13), 1261–1270. https://doi.org/10.1001/jama.2019.11825

French Food Safety Agency (AFSSA) (2001) Report from the AFSSA Expert Committee on Human Nutrition on food fortification by vitamin and mineral: meeting the nutritional and safety needs of the consumers. Maisons Alfort, France: French Food Safety Agency (AFSSA); 8 November 2001 (transcripted version 15 January 2002). http://www.afssa.fr/Documents/NUT2000sa0239Ra.pdf. Accessed February 25, 2025.

Gallo, S., McDermid, J. M., Al-Nimr, R. I., Hakeem, R., Moreschi, J. M., Pari-Keener, M., Stahnke, B., Papoutsakis, C., Handu, D., & Cheng, F. W. (2020). Vitamin D supplementation during pregnancy: An evidence analysis center systematic review and meta-analysis. *Journal of the Academy of Nutrition and Dietetics, 120*(5), 898–924.e4. https://doi.org/10.1016/j.jand.2019.07.002

Global Burden of Disease Study. (2020). Calcium deficiency prevalence estimates. *The Lancet, 396*(10258), 1345–1422. https://doi.org/10.1016/S0140-6736(20)31404-5

Global Fortification Data Exchange. (2021). Technical brief: Global status of food fortification. http://www.fortificationdata.org

Golden, N. H., et al. (2020). Calcium needs in adolescents. *Journal of Pediatrics, 227*, 11–20. https://doi.org/10.1016/j.jpeds.2020.07.005

Green, R., et al. (2021). Maternal folate status and risk of adverse birth outcomes. *Lancet Global Health, 9*(12), e1706–e1715.

Haas, J. D., Rahn, M., Venkatramanan, S., Marquis, G. S., Wenger, M. J., Murray-Kolb, L. E., Wesley, A. S., & Reinhart, G. A. (2014). Double-fortified salt is efficacious in improving indicators of iron deficiency in female Indian tea pickers. *Journal of Nutrition, 144*(6), 957–964. https://doi.org/10.3945/JN.113.183228

Hajimohammadi, M., Shab-Bidar, S., & Neyestani, T. R. (2017). Consumption of vitamin D-fortified yogurt drink increased leptin and ghrelin levels but reduced leptin to ghrelin ratio in type 2 diabetes patients: A single blind randomized controlled trial. *European Journal of Nutrition, 56*(6), 2029–2036. https://doi.org/10.1007/s00394-017-1397-z

Harinarayan, C. V., Akhila, H., & Shanthisree, E. (2021). Modern India and dietary calcium deficiency-half a century nutrition data-retrospect-introspect and the road ahead. *Frontiers in Endocrinology, 12*, 583654. https://doi.org/10.3389/fendo.2021.583654

Hidayat, K., Chen, J. S., Wang, T. C., Liu, Y. J., Shi, Y. J., et al. (2022). The effects of milk supplementation on bone health indices in adults: A meta-analysis of randomized controlled trials. *Advances in Nutrition, 13*(4), 1186–1199. https://doi.org/10.1093/advances/nmab136

Horton, S., Gelband, H., Jamison, D., Levin, C., Nugent, R., & Watkins, D. (2017). Ranking 93 health interventions for low- and middle-income countries by cost-effectiveness. *PLoS One, 12*(8). https://doi.org/10.1371/journal.pone.0182951

Howlett, J., & Binns, N. (2009). Micronutrient landscape of Europe: Comparison of intakes and methodologies with particular regard to higher consumption. *ILSI Europe Report Series*, 1–32.

https://iris.who.int/handle/10665/43781

https://iris.who.int/handle/10665/63322

https://www.ffinetwork.org/plan-standards-and-legislation

https://www.ifpri.org/blog/addressing-global-life-long-health-impacts-zinc-deficiency-call-action/

https://www.who.int/publications/i/item/9241594012

Hurrell, R. F. (2021). Iron fortification practices and implications for iron addition to salt. *Journal of Nutrition, 151*(Suppl 1), 3S–14S. https://doi.org/10.1093/jn/nxaa175

Imdad, A., Mayo-Wilson, E., Haykal, M. R., Regan, A., Sidhu, J., Smith, A., & Bhutta, Z. A. (2022). Vitamin A supplementation for preventing morbidity and mortality in children from six months to five years of age. *Cochrane Database of Systematic Reviews, 3*, CD008524.

Imran, M., Chen, Z., Mehmood, A., Rukh, S., Weixie, W., Asghar, W., & Iftikhar, F. (2023). *Distribution of selenium in soils and human health.* Intech Open. https://doi.org/10.5772/intechopen.110636

Iodine Global Network. (2021). *Global scorecard of iodine nutrition in 2021.* IGN. https://www.ign.org/cm_data/Scorecard_2021_EN.pdf

Jani, R., Salian, N., Udipi, S., Ghugre, P., Lohia, N., Haas, J., & Boy, E. (2015). Folate status and intake of tribal Indian adolescents aged 10 to 17 years. *Food and Nutrition Bulletin, 36*(1), 14–23. https://doi.org/10.1177/156482651503600102

Jayedi, A., Rashidy-Pour, A., Parohan, M., Zargar, M. S., & Shab-Bidar, S. (2021). Dietary antioxidants, circulating antioxidant concentrations, total antioxidant capacity, and risk of all-cause mortality: A systematic review and dose-response meta-analysis of prospective observational studies. *Advances in Nutrition, 9*(6), 2008.

Jiang, Q. (2020). Natural forms of vitamin E and metabolism. *Molecular Nutrition & Food Research, 64*(3), 1900645.

Johnson, L. E. (1997). Oils, fats and margarine: Overview of technology. In *Food fortification to end micronutrient malnutrition: State of the art* (pp. 22–26). The Micronutrient Initiative, International Development Research Centre.

Johnson, J. L., Mistry, V. V., Vukovich, M. D., Hogie-Lorenzen, T., Hollis, B. W., & Specker, B. L. (2005). Bioavailability of vitamin D from fortified process cheese and effects on vitamin D status in the elderly. *Journal of Dairy Science, 88*(7). https://doi.org/10.3168/jds.S0022-0302(05)72907-6

Jones, E., McLean, R., Davies, B., Hawkins, R., Meiklejohn, E., Ma, Z. F., & Skeaff, S. (2016). Adequate iodine status in New Zealand school children post-fortification of bread with iodised salt. *Nutrients, 8*(5), 298. https://doi.org/10.3390/nu8050298

Jones, G. D., Droz, B., Greve, P., Gottschalk, P., Poffet, D., McGrath, S. P., Seneviratne, S. I., Smith, P., & Winkel, L. H. (2017). Selenium deficiency risk predicted to increase under future climate change. *Proceedings of the National Academy of Sciences of the United States of America, 114*(11), 2848–2853. https://doi.org/10.1073/pnas.1611576114

Jones, L., et al. (2022). Impact of "5+ a Day" campaigns on vitamin C intake. *Public Health Nutrition, 25*(6), 1567–1575.

Joosten, E. (2018). Iron deficiency anemia in older adults: A review. *Geriatrics & Gerontology International, 18*(3), 373–379. https://doi.org/10.1111/ggi.13194

Kancherla, V. (2023). Neural tube defects: A review of global prevalence, causes, and primary prevention. *Child's Nervous System, 39*(7), 1703–1710. https://doi.org/10.1007/s00381-023-05910-7

Keats, E. C., Haider, B. A., Tam, E., & Bhutta, Z. A. (2019). Multiple-micronutrient supplementation for women during pregnancy. *The Cochrane Database of Systematic Reviews, 3*(3), CD004905. https://doi.org/10.1002/14651858.CD004905.pub6

Kedir, S., Hassen, K., Mohammed, B., & Ademe, B. W. (2024). Weekly iron-folic acid supplementation and its impact on children and adolescents' iron status, mental health, and school performance: A systematic review and meta-analysis in sub-Saharan Africa. *BMJ Open, 14*(6), e084033. https://doi.org/10.1136/bmjopen-2024-084033

Khadgawat, R., Marwaha, R. K., Garg, M. K., Ramot, R., Oberoi, A. K., Sreenivas, V., et al. (2013). Impact of vitamin D fortified milk supplementation on vitamin D status of healthy school children aged 10–14 years. *Osteoporosis International, 24*, 2335–2343.

Kiouri, D. P., Chasapis, C. T., Mavromoustakos, T., Spiliopoulou, A. C., & Stefanidou, M. E. (2025). Zinc and its binding proteins: Essential roles and therapeutic potential. *Archives of Toxicology, 99*, 23–41. https://doi.org/10.1007/s00204-024-03891-3

Kloosterman, J., Fransen, H. P., de Stoppelaar, V. H., & Rompleberge, C. (2007). Safe addition of vitamins and minerals to foods: Setting maximum levels for fortification in The Netherlands. *Europian Journal of Nutrition, 46*, 220–229. https://doi.org/10.1007/s00394-007-0654-y

Knight, F. A., Ferguson, E. L., Rana, Z. H., Belizán, J. M., Gomes, F., Bourassa, M. W., Dickin, K. L., Weaver, C. M., & Cormick, G. (2023). Including calcium-fortified water or flour in modeled diets based on local foods could improve calcium intake for women, adolescent girls, and young children in Bangladesh, Uganda, and Guatemala. *Annals of the New York Academy of Sciences, 1526*.

Konikoff, M. R., & Denson, L. A. (2006). Role of fecal calprotectin as a biomarker of intestinal inflammation in inflammatory bowel disease. *Inflammatory Bowel Diseases, 12*(6), 524–534. https://doi.org/10.1097/00054725-200606000-00013

Krebs, N. F., Westcott, J. L., & Miller, L. V. (2017). Zinc deficiency in infants and children: A review of its complex and synergistic interactions. *Pediatric Research, 82*(4), 616–620. https://doi.org/10.1038/pr.2017.138

Kujinga, P., Galetti, V., Onyango, E., Jakab, V., Buerkli, S., Andang'o, P., Brouwer, I. D., Zimmermann, M. B., & Moretti, D. (2018). Effectiveness of zinc-fortified water on zinc intake, status and morbidity in Kenyan pre-school children: A randomised controlled trial. *Public Health Nutrition, 21*(15), 2855–2865. https://doi.org/10.1017/S1368980018001441

Lammie, P., Milner, T., & Houston, R. (2007). Unfulfilled potential: Using diethylcarbamazine-fortified salt to eliminate lymphatic filariasis. *Bulletin of the World Health Organization, 85*, 545–549.

Larson, L. M., Cyriac, S., Djimeu, E. W., Mbuya, M. N. N., et al. (2021). Can double fortification of salt with iron and iodine reduce anemia, iron deficiency anemia, iron deficiency, iodine deficiency, and functional outcomes? Evidence of efficacy, effectiveness, and safety. *Journal of Nutrition, 51*. https://doi.org/10.1093/JN/NXAA192

Le Port, A., Bernard, T., Hidrobo, M., Birba, O., Rawat, R., & Ruel, M. T. (2017). Delivery of iron-fortified yoghurt, through a dairy value chain program, increases hemoglobin concentration among children 24 to 59 months old in northern Senegal: A cluster-randomized control trial. *PLoS One, 12*(2), e0172198. https://doi.org/10.1371/journal.pone.0172198

Lee, J. T., Moore, C. E., & Radcliffe, J. D. (2015). Consumption of calcium-fortified cereal bars to improve dietary calcium intake of healthy women: Randomized controlled feasibility study. *PLoS One, 10*(5). https://doi.org/10.1371/JOURNAL.PONE.0125207

Lewis, E. D., Meydani, S. N., & Wu, D. (2019). Regulatory role of vitamin E in the immune system and inflammation. *IUBMB Life, 71*(4), 487–494. https://doi.org/10.1002/iub.1976

Lin, K., Qi, Y., & Sun, J. (2025). Trend and burden of vitamin a deficiency in 1990–2021 and projection to 2050: A systematic analysis for the global burden of disease study 2021. *Nutrients, 17*(3), 572. https://doi.org/10.3390/nu17030572

Lips, P., Cashman, K. D., Lamberg-Allardt, C., Bischoff-Ferrari, H. A., Obermayer-Pietsch, B., Bianchi, M. L., Stepan, J., El-Hajj Fuleihan, G., & Bouillon, R. (2019). Current vitamin D status in European and Middle East countries and strategies to prevent vitamin D deficiency: A position statement of the European calcified tissue society. *European Journal of Endocrinology, 180*(4), 23–54. https://doi.org/10.1530/EJE-18-0736

Lowe, N. M., Hall, A. G., Broadley, M. R., Foley, J., Boy, E., & Bhutta, Z. A. (2024). Preventing and controlling zinc deficiency across the life course: A call to action. *Advances in Nutrition, 15*(3), 100181.

Lykkesfeldt, J., & Carr, A. (2023). Vitamin C – A scoping review for Nordic nutrition recommendations. *Food & Nutrition Research, 67*. https://doi.org/10.29219/fnr.v67.10300

Mahmoud, F. M. (2024). Iron deficiency anemia in infants: Review article. *American Journal of Medical and Clinical Research & Reviews, 3*(10), 1–6. https://doi.org/10.58372/2835-6276.1215

Malik, A., Eggersdorfer, M., & Trilok-Kumar, G. (2021). Vitamin E status in healthy population in Asia: A review of current literature. *International Journal for Vitamin and Nutrition Research, 91*(3–4), 356–369. https://doi.org/10.1024/0300-9831/a000590

Mannar, M. G. V., Garrett, G. S., & Hurrell, R. F. (2018). Future trends and strategies in food fortification. In M. G. Venkatesh Mannar & R. F. Hurrell (Eds.), *Food fortification in a globalized world* (pp. 375–381). Academic Press. https://doi.org/10.1016/B978-0-12-802861-2.00040-7

Manson, J. E., Cook, N. R., Lee, I. M., Christen, W., Bassuk, S. S., Mora, S., Gibson, H., Gordon, D., Copeland, T., D'Agostino, D., Friedenberg, G., Ridge, C., Bubes, V., Giovannucci, E. L., Willett, W. C., Buring, J. E., et al. (2019). Vitamin D supplements and prevention of cancer and cardiovascular disease. *The New England Journal of Medicine, 380*(1), 33–44. https://doi.org/10.1056/NEJMoa1809944

Market Analysis Report. (2025). Food fortifying agents market size, share & trends analysis report by type (Vitamins, minerals, proteins & amino acids, carbohydrates, prebiotics & probiotics, lipids), by application, by region, and segment forecasts, 2024–2030. Available online at: https://www.grandviewresearch.com/industry-analysis/food-fortifying-agents-market-report. Accessed 23 Feb 2025.

Martineau, A. R., Jolliffe, D. A., Hooper, R. L., Greenberg, L., Aloia, J. F., Bergman, P., Dubnov-Raz, G., Esposito, S., Ganmaa, D., Ginde, A. A., Goodall, E. C., Grant, C. C., Griffiths, C. J., Janssens, W., Laaksi, I., Manaseki-Holland, S., Mauger, D., Murdoch, D. R., Neale, R., Rees, J. R., Simpson, S., Stelmach, I., Kumar, G. T., Urashima, M., & Camargo, C. A. (2021). Vitamin D supplementation to prevent acute respiratory infections: Systematic review and meta-analysis of individual participant data. *BMJ, 372*, n232. https://doi.org/10.1136/bmj.n232

Mayo-Wilson, E., Junior, J. A., Imdad, A., Dean, S., Chan, X. H., Chan, E. S., Jaswal, A., & Bhutta, Z. A. (2014). Zinc supplementation for preventing mortality, morbidity, and growth failure in children aged 6 months to 12 years of age. *The Cochrane Database of Systematic Reviews, 5*, CD009384. https://doi.org/10.1002/14651858.CD009384.pub2

McDougall, A. R., Dore, G., Aboud, L., Makama, M., Nguyen, P. Y., Mills, K., Sanderson, B., Hastie, R., Ammerdorffer, A., & Vogel, J. P. (2023). The effect of selenium supplementation in pregnant women on maternal, fetal, and newborn outcomes: A systematic review and meta-analysis. *American journal of obstetrics & gynecology MFM, 5*(11), 101160. https://doi.org/10.1016/j.ajogmf.2023.101160

Micha, R., et al. (2020). Food fortification strategies. *Nutrition Reviews, 78*(12), 1017–1030. https://doi.org/10.1093/nutrit/nuaa048

Mkambula, P., Mbuya, M. N. N., Rowe, L. A., Sablah, M., Friesen, V. M., Chadha, M., Osei, A. K., Ringholz, C., Vasta, F. C., & Gorstein, J. (2020). The unfinished agenda for food fortification in low- and middle-income countries: Quantifying progress, gaps and potential opportunities. *Nutrients, 12*(2), 354. https://doi.org/10.3390/nu12020354

Moghissi, A., Pei, S., & Liu, Y. (2015). Golden rice: Scientific, regulatory and public information processes of a genetically modified organism. *Critical Reviews in Biotechnology, 21*, 1–7. https://doi.org/10.3109/07388551.2014.993586

Mohammed, H., Marquis, G. S., Aboud, F., Bougma, K., & Samuel, A. (2020). Pre-pregnancy iodized salt improved children's cognitive development in randomized trial in Ethiopia. *Maternal & Child Nutrition, 16*(3), e12943. https://doi.org/10.1111/mcn.12943

Moretti, D., Zimmermann, M. B., Muthayya, S., Thankachan, P., Lee, T.-C., Kurpad, A. V., & Hurrell, R. F. (2006). Extruded rice fortified with micronized ground ferric pyrophosphate reduces iron deficiency in Indian schoolchildren: A double-blind randomized controlled trial. *The American Journal of Clinical Nutrition, 84*(4), 1012–1017. https://doi.org/10.1093/ajcn/84.4.822

Moretti, D., Zimmermann, M. B., Muthayya, S., Thankachan, P., Lee, T. C., Kurpad, A. V., & Hurrell, R. F. (2016). Extruded rice fortified with micronized ground ferric pyrophosphate reduces iron deficiency in Indian schoolchildren: A double-blind randomized controlled trial. *American Journal of Clinical Nutrition, 103*, 1249–1256. https://doi.org/10.3945/ajcn.115.126656

Mostafai, R., Mohammadi, R., Nachvak, S. M., Rezaei, M., Pasdar, Y., Abdollahzad, H., Rezvanmadani, F., Moradi, S., Morvaridzadeh, M., Niazi, P., & Adeli, K. (2018). Fortified yogurt with vitamin D as a cost-effective food to prevent diabetes: A randomized double-blind clinical trial. *Journal of Functional Foods, 42*, 137–145. https://doi.org/10.1016/j.jff.2018.01.001

Muthayya, S., Thankachan, P., Hirve, S., Amalrajan, V., Thomas, T., Lubree, H., Agarwal, D., Srinivasan, K., Hurrell, R. F., Yajnik, C. S., et al. (2012a). Iron fortification of whole wheat flour reduces iron deficiency and iron deficiency anemia and increases body iron stores in Indian school-aged children. *Journal of Nutrition, 142*, 1997–2003.

Muthayya, S., Thankachan, P., Hirve, S., Amalrajan, V., Thomas, T., Lubree, H., Agarwal, D., Srinivasan, K., Hurrell, R. F., Yajnik, C. S., & Kurpad, A. V. (2012b). Iron fortification of whole wheat flour reduces iron deficiency and iron deficiency anemia and increases body iron Stores in Indian. *The Journal of Nutrition, 142*(11), 1997–2003. https://doi.org/10.3945/jn.111.155135

Muzembo, B. A., Ngatu, N. R., Januka, K., Huang, H. L., Nattadech, C., Suzuki, T., Wada, K., & Ikeda, S. (2019). Selenium supplementation in HIV-infected individuals: A systematic review of randomized controlled trials. *Clinical Nutrition ESPEN, 34*, 1–7. https://doi.org/10.1016/j.clnesp.2019.09.005

Nalder, L., Zheng, B., Chiandet, G., Middleton, L. T., & de Jager, C. A. (2021). Vitamin B12 and folate status in cognitively healthy older adults and associations with cognitive performance. *The Journal of Nutrition, Health & Aging, 25*(3), 287–294. https://doi.org/10.1007/s12603-020-1489-y

Nazeri, P., Mirmiran, P., Tahmasebinejad, Z., Hedayati, M., Delshad, H., & Azizi, F. (2017). The effects of iodine fortified milk on the iodine status of lactating mothers and infants in an area with a successful salt iodization program: A randomized controlled trial. *Nutrients, 9*(2), 180. https://doi.org/10.3390/nu9020180

Neill, H. R., Gill, C. I. R., McDonald, E. J., McRoberts, W. C., & Pourshahidi, L. K. (2021). The future is bright: Biofortification of common foods can improve vitamin D status. *Critical Reviews in Food Science and Nutrition, 63*(4), 505–521. https://doi.org/10.1080/10408398.2021.1950609

Neufeld, L. M., Aaron, G. J., Garrett, G. S., Baker, S. K., Dary, O., & Van Ameringen, M. (2016). Food fortification for impact: A data-driven approach. *Bulletin of the World Health Organization, 94*, 631–632. https://doi.org/10.2471/BLT.15.164756

Neyestani, T. R., Hajifaraji, M., Omidvar, N., Nikooyeh, B., Eshraghian, M. R., Shariatzadeh, N., & Asadzadeh, S. (2014). Calcium-vitamin D-fortified milk is as effective on circulating bone biomarkers as fortified juice and supplement but has less acceptance: A randomised controlled school-based trial. *Journal of Human Nutrition and Dietetics, 27*(6), 606–616.

Nikooyeh, B., Neyestani, T. R., Farvid, M., Alavi-Majd, H., Houshiarrad, A., Kalayi, A., Shariatzadeh, N., Gharavi, A., Heravifard, S., Tayebinejad, N., & Salekzamani, S. (2011). Daily consumption of vitamin D– or vitamin D + calcium–fortified yogurt drink improved glycemic control in patients with type 2 diabetes: A randomized clinical trial. *The American Journal of Clinical Nutrition, 93*(4), 764–771. https://doi.org/10.3945/ajcn.110.007336

Nikooyeh, B., Hollis, B. W., & Neyestani, T. R. (2021). The effect of daily intake of vitamin D-fortified yogurt drink, with and without added calcium, on serum adiponectin and sirtuins 1 and 6 in adult subjects with type 2 diabetes. *Nutrition & Diabetes, 11*(1), 26. https://doi.org/10.1038/s41387-021-00168-x

Ning, Z., Song, S., Miao, L., Zhang, P., Wang, X., Liu, J., Hu, Y., Xu, Y., Zhao, T., Liang, Y., Wang, Q., Liu, L., Zhang, J., Hu, L., Huo, M., & Zhou, Q. (2016). High prevalence of vitamin D deficiency in urban health checkup population. *Clinical nutrition (Edinburgh, Scotland), 35*(4), 859–863. https://doi.org/10.1016/j.clnu.2015.05.019

Noubactep, C., Kenmogne-Tchidjo, J. F., & Vollmer, S. (2023). Iron-fortified water: A new approach for reducing iron deficiency anemia in resource-constrained settings. *Dental Science Reports, 13*. https://doi.org/10.1038/s41598-023-40600-z

O'Kane, S. M., Pourshahidi, L. K., Mulhern, M. S., Strain, J. J., Mackle, E. M., Koca, D., Schomburg, L., Hill, S., O'Reilly, J., Kmiotek, D., Deitrich, C., Bath, S. C., & Yeates, A. J. (2018). Cow Milk consumption increases iodine status in women of childbearing age in a randomized controlled trial. *The Journal of Nutrition, 148*(3), 401–408.

Olson, R., Gavin-Smith, B., Ferraboschi, C., & Kraemer, K. (2021). Food fortification: The advantages, disadvantages and lessons from sight and life programs. *Nutrients, 13*(4), 1118. https://doi.org/10.3390/nu13041118

Pachón, H., Reynolds, B., Duong, M., Tsang, B. L., Childs, L., Luthringer, C. L., Kang, Y., Vasta, F. C., & Codling, K. (2021). The potential contribution of fortified maize flour, oil, rice, salt, and wheat flour to estimated average requirements and tolerable upper intake levels for 15 nutrients in 153 countries. *Nutrients, 13*(2), 579. https://doi.org/10.3390/nu13020579

Paine, J. A., Shipton, C. A., Chaggar, S., Howells, R. M., Kennedy, M. J., Vernon, G., Wright, S. Y., Hinchliffe, E., Adams, J. L., Silverstone, A. L., & Drake, R. (2015). Improving the nutritional value of Golden Rice through increased pro-vitamin a content. *Nature Biotechnology., 33*, 294–300. https://doi.org/10.1038/nbt.3147

Passarelli, S., Free, C. M., Shepon, A., Beal, T., Batis, C., & Golden, C. D. (2024). Global estimation of dietary micronutrient inadequacies: A modelling analysis. *The Lancet. Global Health, 12*(10), e1590–e1599. https://doi.org/10.1016/S2214-109X(24)00276-6

Pearce, E. N., Lazarus, J. H., Moreno-Reyes, R., & Zimmermann, M. B. (2016). Consequences of iodine deficiency and excess in pregnant women: An overview of current knowns and unknowns. *The American Journal of Clinical Nutrition, 104*(Suppl 3), 918S–923S. https://doi.org/10.3945/ajcn.115.110429

Pieroth, R., Paver, S., Day, S., & Lammersfeld, C. (2018). Folate and its impact on cancer risk. *Current Nutrition Reports, 7*(3), 70–84. https://doi.org/10.1007/s13668-018-0237-y

Pinkaew, S., Udomkesmalee, E., Davis, C. R., & Tanumihardjo, S. A. (2021). Vitamin A-fortified rice increases total body vitamin a stores in lactating Thai women measured by retinol isotope dilution: A double-blind, randomized, controlled trial. *The American Journal of Clinical Nutrition, 113*(5), 1372–1380. https://doi.org/10.1093/ajcn/nqaa418

Pullar, J. M., Carr, A. C., & Vissers, M. C. M. (2017). The roles of vitamin C in skin health. *Nutrients, 9*(8), 866. https://doi.org/10.3390/nu9080866

Ramachandran, P., Prabhakar, K., Kumari, H., & Kalaivani, K. (2023). Impact of the use of iron-fortified iodized salt on hemoglobin levels: A community based open randomized trial. *International Journal of Community Medicine and Public Health, 10*(6), 2197–2207. https://doi.org/10.18203/2394-6040.ijcmph20231703

Ramírez-Luzuriaga, M. J., Larson, L. M., Mannar, V., & Martorell, R. (2018). Impact of double-fortified salt with iron and iodine on hemoglobin, anemia, and iron deficiency anemia: A systematic review and meta-analysis. *Advances in Nutrition, 9*(3), 207–218. https://doi.org/10.1093/advances/nmy008

Rasmussen, S. E., Andersen, N. L., Dragsted, L. O., & Larsen, J. C. (2006). A safe strategy for addition of vitamins and minerals to foods. *Europian Journal of Nutrition, 45*, 123–135. https://doi.org/10.1007/s00394-005-0580-9

Rayman, M. P. (2020). Selenium intake, status, and health: A complex relationship. *Hormones (Athens, Greece), 19*(1), 9–14. https://doi.org/10.1007/s42000-019-00125-5

Ritu, G., & Gupta, A. (2014). Vitamin D deficiency in India: Prevalence, causalities and interventions. *Nutrients, 6*(2), 729–775. https://doi.org/10.3390/nu6020729

Rogers, L. M., Cordero, A. M., Pfeiffer, C. M., Hausman, D. B., Tsang, B. L., De-Regil, L. M., Rosenthal, J., Razzaghi, H., Wong, E. C., Weakland, A. P., & Bailey, L. B. (2018). Global folate status in women of reproductive age: A systematic review with emphasis on methodological issues. *Annals of the New York Academy of Sciences, 1431*(1), 35–57. https://doi.org/10.1111/nyas.13963

Rohner, F., Wirth, J. P., Zeng, W., Petry, N., Donkor, W. E. S., Neufeld, L. M., Mkambula, P., Groll, S., Mbuya, M. N., & Friesen, V. M. (2023). Global coverage of mandatory large-scale food fortification programs: A systematic review and meta-analysis. *Advances in Nutrition, 14*(5), 1197–1210. https://doi.org/10.1016/j.advnut.2023.07.004

Ross, A., & Harrison, E. H. (2007). Vitamin A: Nutritional aspects of retinoids and carotenoids. In *Handbook of vitamins* (pp. 1–40). CRC Press.

Rowe, S., & Carr, A. C. (2020). Global vitamin C status and prevalence of deficiency: A cause for concern? *Nutrients, 12*(7), 2008. https://doi.org/10.3390/nu12072008

Saari, J. (2016). Vitamin a and vision. In *The biochemistry of retinoid signaling II* (Vol. 81, pp. 231–259). Springer. https://doi.org/10.1007/978-94-024-0945-1_9

Salam, A., Briawan, D., Martianto, D., Thaha, A. R., & Virani, D. (2021). Effect of vitamin a supplementation, cooking oil fortification, and nutrition education to postpartum mother on breast milk retinol levels. *Open Access Macedonian Journal of Medical Sciences, 9*, 823–827. https://doi.org/10.3889/OAMJMS.2021.7006

Sandalinas, F., Filteau, S., Joy, E. J. M., Segovia de la Revilla, L., MacDougall, A., & Hopkins, H. (2023). Measuring the impact of malaria infection on indicators of iron and vitamin a status: A systematic literature review and meta-analysis. *The British Journal of Nutrition, 129*(1), 87–103. https://doi.org/10.1017/S0007114522000757

Sandjaja, I. J., Jahari, A. B., Htet, M. K., Tilden, R., Soekarjo, D., Utomo, B., Moench-Pfanner, R., & Korenromp, E. L. (2015). Vitamin A-fortified cooking oil reduces vitamin A deficiency in infants, young children and women: Results from a programme evaluation in Indonesia. *Public Health Nutrition, 18*(14), 2511–2522. https://doi.org/10.1017/S136898001400322X

Sayed, A. R., Bourne, D., Pattinson, R., Nixon, J., & Henderson, B. (2008). Decline in the prevalence of neural tube defects following folic acid fortification and its cost-benefit in South Africa. *Birth Defects Research. Part A, Clinical and Molecular Teratology, 82*(4), 211–216. https://doi.org/10.1002/bdra.20442

Sazawal, S., Dhingra, U., Dhingra, P., Hiremath, G., Sarkar, A., Dutta, A., et al. (2010). Micronutrient fortified milk improves iron status, anemia and growth among children 1–4 years: A double masked, randomized, controlled trial. *PLoS One, 5*(8), e12167. https://doi.org/10.1371/journal.pone.0012167

Sharifan, P., Sahranavard, T., Rashidmayvan, M., Darroudi, S., Fard, M. V., Mohammadhasani, K., Mansoori, A., Eslami, S., Safarian, M., Afshari, A., Bagherniya, M., Tayefi, M., Khadem-Rezaiyan, M., Ghazizadeh, H., Ferns, G., Rezaie, M., & Ghayour-Mobarhan, M. (2024). Effect of dairy products fortified with vitamin d3 on restless legs syndrome in women with premenstrual syndrome, abdominal obesity and vitamin d deficiency: A pilot study. *BMC Women's Health, 24*(1), 434. https://doi.org/10.1186/s12905-024-03159-z

Sharma, A., & Sharma, U. S. (2019). Liposomes in drug delivery: Progress and limitations. *International Journal of Pharmaceutics, 565*, 293–307. https://doi.org/10.1016/j.ijpharm.2019.05.008

Sharma, S., Smith, M. E., Reimer, J., O'Brien, D. B., Brissau, J. M., Donahue, M. C., Carter, C. E., & Michael, E. (2019). Economic performance and cost-effectiveness of using a DEC-salt social enterprise for eliminating the major neglected tropical disease, lymphatic filariasis. *PLOS Neglected Tropical disease, 13*, e0007094.

Skeaff, S. A. (2011). Iodine deficiency in pregnancy: The effect on neurodevelopment in the child. *Nutrients, 3*(2), 265–273. https://doi.org/10.3390/nu3020265

Skeaff, S. A., & Lonsdale-Cooper, E. (2013). Mandatory fortification of bread with iodised salt modestly improves iodine status in schoolchildren. *The British Journal of Nutrition, 109*(6), 1109–1113. https://doi.org/10.1017/S0007114512003236

Solon, F. S. (1997). The progress of wheat flour fortification with vitamin A in The Philippines. In *Food fortification to end micronutrient malnutrition: State of the art* (pp. 66–71). The Micronutrient Initiative, International Development Research Centre.

Sorrenti, S., Baldini, E., Pironi, D., Lauro, A., D'Orazi, V., Tartaglia, F., Tripodi, D., Lori, E., Gagliardi, F., Praticò, M., Illuminati, G., D'Andrea, V., Palumbo, P., & Ulisse, S. (2021). Iodine: Its role in thyroid hormone biosynthesis and beyond. *Nutrients, 13*(12), 4469. https://doi.org/10.3390/nu13124469

Stevens, G. A., Beal, T., Mbuya, M. N. N., Luo, H., & Neufeld, L. M. (2022a). Micronutrient deficiencies among preschool-aged children and women of reproductive age worldwide: A pooled analysis of individual-level data from population-representative surveys. *The Lancet Global Health, 10*(11), e1590–e1599. https://doi.org/10.1016/S2214-109X(22)00367-9

Stevens, G. A., Paciorek, C. J., Flores-Urrutia, M. C., Borghi, E., Namaste, S., Wirth, J. P., Suchdev, P. S., Ezzati, M., Rohner, F., Flaxman, S. R., & Rogers, L. M. (2022b). National, regional, and global estimates of anaemia by severity in women and children for 2000–19: A pooled analysis

of population-representative data. *The Lancet Global Health, 10*(5), e627–e639. https://doi.org/10.1016/S2214-109X(22)00084-5
Straits Research. (2025) Food Fortifying Agents Market Size, Share & Trends Analysis Report by Type (Vitamins, Minerals, Proteins & Amino Acids, Carbohydrates, Prebiotics & Probiotics, Lipids), By Application, By Region, And Segment Forecasts, 2024–2030. May 22, 2024 Available online at: https://straitsresearch.com/report/food-fortification-market#:~:text=The%20global%20food%20fortification%20market,concerns%20about%20micro%2Dnutrient%20deficiency. Accessed on Feb. 23, 2025.
Sudikno, S., Sandjaja, S., et al. (2017). The impact of vitamin a fortified vegetable oil on vitamin a status of children under five years of age: A cohort study. *Health Science Journal of Indonesia, 8*(2), 102–110. https://doi.org/10.22435/HSJI.V8I2.6792.102-110
Szymlek-Gay, E. A., Gray, A. R., Heath, A. M., Ferguson, E. L., Edwards, T., & Skeaff, S. A. (2020). Iodine-fortified toddler milk improves dietary iodine intakes and iodine status in toddlers: A randomised controlled trial. *European Journal of Nutrition, 59*(3), 909–919. https://doi.org/10.1007/s00394-019-01950-5
Ulfberg, J., & Stehlik, R. (2021). Finland's handling of selenium is a model in these times of coronavirus infections. *The British Journal of Nutrition, 125*(12), 1437–1438. https://doi.org/10.1017/S0007114520003827
UNICEF. (2025) *Iodine*. UNICEF Data. Retrieved February 20, 2025, from https://data.unicef.org/topic/nutrition/iodine/
Vannucci, L., Fossi, C., Quattrini, S., Guasti, L., Pampaloni, B., Gronchi, G., Giusti, F., Romagnoli, C., Cianferotti, L., Marcucci, G., & Brandi, M. L. (2018). Calcium intake in bone health: A focus on calcium-rich mineral waters. *Nutrients, 10*(12), 1930. https://doi.org/10.3390/nu10121930
Vishwakarma, S., Genu Dalbhagat, C., Mandliya, S., & Niwas Mishra, H. (2022). Investigation of natural food fortificants for improving various properties of fortified foods: A review. *Food Research International, 156*, 111186. https://doi.org/10.1016/j.foodres.2022.111186
Wang, Y., Zhao, N., Qiu, J., He, X., Zhou, M., Cui, H., Lv, L., Lin, X., Zhang, C., Zhang, H., Xu, R., Zhu, D., Dang, Y., Han, X., Zhang, H., Bai, H., Chen, Y., Tang, Z., Lin, R., Yao, T., et al. (2015). Folic acid supplementation and dietary folate intake, and risk of preeclampsia. *European Journal of Clinical Nutrition, 69*(10), 1145–1150. https://doi.org/10.1038/ejcn.2014.295
Wenger, M. J., Wenger, M. J., Murray-Kolb, L. E., Nevins, J. E. H., Venkatramanan, S., Venkatramanan, S., Reinhart, G. A., Wesley, A. S., & Haas, J. D. (2017). Consumption of a double-fortified salt affects perceptual, attentional, and mnemonic functioning in women in a randomized controlled trial in India. *Journal of Nutrition, 147*(12), 2297–2308. https://doi.org/10.3945/JN.117.251587
World Food Program. (2022) Food fortification: An effective and safe way to fight micronutrient malnutrition and its consequences.
World Health Organization. (2020). *Mobilizing ambitious and impactful commitments for mainstreaming nutrition in health systems: Nutrition in universal health coverage, global nutrition summit*. World Health Organization.
World Health Organization & Food and Agriculture Organization of the United Nations. (2006). *Guidelines on food fortification with micronutrients*. Editted by L. Allen, B. de Benoist, O. Dary, & R. Hurrell. WHO Press.
World Health Organization Executive Board. (2023). Accelerating efforts for preventing micronutrient deficiencies and their consequences, including spina bifida and other neural tube defects, through safe and effective food fortification. Draft decision proposed by Australia, Brazil, Canada, Chile, Colombia, Ecuador, the European Union and its 27 Member States, Israel, Malaysia, and Paraguay. https://apps.who.int/gb/ebwha/pdf_files/EB152/B152_CONF5-en.pdf
Zimmermann, M. B. (2011). The role of iodine in human growth and development. *Seminars in Cell & Developmental Biology, 22*(6), 645–652. https://doi.org/10.1016/j.semcdb.2011.07.009

Microorganisms as Nutrient Factories: Harnessing Bio-fermentation for Food Fortification

5

Patrick Ndovie, Smith G. Nkhata, Nyadani Jali, Gift Chisapo, Marion Sanuka, Lizzie Saka, Kondwani Kammwamba, Sydney Namaumbo, and Justice Munthali

Abstract

Bio-fermentation, a process leveraging the metabolic capabilities of microorganisms, has emerged as a sustainable and efficient strategy for food fortification. Microbial fermentation enhances the nutritional profile of food by synthesizing essential nutrients such as vitamins, amino acids, and bioavailable minerals. This process involves utilizing beneficial microbes, including Lactobacillus, Saccharomyces, and Aspergillus species, to produce fort ified food products that cater to diverse dietary needs and combat micronutrient deficiencies. In addition to nutrient enhancement, bio-fermentation improves food safety, shelf life, and sensory attributes, ensuring consumer acceptance. The approach aligns with global efforts to address malnutrition, particularly in regions with limited access to nutrient-rich foods. This review explores the potential of microorganisms as

P. Ndovie (✉)
Department of Human Nutrition and Health, Lilongwe University of Agriculture and Natural Resources, Lilongwe, Malawi

Department of Agriculture and Food Systems, Natural Resources College, Lilongwe University of Agriculture and Natural Resources, Lilongwe, Malawi
e-mail: pndovie@luanar.ac.mw

S. G. Nkhata · N. Jali · G. Chisapo · M. Sanuka · L. Saka · K. Kammwamba
Department of Agriculture and Food Systems, Natural Resources College, Lilongwe University of Agriculture and Natural Resources, Lilongwe, Malawi

S. Namaumbo
Department of Land and Water Resources, Natural Resources College, Lilongwe University of Agriculture and Natural Resources, Lilongwe, Malawi

J. Munthali
Alliance of Bioversity International & CIAT, Malawi, Agricultural Research Station, Lilongwe, Malawi

G. Sanghvi et al. (eds.), *Sustainable Food Fortification*,
https://doi.org/10.1007/978-981-95-1217-1_5

bio-factories, delving into the mechanisms of nutrient synthesis, advancements in fermentation technology, and applications in fortifying staples like dairy, cereals, and beverages. It also discusses challenges such as strain selection, scalability, and regulatory considerations in implementing bio-fermentation at an industrial scale. Harnessing the power of microbial metabolism could revolutionize food fortification, offering innovative solutions to promote health and nutrition globally.

Keywords

Bio-fermentation · Microbial nutrient synthesis · Food fortification · Nutritional enhancement · Sustainable nutrition

5.1 Introduction

The rising problem of micronutrient deficiencies has made food fortification a key strategy for improving global nutrition (Darnton-Hill & Nalubola, 2002; Dwyer et al., 2015; Faber et al., 2014; García-Casal, 2014; Pritwani & Mathur, 2015). Malnutrition, including a lack of important vitamins and minerals, continues to affect millions of people worldwide, leading to health issues like weaker immune systems, cognitive problems, and a higher risk of diseases (Rajput, 2012; Reinhardt & Fanzo, 2014). Traditional food fortification methods, which usually involve adding synthetic nutrients to food, have been effective but also have some drawbacks. These include lower absorption by the body, high costs, and consumer resistance to foods that are artificially enhanced (Chadare et al., 2019; Ofori et al., 2022).

On the other hand, bio-fermentation, a method that uses the metabolic abilities of microorganisms, has become an innovative and sustainable way to improve food's nutritional value (Praveen & Brogi, 2025a; Sawant et al., 2025; Voidarou et al., 2021). Fermentation is a natural process that has been used for centuries to make food easier to digest, safer to eat, and longer lasting (Siddiqui et al., 2023; Xiang et al., 2019). Microorganisms like *Lactobacillus*, *Saccharomyces*, and *Aspergillus* can produce essential nutrients, such as vitamins, amino acids, and minerals, making them useful for fortifying food (Anumudu et al., 2024; Rusu et al., 2023; Tamang et al., 2016). Unlike synthetic fortification, which simply adds nutrients to food, bio-fermentation increases the nutrient content naturally, making them more stable and easier for the body to absorb (Abbaspour, 2024a, b; Knez et al., 2023; Emmanuel et al., 2024; Siddiqui et al., 2023). Bio-fermentation also provides a cost-effective and environmentally friendly alternative to traditional fortification methods. This process reduces the need for synthetic additives, many of which are made from nonrenewable resources or require energy-intensive production (Bamidele et al., 2025; Kovalchuk, 2021). Furthermore, bio-fermentation fits with the global trend toward natural and minimally processed foods, meeting consumer demand for clean-label products. It also helps improve food digestion by breaking down complex compounds into more absorbable forms and reduces antinutritional factors, like phytates and tannins, that can block nutrient absorption (Anumudu et al., 2024; Siddiqui et al., 2023).

The potential of bio-fermentation to improve food security is also important. In areas where access to nutritious foods is limited due to factors like cost, climate change, or geographic isolation, fermentation can be used to boost the nutritional value of locally available food. Traditional fermented foods, such as tempeh, kimchi, natto, and fortified sourdough bread, show how microbial activity can naturally increase vitamins, minerals, and probiotics (Hedayat & Lapraz, 2019; Sawant et al., 2025; Shah et al., 2023). Moreover, bio-fermentation helps reduce food waste by turning agricultural by-products, like bran, husks, and whey, into nutrient-rich food (Hadj Saadoun et al., 2021; Hedayat & Lapraz, 2019; Sawant et al., 2025; Shah et al., 2023). Technological progress in bio-fermentation has also led to new opportunities in personalized nutrition. By using techniques like genomics, proteomics, and metabolomics, scientists can now optimize microorganisms to produce specific nutrients (Hadj Saadoun et al., 2021; Hedayat & Lapraz, 2019; Sawant et al., 2025; Shah et al., 2023). These advances allow for more precise food fortification that can meet the unique needs of different groups, such as children, pregnant women, and the elderly. Additionally, nutrients produced through fermentation tend to stay stable during food processing and storage, making sure fortified foods keep their nutritional value over time(International Finance Corporation (IFC), 2017; Johnson et al., 2004; Prajapati & Sharma, 2024).

As the global population grows and diets change, there will be a growing need for nutrient-rich, sustainable food sources(Drewnowski et al., 2020; Fanzo et al., 2022; Vermeulen et al., 2020). Bio-fermentation offers a promising solution to malnutrition and food insecurity, combining nutrition, sustainability, and innovation (Materia et al., 2021). This chapter looks at how microorganisms act as natural nutrient producers, the advancements in fermentation technology, and the potential applications for food fortification. By better understanding and using microbial metabolism, we can transform food systems to create healthier, more resilient communities around the world.

5.2 Significant Trends in Bio-fermentation and Its Role in Food Fortification

Bio-fermentation has gained increasing attention as a transformative technology in food fortification, capitalizing on the metabolic processes of microorganisms to enhance the nutritional content of food products(Sundarraj et al., 2018; Sharma et al., 2020). This process involves the use of specific bacteria, yeasts, and molds that ferment substrates, which in turn leads to the synthesis of bioactive compounds, vitamins, minerals, and other nutrients that are often in limited supply in conventional diets (Praveen & Brogi, 2025b; Sarwar et al., 2025). Microorganisms such as *Lactobacillus* and *Saccharomyces cerevisiae* are well-known for their ability to produce essential nutrients like B-vitamins, folate, and certain amino acids (Adesemoye et al., 2025; Guamán et al., 2024; Karina et al., 2017; Linares et al., 2017). This natural enrichment of food products through fermentation can provide a cost-effective and sustainable means of improving public health, particularly in areas where malnutrition is prevalent due to the lack of diverse diets and micronutrient deficiencies (Alfthan et al., 2010; Linares et al., 2017; Thakur et al., 2023; Tura et al., 2025).

The application of bio-fermentation extends beyond simply enhancing the nutrient profile of foods. Fermentation also plays a significant role in improving the bioavailability of essential micronutrients, making them more accessible for absorption by the human body (Nkhata et al., 2018; Samtiya et al., 2020, 2021a). For example, the fermentation of plant-based foods such as legumes and grains can reduce antinutritional factors like phytic acid, which inhibits the absorption of important minerals such as iron and zinc (Manzoor et al., 2021; Nkhata et al., 2018; Samtiya et al., 2020). By breaking down these compounds, fermentation enhances the nutritional value of these foods, making them more beneficial for consumers (Bedi et al., 2023; Siddeeg et al., 2022) (Table 5.1).

Furthermore, fermentation processes are known to increase the digestibility of proteins and other complex carbohydrates, which not only improves the nutritional intake but also aids in gut health by promoting the growth of beneficial microbiota(Lee & Hwang, 2024; Oliphant & Allen-Vercoe, 2019; Rowland et al., 2018).

In addition to enhancing the nutritional and functional properties of food, bio-fermentation can improve the sensory attributes of foods, such as taste, texture, and aroma (Kitessa, 2024; Senanayake et al., 2023; Verni et al., 2024). These changes are particularly important in the development of new food products, especially in plant-based and alternative food markets, where fermentation is used to mimic the

Table 5.1 Role of bio-fermentation in food fortification

Role	Benefits	References
Nutrient enhancement	Microorganisms such as *Lactobacillus* and *Saccharomyces cerevisiae* synthesize essential nutrients like B-vitamins, folate, and amino acids, enriching food naturally	Sundarraj et al. (2018), Sharma et al. (2020), Praveen and Brogi (2025b), Sarwar et al. (2025)
Improved bioavailability	Fermentation reduces antinutritional factors (e.g., phytic acid), enhancing the absorption of key minerals like iron and zinc	Nkhata et al. (2018), Samtiya et al. (2020, 2021a), Manzoor et al. (2021)
Enhanced protein and carbohydrate digestibility	Fermentation breaks down complex proteins and carbohydrates, increasing digestibility and promoting gut health	Bedi et al. (2023), Siddeeg et al. (2022), Lee and Hwang (2024), Oliphant and Allen-Vercoe (2019)
Sensory improvement	Enhances taste, texture, and aroma, making plant-based alternatives more appealing and increasing consumer acceptance	Kitessa (2024), Senanayake et al. (2023), Verni et al. (2024), Abbaspour (Abbaspour, 2024a, b)
Sustainable and cost-effective fortification	Provides a natural, low-cost method for improving food quality, particularly in regions affected by malnutrition and micronutrient deficiencies	Alfthan et al. (2010), Linares et al. (2017), Thakur et al. (2023), Tura et al. (2025)

flavors and textures of traditional animal-based products (Abbaspour, 2024a, b; Boukid et al., 2023; Tachie et al., 2023). Through controlled fermentation, microorganisms can create desirable flavors and textures that appeal to consumers, helping to increase the acceptability and marketability of fortified foods(Ray & Montet, 2014). As demand for fermented foods continues to rise due to their potential health benefits, bio-fermentation will likely play an integral role in the development of fortified foods that meet both the nutritional needs and sensory preferences of global populations(Sharma, 2021).

5.3 Enhancement of Nutritional Content

A significant advancement in bio-fermentation is its role in enhancing the bioavailability of essential nutrients, making them more accessible for human absorption (Bryant et al., 2023; Pasupuleti, 2024; Piskin et al., 2022). Through microbial activity, fermentation transforms food matrices, increasing the concentration of vital micronutrients and improving their digestibility (Castillo-González et al., 2014; Piskin et al., 2022; Seo, 2024). This process is particularly valuable in addressing nutrient deficiencies by naturally enriching foods with essential vitamins and minerals. The growing interest in fermentation as a fortification strategy stems from its ability to modify food composition in a way that benefits both nutrition and overall health (Avnee et al., 2023; Chadare et al., 2019; Fanzo et al., 2023) (Table 5.2).

One of the key benefits of fermentation is its ability to boost the levels of B-complex vitamins, particularly folate, which plays a crucial role in DNA synthesis, cell division, and overall metabolic function (Acevedo-Rocha et al., 2019; Tripathi et al., 2024). Certain strains of probiotic bacteria, such as *Lactobacillus* and

Table 5.2 Role of bio-fermentation in enhancing nutrient availability in food

Food	Nutrient enhanced	Microorganism	References
Fermented dairy (yogurt, kefir)	B-complex vitamins (folate)	*Lactobacillus*, *Bifidobacterium*	García-Burgos et al. (2020), Kaur et al. (2022), Linares et al. (2017), Mahara et al. (2019), Okoniewski et al. (2023), Urrutia-Baca et al. (2024)
Fermented cereals	Iron, zinc, calcium (via phytic acid reduction)	*Lactobacillus*, *Saccharomyces cerevisiae*	Abbas and Ahmad (2021), Gupta et al. (2015), Manzoor et al. (2024), Nissar et al. (2017), Samtiya et al. (2020)
Fermented legumes	Iron, zinc (via phytic acid reduction)	*Bacillus subtilis*, *Rhizopus oligosporus*	Rangaswamaiah (2023), Rizwanuddin et al. (2023), Wu et al. (2024)
Fermented dairy	Protein digestibility, bioactive peptides	*Lactobacillus*, *Streptococcus thermophilus*	Ajomiwe et al. (2024), Yang et al. (2025)
Fermented grains	Antioxidants, immune-boosting compounds	*Saccharomyces cerevisiae*, *Aspergillus oryzae*	Nawaz et al. (2024), Sarıtaş et al. (2024), Seo (2024), Wan-Mohtar et al. (2022)

Bifidobacterium, are known to synthesize folate during the fermentation process, thereby increasing its presence in fermented dairy products like yogurt and kefir (García-Burgos et al., 2020; Kaur et al., 2022; Linares et al., 2017; Mahara et al., 2019; Okoniewski et al., 2023; Urrutia-Baca et al., 2024). This microbial production of folate can serve as a natural means of fortification, particularly for populations at risk of folate deficiency, such as pregnant women and individuals with restricted diets(Homayouni Rad et al., 2016; Ismail et al., 2023).

Beyond vitamin enrichment, fermentation significantly reduces the presence of antinutritional factors that can inhibit the absorption of essential minerals. Phytic acid, commonly found in cereals, legumes, and seeds, is known to bind minerals such as iron, calcium, and zinc, making them less bioavailable to the body (Abbas & Ahmad, 2021; Gupta et al., 2015; Manzoor et al., 2024; Nissar et al., 2017; Samtiya et al., 2020). The enzymatic activity of fermenting microorganisms helps to break down phytic acid, thereby increasing the bio accessibility of these critical nutrients (Rangaswamaiah, 2023; Rizwanuddin et al., 2023; Wu et al., 2024). As a result, fermented foods become a more effective dietary source of essential minerals, which can contribute to improved nutritional status, particularly in populations reliant on plant-based diets (Owusu-Kwarteng et al., 2022; Rezac et al., 2018; Rollán et al., 2019; Tangyu et al., 2019).

In addition to improving mineral bioavailability, fermentation enhances the overall nutritional profile of foods by modifying their protein structure and carbohydrate content(Erem & Kilic-Akyilmaz, 2024; Feng et al., 2023; Kitessa, 2024; Kohli & Singha, 2024; Mefleh et al., 2024). The breakdown of complex proteins into more digestible peptides and amino acids makes fermented foods easier to digest, benefiting individuals with digestive disorders or sensitivities to certain proteins (Ajomiwe et al., 2024; Yang et al., 2025). Moreover, the fermentation process can produce bioactive compounds with antioxidant, antimicrobial, and immune-boosting properties, further increasing the functional benefits of fermented foods (Nawaz et al., 2024; Sarıtaş et al., 2024; Seo, 2024; Wan-Mohtar et al., 2022).

As research in bio-fermentation continues to expand, its potential applications in food fortification are becoming more widely recognized. By leveraging microbial activity, fermentation offers a sustainable and natural method of enhancing food quality without the need for synthetic additives. This approach aligns with the increasing consumer preference for minimally processed, nutrient-dense foods. With ongoing innovations in fermentation technology, the ability to optimize nutrient synthesis and improve dietary outcomes will play a crucial role in addressing global nutrition challenges.

5.4 Utilization of Food by-Products

The growing emphasis on sustainability in the food industry has led to increased interest in utilizing fermentation as a means of transforming agricultural by-products into high-value nutritional resources(Augustin et al., 2024; Enitan-Folami &

Swalaha, 2023; Hashimoto et al., 2024; Salas-Millán & Aguayo, 2024). Through microbial processing, waste materials from fruit and vegetable production can be converted into nutrient-dense ingredients, enriched with proteins, dietary fibers, and bioactive compounds (Aït-Kaddour et al., 2024; Kusumasari et al., 2024). This innovative approach not only minimizes food waste but also enhances the nutritional profile of by-products that would otherwise be discarded. By leveraging fermentation, industries can extract maximum value from raw materials while promoting environmentally friendly food production practices.

Beyond waste reduction, the fermentation of food industry by-products plays a crucial role in the development of functional ingredients for food fortification (Areti et al., 2024; Wibisono et al., 2025). The microbial breakdown of plant-based waste enhances its digestibility and bioavailability, making it suitable for inclusion in fortified foods aimed at addressing micronutrient deficiencies (Rodríguez-Espinosa et al., 2024; Samtiya et al., 2021b; Upadhyay et al., 2024) (Table 5.3).

Additionally, bio-fermentation can produce beneficial compounds with antioxidant, antimicrobial, and probiotic properties, further increasing the health benefits of these repurposed ingredients (Fernandes et al., 2024; Kim et al., 2023; Sharma et al. 2022)). As the demand for sustainable and nutrient-rich foods continues to grow, harnessing fermentation for by-product valorization presents a promising solution to improve both food security and environmental sustainability.

Table 5.3 Bio-fermentation of food by-products for nutritional enhancement and sustainability

Food by-product	Nutrient enriched	Functional benefits	Microorganisms used	References
Fruit and vegetable waste	Proteins, dietary fibers, bioactive compounds	Enhances nutritional profile, reduces food waste	*Lactobacillus*, *Aspergillus*	Aït-Kaddour et al. (2024), Kusumasari et al. (2024)
Cereal bran and husks	Antioxidants, iron, zinc	Improves bioavailability, enhances fortification potential	*Saccharomyces cerevisiae*, *Bacillus subtilis*	Rodríguez-Espinosa et al. (2024), Samtiya et al. (2021b), Upadhyay et al. (2024)
Legume hulls	Digestible proteins, prebiotics	Supports gut health, increases protein quality	*Rhizopus oligosporus*, *Bifidobacterium*	Fernandes et al. (2024), Kim et al. (2023)
Dairy by-products (whey, casein)	Probiotics, bioactive peptides	Enhances gut microbiota, supports immune health	*Lactobacillus*, *Streptococcus thermophilus*	Sharma et al. (2022), Wibisono et al. (2025)
Oilseed meals (soybean, sunflower)	Protein hydrolysates, antimicrobial peptides	Increases protein digestibility, antimicrobial benefits	*Bacillus*, *Aspergillus oryzae*	Areti et al. (2024), Enitan-Folami and Swalaha (2023)

5.5 Development of Plant-Based Alternatives

Fermentation has become an essential tool in the development of plant-based food alternatives, significantly improving their sensory and nutritional qualities (Abbaspour, 2024a, b; Boukid et al., 2023; Harper et al., 2022) (Table 5.4).

By harnessing the metabolic activity of microorganisms, fermentation enhances the texture, taste, and aroma of plant-derived products, making them more palatable and appealing to consumers shifting away from animal-based foods (Day et al., 2022; Dennard, 2019). This natural transformation not only improves the overall eating experience but also boosts the nutritional value by enriching these products with essential vitamins, bioactive compounds, and improved protein digestibility (Abbaspour, 2024a, b).

A key application of fermentation in plant-based foods is its ability to refine the texture and flavor profiles of meat and dairy substitutes (Abbaspour, 2024a, b; Boukid et al., 2023; Tangyu et al., 2023). Microbial fermentation helps mimic the umami-rich taste and fibrous structure of traditional meat, making plant-based proteins more realistic and satisfying(Boukid et al., 2023; Elhalis et al., 2023; Mazumder et al., 2023; Tangyu et al., 2023). Similarly, fermentation is used in dairy alternatives to create complex, creamy textures and tangy flavors, similar to traditional cheese and yogurt (Harper et al., 2022; Pua et al., 2022; Rizzoli & Biver, 2017). As demand for sustainable and health-conscious food options rises,

Table 5.4 Role of fermentation in enhancing plant-based food alternatives

Plant-based food	Nutrient and sensory enhancement	Functional benefits	Microorganisms used	References
Meat substitutes	Improved texture, umami-rich flavor, enhanced protein digestibility	Mimics meat texture, improves taste, enriches protein quality	*Lactobacillus*, *Saccharomyces cerevisiae*, *bacillus*	Abbaspour (2024a, b), Boukid et al. (2023), Tangyu et al. (2023)
Dairy alternatives	Creamy texture, tangy flavors, enhanced bioactive compounds	Mimics cheese and yogurt, improves gut health, boosts probiotic content	*Lactobacillus*, *Streptococcus thermophilus*, *Bifidobacterium*	Harper et al. (2022), Pua et al. (2022), Rizzoli and Biver (2017)
Plant-based proteins	Enriched with vitamins (e.g., B-vitamins), improved protein digestibility	Increases nutritional value, improves palatability	*Rhizopus*, *Aspergillus*	Day et al. (2022), Dennard (2019), Abbaspour (2024a, b)
Plant-based beverages	Enhanced texture, improved flavor profiles, bioactive compounds	Provides health benefits, supports gut microbiota growth	*Saccharomyces cerevisiae*, *lactobacillus*	Boukid et al. (2023), Mazumder et al. (2023)

fermentation-driven innovation continues to shape the future of the plant-based food industry, offering consumers high-quality alternatives without compromising on taste or nutrition.

5.6 Economic and Market Growth

The global market for fermented foods and beverages is undergoing rapid expansion, fuelled by growing consumer recognition of their numerous health benefits (Ibrahim et al., 2023; Patel et al., 2023; Todorovic et al., 2024). As people become more conscious of the role diet plays in overall well-being, there is a heightened demand for naturally fortified foods that support gut health, boost immunity, and enhance nutrient absorption (Aziz et al., 2024; Munteanu & Schwartz, 2022). This surge in awareness has led to an increased preference for probiotic-rich products such as yogurt, kimchi, kombucha, and fermented plant-based alternatives (D'Almeida et al., 2024; García-Barón et al., 2025; Nithya et al., 2023). The shift toward functional and minimally processed foods is driving innovation in the fermentation industry, further solidifying its position as a key player in the evolving food sector.

Consumers are not only seeking traditional fermented products but also exploring novel formulations that integrate fermentation with emerging food trends, such as plant-based diets and personalized nutrition. Additionally, advancements in fermentation technology are expanding the possibilities for nutrient enhancement, improved flavor development, and extended shelf life. As this sector continues to grow, companies investing in fermentation-driven innovations are well-positioned to capitalize on the increasing demand for healthier, more sustainable food options (Boukid et al., 2023; Elhalis et al., 2023; Niyigaba et al., 2025; Teng et al., 2021).

5.7 Conclusion

Bio-fermentation is emerging as a powerful tool in food fortification, offering a natural and sustainable approach to enhancing the nutritional value, bioavailability, and sensory properties of food products. By leveraging microbial activity, fermentation increases essential vitamins and minerals, reduces antinutritional factors, and improves protein digestibility, addressing key nutritional deficiencies, particularly in vulnerable populations. Additionally, the utilization of food by-products through fermentation promotes sustainability, while its role in plant-based alternatives enhances the appeal and functionality of these products. As consumer demand for minimally processed, nutrient-rich foods grow, fermentation-driven innovations will continue to shape the food industry, supporting both health and environmental sustainability.

References

Abbas, Y., & Ahmad, A. (2021). Impact of processing on nutritional and antinutritional factors of legumes: A review. *Revista Mexicana de Ingeniera Quimica, 20*(3), 199–215. https://doi.org/10.24275/rmiq/Alim2433

Abbaspour, N. (2024a). Fermentation's pivotal role in shaping the future of plant-based foods: An integrative review of fermentation processes and their impact on sensory and health benefits. *Applied Food Research, 4*(2), 100468. https://doi.org/10.1016/j.afres.2024.100468

Abbaspour, N. (2024b). Recent progress in the role of fermentation in plant-based foods: A concise review. *Food Science & Nutrition, 10*(175).

Acevedo-Rocha, C. G., Gronenberg, L. S., Mack, M., Commichau, F. M., & Genee, H. J. (2019). Microbial cell factories for the sustainable manufacturing of B vitamins. *Current Opinion in Biotechnology, 56*, 18–29. https://doi.org/10.1016/j.copbio.2018.07.006

Adesemoye, E. T., Sanni, A. I., Spano, G., Capozzi, V., & Fragasso, M. (2025). Lactic acid bacteria diversity in fermented foods as potential bio-resources contributing to alleviate malnutrition in developing countries: Nigeria as a case study. *Fermentation, 11*(2), 1–25. https://doi.org/10.3390/fermentation11020103

Aït-Kaddour, A., Hassoun, A., Tarchi, I., Loudiyi, M., Boukria, O., Cahyana, Y., Ozogul, F., & Khwaldia, K. (2024). Transforming plant-based waste and by-products into valuable products using various "food industry 4.0" enabling technologies: A literature review. *Science of the Total Environment, 955*(June). https://doi.org/10.1016/j.scitotenv.2024.176872

Ajomiwe, N., Boland, M., Phongthai, S., Bagiyal, M., Singh, J., & Kaur, L. (2024). Protein nutrition: Understanding structure, digestibility, and bioavailability for optimal health. *Food, 13*(11), 1771.

Alfthan, G., Aspila, P., Ekholm, P., Eurola, M., Hartikainen, H., Hero, H., Hietaniemi, V., Root, T., Salminen, P., et al. (2010). Nationwide supplementation of sodium selenate to commercial fertilizers: History and 25-year results from the finnish selenium monitoring programme. In *Combating micronutrient deficiencies: Food-based approaches*. https://doi.org/10.1079/9781845937140.0312

Anumudu, C. K., Miri, T., & Onyeaka, H. (2024). Multifunctional applications of lactic acid bacteria: Enhancing safety, quality, and nutritional value in foods and fermented beverages. *Food, 13*(23). https://doi.org/10.3390/foods13233714

Areti, H. A., Muleta, M. D., Abo, L. D., Hamda, A. S., Adugna, A. A., Edae, I. T., Daba, B. J., & Gudeta, R. L. (2024). Innovative uses of agricultural by-products in the food and beverage sector: A review. *Food Chemistry Advances, 5*(April), 100838. https://doi.org/10.1016/j.focha.2024.100838

Augustin, M. A., Hartley, C. J., Maloney, G., & Tyndall, S. (2024). Innovation in precision fermentation for food ingredients. *Critical Reviews in Food Science and Nutrition, 64*(18), 6218–6238. https://doi.org/10.1080/10408398.2023.2166014

Avnee, S., Chaudhary, D. R., Jhorar, P., & Rana, R. S. (2023). Biofortification: An approach to eradicate micronutrient deficiency. *Frontiers in Nutrition, 10*(September), 1–16. https://doi.org/10.3389/fnut.2023.1233070

Aziz, T., Hussain, N., Hameed, Z., & Lin, L. (2024). Elucidating the role of diet in maintaining gut health to reduce the risk of obesity, cardiovascular and other age-related inflammatory diseases: Recent challenges and future recommendations. *Gut Microbes, 16*(1), 1–31. https://doi.org/10.1080/19490976.2023.2297864

Bamidele, M. O., Bamikale, M. B., Cárdenas-Hernández, E., Bamidele, M. A., Castillo-Olvera, G., Sandoval-Cortes, J., & Aguilar, C. N. (2025). Bioengineering in solid-state fermentation for next sustainable food bioprocessing. *Next Sustainability, 6*, 100105.

Bedi, S., Ghosh, S., Bandyopadhyay, B., Bedi, S., & Maity, M. (2023). Fermentation enhanced nutritional quality of food- a review. *Journal of Survey in Fisheries Sciences, 10*, 6139–6145. https://doi.org/10.53555/sfs.v10i1s.2118

Boukid, F., Hassoun, A., Zouari, A., Tülbek, M. Ç., Mefleh, M., Aït-Kaddour, A., & Castellari, M. (2023). Fermentation for designing innovative plant-based meat and dairy alternatives. *Food, 12*(5). https://doi.org/10.3390/foods12051005

Bryant, K. L., Hansen, C., & Hecht, E. E. (2023). Fermentation technology as a driver of human brain expansion. *Communications Biology, 6*(1), 1–11. https://doi.org/10.1038/s42003-023-05517-3

Castillo-González, A. R., Burrola-Barraza, M. E., Domínguez-Viveros, J., & Chávez-Martínez, A. (2014). Rumen microorganisms and fermentation. *Archivos de Medicina Veterinaria, 46*(3), 349–361. https://doi.org/10.4067/S0301-732X2014000300003

Chadare, F. J., Idohou, R., Nago, E., Affonfere, M., Agossadou, J., Fassinou, T. K., Kénou, C., Honfo, S., Azokpota, P., Linnemann, A. R., & Hounhouigan, D. J. (2019). Conventional and food-to-food fortification: An appraisal of past practices and lessons learned. *Food Science and Nutrition, 7*(9), 2781–2795. https://doi.org/10.1002/fsn3.1133

D'Almeida, A. P., Neta, A. A. I., de Andrade-Lima, M., & de Albuquerque, T. L. (2024). Plant-based probiotic foods: Current state and future trends. *Food Science and Biotechnology, 33*. https://doi.org/10.1007/s10068-024-01674-1

Darnton-Hill, I., & Nalubola, R. (2002). Fortification strategies to meet micronutrient needs: Sucesses and failures. *Proceedings of the Nutrition Society, 61*(2), 231–241. https://doi.org/10.1079/pns2002150

Day, L., Altermann, E., Chanyi, R., Hicks, T., Knowles, S., & Mullaney, J. (2022). Fermentation for future food systems-what are the opportunities for New Zealand? *Agreseach White Paper*. http://www.agresearch.co.nz/fermented-foods/

Dennard, R. H. (2019). State of the industry report fermentation. *Gfi, 58*(8), 4–5.

Drewnowski, A., Finley, J., Hess, J. M., Ingram, J., Miller, G., & Peters, C. (2020). Toward healthy diets from sustainable food systems. *Current Developments in Nutrition, 4*(6), nzaa083. https://doi.org/10.1093/CDN/NZAA083

Dwyer, J. T., Wiemer, K. L., Dary, O., Keen, C. L., King, J. C., Miller, K. B., Philbert, M. A., Tarasuk, V., Taylor, C. L., Gaine, P. C., Jarvis, A. B., & Bailey, R. L. (2015). Fortification and health: Challenges and opportunities. *Advances in Nutrition, 6*(1), 124–131. https://doi.org/10.3945/an.114.007443

Elhalis, H., See, X. Y., Osen, R., Chin, X. H., & Chow, Y. (2023). Significance of fermentation in plant-based meat analogs: A critical review of nutrition, and safety-related aspects. *Food, 12*(17). https://doi.org/10.3390/foods12173222

Emmanuel, O. K., Aria, J., Jose, D., & Diego, C. (2024). Fermentation and its role in enhancing nutrient bioavailability.

Enitan-Folami, A. M., & Swalaha, F. M. (2023). Application of biotechnology in the food industry. In *Food science and technology: Fundamentals and innovation* (pp. 235–277). https://doi.org/10.1515/9783111013107-010

Erem, E., & Kilic-Akyilmaz, M. (2024). The role of fermentation with lactic acid bacteria in quality and health effects of plant-based dairy analogues. *Comprehensive Reviews in Food Science and Food Safety, 23*(4), 1–36. https://doi.org/10.1111/1541-4337.13402

Faber, M., Berti, C., & Smuts, M. (2014). Prevention and control of micronutrient deficiencies in developing countries: Current perspectives. *Nutrition and Dietary Supplements, 41*. https://doi.org/10.2147/nds.s43523

Fanzo, J., Rudie, C., Sigman, I., Grinspoon, S., Benton, T. G., Brown, M. E., Covic, N., Fitch, K., Golden, C. D., Grace, D., Hivert, M. F., Huybers, P., Jaacks, L. M., Masters, W. A., Nisbett, N., Richardson, R. A., Singleton, C. R., Webb, P., & Willett, W. C. (2022). Sustainable food systems and nutrition in the 21stcentury: A report from the 22ndannual Harvard nutrition obesity symposium. *American Journal of Clinical Nutrition, 115*(1), 18–33. https://doi.org/10.1093/ajcn/nqab315

Fanzo, J., McLaren, R., Bellows, A., & Carducci, B. (2023). Challenges and opportunities for increasing the effectiveness of food reformulation and fortification to improve dietary and nutrition outcomes. *Food Policy, 119*(August 2023), 102515. https://doi.org/10.1016/j.foodpol.2023.102515

Feng, X., Ng, K., Ajlouni, S., Zhang, P., & Fang, Z. (2023). Effect of solid-state fermentation on plant-sourced proteins: A review. *Food Reviews International, 40*(9), 2580–2617. https://doi.org/10.1080/87559129.2023.2274490

Fernandes, M. T. C., Farinazzo, F. S., Mauro, C. S. I., Ferreira, M., et al. (2024). Effect of fermentation by probiotic bacteria on the bioaccessibility of bioactive compounds from the fruit of the Juçara palm (Euterpe edulis Martius). *Fermentation, 10*(11). https://doi.org/10.3390/fermentation10110541

García-Barón, S. E., Carmona-Escutia, R. P., Herrera-López, E. J., Leyva-Trinidad, D. A., & Gschaedler-Mathis, A. (2025). Consumers' drivers of perception and preference of fermented food products and beverages: A systematic review. *Food, 14*(5), 1–25. https://doi.org/10.3390/foods14050713

García-Burgos, M., Moreno-Fernández, J., Alférez, M. J. M., Díaz-Castro, J., & López-Aliaga, I. (2020). New perspectives in fermented dairy products and their health relevance. *Journal of Functional Foods, 72*(May), 104059. https://doi.org/10.1016/j.jff.2020.104059

García-Casal, M. N. (2014). Planning and implementing food fortification programs to combat micronutrient malnutrition: Iron. *Food and Nutrition Sciences, 05*(10), 880–888. https://doi.org/10.4236/fns.2014.510097

Guamán, L. P., Carrera-Pacheco, S. E., Zúñiga-Miranda, J., Teran, E., Erazo, C., & Barba-Ostria, C. (2024). The impact of bioactive molecules from probiotics on child health: A comprehensive review. *Nutrients, 16*(21). https://doi.org/10.3390/nu16213706

Gupta, R. K., Gangoliya, S. S., & Singh, N. K. (2015). Reduction of phytic acid and enhancement of bioavailable micronutrients in food grains. *Journal of Food Science and Technology, 52*(2), 676–684. https://doi.org/10.1007/s13197-013-0978-y

Hadj Saadoun, J., Bertani, G., Levante, A., Vezzosi, F., Ricci, A., Bernini, V., & Lazzi, C. (2021). Fermentation of agri-food waste: A promising route for the production of aroma compounds. *Food, 10*(4). https://doi.org/10.3390/foods10040707

Harper, A. R., Dobson, R. C. J., Morris, V. K., & Moggré, G. J. (2022). Fermentation of plant-based dairy alternatives by lactic acid bacteria. *Microbial Biotechnology, 15*(5), 1404–1421. https://doi.org/10.1111/1751-7915.14008

Hashimoto, E. H., et al. (2024). Fermentation-mediated sustainable development and improvement of quality of plant-based foods: From waste to a new food. *Systems Microbiology and Biomanufacturing*. https://doi.org/10.1007/s43393-024-00292-6

Hedayat, K. M., & Lapraz, J.-C. (2019). Disorders of intestinal transit. In *The theory of endobiogeny* (pp. 215–235). https://doi.org/10.1016/b978-0-12-816964-3.00010-9

Homayouni Rad, A., Yari Khosroushahi, A., Khalili, M., & Jafarzadeh, S. (2016). Folate bio-fortification of yoghurt and fermented milk: A review. *Dairy Science and Technology, 96*(4), 427–441. https://doi.org/10.1007/s13594-016-0286-1

Ibrahim, S. A., Yeboah, P. J., Ayivi, R. D., Eddin, A. S., Wijemanna, N. D., Paidari, S., & Bakhshayesh, R. V. (2023). A review and comparative perspective on health benefits of probiotic and fermented foods. *International Journal of Food Science and Technology, 58*(10), 4948–4964. https://doi.org/10.1111/ijfs.16619

International Finance Corporation (IFC). (2017). *Global Agribusiness Survey*.

Ismail, S., Eljazzar, S., & Ganji, V. (2023). Intended and unintended benefits of folic acid fortification – A. Narrative review. *Food, 12*(8). https://doi.org/10.3390/foods12081612

Johnson, Q., Mannar, V., & Ranum, P. (2004). *Fortification handbook vitamin and mineral fortification of wheat flour and maize meal* (pp. 1–107).

Karina, T. M.-G., Camila, D. F., Kassiana, T. M., Rosane, F. S., Jorge, A. V. C., & Itaciara, L. N. (2017). Microorganisms in functional food supplementation: A review. *African Journal of Microbiology Research, 11*(8), 319–326. https://doi.org/10.5897/ajmr2017.8445

Kaur, H., Kaur, G., & Ali, S. A. (2022). Dairy-based probiotic-fermented functional foods: An update on their health-promoting properties. *Fermentation, 8*(9). https://doi.org/10.3390/fermentation8090425

Kim, H. J., Kwon, M. S., Hwang, H., Choi, H. S., Lee, W. J., Choi, S. P., Jo, H., & Hong, S. W. (2023). A review of the health benefits of kimchi functional compounds and metabolites. *Microbiology and Biotechnology Letters, 51*(4), 353–373. https://doi.org/10.48022/mbl.2310.10011

Kitessa, D. A. (2024). Review on effect of fermentation on physicochemical properties, anti-nutritional factors and sensory properties of cereal-based fermented foods and beverages. *Annals of Microbiology, 74*(1). https://doi.org/10.1186/s13213-024-01763-w

Knez, E., Kadac-Czapska, K., & Grembecka, M. (2023). Effect of fermentation on the nutritional quality of the selected vegetables and legumes and their health effects. *Life, 13*(3). https://doi.org/10.3390/life13030655

Kohli, V., & Singha, S. (2024). Protein digestibility of soybean: How processing affects seed structure, protein and non-protein components. *Discover Food, 4*(1). https://doi.org/10.1007/s44187-024-00076-w

Kovalchuk, Y. (2021). Bio-based preservatives: A natural alternative to synthetic additives. *International Journal of Science and Research Archive, 01*(January), 56–70.

Kusumasari, S., Syabana, M. A., Pamela, V. Y., & Meindrawan, B. (2024). Potential use of food waste in food processing to add nutritional value. *E3S Web of Conferences, 483*, 1–12. https://doi.org/10.1051/e3sconf/202448302006

Lee, D. B., & Hwang, I. S. (2024). Macronutrient balance determines the human gut microbiome eubiosis: Insights from in vitro gastrointestinal digestion and fermentation of eight pulse species. *Frontiers in Microbiology, 15*(January), 1–15. https://doi.org/10.3389/fmicb.2024.1512217

Linares, D. M., Gómez, C., Renes, E., Fresno, J. M., Tornadijo, M. E., Ross, R. P., & Stanton, C. (2017). Lactic acid bacteria and Bifidobacteria with potential to design natural biofunctional health-promoting dairy foods. *Frontiers in Microbiology, 8*(MAY), 1–11. https://doi.org/10.3389/fmicb.2017.00846

Mahara, F. A., Nuraida, L., & Lioe, H. N. (2019). Fermentation of milk using folate-producing lactic acid bacteria to increase natural folate content: A review. *Journal Of Applied Biotechnology Reports, 6*(4), 129–136. https://doi.org/10.29252/JABR.06.04.01

Manzoor, M., Singh, D., Aseri, G. K., Sohal, J. S., Vij, S., & Sharma, D. (2021). Role of lacto-fermentations in reduction of antinutrients in plant-based foods. *Journal of Applied Biology and Biotechnology, 9*(3), 7–16. https://doi.org/10.7324/JABB.2021.9302

Manzoor, M. F., Ali, A., Ain, H. B. U., Kausar, S., Khalil, A. A., Aadil, R. M., & Zeng, X. A. (2024). Bioaccessibility mechanisms, fortification strategies, processing impact on bioavailability, and therapeutic potentials of minerals in cereals. *Future Foods, 10*(July), 100425. https://doi.org/10.1016/j.fufo.2024.100425

Materia, V. C., Linnemann, A. R., Smid, E. J., & Schoustra, S. E. (2021). Contribution of traditional fermented foods to food systems transformation: Value addition and inclusive entrepreneurship. *Food Security, 13*(5), 1163–1177. https://doi.org/10.1007/s12571-021-01185-5

Mazumder, M. A. R., Panpipat, W., Chaijan, M., Shetty, K., & Rawdkuen, S. (2023). Role of plant protein on the quality and structure of meat analogs: A new perspective for vegetarian foods. *Future Foods, 8*(November), 100280. https://doi.org/10.1016/j.fufo.2023.100280

Mefleh, M., Omri, G., Limongelli, R., Minervini, F., Santamaria, M., & Faccia, M. (2024). Enhancing nutritional and sensory properties of plant-based beverages: A study on chickpea and Kamut® flours fermentation using Lactococcus lactis. *Frontiers in Nutrition, 11*(January), 1–10. https://doi.org/10.3389/fnut.2024.1269154

Munteanu, C., & Schwartz, B. (2022). The relationship between nutrition and the immune system. *Frontiers in Nutrition, 9*(December), 1–23. https://doi.org/10.3389/fnut.2022.1082500

Nawaz, T., Gu, L., Fahad, S., Saud, S., Hassan, S., Harrison, M. T., Liu, K., & Zhou, R. (2024). Unveiling the antioxidant capacity of fermented foods and food microorganisms: A focus on cyanobacteria. *Journal of Umm Al-Qura University for Applied Sciences, 10*(1), 232–243. https://doi.org/10.1007/s43994-023-00095-w

Nissar, J., Ahad, T., Naik, H. R., & Hussain, S. Z. (2017). A review phytic acid: As antinutrient or nutraceutical. *Journal of Pharmacognosy and Phytochemistry, 6*(6), 1554–1560.

Nithya, A., Misra, S., Panigrahi, C., Dalbhagat, C. G., & Mishra, H. N. (2023). Probiotic potential of fermented foods and their role in non-communicable diseases management: An understanding through recent clinical evidences. *Food Chemistry Advances, 3*(June 2022), 100381. https://doi.org/10.1016/j.focha.2023.100381

Niyigaba, T., Küçükgöz, K., Kołożyn-Krajewska, D., Krolikowski, T., & Trząskowska, M. (2025). Advances in fermentation technology: A focus on health and safety. *Applied Sciences, 15*(6), 3001.

Nkhata, S. G., Ayua, E., Kamau, E. H., & Shingiro, J. B. (2018). Fermentation and germination improve nutritional value of cereals and legumes through activation of endogenous enzymes. *Food Science and Nutrition, 6*(8), 2446–2458. https://doi.org/10.1002/fsn3.846

Ofori, K. F., Antoniello, S., English, M. M., & Aryee, A. N. A. (2022). Improving nutrition through biofortification–a systematic review. *Frontiers in Nutrition, 9*(December), 1–20. https://doi.org/10.3389/fnut.2022.1043655

Okoniewski, A., Dobrzyńska, M., Kusyk, P., Dziedzic, K., Przysławski, J., & Drzymała-Czyż, S. (2023). The role of fermented dairy products on gut microbiota composition. *Fermentation, 9*(3), 1–25. https://doi.org/10.3390/fermentation9030231

Oliphant, K., & Allen-Vercoe, E. (2019). Macronutrient metabolism by the human gut microbiome: Major fermentation by-products and their impact on host health. *Microbiome, 7*(1), 1–15. https://doi.org/10.1186/s40168-019-0704-8

Owusu-Kwarteng, J., Agyei, D., Akabanda, F., Atuna, R. A., & Amagloh, F. K. (2022). Plant-based alkaline fermented foods as sustainable sources of nutrients and health-promoting bioactive compounds. *Frontiers in Sustainable Food Systems, 6*(June). https://doi.org/10.3389/fsufs.2022.885328

Pasupuleti, M. K. (2024). Advances in food science and modern technologies. *Food Science and Technology*, 15–33. https://doi.org/10.62311/nesx/97882

Patel, P., Butani, K., Kumar, A., Singh, S., & Prajapati, B. G. (2023). Effects of fermented food consumption on non-communicable diseases. *Food, 12*(4), 1–20. https://doi.org/10.3390/foods12040687

Piskin, E., Cianciosi, D., Gulec, S., Tomas, M., & Capanoglu, E. (2022). Iron absorption: Factors, limitations, and improvement methods. *ACS Omega, 7*(24), 20441–20456. https://doi.org/10.1021/acsomega.2c01833

Prajapati, P., & Sharma, R. (2024). Advances in food fortification: Ensuring a healthier humanity, types, and analytical methods. *Current Trends in Pharmacy and Pharmaceutical Chemistry, 6*(3), 76–83. https://doi.org/10.18231/j.ctppc.2024.021

Praveen, M., & Brogi, S. (2025a). Microbial fermentation in food and beverage industries: Innovations, challenges, and opportunities. *Food, 14*(1). https://doi.org/10.3390/foods14010114

Praveen, M., & Brogi, S. (2025b). Microbial fermentation in food and beverage industries: Innovations, challenges, and opportunities. *Food, 14*(1), 1–24. https://doi.org/10.3390/foods14010114

Pritwani, R., & Mathur, P. (2015). Strategies to combat micronutrient deficiencies: A review. *International Journal of Health Sciences & Research, 5*(2), 362.

Pua, A., Tang, V. C. Y., Goh, R. M. V., Sun, J., Lassabliere, B., & Liu, S. Q. (2022). Ingredients, processing, and fermentation: Addressing the organoleptic boundaries of plant-based dairy analogues. *Food, 11*(6). https://doi.org/10.3390/foods11060875

Rajput, S. (2012). A global issues – Malnutrition a nutritional problem. *Food and Nutrition Journal, 2*(2), 305–316.

Rangaswamaiah, R. (2023). *ScholarWorks @ UARK Use of Microorganisms for the Enrichment of Zn and Se by using Solid-state Fermentation.*

Ray, R. C., & Montet, D. (2014). Microorganisms and fermentation of traditional foods. In *Microorganisms and fermentation of traditional foods*. https://doi.org/10.1201/b17307

Reinhardt, K., & Fanzo, J. (2014). Addressing chronic malnutrition through multi-sectoral, sustainable approaches: A review of the causes and consequences. *Frontiers in Nutrition, 1*(August), 1–11. https://doi.org/10.3389/fnut.2014.00013

Rezac, S., Kok, C. R., Heermann, M., & Hutkins, R. (2018). Fermented foods as a dietary source of live organisms. *Frontiers in Microbiology, 9*(AUG). https://doi.org/10.3389/fmicb.2018.01785

Rizwanuddin, S., Kumar, V., Naik, B., Singh, P., Mishra, S., Rustagi, S., & Kumar, V. (2023). Microbial phytase: Their sources, production, and role in the enhancement of nutritional aspects of food and feed additives. *Journal of Agriculture and Food Research, 12*(September 2022), 100559. https://doi.org/10.1016/j.jafr.2023.100559

Rizzoli, R., & Biver, E. (2017). Yogurt consumption and impact on bone health. In *Yogurt in health and disease prevention*. Elsevier Inc. https://doi.org/10.1016/B978-0-12-805134-4.00029-8

Rodríguez-Espinosa, T., Voukkali, I., Pérez-Gimeno, A., Almendro Candel, M. B., Hernández-Martich, J. D., Zorpas, A. A., Lucas, I. G., & Navarro-Pedreño, J. (2024). Waste as a sustainable source of nutrients for plants and humans: A strategy to reduce hidden hunger. *Sustainability (Switzerland), 16*(16), 1–23. https://doi.org/10.3390/su16167185

Rollán, G. C., Gerez, C. L., & Leblanc, J. G. (2019). Lactic fermentation as a strategy to improve the nutritional and functional values of pseudocereals. *Frontiers in Nutrition, 6*(July). https://doi.org/10.3389/fnut.2019.00098

Rowland, I., Gibson, G., Heinken, A., Scott, K., Swann, J., Thiele, I., & Tuohy, K. (2018). Gut microbiota functions: Metabolism of nutrients and other food components. *European Journal of Nutrition, 57*(1), 1–24. https://doi.org/10.1007/s00394-017-1445-8

Rusu, A. V., Trif, M., & Rocha, J. M. (2023). Microbial secondary metabolites via fermentation approaches for dietary supplementation formulations. *Molecules, 28*(16). https://doi.org/10.3390/molecules28166020

Salas-Millán, J. Á., & Aguayo, E. (2024). Fermentation for revalorisation of fruit and vegetable by-products: A sustainable approach towards minimising food loss and waste. *Food, 13*(22), 1–18. https://doi.org/10.3390/foods13223680

Samtiya, M., Aluko, R. E., & Dhewa, T. (2020). Plant food anti-nutritional factors and their reduction strategies: An overview. *Food Production, Processing and Nutrition, 2*(1), 1–14. https://doi.org/10.1186/s43014-020-0020-5

Samtiya, M., Aluko, R. E., Puniya, A. K., & Dhewa, T. (2021a). Enhancing micronutrients bioavailability through fermentation of plant-based foods: A concise review. *Fermentation, 7*(2), 1–13. https://doi.org/10.3390/fermentation7020063

Samtiya, M., Aluko, R. E., Puniya, A. K., & Dhewa, T. (2021b). Enhancing micronutrients bioavailability through fermentation of plant-based foods: A concise review. *Fermentation, 7*(2). https://doi.org/10.3390/fermentation7020063

Sarıtaş, S., Portocarrero, A. C. M., Miranda López, J. M., Lombardo, M., Koch, W., Raposo, A., El-Seedi, H. R., de Brito Alves, J. L., Esatbeyoglu, T., Karav, S., & Witkowska, A. M. (2024). The impact of fermentation on the antioxidant activity of food products. *Molecules, 29*(16). https://doi.org/10.3390/molecules29163941

Sarwar, R., Ahmad, N., Mahmood, W., & Ahmad, S. (2025). Plant-based fermented foods: A review of microbiological and sensory properties. https://doi.org/10.5281/zenodo.14824653.

Sawant, S. S., Park, H. Y., Sim, E. Y., Kim, H. S., & Choi, H. S. (2025). Microbial fermentation in food: Impact on functional properties and nutritional enhancement – A review of recent developments. *Fermentation, 11*(1). https://doi.org/10.3390/fermentation11010015

Senanayake, D., Torley, P. J., Chandrapala, J., & Terefe, N. S. (2023). Microbial fermentation for improving the sensory, nutritional and functional attributes of legumes. *Fermentation, 9*(7). https://doi.org/10.3390/fermentation9070635

Seo, M. J. (2024). Fermented foods and food microorganisms: Antioxidant benefits and biotechnological advancements. *Antioxidants, 13*(9), 1–5. https://doi.org/10.3390/antiox13091120

Shah, A. M., Tarfeen, N., Mohamed, H., & Song, Y. (2023). Fermented foods: Their health-promoting components and potential effects on gut microbiota. *Fermentation, 9*(2), 1–23. https://doi.org/10.3390/fermentation9020118

Sharma, G. K. (2021). Advances in Fermented Foods and Beverages. https://doi.org/10.59317/9789390591251.

Sharma, R., Garg, P., Kumar, P., Bhatia, S. K., & Kulshrestha, S. (2020). Microbial fermentation and its role in quality improvement of fermented foods. *Fermentation, 6*(4), 1–20. https://doi.org/10.3390/fermentation6040106

Sharma, R., Diwan, B., Singh, B. P., & Kulshrestha, S. (2022). Probiotic fermentation of polyphenols: Potential sources of novel functional foods. *Food Production, Processing and Nutrition, 4*(1). https://doi.org/10.1186/s43014-022-00101-4

Siddeeg, A., Afzaal, M., Saeed, F., Ali, R., Shah, Y. A., Shehzadi, U., Ateeq, H., Waris, N., Hussain, M., Raza, M. A., & Al-Farga, A. (2022). Recent updates and perspectives of fermented healthy super food sauerkraut: A review. *International Journal of Food Properties, 25*(1), 2320–2331. https://doi.org/10.1080/10942912.2022.2135531

Siddiqui, S. A., Erol, Z., Rugji, J., Taşçı, F., Kahraman, H. A., Toppi, V., Musa, L., Di Giacinto, G., Bahmid, N. A., Mehdizadeh, M., & Castro-Muñoz, R. (2023). An overview of fermentation in the food industry – Looking back from a new perspective. *Bioresources and Bioprocessing, 10*(1). https://doi.org/10.1186/s40643-023-00702-y

Sundarraj, A. A. E., et al. (2018). Biotechnology applications in food processing and safety. *Journal of Advanced Pharmacy Education and Research, 10*(4), 598–603.

Tachie, C., Nwachukwu, I. D., & Aryee, A. N. A. (2023). Trends and innovations in the formulation of plant-based foods. *Food Production, Processing and Nutrition, 5*(1). https://doi.org/10.1186/s43014-023-00129-0

Tamang, J. P., Shin, D. H., Jung, S. J., & Chae, S. W. (2016). Functional properties of microorganisms in fermented foods. *Frontiers in Microbiology, 7*(APR), 1–13. https://doi.org/10.3389/fmicb.2016.00578

Tangyu, M., Muller, J., Bolten, C. J., & Wittmann, C. (2019). Fermentation of plant-based milk alternatives for improved flavour and nutritional value. *Applied Microbiology and Biotechnology, 103*(23–24), 9263–9275. https://doi.org/10.1007/s00253-019-10175-9

Tangyu, M., Fritz, M., Tan, J. P., Ye, L., Bolten, C. J., Bogicevic, B., & Wittmann, C. (2023). Flavour by design: Food-grade lactic acid bacteria improve the volatile aroma spectrum of oat milk, sunflower seed milk, pea milk, and faba milk towards improved flavour and sensory perception. *Microbial Cell Factories, 22*(1), 1–21. https://doi.org/10.1186/s12934-023-02147-6

Teng, T. S., Chin, Y. L., Chai, K. F., & Chen, W. N. (2021). Fermentation for future food systems. *EMBO Reports, 22*(5), 1–6. https://doi.org/10.15252/embr.202152680

Thakur, S., Singh, A., Insa, B., & Sharma, S. (2023). Food fortification in India as malnutrition concern: A global approach. *Sustainable Food Technology, 1*(5), 681–695. https://doi.org/10.1039/d3fb00079f

Todorovic, S., Akpinar, A., Assunção, R., Bär, C., Bavaro, S. L., Berkel Kasikci, M., Domínguez-Soberanes, J., Capozzi, V., Cotter, P. D., Doo, E.-H., Gündüz Ergün, B., Guzel, M., Harsa, H. S., Hastaoglu, E., Humblot, C., Hyseni, B., Hosoglu, M. I., Issa, A., Karakaş-Budak, B., et al. (2024). Health benefits and risks of fermented foods – The PIMENTO initiative. *Frontiers in Nutrition, 11*(September), 1–14. https://doi.org/10.3389/fnut.2024.1458536

Tripathi, A., Pandey, V. K., Panesar, P. S., Taufeeq, A., Mishra, H., Rustagi, S., Malik, S., Kovács, B., Suthar, T., & Shaikh, A. M. (2024). Fermentative production of vitamin B12 by Propionibacterium shermanii and Pseudomonas denitrificans and its promising health benefits: A review. *Food Science and Nutrition, 12*, 8675–8691. https://doi.org/10.1002/fsn3.4428

Tura, D. C., Belachew, T., Tamiru, D., & Abate, K. H. (2025). Nutritional enrichment of traditional complementary foods using underutilized nutritious plant foods in sub-Saharan Africa: Their nutritional potential and health benefits: A scoping review. *Applied Food Research, 5*(1), 100726. https://doi.org/10.1016/j.afres.2025.100726

Upadhyay, S. K., Singh, G., Rani, N., Rajput, V. D., Seth, C. S., Dwivedi, P., Minkina, T., Wong, M. H., Show, P. L., & Khoo, K. S. (2024). Transforming bio-waste into value-added products mediated microbes for enhancing soil health and crop production: Perspective views on circular economy. *Environmental Technology and Innovation, 34*(November 2023), 103573. https://doi.org/10.1016/j.eti.2024.103573

Urrutia-Baca, V. H., Hernández-Hernández, S. N., Martínez, L. M., Dávila-Vega, J. P., & Chuck-Hernández, C. (2024). The role of probiotics in dairy foods and strategies to evaluate their functional modifications. *Food Reviews International, 40*(1), 434–456. https://doi.org/10.1080/87559129.2023.2172426

Vermeulen, S. J., Park, T., Khoury, C. K., & Béné, C. (2020). Changing diets and the transformation of the global food system. *Annals of the New York Academy of Sciences, 1478*(1), 3–17. https://doi.org/10.1111/nyas.14446

Verni, M., Holt, S., & Rizzello, C. G. (2024). Editorial: Microbial fermentation for improved sensory properties and functionality of sustainable foods. *Frontiers in Microbiology, 15*. https://doi.org/10.3389/fmicb.2024.1387745

Voidarou, C., Antoniadou, M., Rozos, G., Tzora, A., Skoufos, I., Varzakas, T., Lagiou, A., & Bezirtzoglou, E. (2021). Fermentative foods: Microbiology, biochemistry, potential human health benefits and public health issues. *Food, 10*(1), 1–27. https://doi.org/10.3390/foods10010069

Wan-Mohtar, W. A. A. Q. I., Ilham, Z., Jamaludin, A. A., David, W., & Mohd Zaini, N. A. (2022). Fermented foods as alternative functional foods during post-pandemic in Asia. *Frontiers in Food Science and Technology, 2*(October), 1–7. https://doi.org/10.3389/frfst.2022.1047970

Wibisono, D. A. S., Saw, C. Y., Wu, T. Y., & Chau, C. F. (2025). Advancing industrial food byproduct management: Strategies, technologies, and Progress in waste reduction. *PRO, 13*(1), 1–27. https://doi.org/10.3390/pr13010084

Wu, Q., Kan, J., Cui, Z., Ma, Y., Liu, X., Dong, R., Huang, D., Chen, L., Du, J., & Fu, C. (2024). Understanding the nutritional benefits through plant proteins-probiotics interactions: Mechanisms, challenges, and perspectives. *Critical Reviews in Food Science and Nutrition, 65*, 1–19. https://doi.org/10.1080/10408398.2024.2369694

Xiang, H., Sun-Waterhouse, D., Waterhouse, G. I. N., Cui, C., & Ruan, Z. (2019). Fermentation-enabled wellness foods: A fresh perspective. *Food Science and Human Wellness, 8*(3), 203–243. https://doi.org/10.1016/j.fshw.2019.08.003

Yang, M., Yang, Z., Everett, D. W., Gilbert, E. P., Singh, H., & Ye, A. (2025). Digestion of food proteins: The role of pepsin. *Critical Reviews in Food Science and Nutrition, 1–22*. https://doi.org/10.1080/10408398.2025.2453096

[illegible]

Fortified Gluten-Free Bakery Products: Technological, Sensorial and Nutritional Approaches

6

Karen F. Irigoytia, Demian A. Molina, Camila Carmona, Nancy N. Espósito, Carolina E. Genevois, and Marina F. de Escalada Pla

Abstract

The prevalence of gluten-related disorders (GRDs) is increasing worldwide. The only currently accepted treatment is the strict adherence to a gluten-free (GF) diet lifelong. Several studies have revealed that a GF diet may lead to nutrient deficiencies, especially in dietary fibre, vitamins, and minerals such as iron. Iron deficiency anaemia is commonly present in coeliac disease, recently diagnosed,

K. F. Irigoytia · N. N. Espósito · C. E. Genevois
Instituto de Ciencia y Tecnología de los Alimentos de Entre Ríos (ICTAER), CONICET-Universidad Nacional de Entre Ríos (UNER), Gualeguaychú, Entre Ríos, Argentina

Facultad de Bromatología, UNER, Gualeguaychú, Entre Ríos, Argentina

Consejo Nacional de Investigaciones Científicas y Técnicas (CONICET), Buenos Aires, Argentina

D. A. Molina
Consejo Nacional de Investigaciones Científicas y Técnicas (CONICET), Buenos Aires, Argentina

Facultad de Ciencias Exactas y Naturales (FCEN), Departamento de Industrias, Universidad de Buenos Aires (UBA), Buenos Aires, Argentina

Instituto de Tecnología de Alimentos y Procesos Químicos (ITAPROQ), CONICET – Universidad de Buenos Aires, Buenos Aires, Argentina

C. Carmona
Facultad de Bromatología, UNER, Gualeguaychú, Entre Ríos, Argentina

M. F. de Escalada Pla (✉)
Facultad de Ciencias Exactas y Naturales (FCEN), Departamento de Industrias, Universidad de Buenos Aires (UBA), Buenos Aires, Argentina

Instituto de Tecnología de Alimentos y Procesos Químicos (ITAPROQ), CONICET – Universidad de Buenos Aires, Buenos Aires, Argentina
e-mail: marina@di.fcen.uba.ar

G. Sanghvi et al. (eds.), *Sustainable Food Fortification*,
https://doi.org/10.1007/978-981-95-1217-1_6

and could be persistent despite adherence to a GF diet. GF breads (GFBs) are mainly formulated based on refined flours and starches, which are, in general, not fortified as the gluten-containing counterpart, contributing to an unbalanced diet in people with GRDs. There are several approaches to enhance the iron content; however, the effect of mineral addition on technological, sensory, and nutritional properties should be taken into account. This chapter aims to review different fortified food-based approaches, including traditional strategies such as the addition of different iron compounds; the inclusion of ingredients naturally iron-source such as whole cereals, pseudocereals, and pulses; the application of technological tools for the enhancement the iron-bioaccessibility in GFBs; and finally, to discuss the new challenges and opportunities to be solved.

Keywords

Coeliac disease · Gluten-related disorders · Anaemia · Iron fortification · Bioaccessibility

6.1 Introduction

GRDs include a spectrum of diseases caused by gluten exposure, with a global prevalence of 5%. The CD is the most prevalent among the GRDs, affecting 1% of the population. The adherence to a GF diet is the only treatment available for lifelong (Di Sario et al., 2023); it may also contribute to micronutrient deficiencies (MNDs). These deficiencies affect both developed and developing countries due to a combination of socioeconomic, cultural, and dietary factors. Among MNDs, iron deficiency is the most prevalent, and it is recognised as the main cause of anaemia. Iron deficiency anaemia (IDA), which accounts for approximately 30–50% of cases worldwide (Hurrell, 2022), affects one in four people (Rai et al., 2022; Young et al., 2023). The reduction of IDA is a priority on the global nutrition agenda due to the maternal and child health impact, as well as socioeconomic and country development. All countries are projected to continue with a prevalence of anaemia ≥15% in 2025, which will not meet the World Health Organisation (WHO) target of reducing the prevalence by 50% compared to 2012 levels (WHO, 2012; Hasan et al., 2022).

Beyond the challenges related to iron absorption, individuals following a GF diet may also face additional difficulties due to the nutritional composition of GF products, which often lack essential nutrients as a consequence of their formulation and limited fortification. Understanding the impact of MNDs in this population is crucial for developing strategies to improve the nutritional adequacy of a GF diet.

Considering the high consumption of bread worldwide (>70 kg/per capita/year in Western countries), it is a key vehicle to address these MNDs. Food fortification has been used as a global strategy to address nutritional deficiencies, particularly among vulnerable populations (Olson et al., 2021). Additionally, the incorporation of deficient nutrients through the addition of raw materials that are natural sources of essential components could provide a complementary approach. Although it could be a promising solution to improve the nutritional profile of GF baked goods, it presents new challenges to be solved by the industry and the scientific community (Papagianni et al., 2024). Therefore, it is necessary to study and develop new technological, sensory, and nutritional approaches to ensure that these fortified or enriched products meet either the nutritional needs or the sensory consumers′ expectations.

6.2 GRDs and the GF Diet: An Overview

Cereal grains are the primary food source worldwide, especially in developing countries, where they serve as a crucial part of the diet. These cereals are particularly valued for their cost-effectiveness, adaptability, and versatility in several food applications. Among cereals, wheat has a high production volume. Its global per capita consumption is approximately 67.7 kg/year, making a key contribution to calorie and protein intake (Ibba et al., 2023; FAO, 2024). Furthermore, due to its gluten content, it provides products with unique functional and technological properties.

Gluten is a network developed during kneading due to the union of water molecules with prolamins (storage proteins in the endosperm of cereals). The main function is to provide elasticity and extensibility to the dough, favouring the retention of gases during fermentation and giving the characteristic texture of baked products (Monteiro et al., 2021). The prolamins present in wheat, barley, and rye, called gliadin and glutenin, hordeins and secalins, respectively, are known to have specific amino acid sequences that can trigger an immune response in genetically predisposed individuals, and other GRDs (Sharma et al., 2020). GRDs include a spectrum of diseases caused by gluten exposure, including autoimmune diseases such as CD, dermatitis herpetiformis, and gluten ataxia, and conditions considered to be allergic and non-autoimmune-allergic, such as wheat allergy and non-coeliac gluten sensitivity, with a global prevalence of 5% (Sapone et al., 2012; Tovoli et al., 2015; Taraghikhah et al., 2020). Their prevalence and distribution are related to environmental, cultural, and dietary factors, as well as the emergence of new wheat strains and changes in agricultural practices (Volta et al., 2013). Advances in understanding physio pathogenesis of GRDs have contributed to the effective detection and development of diagnostic methods (Kelly et al., 2015). Despite these advancements and the increasing information available on GRDs, a significant number of cases still go undiagnosed (Choung et al., 2016).

The most common GRDs is the CD. A meta-analysis reported by Singh et al. (2018) revealed that CD prevalence was highest in Europe and Oceania (0.8%), followed by Asia (0.6%), Africa and North America (0.5%), and South America (0.4%). Differences were also observed in the sex and age, with higher prevalence in females with respect to males (0.6% vs. 0.4%), and in children compared to adults (0.9% vs. 0.5%). Intra-regional studies of prevalence suggest that dietary patterns and the increasing adoption of a Western-style diet could play a significant role in the incidence of CD, as has been observed in Asian countries (Kelly, 2016;

Rizzello et al., 2019). For instance, there is a significant geographic variation between northern (1.2%) and southern (0.1%) regions in India. Geographic, genetic, and dietary patterns, particularly high wheat consumption, could contribute significantly to the variations observed (Ramakrishna et al., 2016).

The typical intestinal damage of CD is characterised mainly by histological lesions such as villous atrophy, which leads to malabsorption syndrome and contributes to nutrient deficiencies. Currently, the only effective and safe treatment of CD is the adoption of a lifelong GF diet that could also lead to MNDs and macronutrient excesses, which require special attention (Bianchi et al., 2024).

The level of adherence to a GF diet in adults with CD ranges between 36% and 96%, influenced by factors such as education, socioeconomic status, lack of information about the disease, motivation, symptoms, and social support (Villafuerte-Galvez et al., 2015; Wieser et al., 2021). In young children, parental support is essential, while in adolescents, the situation is more complex, since their greater autonomy, peers, and social environment make adherence more challenging (Garg & Gupta, 2014). Other obstacles include the high cost of GF products, which can be 46–443% higher due to production and certification processes (Myhrstad et al., 2021; Khairuddin & Lasekan, 2021; Estévez et al., 2024), the limited availability, and the lack of clear information on labels (Missbach et al., 2015; Arias-Gastelum et al., 2018; Xhakollari et al., 2019; Cagnasso et al., 2023).

The voluntary or involuntary ingestion of gluten-contaminated foods can trigger symptoms and complications associated with CD. Maintaining a strict diet and ensuring safety through education is important to avoid cross-contamination of GF products (Silvester et al., 2016; Stefanelli et al., 2020; Parsons et al., 2021). In line with this, Di Sario et al. (2023) showed that adherence to a GF diet is crucial for the management of CD and other GRDs, as the continuous ingestion of trace amounts of gluten (>10–50 mg/day) can damage the intestinal mucosa and increase the risk of complications.

The GF diet should be balanced and sufficient in terms of energy and nutritional requirements. Foods allowed in this diet are naturally GF legumes, fruits and vegetables, meat, eggs, and dairy products, as well as products formulated with GF cereals such as baked goods and pastas (El Khoury et al., 2018). GF products generally contain higher saturated and hydrogenated fatty acids, simple carbohydrate, and salt content, lower protein and dietary fibre (DF) content, and are deficient in essential vitamins and minerals such as iron, calcium and magnesium and Vitamin D, E, B12 and folate, with respect to their gluten-containing counterparts (Vici et al., 2016; Cardo et al., 2021).

It is noteworthy that the GF diet is not only adopted by people diagnosed with GRDs. Healthy consumers, family members, and friends of CD can also eat GF products as a lifestyle (Jivraj et al., 2022; de Escalada Pla et al., 2023). For these reasons, the GF food market is expected to experience growth in the coming years, mainly in the bakery products segment (Peñalver et al., 2023). This nutritional imbalance is important, as long-term adherence to a GF diet can increase the risk of MNDs and other health complications.

6.3 MNDs: IDA in People with GRDs

Vitamins and minerals are essential micronutrients that play a fundamental role in all physical and biochemical processes of the body, maintaining homeostasis by acting as cofactors, coenzymes, and regulators in the metabolism (Hernández-Flores et al., 2022). When the daily recommended intake is insufficient, the body cannot carry out these functions efficiently, leading to metabolic and physiological alterations with a negative impact on human health. MNDs affect both developed and developing countries, due to a combination of socioeconomic, cultural, and nutritional factors (de Escalada Pla et al., 2020; Talarico et al., 2021). They are often related to a continuous cycle of malnutrition and can be associated with deficiencies in protein or energy intake (Bailey et al., 2015), being the most prevalent iron, zinc, iodine, and vitamins A, D, and B12 (Espinosa-Salas & Gonzalez-Arias, 2023). Even if the levels were moderately detectable through biochemical or clinical analysis, they could have a significant negative impact on body functions (Genevois et al., 2016).

Among MNDs, iron deficiency is highly prevalent and is recognised as the main cause of anaemia. IDA, which accounts for approximately 30–50% of cases worldwide (Miller, 2013), as was previously discussed in Chap. 2. The reduction of IDA is a priority on the global nutrition agenda due to its impact on maternal and child health, as well as on socioeconomic country development.

Malabsorption syndrome, characteristic of CD, is associated with decreased iron absorption (Lopez et al., 2016; Mahadev et al., 2018). It is important to highlight that CD patients recently diagnosed often present a high prevalence of IDA (10–80%), mainly due to the damaged intestinal epithelium (Efthymakis et al., 2017; Rondanelli et al., 2019). The severity of IDA in these patients depends on the length of time they have had the disease undiagnosed, the intestinal damage, and the degree of malabsorption (Niewinski, 2008). Some studies have shown that, after 8–12 years of following a GF diet, some nutritional deficiencies continue mainly due to the low nutrient intake inherent in the GF diet (Saturni et al., 2010; Skodje et al., 2019; Kreutz et al., 2023). Lack of adherence to a GF diet, cross-contamination, a low-iron diet, or other factors not yet studied, such as alterations in iron absorption, genetic factors, or other underlying mechanisms, could be related to the persistence of IDA (Efthymakis et al., 2017; Stefanelli et al., 2020; Seidita et al., 2022). The limited variety of foods consumed, poor nutritional quality, and limited availability of fortified GF foods could partly explain the high prevalence of IDA in these consumers (Vici et al., 2016; Roman et al., 2019; Khairuddin & Lasekan, 2021).

These findings highlight the importance of reformulating and fortifying GF products to prevent nutritional deficiencies in this population. In turn, iron supplementation should be considered, especially in vulnerable populations (Caruso et al., 2013; Montoro-Huguet et al., 2021; Bianchi et al., 2024). In this context, since bread is a daily staple food in many countries and provides a significant proportion of energy and nutrients, especially in developing regions (González et al., 2022), it represents an opportunity to be used as a nutrient carrier (Collar, 2015).

6.4 Nutritional Profile of GFBs: Current Status and Challenges

Bread is one of the most consumed food products worldwide, given its central role in many diets. Therefore, bread is not only a staple food but also a key source of nutrients. GFBs are often of lower quality than their gluten-containing counterparts, especially in terms of texture, flavour, appearance, nutritional profile, and shelf life. Understanding the current situation and the main challenges of these products is essential for improving their formulations, taking into account consumer needs and expectations.

GF doughs are fluid and viscous and have lower cohesion, leading to difficulties in gas retention and dough stability during proofing and baking (Irigoytia et al., 2025). Consequently, GFBs often present a lower specific volume (SV), a dry, crumbly and compact crumb, a lower crust colour and higher staling rate (Genevois & de Escalada Pla, 2021). To simulate the gluten network and improve the structure and sensorial characteristics, GFBs require the use of functional ingredients and additives such as non-starch polysaccharides, modified starches, proteins, enzymes, emulsifiers, and antimicrobial agents (Bender & Schönlechner, 2020; Šmídová & Rysová, 2022; Sciarini et al., 2023).

The most used raw materials in the formulation of GFBs are rice and maize flours, commonly combined with potato, corn, or cassava starches (Santos et al., 2019; Romão et al., 2021). These ingredients may be the most abundant and the cheapest; however, they are rich in simple carbohydrates that contribute to increasing the product's glycaemic index (GI) and glycaemic load of the meal, consequently, the postprandial blood glucose level (Marciniak et al., 2021). The presence of other components of the formulation, such as fat, protein, DF, hydrocolloids, and/or enzymes such as transglutaminase, could either delay the gastric emptying or interfere with starch digestion, reducing the GI (Sciarini et al., 2023).

The food processing and storage could also modify the starch structure, resulting in the formation of resistant starch (RS) type III when molecules undergo retrogradation, RS type IV when they are chemically modified, and RS type V when amylose-lipid interact forming a complex (Culetu et al., 2021; Srikanlaya et al., 2022; Sciarini et al., 2023). The RS cannot be hydrolysed in the human digestive tract by endogenous enzymes; therefore, dietary interventions using RS and/or the addition of RS in GFB formulation may contribute to delay glucose metabolism and insulin sensitivity and reducing chronic disease risks in consumers with GRDs.

GFBs present a significantly lower protein content, providing between 9% and 16% of the reference nutrient intake per serving size, while traditional wheat bread provides between 17% and 22% (Allen & Orfila, 2018). On the other hand, a recent study showed that the fat content in GFBs ranged from 1.5 to 7.8 $g.100\ g^{-1}$ (w.b.), compared to gluten-containing bread (1.5–3.1 $g.100\ g^{-1}$ w.b.). GFBs were also characterised by a higher variability in their fatty acid composition, that can be explained due to the several oils used in their formulations, for instance coconut, palm, olive, sunflower and soybean oils; meanwhile, traditional breads used mainly sunflower and soybean oils, resulting in a more uniform composition (Tres et al., 2020). The higher content of fats, mainly saturated, observed in commercial GFBs may be due to the emulsifiers that are used to optimise their texture and palatability (Miranda et al., 2014). Besides, fat also stabilises the gas bubbles during proofing and reduces the rate of staling, improving the shelf-life (Pellegrini & Agostoni, 2015).

Regarding DF content, Shepherd and Gibson (2013) observed that GFBs analysed in different studies showed a high variability in DF content, with average values ranging from 0.7 to 9.6 g.100 g^{-1}, explained by the ingredient variations in GFB formulations. These results are consistent with those reported by Djordjević et al. (2021), where variations in DF content in commercially available GFBs are significant, with an average value of 5 ± 3%. In general, gluten-containing breads have significantly higher DF content than GFBs, highlighting the need to improve the nutritional quality of these products (Theethira & Dennis, 2015; Myhrstad et al., 2021; Aguiar et al., 2023).

GF flours used in bread making tend to be low in essential minerals and vitamins, such as iron, manganese, pyridoxine, riboflavin, thiamine, niacin, folate, and vitamin B5 (Melini & Melini, 2019; Larretxi et al., 2019; Bozorg et al., 2024). Additionally, unlike wheat flour, these ingredients are not fortified with essential nutrients, and the levels of these nutrients are lower in rice flour compared to wheat flour, contributing to the nutritional gap between these products (Vici et al., 2016; Diez-Sampedro et al., 2019; Skodje et al., 2019). For instance, a recent study revealed that in 228 commercially available GFBs of different brands around the world, the addition of iron was present in only 7% (Theethira & Dennis, 2015; Roman et al., 2019). In the UK market, only 5% of GFBs were fortified with all four mandatory fortified nutrients, whereas 28% were fortified with calcium and iron (Allen & Orfila, 2018). In GF products such as cereals, breads, flours, and pastas, the iron content was 3 to 4.5-fold lower than gluten-containing counterparts (Jamieson et al., 2018). In addition, another study showed that commercial GFBs contained 67% less iron than traditional breads, contributing 6.9% of the daily value (DV) of iron per 100 g of bread, while gluten-containing breads contributed to 21% of the DV (Jegede et al., 2021).

It is important to address the nutritional challenges associated with GFBs, particularly regarding their micronutrient content. Local and sustainable resources should be considered as fortifying strategy to improve nutritional profile of GFBs ensuring offer, quality and stability in the time.

6.5 Food-Based Approaches for IDA Prevention

Commonly used strategies to prevent and mitigate MNDs in developing countries are supplementation, by providing micronutrients in capsules, tablets, drops, and food-based approaches. These prevention strategies should be part of an integral public health intervention programme, along with education, breastfeeding promotion, and control of infectious and parasitic diseases (FAO and Institute International Life Sciences, 1997a, b). Despite being challenging, food-based approaches are a preventive, cost-effective, and long-term sustainable alternative for addressing

MNDs. Some examples include fortification, biofortification, diversification, and dietary modification.

Food-based approaches play an essential role in addressing IDA in people with GRDs, as they increase the availability and consumption of iron-rich foods in accordance with cultural practices (Bechoff et al., 2023). However, when implementing these strategies, it is necessary to understand their potential effects on the technological and sensory parameters of GFBs to avoid undesirable colours and flavours, provide stability during baking and storage, and ensure the bioaccessibility and bioavailability of minerals (Genevois et al., 2014; de Escalada Pla et al., 2020). Table 6.1 summarises some of the advances and strategies applied to increase iron content and the effects observed on the GFBs' main properties.

6.5.1 Iron Enrichment or Fortification of GFBs

This section explores the use of enrichment or fortification with different iron compounds and their effects on the quality parameters of GFBs. Food fortification or enrichment is considered the most sustainable and cost-effective strategy to compensate for mineral deficiency. According to the Food and Agriculture Organisation (FAO) and WHO, food fortification is defined as the addition of one or more essential nutrients, whether or not it is normally contained in the food, with the aim of preventing or correcting nutritional deficiencies in the population (FAO, 1997a, b). The term enrichment has been commonly used as a synonym for fortification although some authors define it as the intentional increase at the baseline level in the content of some micronutrients that have been removed during processing (Dary & Mora, 2013).

Mass and targeted programme fortification are implemented by government authorities as public health policies in many countries. Meanwhile, market-driven is a voluntary fortification of a commercial food by the manufacturers under the regulatory limits. As mentioned above, GF flours and/or products are excluded from enrichment or fortification programmes in most countries, and their inclusion is voluntary for the food industry (Jastrebova & Jägerstad, 2013; Mitchell et al., 2019; Jeanes et al., 2022).

Several iron compounds are available for fortifying foods, presenting different properties like water solubility, bioavailability, stability, and cost. The WHO recommends the use of ferrous sulfate, ferrous fumarate, ferric pyrophosphate, and electrolytic iron powder for cereal flour fortification. There is a preference for water-soluble iron compounds such as ferrous sulfate due to the highest bioavailability and lower cost (FAO and WHO, 2006; Hurrell, 2022). The addition of iron as a food fortifier represents a special challenge due to its reactivity, limited bioavailability, consumer acceptance, and tendency to interact with components of the food matrix, especially cereal-based foods (de Escalada Pla et al., 2020; Hurrell, 2022).

Several studies have reported the use of different iron compounds to fortify GFBs (Table 6.1). Kisikini et al. (Kiskini et al., 2010) have shown that specific

Table 6.1 Strategies for improving iron content in GFB formulation

		Gluten-free flour formulation base	Main results	References
Food fortification	*Iron compounds*			
	Ferric pyrophosphate, emulsifiers, sodium iron EDTA, ferrous sulfate microencapsulated and electrolytic (elemental) iron	Soy meal, corn starch, amaranth flour	GFB formulation up to 40 mg Fe kg^{-1}. High impact on colour and texture parameters. Electrolytic iron presented less adverse changes to food, but it is usually poorly absorbed	Kiskini et al. (2012)
	Ferric pyrophosphate, emulsifiers, sodium iron EDTA, ferrous sulfate microencapsulated and reduced iron	Soy meal, corn starch, amaranth flour	Fortified-bread characteristics depend on iron type, but not on iron solubility, as well as bread type: Wheat or GFB. Addition of 40 mg kg^{-1} of dry ingredients	Kiskini et al. (2010)
	Electrolytic iron, sodium iron EDTA, ferric pyrophosphate, ferric pyrophosphate with emulsifiers, ferrous bis-glycinate, ferrous gluconate, ferrous lactate and ferrous sulfate	Rice flour, amaranth flour, maize and potato starches	Effects on taste an colour. GFBs with ferric pyrophosphate with emulsifiers, most acceptable and high dialyzable total iron rate ($\approx$ 6%)	Kiskini et al. (2007)
	Iron gluconate (organic compound) and ferrous sulfate heptahydrate (non-organic compound), seeds: Poppy, sunflower and seeds; hazelnuts	Buckwheat or rice and amaranth flour, corn and potato starches	Iron content of GFB with non-organic and organic compounds and seeds: 19.7, 22 and 22.8, mg kg^{-1} rice GFB db, respectively; 22.3, 27.3 and 31.7, mg kg^{-1} buckwheat GFB db, respectively. Rice GFB presented the highest iron bioavailability	Rogaska et al. (2021)

(continued)

Table 6.1 (continued)

		Gluten-free flour formulation base	Main results	References
Food-to-food fortification	*Iron-rich ingredients*			
	Whole buckwheat flour 10%, 20%, 30%, and 40% w/w of corn-starch replacement	Corn and potato starches	The addition of 30% and 40% buckwheat flour improved the specific volume and the size loaf. The 40% buckwheat flour bread contained 54.28 mg Fe kg^{-1}	Krupa-Kozak et al. (2011)
	20% amaranth or Indian ricegrass (*Achnatherum hymenoides*) flour	White rice flour, potato and tapioca starches, soy meal	Compared to commercial GFB, which contains 9.5 mg Fe kg^{-1}, higer content was obtained with FtFF-GFB. GFB with amaranth flour (17.1 mg Fe kg^{-1}) and GFB with Indian ricegrass (12.9 mg Fe kg^{-1}). No significant differences were reported in sensory attributes	Breshears and Crowe (2013)
	Whole milk powder (4%), egg yolk (1%), amaranth flour (3%), hazelnuts (3%), flax (4%), poppy, sunflower and pumpkin seeds (4–5%)	Buckwheat or rice flour, corn and potato starches	Rice-GFB contained 11.9 mg Fe kg^{-1}, increasing to 19.2 with milk and 23.8 with milk and seeds. Buckwheat-GFB contained 20.6 mg Fe kg^{-1}, increasing to 28.4 with milk addition and to 35.1 with milk and seeds. Rice-GFB, without additions, presented the highest iron bioavailability (43.8%)	Regula et al. (2018)
	Green microalgae species (*Nannochloropsis gaditana* L2 and *Chlamydomonas* sp. EL5) addition up to 3.0%.	Rice and buckwheat flours, potato starch	The addition of 3% of the microalgae significantly increased the firmness and adhesiveness and decreased the luminosity. A deep green colour of the supplemented bread was appreciated by consumers. GFBs with 3% w/w microalgae, contained 90 mg kg^{-1}	Khemiri et al. (2020)
	Okara, replacement in the 10%, 20% and 30% formulation	Buckwheat, rice, millet, flours	GFB with 30% okara obtained the highest scores in individual sensory characteristics and very high overall acceptability score (8.71 in a 9-point hedonic scale). Iron content in GFB with 30% okara was 27.71 mg kg^{-1}	Pešić et al. (2023)

Combined strategies	*Fortification + biotechnological processes*			
	Sourdough prepared from buckwheat, quinoa or amarant flours	Rice flour and corn starch	Compared to the control, GFB with 45% replacement of pseudocereal sourdough showed a 45% increase in ash content and a 2.6-fold increase iron content	Yeşil and Levent et al. (2022)
	Flours of hemp, chestnut, carob with lactic acid bacteria (LAB)	Rice, corn and psyllium flours	Hemp and chestnut flour mixture GFBs presented higher specific volume and lower L* values. Iron content increased up to 17.7 times compared to the control bread. Increases were also observed in ca, Mg, P, K, and Zn, as well as in vitamins B1, B3, and B9	Montemurro et al. (2023)
	Ferrous sulfate, soy meal and *L. casei*	Rice flour, coarse and bran, cassava starch	The addition of 30 mg kg^{-1} of iron salt along with acid lactic fermentation improved GFB specific volume, crumb colour, and aeration rising iron bioaccessibility to 85%	Castellanos-Fuentes et al. (2022)
	Spirulina platensis (*Arthrospira platensis*) at 3%. Sourdough (quinoa, amaranth or brown rice, depending on the type of bread)	Corn starch, amaranth, buckwheat and chickpea flours	With the inclusion of spirulina in bread, a significant decrease in the parameters L*, a*, b*. darker and greener-toned breads. Iron content in the GFB with spirulina ranged from 11.74 to 13.06 mg kg^{-1}. GFBs with amaranth sourdough and spirulina presented the highest iron bioaccessibility (~90%)	Peñalver and Nieto (2024)

volume (SV) and relative crumb elasticity of GFBs were not significantly affected by the different iron compounds. However, a greater crust firmness was observed in GFBs containing ferric pyrophosphate, ferric pyrophosphate with emulsifiers, and sodium iron EDTA. Sensory parameters did show significant differences, especially in terms of pore uniformity, moisture smell, metallic taste, and stickiness. Even though iron fortification had a significant impact on the aroma of the samples, those fortified with reduced iron were the most sensorially accepted. On the other hand, microencapsulated iron sources were expected to have a low impact on these sensory parameters, since it has a physical barrier that protects the mineral from interacting with the food matrix (Bryszewska, 2019). Nevertheless, it was found that microencapsulated ferrous sulfate addition negatively affected the flavour and colour of the GFBs, probably due to the instability of the iron compound at high temperatures during baking (Mattar et al., 2022).

The iron fortification with ferric pyrophosphate, ferric pyrophosphate and emulsifiers, and sodium iron EDTA increased the crust firmness of GFBs; meanwhile, no differences were observed in GFB ferrous sulfate and electrolytic iron with respect to the control bread. Descriptive sensory analysis showed that GFBs fortified with ferric pyrophosphate had higher inclusion of air cells in crumb, ferric pyrophosphate with emulsifiers increased the crumb stickiness, and ferrous sulfate had the most intense metallic taste (Kiskini et al., 2012).

Based on an optimised GFB formulation with rice flour:coarse:bran (45:35:20) (Genevois et al., 2020), the authors studied the addition of ferrous sulfate heptahydrate in a GFB formulation, up to 149.28 mg of ferrous sulfate heptahydrate (equivalent to Fe^{2+} content of 30 mg·kg^{-1} of GF flour) (Fig. 6.1). A reduction of SV ≈ 40% was recorded with iron salt addition. Possibly, the presence of the mineral could affect the CO_2 production and/or retention during proofing, reducing the dough expansion and affecting the SV after the baking process (de Gusmão et al., 2016).

Fig. 6.1 Image of GFB loaves and the crumb alveolar structure (cross section). (**a**). M1 (control system), (**b**). M2 (15 mg Fe^{2+}·kg^{-1} of GF flour), (**c**). M3 (30 mg Fe^{2+}·kg^{-1} of GF flour)

Nevertheless, non-significant effects were observed on the parameters of the texture profile analysis, giving average values of hardness 114 N, cohesiveness 1.7, springiness 0.96, gumminess 185 N, chewiness 185 N, and resilience 0.33. Meanwhile, slight and significant colour change on crumb could be observed only with reduction (≈15%) of the a* coordinate, indicating that the maximum iron addition (30 mg·kg^{-1}) reduced red tone intensity (Fig. 6.1). It must be highlighted that the iron content determined on final bread resulted in 63.6 mg kg^{-1} GFB (d.b.). Raw materials, mainly rice bran, are a natural source of iron, as could be observed in the control GFB system, without iron addition, the determined iron content reached 19.2 mg Fe kg^{-1} GFB (d.b.). Additionally, one serving size (50 g) of GFB would contribute to 12% of the iron Recommended Dietary Allowance in adults; meanwhile, this value could be triplicate when ferrous sulfate heptahydrate is added up to 149.28 mg kg $^{-1}$ GF flour. The findings suggest that the selected iron compound and levels added to a basic GFB recipe did not affect relevant quality parameters, such as crumb colour and texture crumb, that could affect the purchase intention in GRDs consumers (Irigoytia et al., 2024), representing an adequate food vehicle for providing iron to the diet (Fig. 6.1).

As regards the bioavailability and bioaccessibility of iron in fortified GFBs, the term bioavailable refers to the efficiency with which a component is used systematically through normal metabolic pathways. However, before becoming bioavailable, the compound must be released from the food matrix during the passage through the gastrointestinal tract. Thus, bioavailability includes the term bioaccessibility, and both terms are part of nutritional effectiveness (de Escalada Pla et al., 2020).

The ferrous and total (ferrous and ferric) iron dialyzable after in vitro gastrointestinal digestion of different iron compounds added to a GFB formulation with 40% of amaranth flour was studied by Kiskini et al. (2007). The results showed that GFBs fortified with ferric pyrophosphate, ferrous bis-glycinate, ferrous gluconate, or ferrous sulfate, presented higher iron dialyzable than GFBs fortified with electrolytic iron, ferrous lactate, or ferric pyrophosphate. Among the iron compounds tested, ferric pyrophosphate with emulsifiers and NaFeEDTA had the highest formation of ferrous and total dialyzable iron. The ferrous dialyzable iron exhibits a better correlation with iron uptake by cells and iron absorption by humans; therefore, it is considered a better index. Additionally, the sensory evaluation of GFBs fortified was also evaluated, showing that ferric pyrophosphate with emulsifiers, ferrous bis-glycinate, or ferric pyrophosphate in GFBs received the highest average quality score (>5), corresponding to a medium/satisfactory quality product. On the contrary, GFBs fortified with ferrous lactate received the lowest score (<3), corresponding to an unacceptable quality with no desirable characteristics.

It is important to highlight that sodium iron EDTA is often used as a fortificant in cereals rich in phytates because it prevents the formation of complexes between these compounds, improving the iron bioavailability compared to ferrous sulfate. Despite this nutritional advantage, the high cost of sodium iron EDTA limits its use in the food industry, and it is usually combined with ferrous sulfate (FAO and WHO, 2006; Kaur et al., 2022).

Castellanos-Fuentes et al. (2022) have shown that fermentation with probiotics, such as *L. casei*, favors intestinal iron solubility and bioaccessibility. In that study, the authors investigated the in vitro bioaccessibility of ferrous sulfate heptahydrate, added at a concentration of 30 mg kg^{-1} in the premix of a GFB formulation containing a soy-based functional ingredient (FI) supporting *L. casei* (ATCC 393™). The GFBs with the addition of the FI and ferrous salt presented the highest SV (2.3 ± 0.2 $cm^3 \cdot g^{-1}$) and the lowest pH (5.69 ± 0.01), compared to the control, suggesting higher acidity and potential improvement in CO_2 retention during leavening. Furthermore, the formulation exhibited significantly higher iron bioaccessibility (85 ± 5%), compared to the control system (76 ± 3%), probably because during the bread fermentation, along with the yeast, the probiotic strain had the ability to reduce pH, generating organic acids that favoured the released of iron from the matrix, increasing its solubility in the in vitro gastrointestinal conditions.

On the other hand, the simultaneous addition of iron as a fortificant and the use of GF raw materials and/or ingredients with high iron content in GFB formulations is another strategy to improve the nutritional profile. Rogaska et al. (2020) carried out a comparative study of iron bioavailability in GFBs based on buckwheat flour or rice flour enriched with natural (poppy seeds and milk) or synthetic (iron gluconate) additives. As part of the experimental design, iron levels were determined in peripheral blood and organs such as the liver, spleen, heart, kidney, and left femur of rats. Increased body weight was observed in rats fed breads with added milk and poppy seeds. No significant differences in femur iron content were detected across the formulations. It was observed that diets containing GFBs enriched with synthetic additives did not improve iron levels in blood, liver, or kidney. Conversely, GFBs incorporating buckwheat with synthetic additives showed significantly higher iron content in the spleen. Moreover, GFBs with natural additives resulted in higher blood iron levels, suggesting improved iron release from the food matrix and enhanced bioavailability. In a following study (Rogaska et al., 2021) GFBs without additives or with natural additives and synthetic additives (organic: iron gluconate; non-organic: ferrous sulfate heptahydrate) were compared. After in vitro digestion, it was observed that breads formulated with rice flour fortified showed a high bioaccessibility, presenting values between 43% and 46%, with no significant differences between their formulations. The GFBs based on buckwheat, the additions of ferrous sulfate heptahydrate showed the highest iron bioaccessibility (40%) compared to the control (17%). This study highlights the importance of the ingredients used in GBF formulation and the type of iron compound that should be taken into account to optimise the mineral release from the food matrix and the absorption.

6.5.2 Incorporation of Iron-Rich Ingredient GBFs: A Sustainable Proposal for Increasing the Iron Dietary Intake

This section explores the incorporation of naturally iron-rich raw materials or ingredients from cereals, pseudocereals, and pulses as a comprehensive and long-term sustainable solution beyond conventional fortification strategies. Recently, a strategy called Food-to-Food Fortification (FtFF) has been reported in the literature (Kruger et al., 2020; Bechoff et al., 2023; Thakur et al., 2023). FtFF involves incorporating micronutrient-rich foods into other staple foods in order to improve their nutritional content and promote dietary diversification. Furthermore, it encourages the use of local raw materials, prioritising the use of neglected and underutilised crops and supporting producers in rural areas to generate nutritional and economic benefits (Desire et al., 2021). This approach offers a sustainable and cost-effective alternative to conventional fortification strategies.

In this sense, Table 6.1 summarises some relevant studies on the incorporation of iron-rich ingredients in GFB formulations, through FtFF approaches. In addition, Fig. 6.2 shows the iron content of emerging GF crops and raw materials commercially used. In general, it is observed that GF flours (millet, rice bran, teff, buckwheat, quinoa, sorghum, brown rice, carob) have significantly higher iron content (mg.100 g^{-1} flour) than fortified wheat flour, which may be relevant for the formulation of GF products with a better contribution of this mineral. However, the presence of iron absorption inhibitors (phytates, polyphenols, calcium, and phosphate) in the food matrix, and the technological processing that maximises the bioavailability and bioaccessibility of the mineral should be considered (Nkhata et al., 2018; Hurrell, 2021; Curti et al., 2023).

Overall, cereals and/or pseudocereals such as buckwheat, amaranth, quinoa, sorghum, and millet have been included in GBF formulations based on corn starch, potato starch, and rice flour to improve the quality of GFBs. Several studies have reported that these raw materials are also characterised by high protein and DF content. Krupa-Kosak et al. (2011) found that GFB formulated with 40% buckwheat had the highest protein content, reaching 7.5% db. According to Breshears and

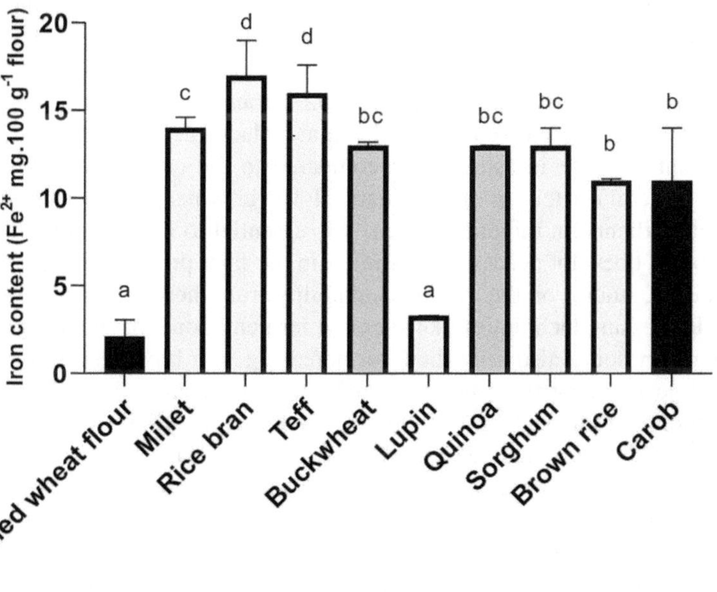

Fig. 6.2 Iron content of GF cereals, pseudocereals, and pulse flours (Fe^{2+} mg.100 g^{-1} flour)

Crowe (2013), bread made with ricegrass flour had a 59% higher DF content than the commercial product, while bread formulated with amaranth showed an approximate 40% increase in DF compared to the commercial reference. Similarly, Khemiri et al. (2020) observed that incorporating microalgae into GFB formulations increased protein, iron, and calcium levels while enriching the fatty acid profile with polyunsaturated fatty acids (ω3 y ω6) and antioxidant compounds. Incorporating these ingredients not only improves the nutritional value of GFBs but also enhances other characteristics such as texture. Krupak-Kosak et al. (2022), reported that the addition of linseed oil cake extract improved the structure and texture of the bread, resulting in greater elasticity compared to the control, probably due to its high protein content.

The use of flours from cereals, pseudocereals, and pulses can improve the nutritional profile of these products and influence their technological and sensory characteristics (Irigoytia et al., 2025). To better understand these relationships, a multiple factor analysis (MFA) was performed using a data matrix containing technological, physicochemical, and sensory data (Irigoytia et al., 2025). A new analysis of the results established novel insight related to the effects observed. Figure 6.3a shows the relationship between the variables studied in the MFA, where 61.48% of the total data variation was explained by the first two factors (F1 and F2). Briefly, as can be observed, the iron content of the GF flours, located in the lower right part of the MFA, presented a very low contribution to the technological and sensory parameters recorded. In other words, the high iron content present naturally in the GF flours does not affect the final quality of GFB; however, iron bioaccessibility and bioavailability should be further studied. In addition, the global acceptability of the GFBs is located in the same quarter of GFB-L, GFB-Q, GFB-S, and GFB-RB. The iron content (mg) could be estimated for a serving size of 50 g GFBs, providing 1.5, 5.2, 5.6, and 6.1, for GFB-L, GFB-Q, GFB-S, GFB-RB, respectively (Figure 6.3b). All these products showed an average level 6–7 in a 9-point hedonic scale, representing a global acceptability of "*like slightly*" and "*like moderately*".

In this regard, Montemurro et al. (2023) found that GFBs formulated with a combination of alternative flours and strains of lactic acid bacteria (LAB) showed a significant increase in ash content compared to the control, which may indicate higher mineral content in the products. However, when considering the impact of these ingredients on mineral content, it is essential to keep in mind that their bioavailability does not necessarily increase in the final product (Regula et al., 2018). In addition, studies on the iron bioavailability from these products should be deepened. Processing techniques like soaking, fermentation, enzymatic hydrolysis, and germination could also contribute to increasing iron bioaccessibility while at the same time reducing the content of antinutritional factors, improving the mineral absorption and bioavailability (de Escalada Pla et al., 2020; Rogaska et al., 2021; Castellanos-Fuentes et al., 2022; Manzoor et al., 2024).

Another biotechnological approach is the fermentation by the addition of sourdough to the GFB formulation as a natural leavening agent. The use of sourdough is the most ancient technological process to obtain leavened breads, and refers to a mixture containing flour and water which is spontaneously fermented by native

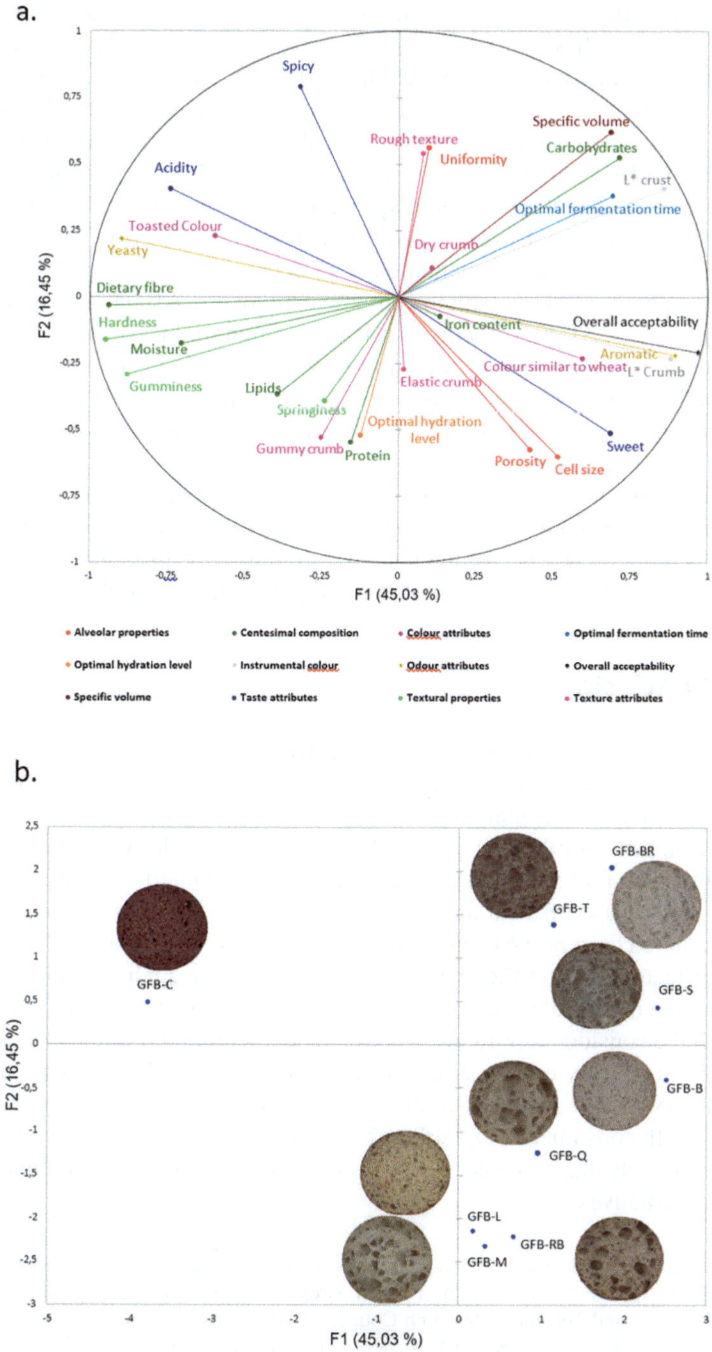

Fig. 6.3 (**a**). Multiple factor analysis for physicochemical properties, textural, alveolar structure, and sensory characteristics of GFBs. (**b**). Representations of the GFB samples in the sensory map. GFB-M, millet; GFB-RB, rice bran, GFB-T, teff; GFB-B, buckwheat; GFB-L, lupin; GFB-Q, quinoa; GFB-S, sorghum; GFB-BR, brown rice; and GFB-C, carob

LAB and yeast (Irigoytia et al., 2023). The sourdough used as a leavening agent for bread fermentation has gained popularity among consumers who demand breads with pronounced flavour and taste, healthy properties, without the addition of additives and an extended shelf life. The use of sourdough to improve iron bioavailability in GFBs has been studied by Peñalver et al. (Peñalver et al., 2023). The results showed that GFBs formulated with sourdough improved the bioavailability of minerals such as calcium, copper, iron, magnesium, potassium, lead, sodium, and zinc, compared to conventionally fermented bread. This effect is attributed to the reduction of phytic acid during fermentation. The decrease in pH activates the grain's endogenous phytases, along with the phytase activity of LAB, leading to the breakdown of phytates and the subsequent release of bound minerals. Germination and fermentation are key processes for improving mineral bioavailability by reducing phytic acid content. These processes also increase the content of protein, amino acids, reducing sugars, and B vitamins, enhancing the nutritional value of breads (Singh et al., 2015).

Therefore, iron intake could be improved by the incorporation of ingredients rich in iron in GFB formulations, and in combination with processing techniques that improve the mineral release from the matrix, minimise the presence of iron absorption inhibitors, and maximise the bioavailability and bioaccessibility of the mineral.

6.6 Conclusion

IDA is highly prevalent in individuals recently diagnosed with CD and could be persistent despite adherence to a GF diet. GF products present lower iron content compared to the gluten-containing counterparts. This makes it difficult to adequately meet the iron needs of this population. Fortified GFBs could represent an opportunity as a carrier for iron intake enhancement. Different food-based approaches, aligned with the use of local and sustainable resources, and processing strategies could be applied to improve its nutritional profile. However, it is essential to explore how these factors impact the technological and sensory quality of GFBs, as they play a key role in overall product acceptance by consumers. Other important aspects that should be considered in future work include integrating the study of GFB technological and sensory quality along with the study of iron bioaccessibility and/or bioavailability. These advances could have a positive social and economic impact, benefiting both consumers and markets by expanding the diversity of commercially available products that can meet the growing demand for healthier and nutritionally balanced alternatives.

Acknowledgements This work was supported by the Universidad Nacional de Entre Ríos (PID UNER 9122; 9175), the Universidad de Buenos Aires (UBACYT-20020170100092BA), and the National Scientific and Technical Research Council of Argentina (CONICET). The authors want to acknowledge ACELA Entre Ríos (Asistencia al Celíaco de la Argentina), Gualeguaychú and Cooperativa Arroceros Villa Elisa for their valuable participation and contribution in our projects.

References

Aguiar, E. V., Santos, F. G., Krupa-Kozak, U., & Capriles, V. D. (2023). Nutritional facts regarding commercially available gluten-free bread worldwide: Recent advances and future challenges. *Critical Reviews in Food Science and Nutrition, 63*, 693–705. https://doi.org/10.1080/10408398.2021.1952403

Allen, B., & Orfila, C. (2018). The availability and nutritional adequacy of gluten-free bread and pasta. *Nutrients, 10*, 1370. https://doi.org/10.3390/NU10101370

Arias-Gastelum, M., Cabrera-Chávez, F., De, M., et al. (2018). The gluten-free diet: Access and economic aspects and impact on lifestyle. *Nutrition and Dietary Supplements, 10*, 27–34. https://doi.org/10.2147/NDS.S143404

Bailey, R. L., West, K. P., & Black, R. E. (2015). The epidemiology of global micronutrient deficiencies. *Annals of Nutrition & Metabolism, 66*(Suppl 2), 22–33. https://doi.org/10.1159/000371618

Bechoff, A., de Bruyn, J., Alpha, A., et al. (2023). Exploring the complementarity of fortification and dietary diversification to combat micronutrient deficiencies: A scoping review. *Current Developments in Nutrition, 7*, 100033. https://doi.org/10.1016/J.CDNUT.2023.100033

Bender, D., & Schönlechner, R. (2020). Innovative approaches towards improved gluten-free bread properties. *Journal of Cereal Science, 91*, 102904. https://doi.org/10.1016/J.JCS.2019.102904

Bianchi, P. I., Aronico, N., Santacroce, G., et al. (2024). Nutritional consequences of celiac disease and gluten-free diet. *Gastroenterol Insights, 15*, 878–894. https://doi.org/10.3390/GASTROENT15040061

Bozorg, S. R., Lee, A. R., Mårild, K., & Murray, J. A. (2024). The economic iceberg of celiac disease: More than the cost of gluten-free food. *Gastroenterology, 167*, 172–182. https://doi.org/10.1053/J.GASTRO.2024.02.051

Breshears, K. L., & Crowe, K. M. (2013). Sensory and textural evaluation of gluten-free bread substituted with Amaranth and Montina™ flour. *Journal of Food Research, 2*, 1. https://doi.org/10.5539/JFR.V2N4P1

Bryszewska, M. A. (2019). Comparison study of iron bioaccessibility from dietary supplements and microencapsulated preparations. *Nutrients, 11*, 273. https://doi.org/10.3390/NU11020273

Cagnasso, C., Marquez, S., & López, L. B. (2023). Regulation and labelling. Methods of analysis for the determination of gluten in foods. In *Designing gluten free bakery and pasta products* (pp. 361–388). https://doi.org/10.1007/978-3-031-28344-4_10

Cardo, A., Churruca, I., Lasa, A., et al. (2021). Nutritional imbalances in adult celiac patients following a gluten-free diet. *Nutrients, 13*, 2877. https://doi.org/10.3390/NU13082877/S1

Caruso, R., Pallone, F., Stasi, E., et al. (2013). Appropriate nutrient supplementation in celiac disease. *Annals of Medicine, 45*, 522–531. https://doi.org/10.3109/07853890.2013.849383

Castellanos-Fuentes, A. P., Cortés, N., Genevois, C. E., et al. (2022). Soybean extruded by-products as substrate for obtaining food ingredients containing probiotics. *International Journal of Food Science and Technology, 57*, 4825–4831. https://doi.org/10.1111/IJFS.15690

Choung, R. S., Larson, S. A., Khaleghi, S., et al. (2016). Prevalence and morbidity of undiagnosed celiac disease from a community-based study. *Gastroenterology, 152*, 830. https://doi.org/10.1053/J.GASTRO.2016.11.043

Collar, C. (2015). Bread and bakery products. In *Handbook of minerals in elementary foods* (pp. 559–572). https://doi.org/10.1002/9781118654316.CH23

Culetu, A., Susman, I. E., Duta, D. E., & Belc, N. (2021). Nutritional and functional properties of gluten-free flours. *Applied Sciences, 11*(14), 6283.

Curti, M. I., Palavecino, P. M., Savio, M., et al. (2023). Sorghum (Sorghum bicolor L. Moench) gluten-free bread: The effect of milling conditions on the technological properties and in vitro bioaccessibility of polyphenols and minerals. *Food, 12*, 3030. https://doi.org/10.3390/FOODS12163030/S1

Dary, O., & Mora, J. O. (2013). Food fortification: Technological aspects. In *Encyclopedia of human nutrition* (Vol. 2–4, pp. 306–314). https://doi.org/10.1016/B978-0-12-375083-9.00120-3

de Escalada Pla, M. F., Flores, S. K., & Genevois, C. E. (2020). Innovative strategies and nutritional perspectives for fortifying pumpkin tissue and other vegetable matrices with iron. *Food Science and Human Wellness, 9*, 103–111. https://doi.org/10.1016/J.FSHW.2020.02.005

de Escalada Pla, M. F., Silva, N. E., Castellanos-Fuentes, A. P., et al. (2023). Gluten free non-fermented bakery. In M. F. de Escalada Pla & C. E. Genevois (Eds.), *Designing gluten free bakery and pasta products* (pp. 211–237). Springer International Publishing.

de Gusmão, R. P., Cavalcanti-Mata, M. E. R. M., Duarte, M. E. M., & Gusmão, T. A. S. (2016). Particle size, morphological, rheological, physicochemical characterization and designation of minerals in mesquite flour (Proposis julifrora). *Journal of Cereal Science, 69*, 119–124. https://doi.org/10.1016/J.JCS.2016.02.017

Desire, M. F., Blessing, M., Elijah, N., et al. (2021). Exploring food fortification potential of neglected legume and oil seed crops for improving food and nutrition security among smallholder farming communities: A systematic review. *Journal of Agriculture and Food Research, 3*, 100117. https://doi.org/10.1016/J.JAFR.2021.100117

Di Sario, F., Monachesi, C., Verma, A. K., & Catassi, C. (2023). Everything that must be known about the relationship of gluten to human health. In M. F. de Escalada Pla & C. E. Genevois (Eds.), *Designing gluten free bakery and pasta products* (pp. 1–17). Springer International Publishing.

Diez-Sampedro, A., Olenick, M., Maltseva, T., & Flowers, M. (2019). A gluten-free diet, not an appropriate choice without a medical diagnosis. *Journal of Nutrition and Metabolism, 2019*, 2438934. https://doi.org/10.1155/2019/2438934

Djordjević, M., Djordjević, M., Šoronja-Simović, D., et al. (2021). Delving into the role of dietary fiber in gluten-free bread formulations: Integrating fundamental rheological, technological, sensory, and nutritional aspects. *Polysaccharides, 3*, 59–82. https://doi.org/10.3390/POLYSACCHARIDES3010003

Efthymakis, K., Milano, A., Laterza, F., et al. (2017). Iron deficiency anemia despite effective gluten-free diet in celiac disease: Diagnostic role of small bowel capsule endoscopy. *Digestive and Liver Disease, 49*, 412–416. https://doi.org/10.1016/J.DLD.2016.12.007

El Khoury, D., Balfour-Ducharme, S., & Joye, I. J. (2018). A review on the gluten-free diet: Technological and nutritional challenges. *Nutrients, 10*, 1410. https://doi.org/10.3390/NU10101410

Espinosa-Salas, S., & Gonzalez-Arias, M. (2023). *Nutrition: Micronutrient intake, imbalances, and interventions*. StatPearls.

Estévez, V., Rodríguez, J. M., Schlack, P., et al. (2024). Persistent barriers of the gluten-free basic food basket: Availability, cost, and nutritional composition assessment. *Nutrients, 16*, 885. https://doi.org/10.3390/NU16060885/S1

FAO. (1997a). Food fortification: Technology and quality control.

FAO. (1997b). *Preventing micronutrient malnutrition: A guide to food-based approaches – A manual for policy makers and programme planners*. International Life Sciences Institute (ILSI).

FAO. (2024). World wheat market at a glance – 2024/25 forecast. Roma.

FAO, WHO. (2006). Guidelines on food fortification with micronutrients A T. Geneva, Switzerland & Rome, Italy.

Garg, A., & Gupta, R. (2014). Predictors of compliance to gluten-free diet in children with celiac disease. *International Scholarly Research Notices, 2014*, 248402. https://doi.org/10.1155/2014/248402

Genevois, C. E., & de Escalada Pla, M. F. (2021). Soybean by-products and modified cassava starch for improving alveolar structure and quality characteristics of gluten-free bread. *European Food Research and Technology, 247*, 1477–1488. https://doi.org/10.1007/s00217-021-03725-x

Genevois, C., Flores, S., & De Escalada, P. M. (2014). Effect of iron and ascorbic acid addition on dry infusion process and final color of pumpkin tissue. *LWT – Food Science and Technology, 58*, 563–570. https://doi.org/10.1016/J.LWT.2014.03.020

Genevois, C., Flores, S., & de Escalada, P. M. (2016). Byproduct from pumpkin (Cucurbita moschata Duchesne ex poiret) as a substrate and vegetable matrix to contain lactobacillus casei. *Journal of Functional Foods, 23*, 210–219. https://doi.org/10.1016/J.JFF.2016.02.030

Genevois, C. E., et al. (2020). Use of different proportions of rice milling fractions as strategy for improving quality parameters and nutritional profile of gluten-free bread. *Journal of Food Science and Technology, 58*, 3913–3923. https://doi.org/10.1007/s13197-020-04852-1

González, M. P., Ballestero-Fernández, C., Fajardo, V., et al. (2022). Gluten-free product contribution to energy and macronutrient intakes in Spanish children and adolescents with celiac disease. *Food, 11*, 3790. https://doi.org/10.3390/FOODS11233790

Hasan, M. M., Soares Magalhaes, R. J., Garnett, S. P., et al. (2022). Anaemia in women of reproductive age in low- and middle-income countries: Progress towards the 2025 global nutrition target. *Bulletin of the World Health Organization, 100*, 196. https://doi.org/10.2471/BLT.20.280180

Hernández-Flores, T. D. J., Pedraza-Brindis, E. J., Cárdenas-Bedoya, J., Ruíz-Carrillo, J. D., Méndez-Clemente, A. S., Martínez-Guzmán, M. A., & Iniguez-Gutierrez, L. (2022). Role of micronutrients and gut microbiota-derived metabolites in COVID-19 recovery. *International Journal of Molecular Sciences, 23*(20), 12324.

Hurrell, R. F. (2021). Iron fortification practices and implications for iron addition to salt. *The Journal of Nutrition, 151*, 3S. https://doi.org/10.1093/JN/NXAA175

Hurrell, R. F. (2022). Ensuring the efficacious iron fortification of foods: A tale of two barriers. *Nutr, 14*, 1609. https://doi.org/10.3390/NU14081609

Ibba, M. I., Palacios-Rojas, N., & Rosales-Nolasco, A. (2023). Screening and use of nutritional and health-related benefits of the main crops. *Dev Sustain Heal Cereal Pseudocereals Conv Mol Breed, 25–55*. https://doi.org/10.1016/B978-0-323-90566-4.00015-1

Irigoytia, K. F., Espósito, N. N., Busch, V. M., et al. (2023). Fermented gluten-free baked goods. In M. F. de Escalada Pla (Ed.), *Designing gluten free bakery and pasta products* (pp. 163–210). Springer International Publishing.

Irigoytia, K. F., Espósito, N. N., de Escalada Pla, M. F., et al. (2024). Technological and multi-sensory analysis approach to holistically understand the quality and consumer perception of gluten-free breads with alternative flours. *International Journal of Food Science and Technology, 59*. https://doi.org/10.1111/IJFS.17557

Irigoytia, K. F., Genevois, C. E., & de Escalada Pla, M. F. (2025). Effect of cornstarch partial substitution by millet flour and agro-industrial by-products on the development of gluten-free doughs. *Future Foods, 11*. https://doi.org/10.1016/J.FUFO.2024.100515

Jamieson, J. A., Weir, M., & Gougeon, L. (2018). Canadian packaged gluten-free foods are less nutritious than their regular gluten-containing counterparts. *PeerJ, 2018*, e5875. https://doi.org/10.7717/PEERJ.5875/SUPP-4

Jastrebova, J., & Jägerstad, M. (2013). Novel fortification strategies for staple gluten-free products. *Handb Food Fortif Heal From Concepts to Public Heal Appl, 1*, 307–320. https://doi.org/10.1007/978-1-4614-7076-2_24

Jeanes, Y., Spitale, A., Nicolini, G., et al. (2022). Calcium and iron content of cereal-based gluten-free products. *Food, 11*, 2001. https://doi.org/10.3390/FOODS11142001/S1

Jegede, O., Enns, A., Kantounia, M., et al. (2021). Cost, nutritional content and number of gluten-free staple foods available in Winnipeg, Manitoba, Canada. *Plant Foods for Human Nutrition, 76*, 196–202. https://doi.org/10.1007/S11130-021-00889-5/METRICS

Jivraj, A., Hutchinson, J. M., Ching, E., et al. (2022). Micronutrient deficiencies are frequent in adult patients with and without celiac disease on a gluten-free diet, regardless of duration and adherence to the diet. *Nutrition, 103–104*, 111809. https://doi.org/10.1016/J.NUT.2022.111809

Kaur, N., Agarwal, A., & Sabharwal, M. (2022). Food fortification strategies to deliver nutrients for the management of iron deficiency anaemia. *Current Research in Food Science, 5*, 2094. https://doi.org/10.1016/J.CRFS.2022.10.020

Kelly, M. (2016). The nutrition transition in developing Asia: Dietary change, drivers and health impacts. In *Eating, drinking: Surviving: The international year of global understanding-IYGU* (pp. 83–90).

Kelly, C. P., Bai, J. C., Liu, E., & Leffler, D. A. (2015). Advances in diagnosis and management of celiac disease. *Gastroenterology, 148*, 1175–1186.

Khairuddin, M. A. N., & Lasekan, O. (2021). Gluten-free cereal products and beverages: A review of their health benefits in the last five years. *Food, 10*, 2523. https://doi.org/10.3390/FOODS10112523

Khemiri, S., Khelifi, N., Nunes, M. C., et al. (2020). Microalgae biomass as an additional ingredient of gluten-free bread: Dough rheology, texture quality and nutritional properties. *Algal Research, 50*, 101998. https://doi.org/10.1016/J.ALGAL.2020.101998

Kiskini, A., Argiri, K., Kalogeropoulos, M., et al. (2007). Sensory characteristics and iron dialyzability of gluten-free bread fortified with iron. *Food Chemistry, 102*, 309–316. https://doi.org/10.1016/J.FOODCHEM.2006.05.022

Kiskini, A., Kapsokefalou, M., Yanniotis, S., & Mandala, I. (2010). Effect of different iron compounds on wheat and gluten-free breads. *Journal of the Science of Food and Agriculture, 90*, 1136–1145. https://doi.org/10.1002/JSFA.3927

Kiskini, A., Kapsokefalou, M., Yanniotis, S., & Mandala, I. (2012). Effect of iron fortification on physical and sensory quality of gluten-free bread. *Food and Bioprocess Technology, 5*, 385–390. https://doi.org/10.1007/S11947-011-0651-2/METRICS

Kreutz, J. M., Heynen, L., & Vreugdenhil, A. C. E. (2023). Nutrient deficiencies in children with celiac disease during long term follow-up. *Clinical Nutrition, 42*, 1175–1180. https://doi.org/10.1016/J.CLNU.2023.05.003

Kruger, J., Taylor, J. R. N., Ferruzzi, M. G., & Debelo, H. (2020). What is food-to-food fortification? A working definition and framework for evaluation of efficiency and implementation of best practices. *Comprehensive Reviews in Food Science and Food Safety, 19*, 3618–3658. https://doi.org/10.1111/1541-4337.12624

KrUpa-KozaK, U., Wronkowska, M., & Soral-ŚMietaNa, M. (2011). Effect of buckwheat flour on microelements and proteins contents in gluten-free bread. *Czech Journal of Food Sciences, 29*(2), 103–108.

Krupa-Kozak, U., et al. (2022). Novel gluten-free bread with an extract from flaxseed by-product: The relationship between water replacement level and nutritional value, antioxidant properties, and sensory quality. *Mol, 27*, 2690. https://doi.org/10.3390/MOLECULES27092690

Larretxi, I., Simon, E., Benjumea, L., et al. (2019). Gluten-free-rendered products contribute to imbalanced diets in children and adolescents with celiac disease. *European Journal of Nutrition, 58*, 775–783. https://doi.org/10.1007/S00394-018-1685-2/TABLES/5

Lopez, A., Cacoub, P., Macdougall, I. C., & Peyrin-Biroulet, L. (2016). Iron deficiency anaemia. *Lancet, 387*, 907–916.

Mahadev, S., Laszkowska, M., Sundström, J., et al. (2018). Prevalence of celiac disease in patients with iron deficiency anemia – A systematic review with meta-analysis. *Gastroenterology, 155*, 374–382. https://doi.org/10.1053/J.GASTRO.2018.04.016

Manzoor, M. F., Ali, A., Ain, H. B. U., et al. (2024). Bioaccessibility mechanisms, fortification strategies, processing impact on bioavailability, and therapeutic potentials of minerals in cereals. *Future Foods, 10*, 100425. https://doi.org/10.1016/J.FUFO.2024.100425

Marciniak, M., Szymczak-Tomczak, A., Mahadea, D., et al. (2021). Multidimensional disadvantages of a gluten-free diet in celiac disease: A narrative review. *Nutrition, 13*, 643. https://doi.org/10.3390/NU13020643

Mattar, G., Haddarah, A., Haddad, J., et al. (2022). New approaches, bioavailability and the use of chelates as a promising method for food fortification. *Food Chemistry, 373*, 131394. https://doi.org/10.1016/J.FOODCHEM.2021.131394

Melini, V., & Melini, F. (2019). Gluten-free diet: Gaps and needs for a healthier diet. *Nutrition, 11*, 170. https://doi.org/10.3390/NU11010170

Miller, J. L. (2013). Iron deficiency anemia: A common and curable disease. *Cold Spring Harbor Perspectives in Medicine, 3*. https://doi.org/10.1101/CSHPERSPECT.A011866

Miranda, J., Lasa, A., Bustamante, M. A., et al. (2014). Nutritional differences between a gluten-free diet and a diet containing equivalent products with gluten. *Plant Foods for Human Nutrition, 69*, 182–187. https://doi.org/10.1007/S11130-014-0410-4/TABLES/3

Missbach, B., Schwingshackl, L., Billmann, A., et al. (2015). Gluten-free food database: The nutritional quality and cost of packaged gluten-free foods. *PeerJ, 2015*, e1337. https://doi.org/10.7717/PEERJ.1337/SUPP-4

Mitchell, S., Gomes, A., Zelig, R., & Parker, A. (2019). Not all grains are created equal: Gluten-free products not included in mandatory folate fortification. *Current Developments in Nutrition, 3*, nzz020. https://doi.org/10.1093/CDN/NZZ020

Monteiro, J. S., Farage, P., Zandonadi, R. P., et al. (2021). A systematic review on gluten-free bread formulations using specific volume as a quality indicator. *Food, 10*, 614. https://doi.org/10.3390/FOODS10030614

Montemurro, M., Beccaccioli, M., Perri, G., et al. (2023). A chestnut-hemp type-II sourdough to improve technological, nutritional, and sensory properties of gluten-free bread. *International Journal of Food Microbiology, 404*, 110322. https://doi.org/10.1016/J.IJFOODMICRO.2023.110322

Montoro-Huguet, M. A., Santolaria-Piedrafita, S., Cañamares-Orbis, P., & García-Erce, J. A. (2021). Iron deficiency in celiac disease: Prevalence, health impact, and clinical management. *Nutrients, 13*, 3437. https://doi.org/10.3390/NU13103437

Myhrstad, M. C. W., Slydahl, M., Hellmann, M., et al. (2021). Nutritional quality and costs of gluten-free products: A case-control study of food products on the Norwegian marked. *Food Nutrition Research, 65*. https://doi.org/10.29219/fnr.v65.6121

Niewinski, M. M. (2008). Advances in celiac disease and gluten-free diet. *Journal of the American Dietetic Association, 108*, 661–672. https://doi.org/10.1016/J.JADA.2008.01.011

Nkhata, S. G., Ayua, E., Kamau, E. H., & Shingiro, J. B. (2018). Fermentation and germination improve nutritional value of cereals and legumes through activation of endogenous enzymes. *Food Science & Nutrition, 6*, 2446–2458. https://doi.org/10.1002/FSN3.846

Olson, R., Gavin-Smith, B., Ferraboschi, C., & Kraemer, K. (2021). Food fortification: The advantages, disadvantages and lessons from sight and life programs. *Nutrients, 13*, 1118. https://doi.org/10.3390/NU13041118

Papagianni, E., Kotsiou, K., Matsakidou, A., et al. (2024). Development of "clean label" gluten-free breads fortified with flaxseed slurry and sesame cake: Implications on batter rheology, bread quality and shelf life. *Food Hydrocolloids, 150*, 109734. https://doi.org/10.1016/J.FOODHYD.2024.109734

Parsons, K., Brown, L., Clark, H., et al. (2021). Gluten cross-contact from common food practices and preparations. *Clinical Nutrition, 40*, 3279–3287. https://doi.org/10.1016/J.CLNU.2020.10.053

Pellegrini, N., & Agostoni, C. (2015). Nutritional aspects of gluten-free products. *Journal of the Science of Food and Agriculture, 95*, 2380–2385. https://doi.org/10.1002/JSFA.7101

Peñalver, R., & Nieto, G. (2024). Developing a functional gluten-free sourdough bread by incorporating quinoa, amaranth, rice and spirulina. *LWT, 201*, 116162. https://doi.org/10.1016/J.LWT.2024.116162

Peñalver, R., Ros, G., & Nieto, G. (2023). Development of functional gluten-free sourdough bread with Pseudocereals and enriched with *Moringa oleifera*. *Food, 12*, 3920. https://doi.org/10.3390/FOODS12213920

Pešić, M. B., Pešić, M. M., Bezbradica, J., et al. (2023). Okara-enriched gluten-free bread: Nutritional, antioxidant and sensory properties. *Molecules, 28*, 4098. https://doi.org/10.3390/molecules28104098

Rai, R. K., Kumar, S. S., Sen Gupta, S., et al. (2022). Shooting shadows: India's struggle to reduce the burden of anaemia. *The British Journal of Nutrition, 129*, 416. https://doi.org/10.1017/S0007114522000927

Ramakrishna, B. S., Makharia, G. K., Chetri, K., et al. (2016). Prevalence of adult celiac disease in India: Regional variations and associations. *The American Journal of Gastroenterology, 111*, 115–123. https://doi.org/10.1038/AJG.2015.398

Regula, J., Cerba, A., Suliburska, J., & Tinkov, A. A. (2018). In vitro bioavailability of calcium, magnesium, iron, zinc, and copper from gluten-free breads supplemented with natural additives. *Biological Trace Element Research, 182*, 140–146. https://doi.org/10.1007/S12011-017-1065-4/TABLES/4

Rizzello, F., Spisni, E., Giovanardi, E., et al. (2019). Implications of the westernized diet in the onset and progression of IBD. *Nutrients, 11*, 1033. https://doi.org/10.3390/NU11051033

Rogaska, A., Reguła, J., Suliburska, J., & Krejpcio, Z. (2020). A comparative study of the bioavailability of Fe, cu and Zn from gluten-free breads enriched with natural and synthetic additives. *Food, 9*, 1853. https://doi.org/10.3390/FOODS9121853

Rogaska, A., Reguła, J., Suliburska, J., & Krejpcio, Z. (2021). Comparison of the in vitro bioavailability of selected minerals from gluten-free breads enriched with grains and synthetic organic and non-organic compounds. *Molecules, 26*, 2085. https://doi.org/10.3390/MOLECULES26072085

Roman, L., Belorio, M., & Gomez, M. (2019). Gluten-free breads: The gap between research and commercial reality. *Comprehensive Reviews in Food Science and Food Safety, 18*, 690–702. https://doi.org/10.1111/1541-4337.12437

Romão, B., Falcomer, A. L., Palos, G., et al. (2021). Glycemic index of gluten-free bread and their main ingredients: A systematic review and meta-analysis. *Food, 10*. https://doi.org/10.3390/FOODS10030506

Rondanelli, M., Faliva, M. A., Gasparri, C., et al. (2019). Micronutrients dietary supplementation advices for celiac patients on long-term gluten-free diet with good compliance: A review. *Medicina (B Aires), 55*, 337. https://doi.org/10.3390/MEDICINA55070337

Santos, F. G., Aguiar, E. V., & Capriles, V. D. (2019). Analysis of ingredient and nutritional labeling of commercially available gluten-free bread in Brazil. *International Journal of Food Sciences and Nutrition, 70*, 562–569. https://doi.org/10.1080/09637486.2018.1551336

Sapone, A., Bai, J. C., Ciacci, C., et al. (2012). Spectrum of gluten-related disorders: Consensus on new nomenclature and classification. *BMC Medicine, 10*, 1–12. https://doi.org/10.1186/1741-7015-10-13/FIGURES/4

Saturni, L., Ferretti, G., & Bacchetti, T. (2010). The gluten-free diet: Safety and nutritional quality. *Nutrients, 2*, 16–34. https://doi.org/10.3390/NU20100016

Sciarini, L., Palavecino, P. M., & Ribotta, P. D. (2023). Use of additives in gluten-free formulations. In *Designing gluten free bakery and pasta products* (pp. 115–161). Springer International Publishing. https://doi.org/10.1007/978-3-031-28344-4_4

Seidita, A., Mansueto, P., Compagnoni, S., et al. (2022). Anemia in celiac disease: Prevalence, associated clinical and laboratory features, and persistence after gluten-free diet. *Journal of Personalized Medicine, 12*, 1582. https://doi.org/10.3390/JPM12101582/S1

Sharma, N., Bhatia, S., Chunduri, V., et al. (2020). Pathogenesis of celiac disease and other gluten related disorders in wheat and strategies for mitigating them. *Frontiers in Nutrition, 7*, 6. https://doi.org/10.3389/FNUT.2020.00006

Shepherd, S. J., & Gibson, P. R. (2013). Nutritional inadequacies of the gluten-free diet in both recently-diagnosed and long-term patients with coeliac disease. *Journal of Human Nutrition and Dietetics, 26*, 349–358. https://doi.org/10.1111/JHN.12018

Silvester, J. A., Weiten, D., Graff, L. A., et al. (2016). Is it gluten-free? Relationship between self-reported gluten-free diet adherence and knowledge of gluten content of foods. *Nutrition, 32*, 777–783. https://doi.org/10.1016/J.NUT.2016.01.021

Singh, A. K., Rehal, J., Kaur, A., & Jyot, G. (2015). Enhancement of attributes of cereals by germination and fermentation: A review. *Critical Reviews in Food Science and Nutrition, 55*, 1575–1589. https://doi.org/10.1080/10408398.2012.706661

Singh, P., Arora, A., Strand, T. A., et al. (2018). Global prevalence of celiac disease: Systematic review and meta-analysis. *Clinical Gastroenterology and Hepatology, 16*, 823–836. https://doi.org/10.1016/J.CGH.2017.06.037

Skodje, G. I., Minelle, I. H., Rolfsen, K. L., et al. (2019). Dietary and symptom assessment in adults with self-reported non-coeliac gluten sensitivity. *Clinical Nutrition ESPEN, 31*, 88–94. https://doi.org/10.1016/J.CLNESP.2019.02.012

Šmídová, Z., & Rysová, J. (2022). Gluten-free bread and bakery products technology. *Food, 11*. https://doi.org/10.3390/FOODS11030480

Srikanlaya, C., Zhou, W., Therdthai, N., & Ritthiruangdej, P. (2022). Effect of hydroxypropyl methylcellulose, protein and fat on predicted glycemic index and antioxidant property of gluten-free bread from rice flour. *Journal of Food Processing & Preservation, 46*, e17087. https://doi.org/10.1111/JFPP.17087

Stefanelli, G., Viscido, A., Longo, S., et al. (2020). Persistent iron deficiency anemia in patients with celiac disease despite a gluten-free diet. *Nutrients, 12*, 2176. https://doi.org/10.3390/NU12082176

Talarico, V., Giancotti, L., Mazza, G. A., et al. (2021). Iron deficiency anemia in celiac disease. *Nutrients, 13*. https://doi.org/10.3390/NU13051695

Taraghikhah, N., Ashtari, S., Asri, N., et al. (2020). An updated overview of spectrum of gluten-related disorders: Clinical and diagnostic aspects. *BMC Gastroenterology, 20*, 1–12. https://doi.org/10.1186/S12876-020-01390-0/FIGURES/2

Thakur, S., Singh, A., Insa, B., & Sharma, S. (2023). Food fortification in India as malnutrition concern: A global approach. *Sustainable Food Technology, 1*, 681–695. https://doi.org/10.1039/D3FB00079F

Theethira, T. G., & Dennis, M. (2015). Celiac disease and the gluten-free diet: Consequences and recommendations for improvement. *Digestive Diseases, 33*, 175–182. https://doi.org/10.1159/000369504

Tovoli, F., Masi, C., Guidetti, E., et al. (2015). Clinical and diagnostic aspects of gluten related disorders. *World Journal of Clinical Cases, 3*, 275. https://doi.org/10.12998/WJCC.V3.I3.275

Tres, A., Tarnovska, N., Varona, E., et al. (2020). Determination and comparison of the lipid profile and sodium content of gluten-free and gluten-containing breads from the Spanish market. *Plant Foods for Human Nutrition, 75*, 344–354. https://doi.org/10.1007/S11130-020-00828-W/TABLES/3

Vici, G., Belli, L., Biondi, M., & Polzonetti, V. (2016). Gluten free diet and nutrient deficiencies: A review. *Clinical Nutrition, 35*, 1236–1241. https://doi.org/10.1016/j.clnu.2016.05.002

Villafuerte-Galvez, J., Vanga, R. R., Dennis, M., et al. (2015). Factors governing long-term adherence to a gluten-free diet in adult patients with coeliac disease. *Alimentary Pharmacology & Therapeutics, 42*, 753–760. https://doi.org/10.1111/APT.13319

Volta, U., Caio, G., Tovoli, F., & De Giorgio, R. (2013). Non-celiac gluten sensitivity: Questions still to be answered despite increasing awareness. *Cellular & Molecular Immunology, 105*(10), 383–392. https://doi.org/10.1038/cmi.2013.28

WHO. (2012). Resolutions and decisions – Sixty-Fifth World Health Assembly. Ginebra.

Wieser, H., Ruiz-Carnicer, Á., Segura, V., et al. (2021). Challenges of monitoring the gluten-free diet adherence in the management and follow-up of patients with celiac disease. *Nutrients, 13*, 2274. https://doi.org/10.3390/NU13072274

Xhakollari, V., Canavari, M., & Osman, M. (2019). Factors affecting consumers' adherence to gluten-free diet, a systematic review. *Trends in Food Science and Technology, 85*, 23–33. https://doi.org/10.1016/J.TIFS.2018.12.005

Yeşil, S., & Levent, H. (2022). The influence of fermented buckwheat, quinoa and amaranth flour on gluten-free bread quality. *LWT – Food Science and Technology, 160*, 113301.

Young, M. F., Luo, H., & Suchdev, P. S. (2023). The challenge of defining the global burden of iron deficiency anaemia. *Lancet Haematology, 10*, e702. https://doi.org/10.1016/S2352-3026(23)00168-0

Marine Algae: A Sustainable Resource for Nutritional Supplements

7

Sara Chadorshabi and Maryam Mahmoudzadeh

Abstract

Micro- and macroalgae have garnered considerable scholarly interest as sustainable and multifunctional resources for the formulation of enhanced nutritional supplements. These aquatic organisms, encompassing microscopic phytoplankton and extensive seaweeds, are abundant in bioactive compounds, which include polysaccharides, proteins, essential fatty acids, minerals, and vitamins. Microalgae, exemplified by *Chlorella* and *Spirulina*, are acclaimed for their elevated protein concentrations and antioxidant pigments, whereas macroalgae, comprising red, green, and brown seaweeds, are distinguished for their polysaccharides such as carrageenan and fucoidan, which demonstrate prebiotic and immunomodulatory attributes. This chapter investigates the nutritional compositions of both micro- and macroalgae, accentuating their health-related advantages, including antioxidant activity, support for cardiovascular health, modulation of the gut microbiome, and anti-inflammatory properties. The escalating interest in algal-derived supplements has prompted significant advancements in cultivation and processing technologies, thereby facilitating the production of products characterized by high quality and bioavailability. Nevertheless, obstacles such as scalability, palatability, safety, and adherence to regulatory frameworks continue to pose challenges. Effectively addressing these issues is imperative for realizing the full potential of algae as a sustainable resource for functional foods and dietary supplements. This chapter offers a

S. Chadorshabi
Student Research Committee, Faculty of Nutrition and Food Science, Department of Food Science and Technology, Tabriz University of Medical Sciences, Tabriz, Iran

M. Mahmoudzadeh (✉)
Faculty of Nutrition and Food Science, Department of Food Science and Technology, Tabriz University of Medical Sciences, Tabriz, Iran
e-mail: mahmoudzadehm@tbzmed.ac.ir

G. Sanghvi et al. (eds.), *Sustainable Food Fortification*,
https://doi.org/10.1007/978-981-95-1217-1_7

thorough examination of current applications, the global market landscape, and prospective research trajectories, emphasizing the transformative impact of algae in fostering health and well-being through sustainable practices.

Keywords

Algae · Gut health · Immune modulation · Cardiovascular disease · Cultivation

7.1 Introduction

Dimensions can be used to categorize aquatic plants. The larger ones are referred to as macroalgae, and the smaller ones are referred to as microalgae. Both play significant ecological roles: microalgae (microscopic, unicellular) and macroalgae (large, multicellular) (Ibañez et al., 2011). Aquatic plants are mainly categorized into three forms: brown, green, and red. Some others are also prokaryotic organisms, commonly known as blue-green, which can also be grouped under this category. Each of the three forms possesses its own nature and functions (Veluchamy & Palaniswamy, 2020). Some of the most prevalent organisms on Earth are marine phytoplankton. Even under the most severe circumstances, they are capable of expanding. They are found in both fresh and saltwater, and they can be found in both terrestrial and aqueous environments (Ścieszka & Klewicka, 2019).

Aquatic plants are an essential part of ecosystems. They absorb atmospheric carbon while they gather sunlight for energy and therefore contribute to the elimination of harmful gases and environmental stability (El-Manaway & Rashedy, 2022; Sahoo et al., 2012). Aquatic organisms play a significant role in maintaining atmospheric oxygen levels. Their photosynthetic activity is essential for sustaining life on the planet, supporting various ecosystems and species (Dissanayake et al., 2023). Also, many species act as protective habitats for a diverse spectrum of plants and animals, contributing to the health of the coastal ecosystem (Omer et al., 2021). Their capacity to take up surplus nutrients and toxins from aquatic ecosystems renders them useful for water filtration and water quality maintenance (Kumar, 2014). They are also an important food source for a variety of other organisms (El-Manaway & Rashedy, 2022).

Recently, the nutritional value of marine algae has gathered significant attention, which has led to their increasing importance in the supplement industry (Bishop & Zubeck, 2012). They have natural products that consist of sugars, amino acids, fats, and pigments. The antimicrobial and protective qualities of these products make foods of higher quality with minimal use of chemical preservatives. Such ingredients guarantee healthier foods and better overall well-being (Ścieszka & Klewicka, 2019). Growing awareness of sea plant benefits has driven their use in health supplements. Global markets for these supplements are experiencing significant growth and demand (Bishop & Zubeck, 2012; Mendes et al., 2022).

The global market for these kinds of health products is expected to almost double. Between the years 2018 and 2027, tremendous financial growth is expected, reflecting increasing consumer demand (Mendes et al., 2022). This growth is fueled by increasing consumer interest in plant-based, sustainable nutrition sources, as well as a deeper understanding of the health benefits associated with algae-derived nutrients (Mendes et al., 2022). The algae supplement industry has responded to growing consumer demand with a variety of innovative products. For example, marine algae, such as *Spirulina*, can be available as capsules or in powder form, which can enhance consumers' diets with nutrient-rich supplements (AlFadhly et al., 2022). Furthermore, there are specialized formulations targeting particular health concerns like cardiovascular health, which can combine with algae to enhance their health benefits (Dini, 2023).

The growing trend toward sustainable and environmentally friendly food sources is another factor driving the increasing demand for products made from algae. Algae farming is a very sustainable substitute for conventional agriculture since it uses few resources, including freshwater and arable land, and can be carried out in controlled settings (Mendes et al., 2022; Enzing et al., 2014). Marine algae are becoming more widely recognized as a sustainable superfood that can help with global issues related to nutrition, food security, and environmental sustainability as consumers become more ecologically conscious and health aware (Zayed et al., 2024).

The current chapter is intended to present an extensive review of marine algae as a renewable source of nutritional supplements. The chapter attempts to bring into perspective the nutritional value, health effects, and uses of algae in the supplement world, as well as challenges and prospects in their commercialization.

7.2 Nutritional Profile of Algae

7.2.1 Microalgae

The global demand for marine algae as dietary components is increasing due to their exceptional nutritional and functional benefits. Microalgae have been esteemed in human diets for ages owing to their abundant bioactive constituents, including proteins, carbs, lipids, vitamins, and minerals (Čmiková et al., 2024). Microalgae are gaining popularity as promising sources of nutraceuticals or functional foods. Their capacity to produce a wide variety of useful products is the explanation for the increasing interest. They contain essential proteins, fatty molecules, pigments, biological catalysts, nutrients, and microelements (Matos, 2017).

Microalgae's impressive protein content is a key factor in their recognition as a promising source of food. Many microalgal species boast high protein levels, like *Spirulina platensis* (55–70% per dry weight) and *Chlorella vulgaris* (42–55% per dry weight) (Matos, 2019). Protein contents of the majority of microalgae are in agreement with international standards. They might, however, be somewhat deficient in a few sulfur-containing amino acids such as cysteine and methionine (Čmiková et al., 2024). *Pseudopediastrum boryanum*, *Chloroidium saccharophilum*, and *Chloromonas* cf. reticulata have high levels of indispensable proteins, 72.6%, 61.2%, and 54.1%, respectively. Glutamic acid was found in all the analyzed species as the most important free amino acid. Yet the next most abundant proteins varied between them, with proline and lysine more dominant in one species and methionine and lysine more dominant in another (Santiago-Díaz et al., 2021).

In addition to proteins, microalgae are rich in carbohydrates, including glucose, starch, and various polysaccharides. These carbohydrates exhibit bioactive properties, including anticoagulant, anticancer, and antihypercholesterolemic activities (Matos et al., 2017). *Porphyridium cruentum*, a unicellular red alga, synthesizes a sulfated galactan exopolysaccharide (EPS) that may serve as a substitute for carrageenan in many applications. In another instance, *Chlamydomonas mexicana* is distinguished by its capacity to excrete around 25% of its entire organic synthesis as extracellular polysaccharides (Matos et al., 2017).

Microalgae show a broad spectrum of lipid and fatty acid composition, reflecting the huge disparity among species. This diversity makes their lipids a unique and precious natural product. Strains capable of producing significant amounts of lipids are especially desired for industrial applications. Further, the capacity of some microalgae to yield lipids reflects their value as a promising resource for future commercial activities (Santiago-Díaz et al., 2021; Michalak et al., 2017).

Experiments show that microalgae's phenolic content is equal to or less than that of land plants. Microalgae phenolics were once thought to be mainly phenolic acids, but more recently, a range of phenolic compounds, including flavonoids like flavonols, flavanones, isoflavones, and dihydrochalcones, have been found (Santiago-Díaz et al., 2021; Safafar et al., 2015). Microalgae contain therapeutic molecules. These molecules promote well-being by inhibiting oxidative stress and inflammation. They are used in nutrition (Silva et al., 2022). Natural pigments derived from microalgae, such as phycobilins, chlorophylls, and carotenoids, are becoming increasingly popular alternatives to synthetic colorants due to their safety and health benefits (Begum et al., 2016). Microalgae also serve as a substantial supply of micronutrients, including vitamins (e.g., C, A, E, biotin, B12, B2, B1, and B6) and minerals (e.g., K, Mg, Na, Ca, Zn, Fe, and trace elements) (Matos et al., 2017). Vitamin B12 is the most common vitamin found in green and red algae (Matos et al., 2017). Table 7.1 delineates the nutritional profiles of microalgae and macroalgae, focusing on their protein, fat, and carbohydrate levels while also summarizing the bioactive chemicals present in each.

7.2.2 Macroalgae

Seaweeds are divided into three broad groups according to color and biochemical characteristics. The first group consists of brown types that include *Laminaria* and *Sargassum* species. The second group contains red types, including *Chondrus* and *Gracilaria*. Finally, green seaweeds include *Ulva* and *Ascophyllum*. These groups have distinct features and functions in marine ecosystems (Holdt & Kraan, 2011; Tavares et al., 2023).

Seaweeds have proteins. They are of varying levels in different species, providing different nutritional values (Healy et al., 2023). Green and red seaweeds have

Table 7.1 Comparison of microalgae and macroalgae nutritional profiles

Algae type	Species	Protein content	Lipid content	Carbohydrate content	Key bioactives	References
Microalgae	*Chlorella vulgaris*	43–61%	5–58%	12–20%	Essential and nonessential amino acids, polyunsaturated fatty acids (PUFA), sulfated polysaccharides, chlorophyll, B-complex vitamins, vitamin A and E, iron	Mendes et al. (2024)
	Chlamydomonas rheinhardii	48%	21%	17%	Carotenoids, chlorophylls, lipids (PUFAs), polysaccharides, proteins	Becker (2013) and Masi et al. (2023)
	Spirulina sp.	67.5 g/100 g	7.5 g/100 g	14.6 g/100 g	Essential amino acids, carotenoids, phycocyanin, chlorophyll, gamma-linolenic acid, iron, vitamin B12	Belay (2013)
	Dunaliella salina	57	6	32	β-Carotene, 3, 3, 5-trimethylheptane	Becker (2013) and Amador-Luna et al. (2024)
	Haematococcus pluvialis	29–45%	20–25%	15–17%	Carotenoids (astaxanthin), proteins, and polyunsaturated fatty acids (PUFAs)	Bueno et al. (2020) and Gherabli et al. (2023)
	Nannochloropsis spp.	19–24% to 44.9%	37–60%	6.4–36.2%	Lipids, proteins, carbohydrates, minerals, and vitamins	Paterson et al. (2023)

(continued)

Table 7.1 (continued)

Algae type	Species	Protein content	Lipid content	Carbohydrate content	Key bioactives	References
Macroalgae	*Laminaria japonica*	17–21%	Up to 4.5%	38%	Polysaccharides, mannitol, vitamins, iodine, and other minerals	Holdt and Kraan (2011) and Jang et al. (2024)
	Porphyra spp.	34.9–37.2%	1.68–2.09%	46.9–49.3%	Proteins, vitamins (e.g., vitamin C and vitamin B family members), minerals, dietary fiber, polyunsaturated fatty acids, carotenoids, saccharides, and mycosporine-like amino acids	Cao et al. (2016)
	Ulva lactuca	10–47%	1–3%	Up to 60%	Polysaccharides, omega-3 and omega-6 fatty acids, phenolic compounds (flavonoids and phenolic acids), chlorophylls, and carotenoids	Putra et al. (2024)
	Gracilaria edulis	14.26 ± 0.88%	0.93 ± 0.00%	32.39 ± 1.90%	Fatty acids, protein, minerals, vitamins, amino acids, and dietary fiber	Debbarma et al. (2016)
	Sargassum spp.	18.21 ± 0.00%	0.73 ± 0.001%	30.32 ± 0.21%	Vitamins, carotenoids, dietary fibers, proteins, minerals, terpenoids, flavonoids, sterols, sulfated polysaccharides, and polyphenols	Debbarma et al. (2016) and Paga and Suek (2024)
	Fucus vesiculosus	12.4 ± 0.01%	3.4 ± 0.2%	30.2 ± 1.3%	Phlorotannins, fucoxanthin, and fucoidans	Nova et al. (2023) and Ferreira et al. (2019)

higher protein concentrations. Brown ones have lower protein concentrations (Healy et al., 2023; Cherry et al., 2019). Seaweeds provide a good range of essential proteins. Seaweeds have high concentrations of different organic compounds with nutritional and health value (Healy et al., 2023).

Seaweeds are fat-free for the most part. It's a healthy food option (Biris-Dorhoi et al., 2020). Brown types have higher fat content compared to the green types. This difference is notable (Biancarosa et al., 2018). Half of the fats in seaweeds are healthy unsaturated kinds. Red and brown types have plenty of these healthy ones. Green types are mostly made up of saturated fats with some unsaturated ones. The ratio of fatty acids in red and brown seaweeds is more friendly to human health. This makes them a superior choice to be included in the diet, contributing to better health status (Biris-Dorhoi et al., 2020).

Seaweeds vary significantly in their carbohydrate content. *Ulva*, for instance, contains low amounts, whereas *Ascophyllum* contains high amounts. *Laminaria* falls in between the two, echoing heterogeneous nutritional content among different species (Biris-Dorhoi et al., 2020; Mahadevan, 2015). Organic compounds can constitute a high proportion of the total weight on a dehydrated weight basis. This highlights nutritional value (Biris-Dorhoi et al., 2020). Seaweeds have unique carbohydrates with numerous health benefits. They are anti-inflammatory and exert protective effect, directly neutralizing harmful molecules in the body (Healy et al., 2023).

The primary natural pigments present in macroalgae include carotenoids, chlorophylls, and phycobiliproteins (Cikoš et al., 2022). Green pigments have a central metal ion which is responsible for light absorption. They are not only responsible for energy conversion during photosynthesis but also as defense molecules that protect algal cells against oxidative damage induced by environmental stress agents (Biris-Dorhoi et al., 2020). Chlorophyll a content in the brown algae species investigated varied from 565 to 2000 mg per kilogram dry weight basis (Holdt & Kraan, 2011). The red or orange pigments called carotenoids assist in photosynthesis by taking in light energy and passing it on to chlorophyll (Biris-Dorhoi et al., 2020). Phycobiliproteins are special, water-soluble pigments that exist in complexes called phycobilisomes. These complexes are associated with thylakoid membranes instead of being embedded in them. They are made up of the pigmented phycobilins, which have linear tetrapyrrole structures that give the distinctive light-absorbing qualities.

Seaweeds are reservoirs of primary nutrients and bioactive compounds. Heat treatment and dehydration can modify their nutritional value, influencing overall quality and health value (Healy et al., 2023). Zinc content ranges widely in various marine organisms. *Mastocarpus stellatus* has high contents, while *Corallina officinalis* has low contents of this essential element (Healy et al., 2023; Stengel et al., 2004). Certain algae have high variation in zinc content. *Mastocarpus stellatus* has high contents, while *Corallina officinalis* has low contents (Healy et al., 2023; Stengel et al., 2004). Beyond zinc, Seaweeds like *Ulva fasciata* and *Gracilaria edulis* are essential providers of iodine (38.8–72.2 mg/100 g), iron (14.8–72 mg/100 g), and calcium (410–870 mg/100 g) (Healy et al., 2023; Ganesan et al., 2020). Macroalgae are also rich in essential vitamins, including B1, B2, B12,

C, beta-carotene, and E (Healy et al., 2023). There are nutrients essential to ensure health. Some vitamins, for example, enhance immunity, and another vitamin safeguards heart health. Some vitamins possess the capability to minimize cancer risk. Adding seaweed to everyday diets has the potential to enhance diet quality. Adding it provides higher-level nutrition to consumers and possible health advantages, and it is a healthful source of food. Adding sea vegetables as a component of healthy eating is a viable alternative (Healy et al., 2023). Microalgae like *Spirulina* and *Chlorella* contain high-quality proteins and fatty acids. Seaweeds, however, have a more varied array of useful compounds and exhibit better antioxidant properties (Agregán et al., 2018).

7.3 Advantages of Nutritional Supplements Derived from Algae

7.3.1 Antioxidant Properties

Reactive molecules are capable of destroying important parts of organisms. Upon accumulation, the molecules cause a state of imbalance known as oxidative stress (Sonani et al., 2017). The state is linked to many serious health conditions, such as malignancies, mental deterioration, metabolic disorders, arthritis, cardiovascular disease, premature aging, and neurodegenerative disorders like Parkinson's and Alzheimer's. Balance is highly desirable for maximum health and longevity (Sonani et al., 2017).

Oceanic plants possess various protective compounds. These include natural pigments, proteins, and essential nutrients that demonstrate notable abilities to combat oxidative damage and support cellular health (Lee et al., 2024). Generally, algae-based antioxidants work by neutralizing ROS and RNS, which are damaging to cells (Pangestuti & Kim, 2011). For example, it is shown that phenolic compounds are potent antioxidants that protect biomolecules like DNA and proteins from oxidative damage (Jimenez-Lopez et al., 2021; Vuolo et al., 2019). Natural compounds effectively counteract harmful molecules. They also inhibit the production of substances that promote inflammation, contributing to overall wellness (Kiokias & Gordon, 2004; Pumas et al., 2011).

Studies identify the function of certain plant substances in regulating cellular communication involved with oxidative damage. One type of these substances in brown seaweeds is capable of inducing a protective pathway that increases the activities of enzymes that protect against oxidative damage (Catarino et al., 2021). Antioxidants found in algae exert potent antioxidant effects against oxidative damage. Algae antioxidants neutralize the damaging molecules and enhance the body's natural defense mechanisms, and hence they are extremely effective against cellular stress (Catarino et al., 2021). Figure 7.1 summarizes the proposed mechanisms for the range of health effects attributed to algae.

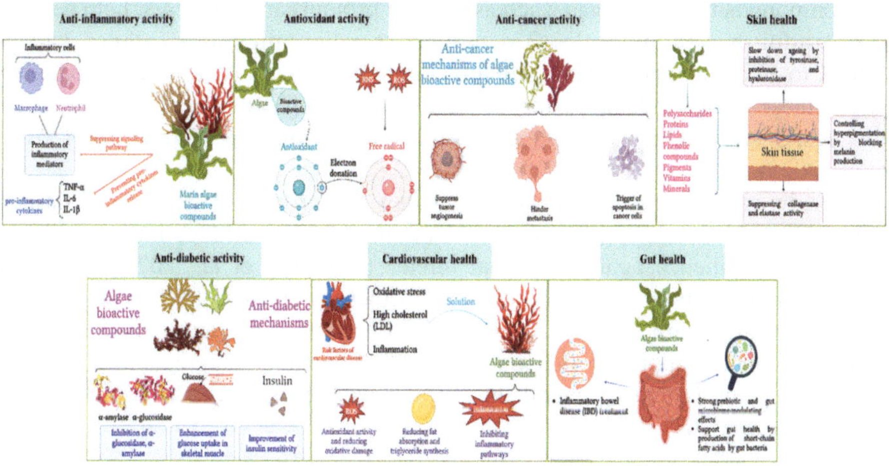

Fig. 7.1 Mechanisms of marine algae bioactives in promoting health and disease prevention

7.3.2 Anti-inflammatory and Immunomodulatory Effects

The immune response triggers a vital process that contributes to the development and advancement of numerous health conditions. This intricate reaction occurs when the body identifies harmful agents, injuries, or infections, leading to protective actions aimed at healing and restoration (Fig. 7.1) (Chen et al., 2017). Acute inflammation is an essential and protective mechanism that facilitates regeneration and fights infections; however, chronic or high levels of inflammation can result in detrimental effects on the body (Pereira & Valado, 2024).

Marine algae have demonstrated significant potential in regulating inflammation and modulating immune responses through various mechanisms (Pereira & Valado, 2024; Conde et al., 2022). Certain compounds from algae interact with components of the immune system, influencing the production and activity of immune regulators. They help maintain a balance between inflammatory and anti-inflammatory processes, reducing excessive immune responses associated with allergic reactions and long-term inflammation (Pereira & Valado, 2024; Elbandy, 2022; Martins et al., 2022).

Algae contain beneficial compounds, such as omega-3 fatty acids and polyphenols, that exhibit properties to reduce inflammation and swelling. These elements can inhibit the production of substances that promote inflammation and block pathways involved in inflammatory responses. Consuming algae-rich foods may help ease symptoms related to allergic reactions and chronic inflammatory disorders (Pereira & Valado, 2024; Elbandy, 2022). Omega-3s influence the processing and signaling of fat molecules. They aid in lowering swelling and reestablishing fat equilibrium during inflammatory states (Pereira & Valado, 2024; De Luca et al., 2021).

Compounds found in algae, such as specific proteins and pigments, help neutralize harmful oxygen molecules. This protective action minimizes cellular damage linked to allergies and chronic inflammation. By shielding cells from oxidative stress, these compounds may support overall well-being and help reduce inflammatory responses, leading to better health outcomes (Pereira & Valado, 2024; Yang et al., 2023).

Certain carbohydrates in algae strengthen the protective barriers in tissues, enhancing their integrity and function. This reinforcement helps improve overall health and resilience against external threats (Reynolds et al., 2021). Including marine plants in the diet can further support these protective mechanisms, potentially preventing harmful substances from entering the body and reducing allergy risks. This natural approach aids in managing inflammation and regulating immune responses, contributing to overall health and wellness (Pereira & Valado, 2024; Reynolds et al., 2021).

7.3.3 Cardiovascular Health

Cardiovascular diseases (CVDs) rank among the foremost causes of mortality globally and are regarded as the primary global killer (Thitame & Aher, 2025). CVDs are responsible for approximately 17.9 million deaths each year, making up nearly 31% of global mortality (WHO 2025). However, proper weight management and lifestyle modifications can substantially diminish the risk of CVDs; maintaining long-term weight loss presents a significant barrier. Consequently, there is increasing interest in the influence of natural dietary constituents, especially those sourced from marine algae, on cardiovascular health enhancement (Thitame & Aher, 2025).

Various types of algae generate a wide range of beneficial compounds that support human health. Supplements made from these organisms are becoming increasingly popular, particularly for weight management (Fernandes et al., 2024). Their appeal comes from their low-calorie content combined with a high concentration of essential nutrients. These supplements work through multiple mechanisms, such as boosting metabolism, regulating blood sugar levels, aiding detoxification, and reducing fat absorption, all due to their unique composition (Fernandes et al., 2024).

Algal omega-3 polyunsaturated fatty acids, such as docosahexaenoic acid and eicosapentaenoic acid, contribute to lipid homeostasis by reducing triglyceride synthesis and enhancing their breakdown. They also upregulate genes responsible for enzymes involved in lipid oxidation, which helps lower circulating cholesterol levels and reduces the risk of atherosclerosis (Fig. 7.1) (Thitame & Aher, 2025). Polysaccharides obtained from red algae, specifically *Porphyra yezoensis*, have demonstrated the ability to lower LDL cholesterol and increase HDL cholesterol in many animal experiments (Amano et al., 2005; Osumi et al., 2002). These findings suggest promising potential for using algae to regulate lipid profiles and improve cardiovascular health in humans (Thitame & Aher, 2025). Compounds like polyphenols and carotenoids found in algae can help reduce inflammatory substances, which may, in turn, lower the risk of CVDs (Thitame & Aher, 2025).

7.3.4 Gut Health and Probiotics

Inflammatory bowel disease (IBD) is a condition that affects the immune system, causing persistent inflammation in the digestive tract (Brizzi et al., 2025). It is classified into two main types: Crohn's disease and ulcerative colitis, each distinguished by specific symptoms, affected areas, and tissue characteristics (Brizzi et al., 2025). A significant imbalance in the gastrointestinal microbiota is commonly observed in individuals with IBD (Brizzi et al., 2025), which is characterized by reduced bacterial diversity, fewer anti-inflammatory bacteria, and more pro-inflammatory bacteria, suggesting that treatments enriched with underrepresented anti-inflammatory bacteria could help reduce intestinal inflammation in IBD patients (Brizzi et al., 2025; Quévrain et al., 2016).

Algae-derived compounds are gaining recognition for their potential health benefits, offering protection against various diseases and promoting overall well-being (Brizzi et al., 2025).

Bioactive substances found in algae, such as certain sugars and proteins, have shown promise in supporting the body's defense mechanisms and reducing inflammation. For example, specific polysaccharides from brown algae exhibit protective properties against oxidative stress and tumors, suggesting their potential role in managing inflammatory bowel diseases. As a result, their therapeutic applications are drawing increasing interest in health research (Brizzi et al., 2025; Bagalagel et al., 2022). The impact of particular polysaccharides on intestinal inflammation has been the subject of numerous investigations. It has been demonstrated that extracts from specific algae have anti-inflammatory and anti-oxidative qualities (Nemoto et al., 2017; Ahmad et al., 2022). These substances have a beneficial effect on the gut microbiota by acting as prebiotics. Beneficial bacteria can ferment them, producing short-chain fatty acids that support digestive health even though they are not broken down in the digestive system. There is evidence that these polysaccharides can improve the integrity of the gut lining, change the structure of the microbial community, and promote the growth of advantageous gut microorganisms (Xie & Cheong, 2022). According to research, polysaccharides derived from algae may enhance the general gut environment and encourage the development of advantageous gut microbes (Guo et al., 2023). Furthermore, in animal studies, some extracts have been shown to improve microbial interactions and strengthen immune responses, suggesting that they may help support digestive health (Wassie et al., 2021).

7.3.5 Other Potential Benefits

High blood sugar is a common symptom of diabetes. It is primarily separated into two types: the first type needs insulin, and the second type doesn't (Ahmed et al., 2024). Certain sugars, phenolic compounds, pigments, and sterols are among the compounds derived from marine plants that have demonstrated promise in the effective management of blood sugar levels. Promising therapeutic benefits for this metabolic disorder may be provided by these natural substances (Ahmed et al., 2024). The body's ability to process sugar is positively impacted by alginate. Fucoidan, on the other hand, lowers blood sugar levels and certain enzyme activity while increasing a hormone that helps regulate glucose, especially in people with type 2 diabetes (Popović-Djordjević et al., 2021; Sakai et al., 2019). Phlorotannins have a number of advantageous effects on controlling blood sugar. They improve insulin response, increase muscle sugar absorption, block specific enzymes, and provide protection against diabetes-related complications (Popović-Djordjević et al., 2021; Lee & Jeon, 2013). Certain enzymes involved in the metabolism of glucose can be inhibited by compounds present in brown seaweeds. In diabetic conditions, this action may help control blood sugar levels (Popović-Djordjević et al., 2021).

Cancer cells in people with diabetes may grow uncontrollably, forming tumors that can spread to other parts of the body (Lee et al., 2021). Algal bioactive compounds exhibit noteworthy anticancer characteristics, underscoring their potential as medicinal agents (Saadaoui et al., 2020). Malignant cells can die when exposed to specific natural substances. Additionally, they stop the development of new blood vessels and stop cancer from spreading to other locations (Saadaoui et al., 2020). Atashrazm, Lowenthal (Atashrazm et al., 2015) reported that fucoidan combats lung cancer by slowing tumor progression, inducing cancer cell death, and enhancing the effectiveness of anticancer chemotherapy drugs when used in combination.

Today, people increasingly prefer natural-ingredient cosmetics over chemical-based ones, as they are safer and free from side effects (Jesumani et al., 2019). Many consumers seek natural products to maintain youthful, healthy-looking skin, prompting the cosmetic industry to focus on ingredients derived from natural sources, such as microbes, plants, algae, and their metabolites. The marine world, rich in diverse species and bioactive compounds, is particularly valuable. Algae are a key source of active compounds with versatile applications (Jesumani et al., 2019; Berthon et al., 2017). Beauty care products emphasize ingredients with antioxidant properties, as well as the ability to inhibit MMPs (extracellular proteinases) and tyrosinase, aiming to counteract ROS generated by UV exposure and slow down skin aging (Jesumani et al., 2019). Tyrosinase is the enzyme responsible for melanin synthesis, the pigment that determines skin color. Hyperpigmentation occurs due to the abnormal accumulation of melanin, often triggered by exposure to ultraviolet (UV) radiation. Tyrosinase inhibitors are promising for controlling hyperpigmentation and achieving skin whitening as they inhibit melanin production. So, algae-derived compounds are valuable for the cosmetic industry due to natural tyrosinase inhibitors (Jesumani et al., 2019; Cha et al., 2011). Also, algae polysaccharides are highly effective in suppressing the activity of collagenase and elastase, which degrade collagen and elastin, respectively, and this leads to wrinkles skin (Jesumani et al., 2019). Furthermore, phlorotannins derived from algae can inhibit hyaluronidase activity, which can degrade hyaluronic acid and lead to the skin aging process (Jesumani et al., 2019).

7.4 Current Applications and Products

7.4.1 Algae in the Supplement Industry

Seaweeds are becoming increasingly popular in nutraceutical supplements because of its rich beneficial compounds (Ghosh et al., 2022). They can produce a wide range of materials that are useful in a variety of industries, including nutrition, medicine, and cosmetics. This versatility improves the economic attractiveness of algae-based energy production and encourages them as an asset for the green industries (Mehta et al., 2017).

Despite the encouraging research on microalgae-derived nutrients, current market products remain short in supply. Food items derived from microalgae fall into two wide categories: (Ibañez et al., 2011) Nutrient-rich dried microalgae of specific species are being marketed as health supplements. In addition, bioactive molecules of such organisms function as beneficial constituents in fortified foods and dietary supplements (Mehta et al., 2017).

Marine algae have garnered considerable attention in the nutritional supplement sector, accessible in diverse forms, including powders, pills, and capsules (Torky et al., 2023). *Chlorella*, a species of green algae, is widely promoted as a health supplement for humans, as well as a feed supplement for fish and a nutritional product, available in various formats, including capsules, pills, and liquids. Dried *Chlorella* biomass consists of roughly 20% carbs, 45% protein, 20% lipids, 5% fiber, 10% minerals, and an array of vitamins (Mehta et al., 2017). These supplements are rich in bioactive substances, including antioxidants, omega-3 fatty acids, and vital vitamins, which are crucial for their health-promoting effects (Nayak et al., 2022). For example, algal oils extracted from species such as *Schizochytrium* and *Crypthecodinium* are especially recognized for their rich docosahexaenoic acid (DHA) content, a nutrient crucial for supporting cardiovascular health and cognitive function (Ansorena & Astiasarán, 2013).

The market for algal-based supplements has grown substantially, with leading products like *Nutrex Hawaii's Spirulina Pacifica*, *Sun Chlorella tablets*, and *Nordic Naturals Algae Omega* driving this expansion (Leu & Boussiba, 2014). For example, *Spirulina Pacifica* is widely recognized for its rich protein levels and antioxidant benefits, whereas *Sun Chlorella tablets* are promoted for their detoxification and immune-enhancing properties (Karkos et al., 2011). Also, *Nordic Naturals Algae Omega*, an oil derived from microalgae, offers a plant-based alternative to fish oil, meeting the increasing demand for vegan omega-3 supplements (Lenihan-Geels et al., 2013). These products utilize the distinctive nutritional properties of algae, positioning them as adaptable components in a wide range of health-focused offerings (Leu & Boussiba, 2014). Growing interest in eco-friendly practices and plant-derived diets positions oceanic flora as vital contributors to the health product sector. They offer unique options to meet the evolving needs of consumers seeking nutritious alternatives (Greene et al., 2022).

7.4.2 Algae-Derived Ingredients in Functional Foods

Although a single definition has yet to be found, health-promoting edibles in general are credited for their further benefits beyond essential nutrition (Cardoso et al., 2015). This new popularity has caused growth in global markets for them, especially across the Asia, Europe, and North America geographies (Ibañez & Cifuentes, 2013).

Producers are increasingly using active compounds from sea plants, which have valuable nutrients. These elements offer cancer, inflammation, and oxidative stress protection, as well as a better overall health profile of products (Admassu et al., 2015). For example, using algae in snacks can restrict calorie, fat, and salt content considerably in relation to conventional versions in addition to offering essential nutrients (Boukid & Castellari, 2021).

Algae are excellent mixing, thickening, and stabilizing food ingredients. Such properties impart rich mouthfeel and food product stability to fulfill the trend of open and natural ingredients in the marketplace (Dominguez, 2013). Additionally, the rising demand for sustainable and eco-friendly food products further complements the use of algae in functional foods. Algae farming requires limited resources, including freshwater and arable land, and can occur in controlled conditions, making it a highly sustainable alternative to conventional agriculture. Incorporating algae into food items enables producers to diminish the carbon footprint of food production while satisfying consumer demand for environmentally sustainable options (Gómez Pinchetti & Martel, 2016).

The adaptability of algae enables their integration into a diverse array of products, including bread, noodles, dairy substitutes, and meat alternatives, expanding their role in the food industry (Healy et al., 2023). Therefore, as global demand for sustainable and nutrient-dense foods increases, algae-derived ingredients are poised to become a key driver in the evolution of functional foods. By effortlessly blending into daily diets, these ingredients help bridge nutritional deficiencies while promoting a shift toward a more sustainable and health-focused food system (Wells et al., 2017).

7.4.3 Challenges in Formulation and Commercialization

Despite the fact that algae have been consumed for generations and have various health benefits, they still face significant barriers that hinder them from transitioning from niche markets to broader market integration (Mendes et al., 2022). New product-to-market development is beset with numerous challenges, including heterogeneity in algal biomass composition (except if produced specifically), product stability issues, formulation issues, and regulatory issues (Michalak & Chojnacka, 2015).

One of the largest challenges in marketing these products is their shelf life. The bioactive compounds that are desired in marine plants may degrade under varying environmental conditions, temperature, light, and exposure to air, ultimately losing their health benefit (Senadheera et al., 2023). All this leads to degradation of the health benefit, as well as the product's shelf life. New technologies like small-scale encapsulation or emulsification are needed to ensure these active compounds are kept effectively (Barrow et al., 2007).

On the other hand, consumer acceptance poses a challenge because of the grassy flavor and fishy odor associated with marine algae. To overcome this barrier, enhancing aroma and taste through careful strain selection and innovative product development can boost market appeal and demand for algae-based supplements and functional foods (Matos et al., 2022).

Scalability presents another hurdle in the commercialization of algae-based supplements. While algae can be cultivated sustainably with minimal environmental impact, scaling up production to meet commercial demands requires significant investment in infrastructure, such as photobioreactors or open pond systems (Barkia et al., 2019). Improving collection and refinement methods is vital for reducing expenses and ensuring the viability of extensive seaweed cultivation (Barkia et al., 2019). Addressing these obstacles with innovative processing methods, protective technologies, and increased public knowledge will be key to unlocking the benefits of oceanic algae in the dietary supplement market (Venugopal, 2008).

7.5 Advances in Cultivation and Processing Technologies

Current advances in the culture of small aquatic plants are directed towards maximizing environmental conditions. Manipulation of light, temperature, and nutrient supply enhances yield and nutritional value. These advances are needed to make microalgae a more economically sound and sustainable source of food (Abreu et al., 2023). For example, photobioreactors (PBRs) have become a highly effective cultivation method, providing controlled conditions that improve photosynthetic efficiency and minimize the risk of contamination (Islam et al., 2024). Closed systems, such as tubular and PBRs, enable precise control of light, temperature, and nutrient levels, resulting in greater productivity compared to conventional open pond systems (Ahmad et al., 2021). Additionally, genetic engineering and selective breeding are being employed to develop microalgae strains with enhanced nutrient profiles, including higher levels of omega-3 fatty acids and carotenoids (Kumari et al., 2024). Macroalgae are being cultivated more effectively using advanced methods. One such approach is integrated multi-trophic aquaculture (IMTA), which involves growing macroalgae alongside fish or shellfish, forming a mutually beneficial system that recycles nutrients and minimizes environmental impact (Angel et al., 2019). Traditionally, macroalgae harvesting has relied on wild collection, a practice that can lead to overharvesting and disrupt ecological balance. So, initiatives have been introduced to explore methods for sustainable harvesting (García-Poza et al., 2022). Wild harvesting of macroalgae typically involves selectively cutting from single-species algae beds, such as rockweeds and kelps, or collecting storm-cast algae fronds, which often include a mix of species along with debris and other contaminants (Mac Monagail et al., 2017). Macroalgae are now being cultivated more effectively using advanced methods. One such method is integrated multi-trophic aquaculture (IMTA), forming a mutually beneficial system that recycles nutrients and lessens environmental harm (Sukhdhane et al., 2018). Besides, seaweed culture is more appealing and scalable. Developments in automated harvesting methods enhance the economic feasibility of large-scale farming activities (Mac Monagail et al., 2017). When harvested, sea plants require processing for release and storage of their precious ingredients, including proteins, omega fatty acids, and antioxidants. More recent technology developments tackle the efficiency and nutritional absorption of these nutrients with the aim that they do not lose their disease-fighting agents upon storage or use (Bürck et al., 2024). Ultrasound extraction involves subjecting high-sound-frequency wave to open algal cells, which enhances substance release. This process not only boosts extraction rates but also decreases processing time and energy consumption, promoting sustainability (Bürck et al., 2024). Alternatively, enzyme-assisted extraction uses particular enzymes that degrade algal cells under the best conditions, enhancing efficiency while decreasing the quantity of harmful solvents (Bürck et al., 2024). This method provides a milder and more precise alternative compared to traditional mechanical techniques (Parsaeimehr & Ozbay, 2024). Microencapsulation is a technology that shields bioactive sensitive ingredients from various steps, including manufacture, transport, and storage. Microencapsulation effectively masks obnoxious flavors without affecting the efficacy of the drugs but enhancing solubility and bioavailability within the body. Through the formation of a protective film, microencapsulation maintains the desirable characteristics of the ingredients and thus leads to improved efficacy and consumer satisfaction (Vieira et al., 2020). Innovations in growth and extraction methods enhance productivity while meeting the increasing need for eco-friendly, premium dietary options (Bürck et al., 2024).

7.6 Regulatory and Safety Considerations

Marine algae offer a sustainable option for third-generation biofuels, avoiding competition with food crops due to their low lignin content (Filote et al., 2021). However, concerns about digestibility and bioavailability arise from anti-nutritional factors like tannins and phytic acid, emphasizing the need for toxicological evaluations to ensure consumer safety (Xie et al., 2024).

One of the most significant safety concerns is the potential for algae to accumulate harmful levels of pollutants, iodine, and heavy metals, which can pose serious health risks to consumers. For example, the iodine content in algae, essential for thyroid hormone synthesis and function, varies widely, with species like *Laminaria* containing over 600 mg/100 g of dry weight (Ferrara, 2022). The European Food Safety Authority advises a daily intake of 150 μg of iodine, with a safe upper limit of 600 μg. Therefore, the ongoing consumption of iodine-dense seaweeds and dietary aids must be closely regulated to prevent possible health complications (Ferrara, 2022).

Despite this, labeling often fails to disclose such contaminants, and regulatory oversight remains insufficient. This lack of transparency highlights the need for stricter regulations and improved monitoring systems to guarantee the security of products derived from algae (Ferrara, 2022). Also, while certain bioactive compounds may possess anti-allergic properties, they can provoke allergic reactions in some individuals (Abril et al., 2024). Techniques like washing and boiling can help reduce risks, but continuous monitoring remains crucial, particularly for algae harvested from polluted environments (Xie et al., 2024).

The European Food Safety Authority suggests 150 μg of iodine daily with an upper safe limit of 600 μg. Therefore, the regular consumption of iodine-containing seaweeds and food supplements must be regulated strictly to prevent any health consequences (Mendes et al., 2022). The FDA regulates food safety in the U.S. It tests the substances to be used by humans. Products that are safe are assigned a certain designation. This procedure protects consumers and encourages public health (Mendes et al., 2022). Canada's standards are the same as the USA, with Health Canada regulating food safety. Novel foods, such as genetically modified foods, are considered new or changed foods and have to go through safety evaluations prior to marketing (Mendes et al., 2022). In the European Union, supplements derived from algae are under the Novel Food Regulation, with thorough safety evaluations and approvals before they can be placed on the market (Mendes et al., 2022).

However, the absence of global regulatory standards poses challenges for manufacturers and consumers. Harmonizing regulations and setting clear guidelines for contaminants and bioactive compounds are essential to ensure the safety and consistency of algae-based supplements. This would not only protect consumers but also facilitate international trade and market expansion for algae-derived products (Eruaga, 2024).

Consumer acceptance remains a significant barrier due to the "grassy" flavor and fishy smell of marine algae. To address this, manufacturers are focusing on improving the taste and aroma of algae products through strain selection and innovative product development (Matos et al., 2022). For example, the Portuguese company *Allmicroalgae* has developed *honey yellow chlorella* powder, a product that eliminates the fishy odor and enhances the taste of food items, making it more appealing to consumers (Matos et al., 2022).

In addition to improving sensory attributes, consumer education plays a crucial role in building trust and awareness. Educating consumers about the health benefits of algae, as well as the sustainable practices involved in their cultivation, can help overcome resistance and increase market acceptance (Proksch, 2013).

By tackling safety issues, complying with regulatory standards, and connecting with consumers, the algae industry can overcome adoption challenges and position itself as a reliable provider of sustainable, high-quality nutritional supplements (Proksch, 2013).

7.7 Future Prospects and Research Directions

The world is facing unprecedented challenges in securing sustainable food and energy supplies, combating climate change, and enhancing public health. As global populations expand, the need for sufficient nutrition intensifies, driving a focus on natural bioactive compounds derived from marine organisms (Ibañez & Cifuentes, 2013). For example, the creation of advanced photobioreactors and closed-loop cultivation systems has dramatically enhanced biomass production and nutrient quality while reducing environmental harm (Antal et al., 2019). Moreover, smart bio-panels and machine learning are being used in algae farming to improve growth conditions, lower production expenses, and increase scalability (Karade et al., 2024). Current advances improve the effectiveness of algae culture. They also enable the manufacturing of uniform, high-quality, and low-cost algal material for food. This innovation is beneficial to both consumers and producers (Mac Monagail et al., 2017).

Genetic alteration of small aquatic organisms is a revolutionary method in environmentally friendly technology. Metabolic adjustment in them leads to increased yield and efficiency, both economically and environmentally friendly. Their application in healthcare products, food industries, animal nutrition, and renewable energy has inspired widespread research in their biochemistry (Ahmad Kamal et al., 2024). Advances in artificial biology and genetic engineering technology can increase the production of beneficial compounds by marine organisms. This enhancement raises the cost-effectiveness of products from these organisms, making them applicable in industries (Qin et al., 2023). Technologies like CRISPR/Cas9 can improve aquatic organisms at a lower cost and a broader range of beneficial compounds for application in food safety and health products (Liu et al., 2022). Nevertheless, addressing moral and regulatory concerns related to modified organisms is essential for building public confidence (Dizon et al., 2016).

The market for products derived from microalgae is developed and has excellent potential for expansion, particularly in Europe. Scientific and technological advances are making increased production and demand possible. In order to improve the market position of such products, aspects of their supply chain need to be optimized (Enzing et al., 2014). The most important ones are operational viability, economic sustainability, consumer acceptance, and regulatory compliance, which are essential for the growth of the market size and opportunities for algae-based products (Rahman, 2020). At present, European nations represent a small share of the world market of microalgae-based food and feed (Enzing et al., 2014). While areas like the United States, Asia, and Oceania are presently at the forefront, Europe can become a major player in this industry within the next 10 years. To achieve this, there need to be concerted efforts and clear goals that will help drive progress and innovation in the sector (Rahman, 2020).

7.8 Conclusion

Oceanic algae, both microscopic and macroscopic, are a healthy and eco-friendly dietary supplement option. They bear beneficial substances, including proteins, essential fatty acids, and natural compounds, that promote overall health. Supermicroscopic algae such as *Spirulina* and *Chlorella* are highly renowned for their protein value and health-giving properties. Conversely, larger forms such as *Laminaria* and *Gracilaria* are prized for their health benefits to the gut and immune system. Although promising, scaling up, palatability, and regulation are present challenges. These are being addressed through breakthroughs in cultivation (photobioreactors, Integrated Multi-Trophic Aquaculture, IMTA) and processing (ultrasound extraction, microencapsulation). Genetic modification further increases bioactive compound yields but needs to be controlled against ethical and regulatory considerations. The global market for algae supplements is expanding rapidly on the back of increasing demand for sustainable, plant-based nutrition. Consistent cooperation among scientists, industries, and regulators is essential to enhance production, ensure security, and create products out of seaweed. This puts aquatic plants in the position as a key tool for addressing international health and environmental issues, providing a stronger and more sustainable future.

References

Abreu, A. P., Martins, R., & Nunes, J. (2023). Emerging applications of Chlorella sp. and Spirulina (Arthrospira) sp. *Bioengineering, 10*(8), 955.

Abril, A. G., Carrera, M., & Pazos, M. (2024). Marine bioactive compounds with functional role in immunity and food allergy. *Nutrients, 16*(16), 2592.

Admassu, H., Zhao, W., Yang, R., Gasmalla, M., & Alsir, E. (2015). Development of functional foods: Seaweeds (algae) untouched potential and alternative resource—A review. *International Journal of Scientific and Technology Research, 4*(1), 108–115.

Agregán, R., Munekata, P. E., Franco, D., Carballo, J., Barba, F. J., & Lorenzo, J. M. (2018). Antioxidant potential of extracts obtained from macro-(Ascophyllum nodosum, Fucus vesiculosus and Bifurcaria bifurcata) and micro-algae (Chlorella vulgaris and Spirulina platensis) assisted by ultrasound. *Medicine, 5*(2), 33.

Ahmad, I., Abdullah, N., Koji, I., Yuzir, A., & Muhammad, S. E. (Eds.). (2021). Evolution of photobioreactors: A review based on microalgal perspective. *IOP Conference Series: Materials Science and Engineering, 1142*, 012004.

Ahmad, T., Ishaq, M., Karpiniec, S., Park, A., Stringer, D., Singh, N., et al. (2022). Oral Macrocystis pyrifera fucoidan administration exhibits anti-inflammatory and antioxidant properties and improves DSS-induced colitis in C57BL/6J mice. *Pharmaceutics, 14*(11), 2383.

Ahmed, N., Sheikh, M. A., Ubaid, M., Chauhan, P., Kumar, K., & Choudhary, S. (2024). Comprehensive exploration of marine algae diversity, bioactive compounds, health benefits, regulatory issues, and food and drug applications. *Measurement: Food, 14*, 100163.

Ahmad Kamal, A. H., Mohd Hamidi, N. F., Zakaria, M. F., Ahmad, A., Harun, M. R., Chandra Segaran, T., et al. (2024). Genetically engineered microalgae for enhanced bioactive compounds. *Discover Applied Sciences, 6*(9), 482.

AlFadhly, N. K., Alhelfi, N., Altemimi, A. B., Verma, D. K., Cacciola, F., & Narayanankutty, A. (2022). Trends and technological advancements in the possible food applications of spirulina and their health benefits: A review. *Molecules, 27*(17), 5584.

Amador-Luna, V. M., Herrero, M., Domínguez-Rodríguez, G., Ibáñez, E., & Montero, L. (2024). Enhancing the bioactivity of Dunaliella salina extracts through ultra-high pressure supercritical fluid extraction (UHP-SFE). *Innovative Food Science & Emerging Technologies, 95*, 103697.

Amano, H., Kakinuma, M., Coury, D. A., Ohno, H., & Hara, T. (2005). Effect of a seaweed mixture on serum lipid level and platelet aggregation in rats. *Fisheries Science, 71*, 1160–1166.

Angel, D., Jokumsen, A., & Lembo, G. (2019). Aquaculture production systems and environmental interactions. In *Organic aquaculture: Impacts and future developments* (pp. 103–118). Springer.

Ansorena, D., & Astiasarán, I. (2013). Development of nutraceuticals containing marine algae oils. In *Functional ingredients from algae for foods and nutraceuticals* (pp. 634–657). Woodhead Publishing.

Antal, T., Konyukhov, I., Volgusheva, A., Plyusnina, T., Khruschev, S., Kukarskikh, G., et al. (2019). Chlorophyll fluorescence induction and relaxation system for the continuous monitoring of photosynthetic capacity in photobioreactors. *Physiologia Plantarum, 165*(3), 476–486.

Atashrazm, F., Lowenthal, R. M., Woods, G. M., Holloway, A. F., & Dickinson, J. L. (2015). Fucoidan and cancer: A multifunctional molecule with anti-tumor potential. *Marine Drugs, 13*(4), 2327–2346.

Bagalagel, A., Diri, R., Noor, A., Almasri, D., Bakhsh, H. T., Kutbi, H. I., et al. (2022). Curative effects of fucoidan on acetic acid induced ulcerative colitis in rats via modulating aryl hydrocarbon receptor and phosphodiesterase-4. *BMC Complementary Medicine and Therapies, 22*(1), 196.

Barkia, I., Saari, N., & Manning, S. R. (2019). Microalgae for high-value products towards human health and nutrition. *Marine Drugs, 17*(5), 304.

Barrow, C. J., Nolan, C., & Jin, Y. (2007). Stabilization of highly unsaturated fatty acids and delivery into foods. *Lipid Technology, 19*(5), 108–111.

Becker, E. W. (2013). Microalgae for human and animal nutrition. In *Handbook of microalgal culture: Applied phycology and biotechnology* (pp. 461–503). Wiley.

Begum, H., Yusoff, F. M., Banerjee, S., Khatoon, H., & Shariff, M. (2016). Availability and utilization of pigments from microalgae. *Critical Reviews in Food Science and Nnutrition, 56*(13), 2209–2222.

Belay, A. (2013). Biology and industrial production of Arthrospira (spirulina). In *Handbook of microalgal culture: Applied phycology and biotechnology* (pp. 339–358). Wiley.

Berthon, J.-Y., Nachat-Kappes, R., Bey, M., Cadoret, J.-P., Renimel, I., & Filaire, E. (2017). Marine algae as attractive source to skin care. *Free Radical Research, 51*(6), 555–567.

Biancarosa, I., Belghit, I., Bruckner, C. G., Liland, N. S., Waagbø, R., Amlund, H., et al. (2018). Chemical characterization of 21 species of marine macroalgae common in Norwegian waters: Benefits of and limitations to their potential use in food and feed. *Journal of the Science of Food and Agriculture, 98*(5), 2035–2042.

Biris-Dorhoi, E.-S., Michiu, D., Pop, C. R., Rotar, A. M., Tofana, M., Pop, O. L., et al. (2020). Macroalgae—A sustainable source of chemical compounds with biological activities. *Nutrients, 12*(10), 3085.

Bishop, W. M., & Zubeck, H. M. (2012). Evaluation of microalgae for use as nutraceuticals and nutritional supplements. *Journal of Nutrition & Food Sciences, 2*(5), 1–6.

Boukid, F., & Castellari, M. (2021). Food and beverages containing algae and derived ingredients launched in the market from 2015 to 2019: A front-of-pack labeling perspective with a special focus on Spain. *Foods, 10*(1), 173.

Brizzi, A., Rispoli, R. M., Autore, G., & Marzocco, S. (2025). Anti-inflammatory effects of algae-derived biomolecules in gut health: A review. *International Journal of Molecular Sciences, 26*(3), 885.

Bueno, M., Gallego, R., Chourio, A. M., Ibáñez, E., Herrero, M., & Saldaña, M. D. (2020). Green ultra-high pressure extraction of bioactive compounds from Haematococcus pluvialis and Porphyridium cruentum microalgae. *Innovative Food Science & Emerging Technologies, 66*, 102532.

Bürck, M., Ramos, S. P., & Braga, A. R. C. (2024). Enhancing the biological effects of bioactive compounds from microalgae through advanced processing techniques: Pioneering ingredients for next-generation food production. *Foods, 13*(12), 1811.

Cao, J., Wang, J., Wang, S., & Xu, X. (2016). Porphyra species: A mini-review of its pharmacological and nutritional properties. *Journal of Medicinal Food, 19*(2), 111–119.

Cardoso, S. M., Pereira, O. R., Seca, A. M., Pinto, D. C., & Silva, A. M. (2015). Seaweeds as preventive agents for cardiovascular diseases: From nutrients to functional foods. *Marine Drugs, 13*(11), 6838–6865.

Catarino, M. D., Amarante, S. J., Mateus, N., Silva, A. M., & Cardoso, S. M. (2021). Brown algae phlorotannins: A marine alternative to break the oxidative stress, inflammation and cancer network. *Foods, 10*(7), 1478.

Cha, S. H., Ko, S. C., Kim, D., & Jeon, Y. J. (2011). Screening of marine algae for potential tyrosinase inhibitor: Those inhibitors reduced tyrosinase activity and melanin synthesis in zebrafish. *The Journal of Dermatology, 38*(4), 354–363.

Chen, L., Deng, H., Cui, H., Fang, J., Zuo, Z., Deng, J., et al. (2017). Inflammatory responses and inflammation-associated diseases in organs. *Oncotarget, 9*(6), 7204.

Cherry, P., O'Hara, C., Magee, P. J., McSorley, E. M., & Allsopp, P. J. (2019). Risks and benefits of consuming edible seaweeds. *Nutrition Reviews, 77*(5), 307–329.

Cikoš, A.-M., Šubarić, D., Roje, M., Babić, J., Jerković, I., & Jokić, S. (2022). Recent advances on macroalgal pigments and their biological activities (2016–2021). *Algal Research, 65*, 102748.

Čmiková, N., Kowalczewski, P. Ł., Kmiecik, D., Tomczak, A., Drożdżyńska, A., Ślachciński, M., et al. (2024). Characterization of selected microalgae species as potential sources of nutrients and antioxidants. *Foods, 13*(13), 2160.

Conde, T., Lopes, D., Łuczaj, W., Neves, B., Pinto, B., Maurício, T., et al. (2022). Algal lipids as modulators of skin disease: A critical review. *Metabolites, 12*(2), 96.

De Luca, M., Pappalardo, I., Limongi, A. R., Viviano, E., Radice, R. P., Todisco, S., et al. (2021). Lipids from microalgae for cosmetic applications. *Cosmetics, 8*(2), 52.

Debbarma, J., Rao, B. M., Murthy, L. N., Mathew, S., Venkateshwarlu, G., & Ravishankar, C. (2016). Nutritional profiling of the edible seaweeds Gracilaria edulis, Ulva lactuca and Sargassum sp. *Indian Journal Of Fisheries, 63*(3), 81–87.

Dini, I. (2023). The potential of algae in the nutricosmetic sector. *Molecules, 28*(10), 4032.

Dissanayake, W., Hailstone, R., Castillo, L., Bao, M., Wang, R.-Q., Gates, C., et al. (2023). The effects of CdSe/ZnS quantum dots on the photosynthesis rate of the chlorella vulgaris beads. *Algal Research, 72*, 103095.

Dizon, F., Costa, S., Rock, C., Harris, A., Husk, C., & Mei, J. (2016). Genetically modified (GM) foods and ethical eating. *Journal of Food Science, 81*(2), R287–RR91.

Dominguez, H. (2013). *Functional ingredients from algae for foods and nutraceuticals*. Elsevier.

El-Manaway, I. M., & Rashedy, S. H. (2022). The ecology and physiology of seaweeds: An overview. In *Sustainable global resources of seaweeds volume 1: Bioresources, cultivation, trade and multifarious applications* (pp. 3–16). Springer.

Elbandy, M. (2022). Anti-inflammatory effects of marine bioactive compounds and their potential as functional food ingredients in the prevention and treatment of neuroinflammatory disorders. *Molecules, 28*(1), 2.

Enzing, C., Ploeg, M., Barbosa, M., & Sijtsma, L. (2014). *Microalgae-based products for the food and feed sector: An outlook for Europe* (pp. 1017–1024). Institute for Prospective Technological Studies fashion Nutraceuticals.

Eruaga, M. (2024). Enhancing global food safety standards through international collaboration and policy harmonization. *International Journal of Scientific Research in Multidisciplinary Studies, 4*, 20–32.

Fernandes, F., Martins, R., Barbosa, M., & Valentão, P. (2024). Algae-based supplements claiming weight loss properties: Authenticity control and scientific-based evidence on their effectiveness. *Marine Drugs, 22*(3), 123.

Ferrara, L. (2022). Issues regarding toxicity and safety of foods from seaweeds. In *Sustainable global resources of seaweeds volume 2: Food, pharmaceutical and health applications* (pp. 293–308). Springer.

Ferreira, R. M., Ramalho Ribeiro, A., Patinha, C., Silva, A. M., Cardoso, S. M., & Costa, R. (2019). Water extraction kinetics of bioactive compounds of Fucus vesiculosus. *Molecules, 24*(18), 3408.

Filote, C., Santos, S. C., Popa, V. I., Botelho, C. M., & Volf, I. (2021). Biorefinery of marine macroalgae into high-tech bioproducts: A review. *Environmental Chemistry Letters, 19*, 969–1000.

Ganesan, A. R., Subramani, K., Shanmugam, M., Seedevi, P., Park, S., Alfarhan, A. H., et al. (2020). A comparison of nutritional value of underexploited edible seaweeds with recommended dietary allowances. *Journal of King Saud University-Science, 32*(1), 1206–1211.

García-Poza, S., Pacheco, D., Cotas, J., Marques, J. C., Pereira, L., & Gonçalves, A. M. (2022). Marine macroalgae as a feasible and complete resource to address and promote Sustainable Development Goals (SDGs). *Integrated Environmental Assessment and Management, 18*(5), 1148–1161.

Gherabli, A., Grimi, N., Lemaire, J., Vorobiev, E., & Lebovka, N. (2023). Extraction of valuable biomolecules from the microalga Haematococcus pluvialis assisted by electrotechnologies. *Molecules, 28*(5), 2089.

Ghosh, S., Sarkar, T., Pati, S., Kari, Z. A., Edinur, H. A., & Chakraborty, R. (2022). Novel bioactive compounds from marine sources as a tool for functional food development. *Frontiers in Marine Science, 9*, 832957.

Gómez Pinchetti, J. L., & Martel, Q. A. (2016). Algae production and their potential contribution to a nutritional sustainability. *Journal of Environmental Health Science and Engineering, 2*(3), 1–3.

Greene, C., Scott-Buechler, C., Hausner, A., Johnson, Z., Lei, X. G., & Huntley, M. (2022). Transforming the future of marine aquaculture: A circular economy approach. *Oceanography, 35*(2). https://doi.org/10.5670/oceanog.2022.213

Guo, Q., Li, Y., Dai, X., Wang, B., Zhang, J., & Cao, H. (2023). Polysaccharides: The potential prebiotics for metabolic associated fatty liver disease (MAFLD). *Nutrients, 15*(17), 3722.

Healy, L. E., Zhu, X., Pojić, M., Sullivan, C., Tiwari, U., Curtin, J., et al. (2023). Biomolecules from macroalgae—Nutritional profile and bioactives for novel food product development. *Biomolecules, 13*(2), 386.

Holdt, S. L., & Kraan, S. (2011). Bioactive compounds in seaweed: Functional food applications and legislation. *Journal of Applied Phycology, 23*, 543–597.

Ibañez, E., & Cifuentes, A. (2013). Benefits of using algae as natural sources of functional ingredients. *Journal of the Science of Food and Agriculture, 93*(4), 703–709.

Ibañez, E., Herrero, M., Mendiola, J. A., & Castro-Puyana, M. (2011). Extraction and characterization of bioactive compounds with health benefits from marine resources: Macro and micro algae, cyanobacteria, and invertebrates. In *Marine bioactive compounds: Sources, characterization and applications* (pp. 55–98). Springer.

Islam, M. M., Alam, H., Acharjee, A., & Mozumder, M. S. I. (2024). Establishment of an effective photobioreactor for growing microalgae: A review. *International Journal of Agricultural Research, Innovation and Technology, 14*(2), 153–162.

Jang, H., Lee, J., Park, Y.-K., & Lee, J.-Y. (2024). Exploring the health benefits and concerns of brown seaweed consumption: A comprehensive review of bioactive compounds in brown seaweed and its potential therapeutic effects. *Journal of Agriculture and Food Research, 17*, 101215.

Jesumani, V., Du, H., Aslam, M., Pei, P., & Huang, N. (2019). Potential use of seaweed bioactive compounds in skincare—A review. *Marine Drugs, 17*(12), 688.

Jimenez-Lopez, C., Pereira, A., Lourenço-Lopes, C., García-Oliveira, P., Cassani, L., Fraga-Corral, M., et al. (2021). Main bioactive phenolic compounds in marine algae and their mechanisms of action supporting potential health benefits. *Food Chemistry, 341*, 128262.

Karade, N. R., Lohar, S. D., Patil, R. S., & Desai, S. R. (2024). Integrating smart bio-panels and machine learning for enhanced microalgae cultivation and carbon reduction. *Asian Journal of Engineering and Applied Technology, 13*(2), 36–43.

Karkos, P., Leong, S., Karkos, C., Sivaji, N., & Assimakopoulos, D. (2011). Spirulina in clinical practice: Evidence-based human applications. *Evidence-Based Complementary and Alternative Medicine, 2011*(1), 531053.

Kiokias, S., & Gordon, M. H. (2004). Antioxidant properties of carotenoids in vitro and in vivo. *Food Reviews International, 20*(2), 99–121.

Kumar, N. (2014). Biopurification of mine wastewater through aquatic plants – A review. *International Journal of Engineering and Technical Research, 2*(6), 186–188.

Kumari, A., Pabbi, S., & Tyagi, A. (2024). Recent advances in enhancing the production of long chain omega-3 fatty acids in microalgae. *Critical Reviews in Food Science and Nutrition, 64*(29), 10564–10582.

Lee, S.-H., & Jeon, Y.-J. (2013). Anti-diabetic effects of brown algae derived phlorotannins, marine polyphenols through diverse mechanisms. *Fitoterapia, 86*, 129–136.

Lee, H., Selvaraj, B., & Lee, J. W. (2021). Anticancer effects of seaweed-derived bioactive com pounds. *Applied Sciences, 11*(23), 11261.

Lee, J. Y., Wong, C. Y., Koh, R. Y., Lim, C. L., Kok, Y. Y., & Chye, S. M. (2024). Natural bioactive compounds from macroalgae and microalgae for the treatment of Alzheimer's disease: A review. *The Yale Journal of Biology and Medicine, 97*(2), 205.

Lenihan-Geels, G., Bishop, K. S., & Ferguson, L. R. (2013). Alternative sources of omega-3 fats: Can we find a sustainable substitute for fish? *Nutrients, 5*(4), 1301–1315.

Leu, S., & Boussiba, S. (2014). Advances in the production of high-value products by microalgae. *Industrial Biotechnology, 10*(3), 169–183.

Liu, H., Chen, W., Li, Y., Sun, L., Chai, Y., Chen, H., et al. (2022). CRISPR/Cas9 technology and its utility for crop improvement. *International Journal of Molecular Sciences, 23*(18), 10442.

Mac Monagail, M., Cornish, L., Morrison, L., Araújo, R., & Critchley, A. T. (2017). Sustainable harvesting of wild seaweed resources. *European Journal of Phycology, 52*(4), 371–390.

Mahadevan, K. (2015). Seaweeds: A sustainable food source. In *Seaweed sustainability* (pp. 347–364). Elsevier.

Martins, B., Vieira, M., Delerue-Matos, C., Grosso, C., & Soares, C. (2022). Biological potential, gastrointestinal digestion, absorption, and bioavailability of algae-derived compounds with neuroprotective activity: A comprehensive review. *Marine Drugs, 20*(6), 362.

Masi, A., Leonelli, F., Scognamiglio, V., Gasperuzzo, G., Antonacci, A., & Terzidis, M. A. (2023). Chlamydomonas reinhardtii: A factory of nutraceutical and food supplements for human health. *Molecules, 28*(3), 1185.

Matos, Â. P. (2017). The impact of microalgae in food science and technology. *Journal of the American Oil Chemists' Society, 94*, 1333–1350.

Matos, Â. P. (2019). *Microalgae as a potential source of proteins. Proteins: Sustainable source, processing and applications* (pp. 63–96). Elsevier.

Matos, J., Cardoso, C., Bandarra, N. M., & Afonso, C. (2017). Microalgae as healthy ingredients for functional food: A review. *Food & Function, 8*(8), 2672–2685.

Matos, Â. P., Novelli, E., & Tribuzi, G. (2022). Use of algae as food ingredient: Sensory acceptance and commercial products. *Frontiers in Food Science and Technology, 2*, 989801.

Mehta, P., Singh, D., Saxena, R., Rani, R., Gupta, R. P., Puri, S. K., et al. (2017). High-value coproducts from algae—An innovational way to deal with advance algal industry. In *Waste to wealth* (pp. 343–363). Springer.

Mendes, M. C., Navalho, S., Ferreira, A., Paulino, C., Figueiredo, D., Silva, D., et al. (2022). Algae as food in Europe: An overview of species diversity and their application. *Foods, 11*(13), 1871.

Mendes, A. R., Spínola, M. P., Lordelo, M., & Prates, J. A. (2024). Chemical compounds, bioactivities, and applications of Chlorella vulgaris in food, feed and medicine. *Applied Sciences, 14*(23), 10810.

Michalak, I., & Chojnacka, K. (2015). Algae as production systems of bioactive compounds. *Engineering in Life Sciences, 15*(2), 160–176.

Michalak, I., Chojnacka, K., & Saeid, A. (2017). Plant growth biostimulants, dietary feed supplements and cosmetics formulated with supercritical CO2 algal extracts. *Molecules, 22*(1), 66.

Nayak, G., Bhuyan, R., & Sahu, A. (2022). Review on biomedical applications of marine algae-derived biomaterials. *Universal Journal of Public Health, 10*(1), 15–24.

Nemoto, M., Kuda, T., Eda, M., Yamakawa, H., Takahashi, H., & Kimura, B. (2017). Protective effects of mekabu aqueous solution fermented by Lactobacillus plantarum Sanriku-SU7 on human enterocyte-like HT-29-luc cells and DSS-induced murine IBD model. *Probiotics and Antimicrobial Proteins, 9*, 48–55.

Nova, P., Pimenta-Martins, A., Maricato, É., Nunes, C., Abreu, H., Coimbra, M. A., et al. (2023). Chemical composition and antioxidant potential of five algae cultivated in fully controlled closed systems. *Molecules, 28*(12), 4588.

Omer, M. Y., Abd El-Wakeil, K. F., Hussein, H. N. M., & Rashedy, S. H. (2021). Invertebrate assemblages associated with seaweeds at different locations in the Red Sea, Egypt. *Egyptian Journal of Aquatic Biology and Fisheries, 25*(1), 407–421.

Osumi, Y., Kawai, M., Amano, H., & Noda, H. (2002). Physiological activities of oligosaccharides derived from marine algae Porphyra yezoensis Porphyran. *Fisheries Science, 68*(sup2), 1441–1444.

Paga, A., & Suek, F. S. (2024). Chemical composition and carbohydrates of seaweed Sargassum sp. with different drying methods. *Journal of Tropical Animal & Veterinary Sciences/Jurnal Ilmu Peternakan dan Veteriner Tropis, 14*(4), 190–196.

Pangestuti, R., & Kim, S.-K. (2011). Biological activities and health benefit effects of natural pigments derived from marine algae. *Journal of Functional foods, 3*(4), 255–266.

Parsaeimehr, A., & Ozbay, G. (2024). Enzymatic processing of algae for food applications. *Biocatalysis and Agricultural Biotechnology, 56*, 103042.

Paterson, S., Gómez-Cortés, P., de la Fuente, M. A., & Hernández-Ledesma, B. (2023). Bioactivity and digestibility of microalgae Tetraselmis sp. and Nannochloropsis sp. as basis of their potential as novel functional foods. *Nutrients, 15*(2), 477.

Pereira, L., & Valado, A. (2024). Exploring the therapeutic potential of algae derived food and diet factors in allergy and inflammation. *Exploration of Asthma & Allergy, 2*(2), 127–147.

Popović-Djordjević, J. B., Katanić Stanković, J. S., Mihailović, V., Pereira, A. G., Garcia-Oliveira, P., Prieto, M. A., et al. (2021). Algae as a source of bioactive compounds to prevent the development of type 2 diabetes mellitus. *Current Medicinal Chemistry, 28*(23), 4592–4615.

Proksch, G. (2013). Growing sustainability-integrating algae cultivation into the built environment. *Edinburgh Architecture Research Journal, 33*, 147–162.

Pumas, C., Vacharapiyasophon, P., Peerapornpisal, Y., Leelapornpisid, P., Boonchum, W., Ishii, M., et al. (2011). Thermostablility of phycobiliproteins and antioxidant activity from four thermotolerant cyanobacteria. *Phycological Research, 59*(3), 166–174.

Putra, N. R., Fajriah, S., Qomariyah, L., Dewi, A. S., Rizkiyah, D. N., Irianto, I., et al. (2024). Exploring the potential of Ulva Lactuca: Emerging extraction methods, bioactive compounds, and health applications – A perspective review. *South African Journal of Chemical Engineering, 47*(1), 233–245.

Qin, J., Kurt, E., LBassi, T., Sa, L., & Xie, D. (2023). Biotechnological production of omega-3 fatty acids: Current status and future perspectives. *Frontiers in Microbiology, 14*, 1280296.

Quévrain, E., Maubert, M., Michon, C., Chain, F., Marquant, R., Tailhades, J., et al. (2016). Identification of an anti-inflammatory protein from Faecalibacterium prausnitzii, a commensal bacterium deficient in Crohn's disease. *Gut, 65*(3), 415–425.

Rahman, K. M. (2020). Food and high value products from microalgae: Market opportunities and challenges. In *Microalgae biotechnology for food, health and high value products*. (pp. 3–27). Springer.

Reynolds, D., Huesemann, M., Edmundson, S., Sims, A., Hurst, B., Cady, S., et al. (2021). Viral inhibitors derived from macroalgae, microalgae, and cyanobacteria: A review of antiviral potential throughout pathogenesis. *Algal Research, 57*, 102331.

Saadaoui, I., Rasheed, R., Abdulrahman, N., Bounnit, T., Cherif, M., Al Jabri, H., et al. (2020). Algae-derived bioactive compounds with anti-lung cancer potential. *Marine Drugs, 18*(4), 197.

Safafar, H., Van Wagenen, J., Møller, P., & Jacobsen, C. (2015). Carotenoids, phenolic compounds and tocopherols contribute to the antioxidative properties of some microalgae species grown on industrial wastewater. *Marine Drugs, 13*(12), 7339–7356.

Sahoo, D., Elangbam, G., & Devi, S. S. (2012). Using algae for carbon dioxide capture and biofuel production to combat climate change. *Phykos, 42*(1), 32–38.

Sakai, C., Abe, S., Kouzuki, M., Shimohiro, H., Ota, Y., Sakinada, H., et al. (2019). A randomized placebo-controlled trial of an oral preparation of high molecular weight fucoidan in patients with type 2 diabetes with evaluation of taste sensitivity. *Yonago Acta Medica, 62*(1), 014–023.

Santiago-Díaz, P., Rivero, A., Rico, M., & Gómez-Pinchetti, J. L. (2021). Characterization of novel selected microalgae for antioxidant activity and polyphenols, amino acids, and carbohydrates. *Marine Drugs, 20*(1), 40.

Ścieszka, S., & Klewicka, E. (2019). Algae in food: A general review. *Critical Reviews in Food Science and Nutrition, 59*(21), 3538–3547.

Senadheera, T. R., Hossain, A., & Shahidi, F. (2023). Marine bioactives and their application in the food industry: A review. *Applied Sciences, 13*(21), 12088.

Silva, M., Kamberovic, F., Uota, S. T., Kovan, I.-M., Viegas, C. S., Simes, D. C., et al. (2022). Microalgae as potential sources of bioactive compounds for functional foods and pharmaceuticals. *Applied Sciences, 12*(12), 5877.

Sonani, R., Rastogi, R., & Madamwar, D. (2017). Natural antioxidants from algae: A therapeutic perspective. In *Algal Green chemistry* (pp. 91–120). Elsevier.

Stengel, D. B., Macken, A., Morrison, L., & Morley, N. (2004). Zinc concentrations in marine macroalgae and a lichen from western Ireland in relation to phylogenetic grouping, habitat and morphology. *Marine Pollution Bulletin, 48*(9–10), 902–909.

Sukhdhane, K. S., Kripa, V., Divu, D., Vase, V. K., & Mojjada, S. K. (2018). Integrated multitrophic aquaculture systems: A solution for sustainability. *Aquaculture Asia, 22*(4), 26–29.

Tavares, J. O., Cotas, J., Valado, A., & Pereira, L. (2023). Algae food products as a healthcare solution. *Marine Drugs, 21*(11), 578.

Thitame, S. N., & Aher, A. A. (2025). Algal biomolecules in cardiovascular disease: A review of current evidence and emerging therapeutic avenues. *Journal of Pharmacy and Bioallied Sciences*. https://doi.org/10.4103/jpbs.jpbs_1760_24

Torky, A., Saad, S., & Eltanahy, E. (2023). Microalgae as dietary supplements in tablets, capsules, and powder. In *Handbook of food and feed from microalgae* (pp. 357–369). Elsevier.

Veluchamy, C., & Palaniswamy, R. (2020). A review on marine algae and its applications. *Asian Journal of Pharmaceutical and Clinical Research, 13*(3), 21–27.

Venugopal, V. (2008). *Marine products for healthcare: Functional and bioactive nutraceutical compounds from the ocean*. CRC Press.

Vieira, M. V., Pastrana, L. M., & Fuciños, P. (2020). Microalgae encapsulation systems for food, pharmaceutical and cosmetics applications. *Marine Drugs, 18*(12), 644.

Vuolo, M. M., Lima, V. S., & Junior, M. R. M. (2019). Phenolic compounds: Structure, classification, and antioxidant power. In *Bioactive compounds* (pp. 33–50). Elsevier.

Wassie, T., Lu, Z., Duan, X., Xie, C., Gebeyew, K., Yumei, Z., et al. (2021). Dietary Enteromorpha polysaccharide enhances intestinal immune response, integrity, and caecal microbial activity of broiler chickens. *Frontiers in Nutrition, 8*, 783819.

Wells, M. L., Potin, P., Craigie, J. S., Raven, J. A., Merchant, S. S., Helliwell, K. E., et al. (2017). Algae as nutritional and functional food sources: Revisiting our understanding. *Journal of Applied Phycology, 29*, 949–982.

WHO. (2025). *Cardiovascular diseases (CVDs)*. Fact Sheet. Available from: https://www.who.int/news-room/fact-sheets/detail/cardiovascular-diseases-(cvds). Last accessed on 31 July 2025.

Xie, X.-T., & Cheong, K.-L. (2022). Recent advances in marine algae oligosaccharides: Structure, analysis, and potential prebiotic activities. *Critical Reviews in Food Science and Nutrition, 62*(28), 7703–7717.

Xie, C., Lee, Z. J., Ye, S., Barrow, C. J., Dunshea, F. R., & Suleria, H. A. (2024). A review on seaweeds and seaweed-derived polysaccharides: Nutrition, chemistry, bioactivities, and applications. *Food Reviews International, 40*(5), 1312–1347.

Yang, N., Zhang, Q., Chen, J., Wu, S., Chen, R., Yao, L., et al. (2023). Study on bioactive compounds of microalgae as antioxidants in a bibliometric analysis and visualization perspective. *Frontiers in Plant Science, 14*, 1144326.

Zayed, A. M., El-Seadawy, H., Attia, E. Z., Rushdi, I. M., & Abdelmohsen, U. R. (2024). Adopting biorefinery and a circular bioeconomy for extracting and isolating natural products from marine algae. *Frontiers in Natural Products, 3*, 1425242.

Plant-Based Approach for Nutrient Enhancement and Quality Assurance in Food

8

Mukesh N. Kher, Ashish Warghane, Sandip P. Dholakia, Vidhi V. Shah, and Vaibhav D. Bhatt

Abstract

In order to address micronutrient shortfalls across the globe, food fortification is essential, especially for people who have limited access to a variety of nutrient-dense foods. A growing number of consumers are switching to a more plant-based diet due to ethical, environmental, and health concerns. Utilizing the inherent nutrient profiles of plants, plant-based food fortification technologies provide a bioavailable and sustainable substitute for traditional fortification techniques, increasing the nutritious content of frequently consumed meals. This strategy entails adding vitamins, minerals, vital fatty acids, and bioactive compounds derived from plants to everyday diets. Additionally, the increasing demand for vegetarian diets presents an opportunity to fortify conventional food products with plant-based ingredients, enhancing public health outcomes and advancing sustainable agriculture. At the same time, there is growing concern about food adulteration, particularly with regard to the integrity of phytoconstituents. When synthetic or harmful substances are added to plant-based foods, their nutritional value is compromised and health hazards are exacerbated. Therefore, in order to ensure the safety and authenticity of plant-based fortified foods, reliable techniques for detecting adulteration are essential. With an emphasis on their phytochemical content, nutritional advantages, difficulties in putting plant-based fortification strategies into practice, and future potential as a

M. N. Kher (✉) · S. P. Dholakia
Department of Pharmaceutical Chemistry and Quality Assurance, L. M. College of Pharmacy, Ahmedabad, Gujarat, India

A. Warghane · V. D. Bhatt
School of Applied Sciences and Technology, Gujarat Technological University, Ahmedabad, Gujarat, India

V. V. Shah
Department of Pharmaceutics, L. M. College of Pharmacy, Ahmedabad, Gujarat, India

G. Sanghvi et al. (eds.), *Sustainable Food Fortification*,
https://doi.org/10.1007/978-981-95-1217-1_8

cutting-edge tactic to enhance global nutrition and health outcomes, this chapter examines a variety of plant-based sources that can be used to fortify foods. Additionally, this chapter explores techniques for detecting adulteration and confirming the purity of phytoconstituents in fortified food products, including spectroscopic approaches and chromatographic analysis.

Keywords

Plant · Fortification · Sustainable food solution · Adulteration analysis

8.1 Introduction

The technique of adding essential nutrients to frequently consumed food in order to prevent or address nutrient deficiencies in a population is known as food fortification. This public health approach aims to enhance the nutritional value of foods, particularly for those that might not have access to a varied or nutrient-dense diet (Miller & Welch, 2013). Fortification can be done by adding nutrients to staple foods and processed foods. In order to combat malnutrition, particularly in developing nations, and to avoid common deficiencies like iron-deficiency anemia or vitamin D deficiency, food fortification is widely utilized (Singh, 2016). Fortifications can be broadly classified into two basic types based on nutritional requirements: micronutrient fortification, which involves adding different micronutrients, such as vitamins and minerals, to foods. For instance, adding calcium to orange juice, iodine to salt, vitamin D to milk, folic acid to flour, etc., and macronutrient fortification, which involves adding macronutrients including proteins, lipids, and fatty acids. Examples include adding omega-3 fatty acids to foods like margarine, milk, or eggs, and fortifying soy protein into processed foods or drinks (Bechoff et al., 2023). The following are the main ways that food fortification can be implemented, depending on the target populations and specific health goals for improving public health: (a) Mandatory (Universal) fortification: This refers to the mandatory addition of nutrients to foods by law. When public health concerns are identified, governments frequently mandate the fortification of specific foods. For example, many countries require the fortification of salt with iodine to prevent iodine deficiency disorders. (b) Targeted fortification: Adding minerals to particular foods those are consumed by high-risk groups, such as iron-fortified cereals for kids or expectant mothers. (c) Voluntary fortification: In this case, food manufacturers voluntarily add nutrients to their products to improve the nutritional profile or cater to consumer demand. Products like breakfast cereals, juices, and dairy items may be fortified with additional vitamins or minerals. (d) Home fortification: This entails incorporating micronutrients into food at the moment of consumption, usually by sprinkling meals with vitamin powders to make up for deficits. (e) Biofortification: The process of fortifying crops by breeding them to naturally contain higher levels of micronutrients. Golden rice, for example, has been genetically modified to contain more vitamin A (Kurpad et al., 2021; Lewis et al., 2013; Pachon, 2023).

8.1.1 Global Issues with Food Quality and Nutritional Deficiencies

In the worldwide battle against malnutrition, food fortification is still an essential weapon. Food fortification techniques will continue to develop as new issues emerge, including shifting dietary habits, climate change, and the rising prevalence of non-communicable diseases worldwide (Jha et al., 2024). Some of the world's nutritional problems might have long-term solutions and thanks to innovations like biofortification. Despite improvements in food supply, nutritional shortages continue to be a major global health concern. These shortages can have serious effects on physical health, cognitive development, and economic growth, and they impact billions of people, especially in low- and middle-income nations. Accessibility, cost, and consumer acceptance are global obstacles to using food fortification to address vitamin deficiencies, especially in low-income areas. The efficacy of food fortification schemes may be compromised by inconsistent fortification standards and inadequate quality control among nations (Bouis et al., 2024). Concerns about nutrient interactions and bioavailability can arise because the body may not properly absorb some additional nutrients. Furthermore, nutritional stability during processing and storage can undermine the benefits of fortification, and over-fortification poses toxicity hazards. To address the problems of nutrient shortages and food quality globally, a multifaceted strategy incorporating fortification, sustainable agriculture, education, and international cooperation is required (Sandhu et al., 2023).

8.1.2 Trends in Plant-Based Diets Globally

Growing awareness of brutality toward animals, environmental sustainability, and health has led to a major global change in diets in recent years toward plant-based diets. The emergence of plant-based food items, including dairy and meat substitutes as well as plant-based packaged foods, which have become widely accepted in many areas, is indicative of this trend (Jareonsin et al., 2024). The trend toward plant-based diets has drawn flexitarians, or people who limit their intake of animal products without completely giving them up, in addition to vegans and vegetarians. The global plant-based food market is expected to grow significantly in the upcoming years, with North America and Europe leading the way in consumption. Industry reports predict that the market will reach nearly $74.2 billion by 2027. This movement is being driven by research that shows the benefits of plant-based diets, such as a lower risk of chronic diseases like diabetes, heart disease, and some types of cancer. Additionally, environmental concerns, such as reducing greenhouse gas emissions, conserving water, and protecting biodiversity, are driving many to adopt plant-based lifestyles (McClements & McClements, 2023). Table 8.1 summarize the various plantbased fortified food products available in the global market (Singh et al., 2025b; Moshtaghian et al., 2024; Xavier et al., 2025; Embling et al., 2024; Shi et al., 2024; Muleya et al., 2024; Perez-Rodriguez et al., 2023; Grau-Fuentes et al., 2023).

Table 8.1 Plant-based fortified food products available in the global market

Sr. no.	Product	Fortified nutrients	Product type	Examples
1	Plant-based milk	Calcium, vitamin D, vitamin B12, vitamin A, vitamin D2, iodine	Beverage (milk alternative)	Soy milk, almond milk, oat milk, rice milk
2	Plant-based yogurt	Calcium, vitamin D, vitamin B12	Yogurt alternative	Almond yogurt, soy yogurt, coconut yogurt, oat yogurt
3	Plant-based cereals	Iron, vitamin B12, vitamin D, calcium, folic acid	Cereal	Breakfast cereals, muesli, granola
4	Plant-based protein powders	Vitamin B12, vitamin D, iron, calcium, omega-3 fatty acids	Protein powder	Pea protein, rice protein, hemp protein, soy protein
5	Plant-based frozen meals	Vitamin B12, vitamin D, iron, calcium, folate	Cheese-less pizzas/burgers	Vegan pizzas, vegan burgers
6	Plant-based meat substitutes	Vitamin B12, vitamin D, iron, calcium, omega-3 fatty acids	Meat alternative	Plant-based burgers, sausages, nuggets
7	Plant-based cheese	Vitamin B12, calcium, vitamin D	Cheese alternative	Vegan cheese slices, shredded vegan cheese, cheese spreads
8	Fortified tofu	Calcium, iron, vitamin B12, vitamin D	Tofu	Soybeans, fortified with minerals
9	Plant-based drinks	Vitamin B12, calcium, vitamin D, omega-3 fatty acids	Energy drink	Fortified smoothies, juices, protein shakes
10	Plant-based spreads	Vitamin B12, iron, calcium, omega-3 fatty acids, iodine	Butter alternative	Nut butters (almond, peanut), hummus
11	Fortified plant-based snacks	Iron, vitamin B12, vitamin D, omega-3 fatty acids, zinc	Snack bar	Granola bars, protein bars, chips
12	Fortified plant-based margarine	Vitamin D, omega-3 fatty acids, vitamin B12	Margarine/ butter alternative	Plant oils, fortified with omega-3, vitamins

8.1.3 Role and Significance of Plant-Based Nutrition

Plant-based nutrition plays a crucial role in promoting health and well-being. Rich in essential nutrients such as fiber, vitamins, minerals, and antioxidants, plant-based diets contribute to improved digestion, enhanced immune function, and better weight management. Moreover, these diets are naturally low in saturated fats and cholesterol, which are linked to various health issues (Prakash et al., 2023). From an environmental perspective, plant-based diets are more sustainable. Producing plant-based foods generally requires fewer resources, including land and water, compared to animal-based products. Reducing meat consumption can significantly lower carbon footprints, contributing to the fight against climate change. Furthermore, plant-based nutrition supports ethical considerations regarding animal welfare, aligning with the growing trend of conscious consumerism. By choosing plant-based options, individuals contribute to more humane and sustainable food systems (Boukid et al., 2024).

8.2 Plant-Based Nutrient Enhancement Strategies

The worldwide agricultural landscape faces immense challenges in addressing food security, climate change, and the increasing demand for more sustainable farming practices (Karabulut et al., 2014). Improving the nutritional value of plant-based meals is one of the most important research areas (Minich et al., 2019). The term "plant-based nutrition" describes the essential nutrients, such as proteins, vitamins, minerals, and other bioactive compounds that plants can offer for human health (Samtiya et al., 2021). However, the ability of many plant-based diets to provide all of the nutritional requirements of humans is frequently limited. Consequently, research in the fields of agriculture and biotechnology is showing promise in the development of plant-based nutrition augmentation techniques. By increasing the nutrient density of crops, these techniques hope to make them healthier for people and more sustainable to produce (Alcorta et al., 2021).

Plant-based diets frequently lack adequate amounts of several nutrients, including protein, vital amino acids, iron, zinc, calcium, and vitamins B12, D, and A (Hever & Cronise, 2017). The main source of this issue is that many crops lack some vital nutrients naturally, and antinutritional agents or inadequate nutrient concentration frequently impair their bioavailability. For instance, heme iron, which comes from animals, has a higher bioavailability than non-heme iron, which comes from plants. Similarly, phytic acid and oxalates can hinder mineral absorption and affect the bioavailability of calcium and zinc in plant meals (Hever & Cronise, 2017). Given the transition toward plant-based diets due to environmental sustainability and health-conscious tendencies, these issues call for innovative approaches to improve the nutritional value of plant-based foods. Further significant plant-based nutrition augmentation techniques are discussed to satisfy human requirements (Samtiya et al., 2021) (Fig. 8.1).

(a) Conventional breeding and selection: The centuries-old methods of conventional plant breeding continue to be important for improving the nutritional value of plants. The nutritional profiles of crops can be progressively improved by choosing and crossing plants with naturally occurring genetic variants that contain higher concentrations of particular nutrients. Biofortified crops are among the greatest examples (Kaiser et al., 2020). Golden rice, which has been genetically altered to produce higher levels of provitamin A (beta-carotene) in the endosperm, is one of the nutrient-dense crops that have expanded as a result of biofortification through conventional breeding. The production of iron-rich beans via selective breeding serves as another illustration. These crops have the

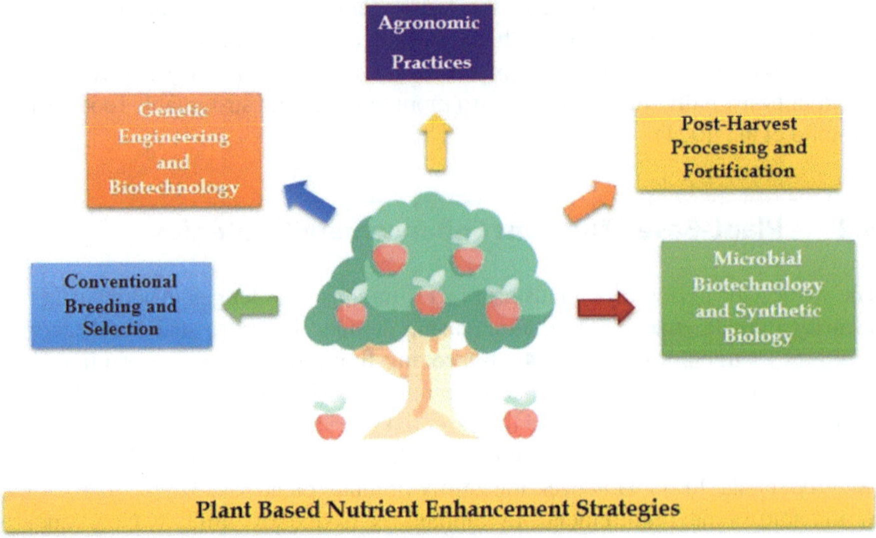

Fig. 8.1 Plant-based nutrient enhancement strategies

potential to help fight hunger worldwide, particularly in regions where traditional diets are deficient in vital nutrients. Even while it works, traditional breeding is frequently sluggish and might take a long time to develop new crop types with the right nutrient profiles. Dealing with crops that are genetically complex or lack naturally occurring genetic variety for specific nutrients presents additional difficulties (Naik et al., 2024).

(b) Genetic engineering and biotechnology: More accurate and rapid methods of improving the nutritional profile of plants are provided by genetic modification (GM) and gene-editing technologies like CRISPR/Cas9 (Warghane et al., 2024). These methods enable the targeted alteration of plant genes involved in the transport or production of nutrients. Golden rice, as previously said, is a genetically modified crop that is intended to produce more beta-carotene, a form of provitamin A (Hamdan & Tan, 2024). Beta-carotene synthesis in the rice endosperm was made possible by the insertion of genes from *Erwinia* and *daffodils* into the rice genome. This has had significant potential effects on lowering vitamin A deficiency, which is a major contributor to childhood mortality and blindness in developing nations (Al-Babili et al., 2002). Researchers have used CRISPR/Cas9 to target genes that produce phytates, or phytic acid, which lowers the bioavailability of minerals like iron and zinc (Warghane et al., 2024). Mineral shortages can be lessened by developing crops with increased mineral bioavailability by the suppression of these genes' expression. Compared to conventional genetic alteration, gene editing is less contentious and is widely regarded as a more accurate method of obtaining desired characteristics. However, public perceptions and regulatory barriers still pose challenges to being widely adopted GM and gene-edited crops (Song et al., 2022).

(c) Agronomic practices: In addition to genetic modification, agronomic techniques can be modified to improve plant nutrient availability. Crop rotation, irrigation methods, and soil fertility can all have a big impact on the nutrients that crops contain (Tamara et al., 2024). Plant nutrient content can be enhanced by maximizing soil health through balanced fertilization. For instance, plants can be guaranteed to have access to the minerals required for their growth through the use of organic matter, micronutrient fertilizers, and biofertilizers. Additionally, techniques like foliar fertilization can facilitate direct nitrogen uptake through plant leaves, occasionally avoiding soil deficits (Jose et al., 2024).

Crop rotation and intercropping are two sustainable agricultural practices that can improve nutrient uptake. Legumes are among the plants that can fix nitrogen in the soil, increasing the amount of nitrogen in crops that grow after them. When cultivated in rotation with legumes, this can therefore raise the total protein content of food crops like wheat and maize. Crops under water stress may become less nutritious. Plants are guaranteed to receive the required water supply during crucial growth phases when irrigation techniques, such as drip irrigation or rainwater harvesting, are optimized. Fruits and seeds' ability to accumulate nutrients like sugars, amino acids, and vitamins can be directly impacted by proper irrigation (Chamkhi et al., 2022).

(d) Postharvest processing and fortification: Plant-based meals can have their nutrient density increased by optimizing postharvest processing techniques like fermentation, heating, and milling. By breaking down antinutritional substances like phytates, fermentation is a conventional method that can increase the bioavailability of nutrients, particularly minerals like iron and zinc. For example, foods made from fermented plants, including tempeh, miso, and sauerkraut, have higher amounts of accessible minerals, amino acids, and B vitamins (Tolulope et al., 2025). The addition of extra nutrients to plant-based foods is another postharvest tactic. For instance, iron, folic acid, and vitamin D can be added to rice, wheat flour, and maize flour. Similarly, in order to ensure that people on plant-based diets have access to sufficient levels of calcium and vitamin B12, plant-based milk substitutes like almond or soy milk are frequently fortified with these elements (Gupta, 2014). The nutritional value of plant meals can also be preserved or even increased with innovative drying and preservation methods. Fruits and vegetables that are sun-dried retain more of their vitamin C content than those that are dried in an industrial setting using higher temperatures (Giannakourou & Taoukis, 2021).

(e) Microbial biotechnology and synthetic biology: New techniques to enhance the nutritional profile of plant-based foods are being made possible by developments in synthetic biology and microbial biotechnology. Plants can absorb more nutrients, especially minerals like phosphorus, zinc, and iron, when modified microbes are used in the rhizosphere, or the soil-root interface (Mahmud et al., 2021). Utilizing beneficial microorganisms like plant-growth-promoting rhizobacteria (PGPR), mycorrhizal fungi, and nitrogen-fixing bacteria can improve the nutrient content of crops and increase the availability of nutrients in the soil. By developing symbiotic interactions with plant roots, these microbes can improve plant growth, improve nutrient density, and increase nutrient uptake. Designing microbes that can synthesize vital nutrients or bioactive substances that plants might not have is made possible by synthetic biology (Vejan et al., 2016).

8.3 Quality Assurance in Plant-Based Foods

The illegal or immoral practice of adding hazardous substances to food items in order to either stretch profit margins or conceal substandard products is known as food adulteration in fortification, and it is an important concern in the worldwide food industry. However, the contamination or dilution of additional micronutrients meant to increase nutritional value can also be referred to as adulteration in the context of food fortification. This problem jeopardizes the desired health advantages and may negatively impact the well-being of customers (Czepielewska et al., 2018). Adulterated fortification, on the other hand, not only does not offer the anticipated nutritional advantages but also presents certain health hazards. In this context, adulteration can take many different forms, such as: (a) Dilution of fortifiers: Inadvertently or purposely, fortifying nutrients are occasionally administered at lower-than-recommended amounts. For instance, purposefully lowering vital elements like iron, iodine, or vitamins in order to save cost jeopardizes the efficacy of the fortification and may exacerbate diseases like goiter, rickets, or anemia. (b) Using inferior fortifiers: Low-quality fortifying micronutrients may have detrimental consequences or decreased bioavailability and efficiency. For instance, synthetic vitamin forms and substandard iodine are less effective than their natural counterparts. (c) Contamination with hazardous ingredients: Heavy metals, allergies, or industrial chemicals may be contaminated during the fortification process. This can result in toxicity, which can cause long-term health problems like neurological damage, kidney dysfunction, or developmental delays in children. (d) Fraudulent claims or mislabeling: Food items may occasionally be erroneously marketed as "fortified" even though the fortification is either nonexistent or of poor quality (Pastor et al., 2019; Kaur et al., 2022).

Food adulteration in fortification has serious repercussions. Fortification programs are intended to alleviate nutritional deficiencies, and they only exacerbate them when the micronutrients provided are inadequate or have low bioavailability. Furthermore, consumers, especially vulnerable groups like children, the elderly, and pregnant women, may suffer direct harm when fortification agents are tainted or adulterated with dangerous compounds. Additionally, poor fortification might cause the public to lose faith in food safety systems, which can undermine the effectiveness of large-scale fortification projects and raise doubts about their efficacy (Meena et al., 2024; Garrett, 2018). Authorities can keep an eye on and regulate the quality of fortified foods through analytical techniques like chromatography, spectroscopy, and microbiological examination, which are crucial in identifying and detecting cases of adulteration. To ensure that fortification is safe and effective, regulatory frameworks established by agencies such as the Food Safety and Standards Authority of India (FSSAI), Food and Agriculture Organization (FAO), Codex Alimentarius Commission, and the US Food and Drug Administration (FDA) assist in establishing guidelines for fortification practices. Improving the safety of fortified foods is only one aspect of addressing food adulteration in fortification; another is increasing the overall efficacy of global nutrition initiatives meant to lower malnutrition and promote public health (Abedini et al., 2023; Ghosh & Ghosh, 2023).

8.3.1 Methods for Detecting Adulteration and Standardization of Fortified Foods

The integrity of fortification programs and public health depends on the detection of adulteration and the uniformity of fortified foods. Chromatography, spectroscopy, hyphenated analytical techniques, and microbiological examinations are efficient techniques used to standardize fortified foods and detect and recognize cases of adulteration. By precisely identifying and measuring fortifying chemicals, pollutants, and adulterants, chromatographic techniques are essential for confirming the quality of fortified food. Unauthorized synthetic versions of vitamins or additional additives that are designed to mimic higher fortification levels may be incorporated in some fortified foods. Chromatographic techniques can identify chemical preservatives (such as sulfites and benzoates) that may be present in food due to processing and can differentiate between natural and synthetic forms of nutrients (Aslam et al., 2024). Chromatography techniques are useful not only for detecting adulterants and fortified foods but also for tracking the stability of fortifying substances over time. Because some vitamins and minerals might deteriorate during processing or storage, nutrient stability is a major problem for fortified foods. For instance, the levels of vitamin C in a product, such as a fortified beverage, can gradually drop due to its extreme sensitivity to heat and light. Manufacturers can evaluate the shelf life of fortified foods and make sure that their nutritional content is maintained over the course of the product's life by using chromatography techniques to monitor the degradation of these nutrients (Abedini et al., 2023). Additionally, these techniques enable the simultaneous examination of various components, which is particularly helpful for confirming that foods have been fortified with a variety of vitamins, minerals, and other nutrients because it allows for comprehensive analysis without requiring several tests.

(a) High-performance liquid chromatography (HPLC): One of the most popular methods for identifying adulteration in fortified foods is high-performance liquid chromatography (HPLC), which is especially useful for figuring out how much vitamin and other nutrients are added during fortification. HPLC offers insights into non-volatile substances. Proteins and vitamins can be separated, identified, and quantified using HPLC. HPLC can offer accurate measures of the product's actual nutrient content in cases of adulteration, where nutrients may be depleted or substituted with inferior forms, detecting differences between the reported and actual levels of fortification (Russo et al., 2016).

(b) Gas chromatography (GC): Usually, gas chromatography is used to examine volatile substances like vitamins or pollutants like solvents or pesticides in fortified meals. Spotting dangerous substances or impurities that should not be in fortified products, it can help in the detection of adulteration. The technique allows for the identification of impurities or deficiencies in fortifying agents by vaporizing the sample and separating its constituent parts according to how they interact with a stationary phase (Chen et al., 2024; Zou et al., 2024).
(c) High-performance thin-layer chromatography (HPTLC): TLC and HPTLC are used as fingerprinting tools for identifying the fortified content and detecting the presence of adulterants in fortified foods. TLC is a cost-effective and relatively simple technique, while HPTLC, an advanced version of TLC, offers enhanced sensitivity, reproducibility, and resolution, making it ideal for more complex analyses. HPTLC uses automated systems for sample application, solvent development, and detection, as well as a more refined and optimized stationary phase (silica gel or alumina).(Chen et al., 2020).

Spectroscopic methods enable the quick detection and quantification of unauthorized ingredients in food items, including artificial coloring agents, preservatives, and less expensive alternatives to premium ingredients. One of the biggest benefits of spectroscopic techniques is their adaptability; they can identify both organic and inorganic substances and assess a wide variety of food products, including liquids, solids, and powders (Jamwal et al., 2021). They are perfect for routine testing in both production facilities and regulatory environments because they provide high-throughput, quick analysis with minimal sample preparation. Additionally, on-site testing is made possible by portable spectroscopic instruments, which are particularly helpful for supply chain inspections or quality control in remote locations (Mohammed et al., 2021).

(a) Atomic absorption spectroscopy (AAS): The content of particular metals in fortified meals can be measured using the AAS instrumental approach. When it comes to identifying heavy metal contamination like lead, mercury, or cadmium that can be present in tainted fortified foods, this technique is especially helpful. Because AAS offers sensitive detection, regulatory agencies can make sure that fortified foods do not contain potentially hazardous levels of these metals, which may pose major health hazards.
(b) Inductively coupled plasma mass spectrometry (ICP-MS): For the detection of trace amounts of minerals and heavy metals in fortified meals, ICP-MS is an extremely sensitive technique. This method can be used to determine the amount of fortifying minerals (such as iron, calcium, or zinc) present as well as to look for heavy metal contamination that might have occurred during packaging or processing. According to Kwon et al. (2024), it is especially useful for making sure that fortified foods fulfill the safety requirements established by regulatory bodies.
(c) Infrared spectroscopy (IR): The presence of foreign oils or adulteration with less expensive oils or other vegetable oils or fats in food products can be found using infrared (IR) and near-infrared (NIR) spectroscopy. The infrared spec-

trum can be used to identify artificial food additives, preservatives, or flavor enhancers. It is feasible to detect contaminants or adulterants by comparing the spectra of contaminated foods with those of recognized pure standards (Badaro et al., 2022; Jamwal et al., 2021).

(d) Ultraviolet-visible (UV-Vis) spectroscopy: The quantity of specific vitamins and other fortifying agents in fortified meals is frequently examined using UV-Vis spectroscopy. The foundation of this technique is the target molecules' ability to absorb light at particular wavelengths. Through the identification of discrepancies in nutrient levels, this quick and affordable technique can assist in the detection of food adulteration. Similar to IR spectroscopy, adulteration can be detected by comparing the absorption spectra of suspected adulterants with those of established standards (Faez et al., 2021).

(e) Mass spectrometry (MS): The molecular weight and structure of the chemicals in the sample are revealed in tremendous clarity by MS. In fortified foods, MS can detect and precisely measure the amounts of adulterants and micronutrients, including synthetic vitamins, prohibited preservatives, and foreign compounds. It works especially well with complicated adulterant combinations (Chen et al., 2025).

The hyphenated techniques describe the integration of two or more analytical techniques that enable the study of samples either sequentially or simultaneously. In food analysis, the following hyphenated procedures are most frequently employed: gas chromatography-mass spectrometry (GC-MS), high-performance liquid chromatography-mass spectrometry (HPLC-MS), inductive coupled plasma mass spectrometry (ICP-MS), gas chromatography-tandem mass spectrometry (GC-MS/MS), liquid chromatography-tandem mass spectrometry (LC-MS/MS), Fourier transform infrared spectroscopy (FTIR) in conjunction with other methods. Due to their high sensitivity and specificity, GC-MS/MS and LC-MS/MS are ideal for detecting traces of adulterants that may be present in fortified food products, such as pesticides or illegal additives. The precise measurement of nutrients in fortified foods using hyphenated procedures guarantees that the necessary fortification levels are met. When evaluating the stability of fortified nutrients under various circumstances, hyphenated approaches are useful. Additionally, it offers comprehensive details regarding the chemical composition of substances, which facilitates the identification of intricate adulterants. Hyphenated procedures can be used for regulatory compliance testing, confirming that food items fulfill the set fortification standards, and guaranteeing consumer safety, according to a number of national and international standards. Hyphenated techniques, like FTIR and ICP-MS, are perfect for quality control in the food industry since they enable the analysis of food samples without changing their composition. With continued advancements in analytical technology, hyphenated techniques will remain at the forefront of ensuring the integrity of fortified food products worldwide (Kher et al., 2024).

Microbiological methods for probiotics and fortified microorganisms: Microbiological assays are used to assess the viability and concentration of microorganisms added to fortified meals, such as probiotics. Reduced health benefits may arise from adulteration, which may involve the use of inferior or ineffective strains of microorganisms. Enzyme-linked immunosorbent assay (ELISA) and polymerase chain reaction (PCR) are two examples of biochemical tests that can be performed to confirm the legitimacy of the microorganisms used in the fortification process and make sure they adhere to legal requirements (Mudgil et al., 2022).

8.4 Sustainable Food Solution

Traditional animal-based sources of protein are facing increasing difficulties with regard to resource depletion, environmental sustainability, and ethical issues as the world's protein needs continue to expand. Diversifying plant-based protein sources and cutting down on food waste and nutrient loss are crucial steps in creating a sustainable food solution in response to these issues (Anyiam et al., 2025). A plant-based protein source has become crucial to lowering reliance on animal-based proteins and enhancing sustainability. Some tactics include:

(a) Exploring alternative grains and legumes: Although peas, beans, and lentils are popular plant-based protein sources, numerous more neglected choices can be included in diets, for instance (Joseph, 2024; Romano et al., 2023), chickpeas, mung beans, and fava beans: high in protein and widely adaptable in recipes. Quinoa and buckwheat: complete proteins containing all nine essential amino acids. Spirulina: a blue-green algae that is a complete protein and contains a wide array of micronutrients. Millets and sorghum: both millets and sorghum are drought-resistant, require less water, and are rich in protein and essential micronutrients. These grains are well-suited for regions facing water scarcity, contributing to global food security.

(b) Incorporating nuts, seeds, and soy products: Nuts provide not only protein but also heart-healthy fats, antioxidants, and micronutrients. They are versatile for snacking or as ingredients in protein powders and dairy alternatives. Nuts and seeds are packed with protein and healthy fats. For instance, hemp, chia, and flaxseeds are excellent sources of plant protein and omega-3 fatty acids. Soy products like tofu, tempeh, and edamame are versatile protein-rich foods (Balakrishnan & Ishwarya, 2025).

(c) Cultivating protein-rich vegetables: Several vegetables are high in protein content and can be easily added to a balanced diet. Broccoli, spinach, and brussels sprouts are great sources of plant protein that also provide vitamins and fiber. Edamame (young soybeans) is also an example of complete protein and an excellent snack or addition to meals (Karabulut et al., 2025).

(d) Mushrooms and algae as protein sources: Fungi like shiitake, maitake, and oyster mushrooms are emerging as excellent protein sources with unique flavors and textures, making them ideal substitutes for meat in various dishes. Algae-based proteins like *Spirulina* and *Chlorella* not only are rich in protein but also contain vitamins, minerals, and antioxidants. They offer a high-density nutrient profile in small quantities, contributing to both sustainable and nutritious diets (Batubara et al., 2025).

However, to ensure long-term sustainability, it is crucial not only to diversify plant-based protein sources but also to reduce food waste and nutrient loss throughout the food supply chain, and reducing both is key to sustainability (Beuving et al., 2024). The food system is plagued by significant losses in both food and nutrients at multiple stagesfrom production and harvest to processing, distribution, and consumption. The following strategies can be implemented at various stages of food production, storage, and consumption (Singh et al., 2025a, Manzoor et al., 2024).

(a) Improving harvest and postharvest practices: Enhancing agricultural practices, such as optimizing harvesting times, reducing spoilage, and improving storage conditions, can significantly cut down postharvest losses. Using technologies like controlled atmosphere storage or cold chain logistics helps preserve food quality. Farmer training programs that teach efficient crop management and postharvest handling techniques are essential to minimizing waste at the initial stages of food production.
(b) Valorizing food waste as a resource: Food by-products, such as peels, stems, and seeds, often discarded in conventional food production, can be repurposed into valuable ingredients. For example, pulp from fruit and vegetable processing can be used in smoothies or baked goods, while leftover grain husks can be transformed into animal feed or high-fiber snacks. Circular economy models are gaining traction, where food waste is treated as a valuable resource for other products, such as protein powder, biofuels, or compost, reducing waste and conserving resources.
(c) Innovative processing techniques to retain nutrients: Traditional food processing methods can cause nutrient degradation, but modern techniques like freeze-drying, vacuum sealing, and cold pressing help retain more nutrients while extending the shelf life of plant-based foods. When possible, eating whole foods and minimizing food processing helps retain more nutrients. For instance, steaming and roasting are better ways to maintain vitamins than boiling. Heat-sensitive vitamins, such as vitamin C and folate, can be preserved by using the proper cooking temperature and duration. By adding probiotics and boosting nutrient bioavailability, fermentation is another method that can improve the nutritional profile of plant-based foods.
(d) Waste reduction at the consumer level: Education initiatives aimed at consumers can reduce household food waste by increasing knowledge about portion control, appropriate food storage, and the value of using leftovers. There are creative methods to repurpose food that would otherwise go to waste due to a variety of digital apps and platforms that link customers with excess food from grocery shops and restaurants
(e) Eco-friendly packaging: Using compostable or biodegradable packaging instead of plastic helps protect the environment while maintaining the nutrition and freshness of food.

8.4.1 Synergies Between Diversifying Plant-Based Proteins and Reducing Food Waste

Strategies for reducing food waste can be combined with a diverse plant-based protein system. Utilizing underutilized crops, such as chickpeas, lentils, and ancient grains, for example, frequently requires less resource-intensive cultivation, which lowers waste during the production stage. The entire food production system is made more sustainable and resource-efficient by including food waste valorization techniques, such as converting vegetable scraps into protein powder or putting protein-rich byproducts into animal feed (Pontonio et al., 2024).

8.5 Challenges, Opportunities, and Future Trends in Plant-Based Nutrition

Due to the rising demand for nutrients derived from plants, the transition to more sustainable agricultural systems, and growing awareness of the health advantages of plant-based diets, plant-based approaches to food fortification have become increasingly popular in recent years. The challenges, opportunities, and future prospects in the plant-based fortification strategy are highlighted in the following section.

One of the biggest obstacles to fortifying plant-based diets is the standardization of plant-based ingredients. The nutrient composition of plant-based raw materials can vary significantly depending on a number of variables, including climate, soil quality, geographic location, and cultivation methods. The standardization of fortification procedures may be made more difficult by this diversity (Sharma & Yadav, 2023). It can be resource-intensive to closely monitor ingredient quality and make adjustments to fortification levels in order to guarantee that plant-based foods continuously satisfy nutritional requirements. Another major issue with fortified food is the lack of universal regulatory frameworks for plant-based food fortification, particularly for novel or emerging components. Different nations have different fortification regulations, which result in variations in safety standards, labeling, and nutrient levels. Global food producers seeking to guarantee that products meet local legislation and the fortification requirements required for nutritional advantages may also face difficulties due to inconsistent regulations among nations (Mathanghi & Reddy, 2025). Another issue is the stability and bioavailability of added nutrients. Compounds like phytates, oxalates, and fiber found in many plant-based diets might hinder the absorption of minerals, including calcium, zinc, and iron. For instance, although there are plant-based sources of iron, heme iron from animal products is more easily absorbed than non-heme iron from plants. Fortification techniques are therefore essential, but choosing nutritious forms that can get beyond these obstacles is a difficult task. Furthermore, a small number of plant-based nutrients, such as some vitamins and thermolabile substances, are frequently prone to deterioration as a result of processing, storage, and climatic conditions. One of the biggest challenges for quality assurance is making sure that these nutrients are stable and bioavailable during the product's shelf life (Xie et al., 2024).

Plant-based diets are becoming more and more popular, which presents an enormous fortification potential. Fortified plant-based products that can supply vital nutrients frequently lacking in a vegan or vegetarian diet are in greater demand as more people choose plant-based diets for ethical, health, or environmental reasons (Rogers et al., 2024). An opportunity to bridge dietary shortages and enhance public health outcomes is presented by fortifying plant-based diets with nutrients like iron, vitamin B12, and omega-3 fatty acids. Furthermore, plant-based fortifiers like fortified yeast and omega-3s derived from algae are gaining popularity and providing worthy replacements for artificial nutrients (Boukid, 2024). The stability, bioavailability, and controlled release of nutrients in fortified plant-based meals can be improved owing to new developments in food processing and fortification technologies, including microencapsulation and nanotechnology. By enhancing nutrient absorption, ensuring a long shelf life, and reducing nutrient loss during processing, these technologies can enhance the quality assurance procedure (Soni et al., 2022). The growing need for customized fortification presents yet another noteworthy potential. Certain populations' demands can be catered to by fortification (Calderon, 2025). For example, vegans and vegetarians can have their needs met by adding vitamin B12, iodine, or calcium to plant-based milks or plant-based meat substitutes. With the developments in genetics and biotechnology, personalized nutrition presents the chance to develop foods that are tailored to the unique nutritional needs of certain populations, such as youngsters, athletes, and expectant mothers. The emergence of plant-based food substitutes has spurred cooperation between a number of industries, including biotechnology, food science, and agriculture. These partnerships offer chances for innovation in ingredient sourcing and fortification methods. The chance for new and established food companies to expand into new areas and satisfy customer demand for sustainable and fortified food items is increasing, along with the need for plant-based products (Aggarwal et al., 2024).

Genetic engineering and biofortification are two major upcoming themes in plant-based food fortification. A sustainable method of organically fortifying meals is biofortification, which involves growing plants to have higher nutritional content. The production of plant varieties that are naturally higher in particular nutrients, such as vitamin D, B12, or iron, may be made possible by genetic engineering and CRISPR technology, which would lessen the need for external fortification. Genetically modified plant-based foods might be developed in the future to fill certain nutritional shortages in the world. (Bouis et al., 2024). Personalized nutrition is another trend in fortified foods. Fortification techniques may change to become more customized as the field of personalized nutrition develops. The creation of customized plant-based fortified meals according to consumers' dietary preferences, genetic profiles, and particular health requirements may be one of the future trends. The emphasis on integrating artificial intelligence (AI) into quality assurance systems is one emerging trend. With AI technologies that can track food manufacturing processes, identify contaminants, analyze nutritional content, and forecast product quality, the use of AI in fortified food quality assurance is growing. Future fortified plant-based food products may be monitored in real-time by AI-powered systems, which would enable more effective quality control and guarantee that the fortified meals are safe to eat and meet necessary nutritional standards (Yu et al., 2025).

8.6 Conclusion

Adopting a plant-based strategy to improve nutrients offers a viable alternative to meet the rising demand for more sustainable and healthful food options worldwide. Genetics, agronomy, biotechnology, and postharvest processing must all be integrated to address the complex problem of improving the nutritional value of plant-based diets. Nutrient shortages in plant-based diets can be addressed using a variety of techniques, including fermentation, fortification, agronomic methods, genetic engineering, and conventional breeding. We can enhance public health outcomes by incorporating fortified plant-based foods into everyday diets, especially for groups that depend on plant-based diets or have limited access to a variety of food options, or in the worldwide fight against hunger. In order to preserve the nutritional integrity and safety of these products, techniques for identifying adulteration and standardizing these fortified foods are equally important. Developments in biochemical, chromatographic, and spectroscopic techniques provide trustworthy ways to detect adulteration and confirm the legitimacy of fortified foods. Additionally, standardizing fortification procedures guarantees uniformity in nutrient amounts and quality, giving customers safe and efficient products. When combined, these strategies provide a route to a food system that is more nourishing, sustainable, and equitable.

References

Abedini, A., Salimi, M., Mazaheri, Y., Sadighara, P., AlizadehSani, M., Assadpour, E., & Jafari, S. M. (2023). Assessment of cheese frauds, and relevant detection methods: A systematic review. *Food Chem X, 19*, 100825.

Aggarwal, B., Rajora, N., Raturi, G., Dhar, H., Kadam, S. B., Mundada, P. S., Shivaraj, S. M., Varshney, V., Deshmukh, R., Barvkar, V. T., Salvi, P., & Sonah, H. (2024). Biotechnology and urban agriculture: A partnership for the future sustainability. *Plant Science, 338*, 111903.

Al-Babili, P., Ye, X., Lucca, P., Schaub, P., Welsch, R., & Potrykus, I. (2002). Golden rice: Introducing the beta-carotene biosynthesis pathway into rice endosperm by genetic engineering to defeat vitamin a deficiency. *The Journal of Nutrition, 132*(3), 506S–510S.

Alcorta, A., Porta, A., Tarrega, A., Alvarez, M. D., & Vaquero, M. P. (2021). Foods for plant-based diets: Challenges and innovations. *Foods, 10*(2), 293.

Amiri, S., NezamdoostSani, N., Mostashari, P., McClements, D. J., Marszałek, K., & MousaviKhaneghah, A. (2022). Effect of the molecular structure and mechanical properties of plant-based hydrogels in food systems to deliver probiotics: An updated review. *Critical Reviews in Food Science and Nutrition, 64*, 1–27.

Anyiam, P. N., Phongthai, S., Grossmann, L., Jung, Y. H., Sai-Ut, S., Onsaard, E., & Rawdkuen, S. (2025). Potential plant proteins for functional food ingredients: Composition, utilization and its challenges. *Nutrition & Food Science, 38*, 100216.

Aslam, N., Fatima, R., Altemimi, A. B., Ahmad, T., Khalid, S., Hassan, S. A., & Aadil, R. M. (2024). Overview of industrial food fraud and authentication through chromatography technique and its impact on public health. *Food Chemistry, 460*(1), 140542.

Badaro, A. T., Tavares, J. P. H., Blasco, J., Aleixos-Borras, N., & Barbin, D. F. (2022). Near infrared techniques applied to analysis of wheat-based products: Recent advances and future trends. *Food Control, 140*, 109115.

Balakrishnan, G., & Ishwarya, S. P. (2025). Seed and nut protein-applications and health benefits. In *Plant-based proteins sources, extraction, applications, value-chain and sustainability* (pp. 267–302). Academic Press.

Batubara, I., Anisa, A. S., Khotib, M., Achmadi, S. A., Witasari, L. D., & Santoso, D. (2025). Heimioporus sp. (Pelawan mushroom) as source of nutrition and antioxidant. *Food Chemistry Advances, 6*, 100946.

Bechoff, A., de Bruyn, J., Alpha, A., Wieringa, F., & Greffeuille, V. (2023). Exploring the complementarity of fortification and dietary diversification to combat micronutrient deficiencies: A scoping review. *Current Developments in Nutrition, 7*(2), 100033.

Beuving, M., McNabb, W. C., & Smith, N. W. (2024). Global nutrient content embedded in food losses and waste: Identifying the sources and magnitude along the food supply chain. *Sustainable Production and Consumption, 51*, 519–531.

Bouis, H., Foley, J., Lividini, K., Jumrani, J., Reinke, R., Van Der Straeten, D., et al. (2024). Biofortification: Future challenges for a newly emerging technology to improve nutrition security sustainably. *Current Developments in Nutrition, 8*(12), 104478.

Boukid, F. (2024). Holistic benefits of plant-based foods for sustainable agrifood systems. *Current Opinion In Food Science, 58*, 101184.

Calderon, P. (2025). Personalized nutrition: Fruit and vegetables consumption in obesity prevention. In *Improving health and nutrition through bioactive compounds benefits and applications* (pp. 25–46). Academic Press.

Chamkhi, I., Cheto, S., Geistlinger, J., Zeroual, Y., Kouisni, L., Bargaz, A., & Ghoulam, C. (2022). Legume-based intercropping systems promote beneficial rhizobacterial community and crop yield under stressing conditions. *Industrial Crops and Products, 183*, 114958.

Chen, Y., Caihong, H., Zhengyu, J., Xu, X., Cai, Y., & Yuxiang, B. (2020). HPTLC-bioautography/SERS screening nifedipine adulteration in food supplement based on Ginkgo biloba. *Microchemical Journal, 154*, 104647.

Chen, J., Tang, H., Wang, M., Wei, H., & Ou, C. (2024). Explorative study for the rapid detection of adulterated surimi using gas chromatography-ion mobility spectrometry. *Food Chemistry, 439*, 138083.

Chen, M., Peng, M., Yuan, M., Huang, C., Liu, J., Wu, Z., Chen, W., Hu, S., Liu, Q., Dong, J., & Ling, L. (2025). Detection of Salmonella enterica in food using targeted mass spectrometry. *Food Chemistry, 465*(15), 141985.

Czepielewska, E., MakarewiczWujec, M., Rozewski, F., Wojtasik, E., & Kozłowska-Wojciechowska, M. (2018). Drug adulteration of food supplements: A threat to public health in the European Union? *Regulatory Toxicology and Pharmacology, 97*, 98–102.

Embling, R., Neilson, L., Mellor, C., Durodola, M., Rouse, N., Haselgrove, A., Shipley, K., Tales, A., & Wilkinson, L. (2024). Exploring consumer beliefs about novel fortified foods: A focus group study with UK-based older and younger adult consumers. *Appetite, 193*, 107139.

Faez M., Dominique G., Jon W., Nada A., (2021). Analytical methods to detect adulteration of argan oil: A critical review.*Microchemical Journal. 168*, 106501.

Garrett, G. S. (2018). National mandated food fortification programs. In *Food fortification in a globalized world* (pp. 53–62). Academic Press.

Ghosh, P., & Ghosh, P. K. (2023). Regulatory guidelines and protocols for food fortification and enrichment by liposomes. In *Liposomal encapsulation in food science and technology* (pp. 255–268). Academic Press.

Giannakourou, M. C., & Taoukis, P. S. (2021). Effect of alternative preservation steps and storage on vitamin C stability in fruit and vegetable products: Critical review and kinetic modelling approaches. *Foods, 10*(11), 2630.

Grau-Fuentes, E., Rodrigo, D., Garzon, R., & Rosell, C. M. (2023). Understanding the marketed plant-based beverages: From ingredients technological function to their nutritional value. *Journal of Functional Foods, 106*, 105609.

Gupta, G. R. (2014). Fortification of foods with vitamin D in India. *Nutrients, 6*(9), 3601–3623.

Hamdan, M. F., & Tan, B. C. (2024). Genetic modification techniques in plant breeding: A comparative review of CRISPR/Cas and GM technologies. *Horticultural Plant Journal*. https://doi.org/10.1016/j.hpj.2024.02.012

Hever, J., & Cronise, R. J. (2017). Plant-based nutrition for healthcare professionals: Implementing diet as a primary modality in the prevention and treatment of chronic disease. *Journal of Geriatric Cardiology, 14*(5), 355–368.

Jamwal, R., Amit, K. S., Sharma, S., Kelly, S., Cannavan, A., & Singh, D. K. (2021). Recent trends in the use of FTIR spectroscopy integrated with chemometrics for the detection of edible oil adulteration. *Vibrational Spectroscopy, 113*, 103222.

Jareonsin, S., Pumas, C., Jaitiang, D., & Uttarotai, T. (2024). Green fusion proteins: An approach to sustainable nutrition blending plant and algae-based proteins for a circular food system. *Future Foods, 10*, 100415.

Jha, R., Zhang, K., He, Y., Mendler-Drienyovszki, N., Magyar Tabori, K., Quinet, M., et al. (2024). Global nutritional challenges and opportunities: Buckwheat, a potential bridge between nutrient deficiency and food security. *Trends in Food Science and Technology, 145*, 104365.

Jose, T. S., Tindwa, H. J., Alovisi, A. M. T., Shitindi, M. J., & Semoka, J. M. (2024). Enhancing sustainable crop production through integrated nutrient management: A focus on vermicompost, bio-enriched rock phosphate, and inorganic fertilisers—A systematic review. *Frontiers in Agronomy, 6*, 100000.

Joseph, M. (2024). Grains and legumes: By-product generation and their utilization. In *Unit operations in food grain processing* (pp. 473–512). Academic Press.

Kaiser, N., Douches, D., Dhingra, A., Glenn, K. C., Herzig, P. R., Stowe, E. C., & Swarup, S. (2020). The role of conventional plant breeding in ensuring safe levels of naturally occurring toxins in food crops. *Trends in Food Science and Technology, 100*, 51–66.

Karabulut, G., Goksen, G., & MousaviKhaneghah, A. (2014). Plant-based protein modification strategies towards challenges. *Journal of Agriculture and Food Research, 15*, 101017.

Karabulut, G., Subasi, B. G., Ivanova, P., Goksen, G., Chalova, V., & Capanoglu, E. (2025). Towards sustainable and nutritional-based plant protein sources: A review on the role of rapeseed. *Food Research International, 202*, 115553.

Kaur, N., Agarwal, A., & Sabharwal, M. (2022). Food fortification strategies to deliver nutrients for the management of iron deficiency anaemia. *Current Research in Food Science, 5*, 2094–2107.

Kher, M., Shah, R., Gajjar, A., Chhabria, M., & Rakholiya, K. (2024). Analytical characterization of herbal biomolecules using hyphenated techniques. In *Herbal formulation, phytochemistry and pharmacognosy* (pp. 241–253). Elsevier.

Kurpad, A. V., Ghosh, S., Thomas, T., Bandyopadhyay, S., Goswami, R., Gupta, A., et al. (2021). Perspective: When the cure might become the malady: The layering of multiple interventions with mandatory micronutrient fortification of foods in India. *The American Journal of Clinical Nutrition, 114*(4), 1261–1266.

Kwon, S. Y., Kim, Y. I., Kim, Y. K., Lee, Y. B., & Mok, J. H. (2024). Microwave-assisted sample preparation for screening of heavy metal elements in food additives by ICP-MS. *LWT - Food Science and Technology, 208*, 116708.

Lewis, M. J., Abdul Hamid, N. K., Alhazzaa, R., Hermon, K., Donald, J. A., Sinclair, A. J., et al. (2013). Targeted dietary micronutrient fortification modulates n − 3 LC-PUFA pathway activity in rainbow trout (Oncorhynchusmykiss). *Aquaculture, 412–413*, 215–222.

Mahmud, K., Missaoui, A., Lee, K., Ghimire, B., Presley, H. W., & Makaju, S. (2021). Rhizosphere microbiome manipulation for sustainable crop production. *Current Plant Biology, 27*, 100210.

Manzoor, S., Fayaz, U., Dar, A. H., Dash, K. K., Shams, R., Bashir, I., Pandey, V. K., & Abdi, G. A. (2024). Sustainable development goals through reducing food loss and food waste: A comprehensive review. *Future Foods, 9*, 100362.

Mathanghi, S. K., & Reddy, M. K. (2025). Sustainable plant-based meat protein alternatives – Regulatory aspects and consumer acceptance. In *Plant-based proteins – Sources, extraction, applications, value-chain and sustainability* (pp. 377–401). Academic Press.

McClements, I. F., & McClements, D. J. (2023). Designing healthier plant-based foods: Fortification, digestion, and bioavailability. *Food Research International, 169*, 112853.

Meena, S., Raj, A., Meena, J., Reddy, B. K., Rai, D. C., Duary, R. K., & Singh, G. P. (2024). Study on curcumin fortification to improve the physicochemical properties and anti-oxidative potentialities of butteroil (ghee). *Food and Humanity, 3*, 100316.

Miller, D. D., & Welch, R. M. (2013). Food system strategies for preventing micronutrient malnutrition. *Food Policy, 42*, 115–128.

Minich, D. M. (2019). A review of the science of colorful, plant-based food and practical strategies for "eating the rainbow". *Journal of Nutrition and Metabolism, 2019*, 2125070.

Mohammed, F., Guillaume, D., Warland, J., & Abdulwali, N. (2021). Analytical methods to detect adulteration of argan oil: A critical review. *Microchemical Journal, 168*, 106501.

Moshtaghian, H., Hallstrom, E., Bianchi, M., & Bryngelsson, S. (2024). Nutritional profile of plant-based dairy alternatives in the Swedish market. *Current Research in Food Science, 8*, 100712.

Mudgil, P., Aldhaheri, F., Hamdi, M., Punia, S., & Maqsood, S. (2022). Fortification of Chami (traditional soft cheese) with probiotic-loaded protein and starch microparticles: Characterization, bioactive properties, and storage stability. *LWT - Food Science and Technology, 158*, 113036.

Muleya, M., Bailey, E. F., & Bailey, E. H. (2024). A comparison of the bioaccessible calcium supplies of various plant-based products relative to bovine milk. *Food Research International, 175*, 113795.

Naik, B., Kumar, V., Rizwanuddin, S., Mishra, S., Kumar, V., Saris, P. E. J., Khanduri, N., Kumar, A., Pandey, P., Gupta, A. K., Khan, J. M., & Rustagi, S. (2024). Biofortification as a solution for addressing nutrient deficiencies and malnutrition. *Heliyon, 10*(9), e30595.

Pachon, H. (2023). Food fortification programs. In *Encyclopedia of human nutrition* (4th ed., pp. 288–300). Elsevier.

Pastor, K., Acanski, M., & Vujic, D. (2019). A review of adulteration versus authentication of flour. In *Flour and breads and their fortification in health and disease prevention* (2nd ed., pp. 21–35). Academic Press.

Perez-Rodriguez, M. L., Serrano-Carretero, A., Garcia-Herrera, P., Camara-Hurtado, M., & Sanchez-Mata, M. C. (2023). Plant-based beverages as milk alternatives? Nutritional and functional approach through food labeling. *Food Research International, 173*, 113244.

Pontonio, E., Stringari, A., Di Cagno, R., Filannino, P., Rizzello, C. G., Polo, A., Nikoloudaki, O., & Gobbetti, M. (2024). Plant-derived food waste management, valorization, and recycling through sourdough fermentation. *Trends in Food Science and Technology, 150*, 104589.

Prakash, S., Gaiani, C., & Bhandari, B. R. (2023). Plant-based food as a sustainable source of food for the future. In *Engineering plant-based food systems* (pp. 1–12). Academic Press.

Rogers, H., Dora, M., Tsolakis, N., & Kumar, M. (2024). Plant-based food supply chains: Recognising market opportunities and industry challenges of pea protein. *Applied Food Research, 4*(2), 100440.

Romano, G., Tufariello, M., Calabriso, N., Del Coco, L., Fanizzi, F. P., Blanco, A., Carluccio, M. A., Grieco, F., & Laddomada, B. (2023). Pigmented cereals and legume grains as healthier alternatives for brewing beers. *Food Bioscience, 52*, 102463.

Russo, G., Barbato, F., & Grumetto, L. (2016). A validated LC/UV method for the determination of four adulterating drugs in herbal slimming capsules. *Journal of Pharmaceutical and Biomedical Analysis, 117*(5), 436–445.

Samtiya, M., Aluko, R. E., Dhewa, T., & Moreno-Rojas, J. M. (2021). Potential health benefits of plant food-derived bioactive components: An overview. *Foods, 10*(4), 839.

Sandhu, R., Chaudhary, N., Bindia, Shams, R., Singh, K., & Pandey, V. K. (2023). A critical review on integrating bio fortification in crops for sustainable agricultural development and nutritional security. *Journal of Agriculture and Food Research, 14*, 100830.

Sharma, B., & Yadav, D. K. (2023). Chromatography and hyphenated techniques in quality-based standardization of medicinal plants: Current scenario and future perspectives. *South African Journal of Botany, 157*, 467–483.

Shi, H., Nolan, J. M., Pryor, C., Power, T., & Prado-Cabrero, A. (2024). Development and evaluation of a lutein-fortified yoghurt: Stability assessment, sensory properties and human bioavailability study. *Journal of Functional Foods, 121*, 106430.

Singh, J. (2016). Biofortification of food legumes and bioavailability of nutrients. In *Biofortification of food crops* (pp. 51–60).

Singh, A., Prasad, S., Singh, R., Younis, K., & Yousuf, O. (2025a). Revolutionizing the supply chain: Cutting-edge strategies and technologies for food waste reduction. *Bioresource Technology Reports, 29*, 102047.

Singh, M., Mal, N., Trivedi, D., Krishnamoorthy, S., Behera, C., Krishnan, C., Naik, S., & Kuppam, C. (2025b). An overview of the role of algae-fortified foods in nutraceutical industries: Synthesis pathway of value-added bioproducts and co-products. *Food Bioscience, 63*, 105568.

Song, J. H., Shin, G., Kim, H. J., Lee, S. B., Moon, J. Y., Jeong, J. C., Choi, H. K., Kim, I. A., Song, H. J., Kim, C. Y., & Chung, Y. S. (2022). Mutation of GmIPK1 gene using CRISPR/Cas9 reduced phytic acid content in soybean seeds. *International Journal of Molecular Sciences, 23*(18), 10583.

Soni, M., Maurya, A., Das, S., Prasad, J., Yadav, A., Singh, V. K., Singh, B. K., Dubey, N. K., & Dwivedy, A. K. (2022). Nanoencapsulation strategies for improving nutritional functionality, safety and delivery of plant-based foods: Recent updates and future opportunities. *Plant Nano Biology, 1*, 100004.

Tolulope, J. A., Laszlo V. B., (2025). Nutritional and functional aspects of European cereal-based fermented foods and beverages. *Food Research International, 209*, 116221.

Vejan, P., Abdullah, R., Khadiran, T., Ismail, S., & Nasrulhaq Boyce, A. (2016). Role of plant growth promoting rhizobacteria in agricultural sustainability-a review. *Molecules, 21*(5), 573.

Warghane, A., Paserkar, N. G., & Bhose, S. (2024). Genome editing against bacterial plant pathogens. In P. M. Halami & A. Sundararaman (Eds.), *Genome editing in bacteria* (pp. 43–67). Bentham Science.

Xavier, J. R., Shashikumar, S. H., Vats, D., & Chauhan, O. P. (2025). Future trends in plant-based meat: Consumer perception, market growth and health benefits. *Future Foods, 11*, 100551.

Xie, Y., Cai, L., Zhou, G., & Li, C. (2024). Comparison of nutritional profile between plant-based meat analogues and real meat: A review focusing on ingredients, nutrient contents, bioavailability, and health impacts. *Food Research International, 187*, 114460.

Yu, W., Ouyang, Z., Zhang, Y., Lu, Y., Wei, C., Tu, Y., & He, B. (2025). Research progress on the artificial intelligence applications in food safety and quality management. *Trends in Food Science and Technology, 156*, 104855.

Zou, Z., Wang, Q., Li, M., Zhen, J., Yuan, D., Zhang, Q., Cui, Q., Li, M., Zhou, M., Yin, S., & Xu, L. (2024). Combination of gas chromatography-mass spectrometry and hyperspectral imaging for identification of adulterated safflower seed oil. *Journal of Food Composition and Analysis, 135*, 106593.

Emerging Biotechnological Trends and Techniques in Fungal and Lichen Fortification

9

Vidya Patil-Patankar, Mayuri Pawar, and Prajakta Kulkarni

Abstract

Fungal and lichenized symbiotic systems, which include between 2.2 and 3.8 million fungal species and more than 20,000 lichen species globally, are emerging as innovative biotechnological platforms for nutritional fortification. Lichens, which form mutualistic partnerships between fungi and photosynthetic partners (algae or cyanobacteria), excel at nutrient accumulation, stress resistance, and bioactive chemical production. Their peculiar structural organisation facilitates to optimise micronutrient concentration, including rare vitamins, minerals, and functional metabolites. Fungi and lichens can be transformed into specific nutrient manufacturing systems using advanced biotechnology techniques, such as metabolic pathway engineering and (CRISPR)-Clustered Regularly Interspaced Short Palindromic Repeats Cas-9 genome editing. This chapter provides thorough insights into the revolutionary potential of fungal and lichen species for generating innovative functional food solutions by critically analysing advanced techniques for nutritional optimisation, bioactive chemical synthesis, and targeted micronutrient accumulation. This chapter presents fungal-lichen biotechnology as an innovative strategy for dealing with global nutritional security, ecological challenges, and personalised nutrition requirements through advanced biological engineering.

Keywords

Fungi · Lichen · Fortification · Symbiotic systems · Metabolites · Biotechnology

V. Patil-Patankar (✉) · M. Pawar · P. Kulkarni
Department of Botany, PDEA'S Waghire College, Pune, Maharashtra, India

G. Sanghvi et al. (eds.), *Sustainable Food Fortification*,
https://doi.org/10.1007/978-981-95-1217-1_9

9.1 Introduction

The intensification of global food production represents a significant challenge to achieving food security. The majority of surveys and research initiatives in agriculture concentrate on crop production. These studies do not look at how yields change over time or how cereal production changes and stays the same from year to year (Knapp & van der Heijden, 2018). With the ongoing increase in global population, agricultural production must correspondingly expand (Fróna et al., 2019). One of the greatest challenges in the contemporary world is to provide sufficient nutrition to a rapidly increasing world population. Consequently, the imperative for sustainable food production is of utmost importance. Along with quantity, we need to focus on the quality of food production. Food or crops, which are micronutrient-deficient, are prevalent and frequently referred to as 'hidden hunger' (Lockyer et al., 2018). Micronutrients are critical for the body, so not getting enough of them can have many bad effects, such as slow growth, problems with thinking and memory, and ultimately a higher risk of illness and death, depending on which micronutrients are missing. Eradicating hunger in all its manifestations is a principal global concern and constitutes one of the United Nations' Sustainable Development Goals for 2030. Consumption of age- and activity-appropriate balanced meals, including enough amounts of micronutrient-rich plant and animal-source foods, is the optimal strategy for the eradication of micronutrient deficiencies. Especially in rural parts of developing nations, many households may not have access to these ideal, balanced diets because they are either too expensive or too seasonal. Three effective therapies that work together across time and geography include biofortification, fortification, and supplementation in the absence of balanced meals (Lockyer et al., 2018). Fortifying food is cost-effective and beneficial to health, the economy, and society. Despite global and national debates about food fortification's performance and safety, all three main vehicles—large-scale food fortification, biofortification, and point-of-use or home fortification—reduce nutritional deficiencies and benefit societies and economies (Olson et al., 2021). In light of changing food systems and fighting malnutrition and micronutrient deficiencies, biobased fortification is becoming a credible answer. The following framework specifies the term biobased fortification and discusses its universal significance, technology development, and the pivotal shift from conventional techniques to new ways (Bouis et al., 2011).

9.1.1 Global Context of Nutritional Issues

The world's nutritional issues are very diverse. Stunting, wasting, and an absence of vital micronutrients such as iodine, zinc, and iron are symptoms of malnutrition in developing nations. Overnutrition, obesity, and related lifestyle diseases, such as diabetes and cardiovascular disease, are on the rise in affluent nations. Around 38.2 million children younger than 5 were predicted to be overweight or obese in 2019, as per research by the World Health Organisation. Among adults, 39% are overweight and 13% are obese; being overweight or obese is a bigger problem than being underweight in most countries. The prevalence of overweight and obesity is increasing in low- and middle-income nations, challenging the notion that these conditions primarily affect high-income nations. More than 340 million kids and teens (those between the ages of 5 and 19) were overweight or obese in 2016. The percentage of children and adolescents who were overweight or obese rose sharply from 4% in 1975 to slightly more than 18% in 2016. Among adults, over 1.9 billion were overweight in 2016, with 600 million falling into the obese category (WHO, 2020). According to research by Harastani et al. (2020), four interrelated elements contribute to obesity: one's biological makeup, level of physical activity, psychological state, and access to unhealthy food. Two people are overweight or obese for every person who is malnourished; this is an unsustainable paradox in our world (Onyeaka et al., 2023). Approximately 2 billion individuals worldwide experience micronutrient deficiencies, referred to as "hidden hunger," whereas 39% of adults are classified as overweight (WHO, 2021). These divergent situations underscore the pressing necessity for customised, novel approaches such as biobased fortification to comprehensively tackle both undernutrition and overnutrition.

9.1.2 Malnutrition in Developing and Developed Nations

Micronutrient malnutrition, which impacts almost 2 billion people worldwide, is caused by an inadequate diet or impaired absorption as a result of infections, illnesses, or inflammation Kiani et al., 2022. Physiological function and, by implication, health, are compromised because these individuals do not obtain adequate amounts of vitamins and minerals. Many micronutrient deficiencies can be remedied with supplements, but some have irreversible consequences; for instance, a child delivered to a woman who does not get enough iodine during pregnancy may have intellectual impairment (Bailey et al., 2015). Iron deficiency is the most prevalent and widespread public health concern, followed by inadequate levels of iodine, zinc, and vitamin A (WHO, 2021).

Malnutrition has varying effects on developing and prosperous countries, creating a complex issue. Undernutrition is the most common form of malnutrition in developing nations, and it causes stunted growth, impaired immunity, and a higher likelihood of child mortality. Inadequate dietary diversity and an overreliance on carbohydrate-based foods lead to micronutrient deficiencies in South Asia and Sub-Saharan Africa (Development Initiatives, 2022).

Obesity and diet-related diseases like diabetes and cardiovascular disease are on the rise in industrialised countries, where malnutrition often takes the form of overnutrition. Public health concerns related to insufficient amounts of vitamin A, zinc, iodine, and iron are of utmost importance; iron deficiency, in particular, is widespread and prevalent, and it poses health risks. There are no specific dietary reference values (DRVs) for pregnant women, but micronutrient deficiencies can have devastating effects on both the mother and her unborn child, regardless of whether the country is developing (like India) or developed (like the United Kingdom) (Lockyer et al., 2018). But these negative effects can happen to anyone, even individuals who are not pregnant. Poor dietary quality can be indicated by insufficient levels of multiple micronutrients, not just one. Ironically, vitamin shortages and conditions like anaemia still impact certain demographics, especially the elderly and low-income communities. Two problems of malnutrition need to be fixed at the same time: fortified staples for people who are not getting enough food and biofortified, nutrient-rich crops for people who are getting enough food in richer countries.

9.1.3 The Utility of Fortification in Densely Populated Nations

A practical and scalable solution to combat malnutrition in densely populated countries like Nigeria, China, and India is fortification. These countries face unique challenges, including limited access to healthcare, high poverty rates, and dependence on staple crops for subsistence. By fortifying food sources, we can ensure that large populations get plenty of essential nutrients without having to drastically change their eating habits. For instance, in India, iron-enriched wheat flour helps reduce the prevalence of anaemia, whereas in China, fortified rice programmes address vitamin A insufficiency (Field et al., 2020). Countries with large populations can significantly reduce the frequency of nutritional shortages by including fortification in public health programmes.

9.1.4 Food Fortification

Fortification and biofortification represent distinct methodologies in the enhancement of food. The former entails the direct addition of nutrients during the processing phase, whereas the latter focuses on the incorporation of these nutrients at the production stage of the food itself. Biofortification is a planned way to make nutrients more concentrated and available in plants. This can be done using genetic engineering or traditional plant breeding methods, and it should be done in a way that keeps costs low. It denotes the enhancement of food and crops with vital nutrients by biological techniques, including genetic alteration, selective breeding, and microbial treatments. Biofortification is the method of enhancing the concentration of vitamins and minerals in a crop via plant breeding, genetic engineering, or agricultural practices (Naik et al., 2024; Bhardwaj et al., 2022). It is hoped that combining agricultural and nutritional sciences will help fight malnutrition by making staple foods higher in nutrients like iron, provitamin A, and zinc without making people spend more money. Most staple crops exhibit a deficiency in the essential nutrients necessary for optimal growth. Regular consumption of biofortified staple crops will yield quantifiable enhancements in human health and nutrition. In contrast to traditional fortification, which entails the incorporation of nutrients into processed meals, biobased fortification improves the nutritional quality of crops throughout their cultivation. For instance, biofortified rice enhanced with vitamin A or iron-enriched beans is cultivated using sustainable agriculture methods (Bouis & Saltzman, 2017; Lockyer et al., 2018). This method combines biotechnology with natural processes to tackle micronutrient deficiencies at their origin, hence enhancing nutritional accessibility for at-risk populations. There are three categories of fortification strategies: traditional, industrial, and biobased methods.

1. *Traditional fortification:* The incorporation of nutrients into staple foods during processing, exemplified by the addition of iodine to salt or folic acid to flour.
2. *Industrial fortification:* Enhancing nutritional incorporation for large-scale production, such as fortified cereals or beverages.
3. *Biobased fortification:* Naturally or biotechnologically enhancing crops, exemplified by zinc-enriched wheat or beta-carotene-rich sweet potatoes. The term "biofortification" describes the process of breeding crops to have a higher concentration of a certain micronutrient by harvest compared to conventional crops.

Each type helps with different nutritional issues in its own way, and biobased fortification has the added benefit of being sustainable and able to adapt to different farming situations (Bouis & Saltzman, 2017). The Joint Food and Agriculture Organisation (FAO)/WHO Codex Committee on Nutrition and Foods for Special Dietary Uses lists three ways biofortification can happen. This committee is currently working on an official definition of biofortification (Codex Alimentarius Commission: 17-22 July 2017, n.d.; Lockyer et al., 2018).

1. *Traditional biofortification:* this method involves picking plants that have higher amounts of a certain micronutrient and then breeding them in traditional ways to make staple crops that are better for your health and the environment.
2. *Agricultural/farming biofortification:* Use of micronutrient-rich sprays or fertilisers is the second method. The crop's edible part absorbs these nutrients for a short period.
3. *Transgenic biofortification:* This practice involves inserting genes into a crop to make it accumulate a micronutrient that would not be present in that crop normally (either completely or in a bioavailable form).

9.1.5 Shift from Conventional to Sophisticated Biobased Methods

The transition from conventional fortification techniques to biobased technologies signifies a crucial advancement in nutritional research. Conventional fortification methods, such as the addition of iodine to salt or vitamin D to milk, have demonstrated efficacy but frequently depend on industrial processing and uniform consumer practices. On the other hand, advanced biobased fortification tries to increase the natural nutritional content of crops, which adds nutrition to the food system. This transformation utilises advanced technologies such as CRISPR, metabolomics, and synthetic biology to create nutrient-rich crops that flourish in various conditions, addressing the requirements of diverse populations (Naik et al., 2024). Food fortification has become an essential strategy for addressing malnutrition, bolstered by ongoing advancements in technologies that enhance its efficacy and sustainability. Nanoencapsulation, double fortification, biofortification, and encapsulated micronutrients are some of the new ways that nutrients are added to foods that have greatly changed how these nutrients are delivered (Pateiro et al., 2021). As a result of new technologies, nutrients are more stable, bioavailable, and pleasant to taste. This makes fortified foods more effective and popular with consumers.

9.1.5.1 Nanoencapsulation: Enhancing the Absorption and Stability of Nutrients

A contemporary technique called nanoencapsulation encapsulates nutrients into minuscule particles, improving their ability to function in meals. The following are some major advantages of the technology, which greatly improve the effectiveness of nutrient delivery:

1. *Enhanced stability:* Over time, nutrients may deteriorate due to their susceptibility to environmental elements, including heat, light, and oxygen. These nutrients' longevity and effectiveness are increased by the protective layer that nanoencapsulation provides (Sajid & Fessi, 2012).
2. *Improved bioavailability:* Encapsulated particles' small size allows for easy passage through the digestive tract. This process makes it easier for the body to absorb the nutrients, which enhances overall health (Alleleyn et al., 2018).
3. *Masking unpleasant flavours:* Nanoencapsulation may modify the sensory qualities of the food by increasing the consumer's attraction to the strong flavours of iron and omega-3 fatty acids, two essential nutrients. The consumer's attraction to fortified foods is enhanced by nanoencapsulation's successful concealment of these flavours (Sajid & Fessi, 2012).

9.1.5.2 Biofortification: An Enduring Agricultural Remedy

Biofortification is an innovative method aimed at enhancing the nutritional composition of crops via conventional breeding or genetic alteration. Unlike traditional food fortification, which adds nutrients after harvest, biofortification aids in crop growth, offering numerous benefits:

1. *Long-term effects on nutrition:* Biofortifying staple foods like rice, wheat, and maize to make them higher in nutrients is a very effective way to fight malnutrition, especially in places where fortified foods are difficult to find (Bouis et al., 2011).
2. *Keeping cultural food traditions alive:* Unlike artificial fortification methods that change the way food tastes, looks, and feels, biofortified crops keep the same taste, texture, and visual appeal as regular varieties, making them more acceptable to consumers (de Moura et al., 2014).
3. *Targeted nutrient fortification:* Specific nutrients are incorporated into staple foods to rectify common deficits. Changing the genes of orange sweet potatoes, beans, and wheat to add zinc or iron has been done to help vulnerable groups that

are lacking in certain micronutrients (Bouis & Saltzman, 2017). The addition of nutritional supplements to agricultural production through biofortification is a cost-effective and preventative way to improve the quality of food around the world.

9.1.5.3 Double Fortification: An Integrated Approach to Addressing Nutrient Insufficiency

Double fortification is an efficient way to address multiple nutrient shortages at once since it adds two or more important nutrients to a single diet. According to Hurrell et al. (Hurrell, 2020), it is especially advantageous for groups with minimal dietary diversity. Fortification with Iron and Vitamin A: De Moura et al. (2014) say that adding iron and vitamin A to staple foods like rice and wheat flour stops anaemia and treats vitamin A deficiency, both of which are important for the immune system and the eyes. Better bone health is ensured by fortifying milk products and beverages with both calcium and vitamin D, especially for people who do not get much sun exposure (Bouis & Saltzman, 2017). By fortifying items that many consume, double fortification improves total nutrition without requiring drastic dietary changes.

9.1.5.4 Encapsulated Micronutrients: Optimizing Nutrient Delivery

Embedding essential vitamins and minerals within edible proteins, lipids, or carbohydrates serves the dual purpose of protecting encapsulated micronutrients while also increasing their delivery efficiency. This technique addresses common issues encountered in food fortification. Encapsulation permits the gradual release of nutrients during digestion, ensuring maximum absorption. This effective method results in incredible health benefits. The protective layer also functions as a shield from environmental degradation, improving the shelf life of the nutrients and ensuring their efficacy (Hotz & Gibson, 2007). Encapsulation serves as a solution to sensory issues posed by certain micronutrients, like iron or zinc, which can impart an unpleasant taste or strong odour to fortified foods. This makes the food more palatable and, therefore, more acceptable to consumers (Sajid & Fessi, 2012). Consumers tend to exaggerate the great taste snatched away by fortified beverages, baked goods, and confectionery products and forget the nutrition that now complements these products.

Advancements in food fortification are essential in meeting needs that relate to nutrition and health on a global scale. Nanoencapsulation, double fortification, biofortification, and encapsulated micronutrients are just a few of the new methods that have made fortified foods much more effective and popular. These breakthroughs guarantee that critical micronutrients are added to the daily diet of people across the globe by ensuring nutrients are made more stable, bioavailable, and appealing to the senses (Hurrell, 2020). Further advancement in food fortification science will enable many opportunities in combating malnutrition and improving health in a productive and truly sustainable manner.

9.1.6 Importance of Innovative Fortification Technologies

Innovative fortification technologies connect old and sophisticated methods, providing specific answers to intricate nutritional challenges. The foundation of this effort is fortification, which makes it possible to successfully correct deficiencies by incorporating vitamins into staple foods. By adding or incorporating vitamins and minerals into food, this technique improves its nutritional profile and makes up for any innate inadequacies. Several different fortification techniques use either synthetic or natural sources of the targeted micronutrients (Srivastav & Wadhawan, 2025). Methods such as biofortification, nanotechnology, and precision agriculture are transformative. Biofortification enriches crops with nutrients at the genetic level, whereas nanotechnology optimises the delivery and absorption of these nutrients within the body. Nanoencapsulation of iron or zinc improves bioavailability, even in people who do not absorb nutrients well (Kaur et al., 2020). These advances provide sustainable, economical, and significant interventions, especially in resource-constrained environments (Fig. 9.1).

9.1.7 Significance of Biofortification and Nutritional Requirements

Biofortification is a proactive strategy to address the increasing nutrient requirements of the world population. In contrast to traditional fortification, biofortification adds important nutrients to plants as they grow, so the food is already high in nutrients. This process is especially vital in areas with inferior soil quality, where diets

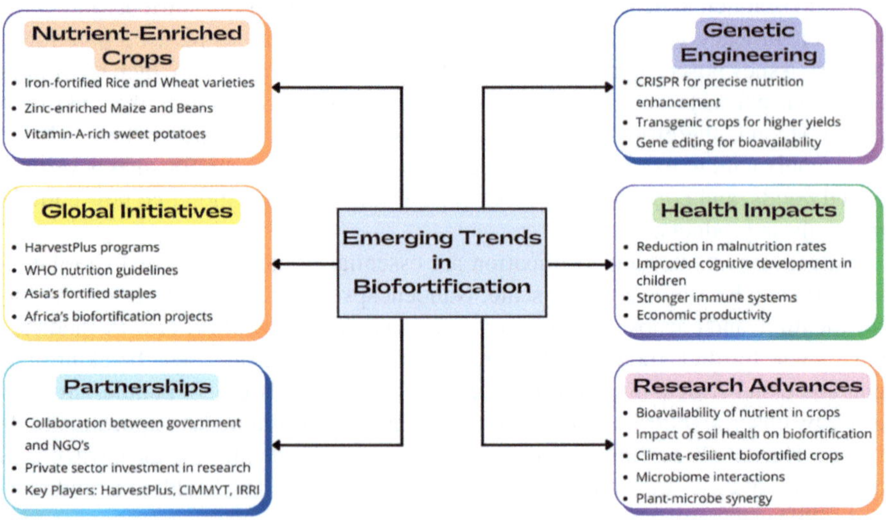

Fig. 9.1 Emerging trends in biofortification

predominantly depend on a singular staple crop. Iron-fortified beans and zinc-enriched rice serve communities with restricted access to varied food options. Biofortification is a long-term way to improve food security and health because it meets growing nutritional needs caused by changing lifestyles and environmental factors (Bouis et al., 2011).

9.2 Fungal and Lichen-Based Fortification

9.2.1 Fungi as Biofortification Agents

Taxonomic and Physiological Diversity The Fungi kingdom comprises an estimated 2.2–3.8 million species with exceptional morphological, ecological, and physiological diversity (Hawksworth & Lücking, 2017). This diversity is also seen in the ways that different fungal species acquire nutrients, with each showing unique methods for absorbing, storing, and breaking down vital elements. The bioaccumulation patterns of microfungi, such as different moulds and yeasts, and macrofungi, mainly basidiomycetes and some ascomycetes, are different and can be used for biofortification (Falandysz & Borovička, 2012).

9.2.2 Mediators of Plant-Soil Nutrient Dynamics

Mycorrhizal Fungi Mycorrhizal fungi establish mutualistic relationships with approximately 90% of terrestrial plant varieties, significantly impacting the nutrient uptake capabilities of their host plants (Diagne et al., 2020). These symbiotic relationships differ in that they involve the exchange of nutrients between the organisms. The plant provides the fungus with carbon compounds produced by photosynthesis, while the fungus facilitates nutrient absorption for the plant through extensive extraradical hyphal networks. Arbuscular mycorrhizal fungi (AMF), classified within the phylum Glomeromycota, represent the predominant form of mycorrhizal association. The AMF mycelium can reach 20–100 cm outside the rhizosphere and get nutrients from a far wider area of soil than non-mycorrhizal roots can reach (Lehmann & Rillig, 2014). The molecular mechanisms behind AMF's nutrition acquisition have been clarified by transcriptomic and proteomic investigations, which have shown that the extraradical mycelium expresses high-affinity phosphate transporters (like GintPT) and zinc transporters (like GintZnT) (Tamayo et al., 2014).

The effectiveness of AMF inoculation in raising the micronutrient content of commonly grown crops has been shown in field tests carried out in a variety of agroecological zones. AMF colonisation can raise the concentration of iron and zinc in wheat (*Triticum aestivum*), by 8–15% and 10–25%, respectively, according to meta-analytical studies. The extent of the enhancement varies depending on the genotype of the plant, the fungal species, and the soil characteristics (Lehmann et al., 2013).

Similarly, it has been demonstrated that in field conditions, an AMF inoculation of rice (*Oryza sativa*) might increase grain zinc concentration by as much as 32% (Wang et al., 2018).

Mycorrhizal fungi promote micronutrient absorption through the following mechanisms:

1. *Expression of high-affinity transporters:* Molecular research has identified a number of metal transporters in mycorrhizal fungi, including members of the ZIP (Zinc-Iron Permease) family and heavy metal ATPases. According to González-Guerrero et al. (2016), they aid in the copper, iron, and zinc uptake from soil solutions.
2. *Siderophore sequestration:* Mycorrhizal fungi increase iron availability and solubility in the rhizosphere by producing low-molecular-weight molecules that have a strong affinity for ferric iron (Fe^{3+}). Among them are hydroxamate siderophores that have been found in the extraradical mycelium of different ectomycorrhizal fungi, including ferricrocin and ferrichrome (Haselwandter et al., 2013).
3. *Production of organic acids:* Mycorrhizal fungi secrete three-carbonylic acid cycle intermediates, such as citric, malic, and oxalic acids, which aid in the breakdown of mineral-bound phosphorus and micronutrients by means of acidification and chelation processes (Andrino et al., 2021).
4. *Phosphatase activity:* Mycorrhizal fungi secrete acid phosphatases that hydrolyse organic phosphorus molecules in soil, releasing phosphate and related micronutrients that might form a complex with phytic acid (Joner et al., 2000).
5. *Edible macrofungi:* Direct vectors for nutrient delivery: Edible mushrooms are a direct vector for providing humans with bioavailable micronutrients. Essential minerals are naturally accumulated by these species at higher concentrations than in plant-based food (Kalač, 2012). Mushroom nutritional composition is regulated by species-specific variables, substrate composition, and natural conditions, creating prospects for customised biofortification techniques (Falandysz, 2008). Due to the vital function that selenium plays in thyroid hormone metabolism and antioxidant defence mechanisms, selenium-enriched mushrooms have caught particular interest. *Pleurotus ostreatus* (oyster mushroom) can accumulate selenium when grown on selenium-enriched substrates, mainly in the form of selenomethionine and selenocysteine, which are more bioavailable to humans than inorganic forms of selenium (Da Silva et al., 2011). Metabolomic investigations have demonstrated that selenium is integrated into a number of fungal metabolic pathways, such as the creation of proteins and amino acids that contain selenium (Turło et al., 2010). Zinc-enriched *Agaricus bisporus*, also known as white button mushrooms, was developed by adding zinc sulphate to the substrate. Zinc is primarily associated with proteins and low-molecular-weight components in these mushrooms, which increases its bioavailability when consumed (Bhatia et al., 2014). The bioaccessibility of zinc from these enhanced mushrooms is 65–78%, according to in vitro gastrointestinal digestion models, which is significantly greater than that of the majority of plant-based nourishment (Fernández-Fueyo et al., 2014) (Fig. 9.2).

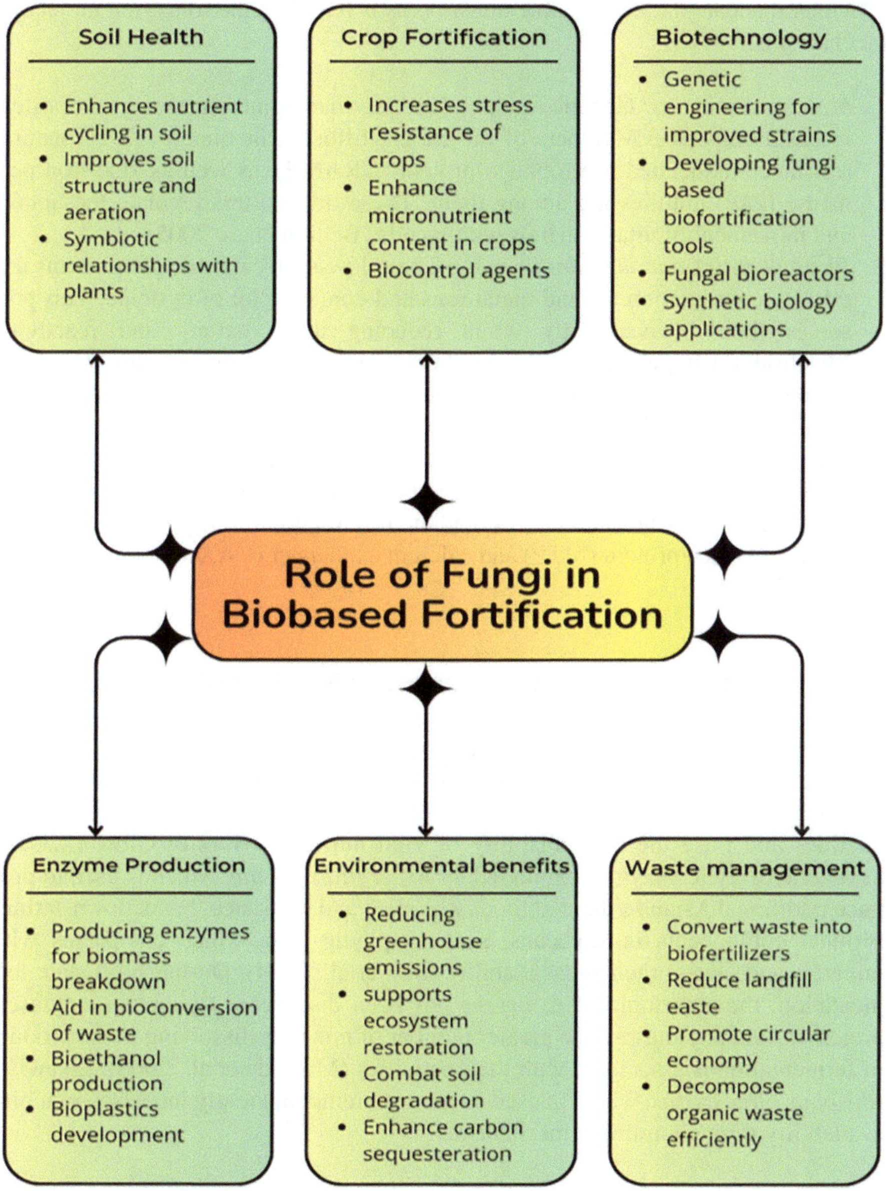

Fig. 9.2 Role of fungi in biobased fortification

Mineral accumulation in edible mushrooms is based on the following molecular mechanisms:

1. *Metal transporters:* Genetic research has identified numerous metal transporters in basidiomycetes. Members of the cation diffusion facilitator (CDF), natural resistance-associated macrophage protein (NRAMP), as well as zinc-iron permease (ZIP) families are among them. These transporters regulate the uptake and movement of vital minerals within cells (Bellion et al., 2005).
2. *Phytochelatins and metallothioneins:* These low-molecular-weight proteins use thiolate coordination to bind metal ions and contain a lot of cysteine. This preserves their bioavailability while reducing their toxicity and reactivity. Metallothionein gene expression has been found to increase in *Lentinula edodes* (shiitake) in response to increased amounts of copper and zinc in the growing medium (Hložková et al., 2015).
3. *Vacuolar sequestration:* The high capacity of fungi to accumulate minerals without compromising cellular functions is facilitated by the compartmentalisation of minerals in vacuoles, which is mediated by vacuolar transporters such as the metal tolerance protein (MTP) and calcium exchanger (CAX) families (Becquer et al., 2014).

9.2.3 Fermentative Fungi: Enhancers of Nutrient Bioavailability

Yeasts and filamentous fungi-mediated fermentation processes are a traditional yet advanced method of raising the nutritional content of meals obtained from plants. Through a variety of biochemical changes, these procedures enhance organoleptic qualities and raise the bioavailability of vital nutrients (Hotz & Gibson, 2007). Several hydrolytic enzymes expressed by *Aspergillus oryzae*, which is used to produce traditional Asian fermented foods like miso and soy sauce, break down antinutritional components in soybeans, such as phytic acid, which complexes with minerals and lowers their bioavailability (Sun et al., 2024). During solid-state fermentation, the secretome of *A. oryzae* has been discovered to contain amylases, proteases, and phytotases. The greater number of minerals dissolving in the product of fermentation is associated with their activities (Machida et al., 2008). Likewise, *Rhizopus oligosporus*, which is used to produce tempeh, greatly increases iron bioavailability through multiple mechanisms:

1. Phytate hydrolysis: *R. oligosporus* produces phytase enzymes, which break down phytic acid and liberate bound iron and zinc. According to in vitro research, tempeh fermentation can boost iron dialysability by two to three times while lowering phytate level by 57–69% (Astuti et al., 2000).
2. Production of organic acids: According to metabolomic analyses, tempeh accumulates lactic, acetic, and citric acids, among other organic acids. These acids form soluble complexes with iron, preventing precipitation at intestinal pH and improving absorption (Starzyńska-Janiszewska et al., 2015).

3. Microbial siderophores: *R. oligosporus* secretes hydroxamate siderophores during fermentation, which chelate iron, preserving its solubility and promoting its absorption in the gastrointestinal system (Zielińska-Dawidziak, 2015). Beyond its conventional use in the manufacturing of bread and alcoholic beverages, *Saccharomyces cerevisiae* shows promise as a vector for mineral biofortification. *S. cerevisiae* can accumulate zinc, selenium, and chromium when grown in mineral-enriched conditions; these elements are mostly linked to internal organelles and cell wall mannoproteins (Kieliszek et al., 2015) (Table 9.1).

Table 9.1 List of fungi suitable for biofortification

Sr. no.	Name of fungi	Properties	References
1	*Aspergillus*	Enhances bioavailability through fermentation	Hotz and Gibson (2007)
2	*Agaricus bisporus*	Bioaccumulation of selenium	Hu et al. (2018)
3	*Ganoderma lucidium*	Ability to bioaccumulate lithium	Mleczek et al. (2016)
4	*Ganoderma*	Bioaccumulation of se and other minerals	Zhao et al. (2004), Falandysz (2008)
5	*Cordyceps*	Potential for accumulation of se and Zn	Dong et al. (2012), Kalac (2000)
6	*Hericium erinaceus*	Bioaccumulation of selenium	Wang et al. (2002), Vetter (2003)
7	*Saccharomyces*	Extensively used for se biofortification	Kieliszek et al. (2015), Perez-Corona et al. (2010)
8	*Auricularia*	Capable of accumulating se and Fe	Falandysz and Borovicka (2012)
9	*Volvariella*	Shows potential for mineral enrichment	Falandysz and Borovicka (2012)
10	*Pleurotus eryngii*	High metal accumulation capacity	Budzynska et al. (2022)
11	*Pleurotus nameko*	High metal accumulation capacity	Budzynska et al. (2022)
12	*Pleurotus osteratus*	Metal accumulation capacity, high nutritional value	Budzynska et al. (2022)
13	*Lentinus crinitus*	Ability to bioaccumulate lithium	Bertéli et al. (2021)
14	*Lentinus edodes*	Bioaccumulation of selenium	Hu et al. (2018)
15	*Lentinula*	Ability to accumulate se and other minerals	Zhao et al. (2004)
16	*Pleurotus Florida*	Bioaccumulation of selenium	Hu et al. (2018)
17	*Pleurotus sajor*	Bioaccumulation of se and Zn	Madaan et al. (2023)
18	*Pleurotus dijamor*	Bioaccumulation of Se and Zn	Madaan et al. (2023)
19	*Schizophyllum commune*	Bioaccumulation of Fe	Scheid et al. (2020)
20	*Agaricus subrufescens*	Bioaccumulation of Fe	Scheid et al. (2020)

9.2.4 Lichens as Biofortification Agents

Ecological Significance and Symbiotic Nature:

Lichens comprise what is made of fungi (mycobionts) and photosynthetic partners (photobionts), which might be green algae, cyanobacteria, or a combination of the two. An estimated 20,000 varieties of lichens have been described worldwide, and their symbiotic relationship has allowed them to colonise a variety of terrestrial settings, from desert ecosystems to the Arctic tundra (Lücking et al., 2017). Lichens are characterised by their stratified thallus, in which the photobiont is usually embedded in a matrix of fungal hyphae. This configuration promotes nutrition exchange between the symbionts: the mycobiont assists the photobiont to acquire nutrients and provides structural support and protection, while the photobiont gives the mycobiont carbon compounds produced from photosynthetic processes (Grube et al., 2014) (Fig. 9.3).

Lichens have no root systems or cuticles, so they obtain water and nutrients directly from atmospheric deposition throughout the entire thallus surface. Along with their perennial nature and lagging growth rates, this trait leads to exceptional bioaccumulation capacities, especially for trace elements (Garty, 2001). The following are the mechanisms that underlie mineral accumulation in lichens:

1. *Sites for extracellular binding:* Functional groups such as carboxyl, phosphoryl, and hydroxyl moieties are found in the cell walls of lichen-forming fungi and act as cation exchange sites. These groups have different affinities for binding metal ions, and the specific chemical characteristics of the metal, pH, and competing ions all affect the binding capacity (Sarret et al., 1998).
2. *Intracellular sequestration:* Metals may be internalised and stored in cellular compartments such as cytoplasmic vesicles and vacuoles after external adsorption. Metals are frequently combined with organic ligands such as organic acids, metallothioneins, and phytochelatins during this process. Metals become less poisonous and reactive when these ligands are present (Pawlik-Skowrońska & Bačkor, 2010).
3. *Secondary metabolite complexity:* Lichens generate a variety of secondary metabolites, such as derivatives of usnic acid, dibenzofurans, depsides, and depsidones. Many of them have the ability to chelate metals. These substances have the ability to combine with metals to create stable complexes that affect the metals' mobility and bioavailability inside the thallus (Manojlović et al., 2010). Lichens' exceptional ability to accumulate a variety of essential nutrients has been shown by quantitative research. For example, *Hypogymnia physodes* can accumulate zinc at concentrations 100–150 times higher than those found in the environment; similar trends have been noted for iron, manganese, and copper (Garty, 2001). Reindeer rely heavily on *Cladonia rangiferina* as a food source in Arctic regions. According to Lavu et al. (2015), it mostly stores selenium in organic forms like selenomethionine and selenocysteine, which mammals can absorb more readily than inorganic forms.

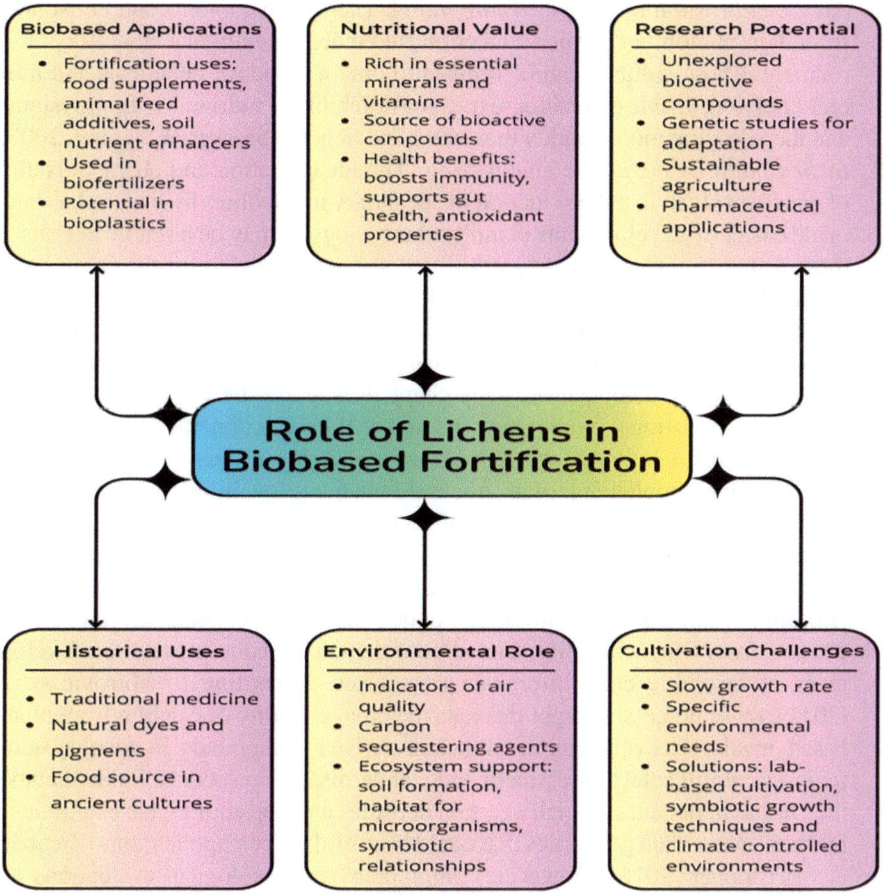

Fig. 9.3 Role of lichens in biobased fortification and in bioaccumulation

9.2.5 Applications of Lichens in Biofortification Strategies

Lichens are not commonly consumed by humans due to their slow growth rates, potential irritating chemical content, and ecological conservation concerns. Several novel technologies use the biochemical features of lichens for biofortification applications:

1. *Isolation of bioactive compounds:* Molecules formed from lichens that have the ability to chelate metals, such as usnic acid and its derivatives, can be separated and examined for possible use in food fortification systems. By forming soluble complexes that prevent minerals from adhering to dietary inhibitors, these substances may increase the bioavailability of minerals (Fernández-Moriano et al., 2015).

2. *Axenic cultivation of lichen-forming fungi:* Lichen mycobionts can be isolated from their photobiont partners and cultivated independently in a laboratory environment. These cultures can be utilised to obtain bioactive chemicals and mineral-chelating agents that increase the bioavailability of nutrients in food systems, and they develop more quickly than complete lichens (Stocker-Wörgötter, 2007).
3. *Lichen pathway metabolic engineering:* The identification and characterisation of genes involved in the production of secondary metabolites by lichens has been made easier by developments in molecular biology. It may be possible to transfer these routes to more frequently cultivated species, allowing for the commercial production of bioactive chemicals produced from lichens (Calcott et al., 2017).
4. *Biological markers for initiatives involving biofortification:* Lichens are useful bioindicators for tracking environmental metal concentrations because of their sensitivity to air composition. This characteristic can be used to detect areas where crop nutritional quality may be impacted by soil mineral deficits, guiding focused biofortification efforts (Giordani et al., 2012). Lichen extracts' potential as nutritional supplements with mineral-enhancing qualities has been investigated recently. Due to the presence of metal-chelating chemicals such as protocetraric acid and atranorin, *Parmelia sulcata* water-soluble extracts have been shown in vitro to improve iron dialysability from cereal-based diets by 25–30% (Kosanić et al., 2014). Additionally, under simulated intestinal circumstances, extracts from Cladonia foliacea have been demonstrated to prevent the development of insoluble calcium-phytate complexes. According to Mitrović et al. (2011), these processes might increase the bioavailability of calcium from plant-based meals. The application of lichen-derived compounds in food systems requires careful safety assessment and standardisation protocols. Some metabolites found in lichen, especially certain depsides and depsidones, have antimicrobial and antioxidant properties that could be useful in food applications (Zambare & Christopher, 2012). However, comprehensive toxicological evaluations are necessary to establish acceptable daily intake levels and ensure consumer safety (Table 9.2).

9.2.6 Comparative Analysis of Fungi and Lichens Biofortification Potential

9.2.6.1 Bioaccumulation Efficiency

Both fungi and lichens have excellent mineral accumulation capacity, albeit through different mechanisms according to their ecological niches. Lichens exhibit greater bioconcentration factors for metals, including zinc, copper, and manganese, that are deposited in the surroundings, according to comparative research. This is because they lack cuticles and have a high surface area-to-volume ratio (Garty, 2001). In contrast, fungi usually exhibit better uptake of soil-derived materials due to their large hyphal networks and unique uptake processes (Gadd, 2007). These organisms also differ greatly in the temporal dynamics of bioaccumulation. Long-term mineral accumulation by lichens, which frequently lasts for decades, raises steady-state

Table 9.2 List of lichens suitable for biofortification

Sr. no.	Name of Lichen	Properties	References
1	*Cetraria islandica*	Metal accumulation	Bjerke et al. (2004)
2	*Lobaria*	Ability to accumulate Zn and Mn	Gauslaa et al. (2015), Mateos and Gonzalez (2015)
3	*Umbilicaria*	Significant Fe and Zn accumulation capacities	Bačkor and Loppi (2009)
4	*Xanthoria*	High metal accumulation capacity, particularly for Zn and cu	Cuny et al. (2004)
5	*Parmelia*	Exhibits Zn and Fe bioaccumulation	Bačkor et al. (2009)
6	*Cladonia*	Some species accumulate Se and other minerals	Lawrey and Rudolph (1975)
7	*Evernia*	Shows capacity to accumulate minerals	Paoli et al. (2019)
8	*Ramalina*	Selective mineral accumulation	Godinho et al. (2008), Garty et al. (2001)
9	*Usnea*	Metal chelating property	Conti and Cecchetti (2001), Bergamaschi et al. (2007)

concentrations but slows their reaction to changes in the environment. Alternatively, fungi absorb minerals more rapidly, reacting to substrate changes in a matter of days to weeks. This enhances them for applications involving regulated biofortification (Falandysz & Borovička, 2012).

9.2.6.2 Bioavailability Enhancement

The bioavailability of minerals stored by fungi and lichens varies according to their chemical composition, the way they bind to cell components, and whether the food matrix contains enhancers or inhibitors. Minerals collected by edible fungi often have higher bioaccessibility than those in lichens, according to in vitro gastrointestinal digestion studies. This is thought to be due to variations in binding ligands and cellular compartmentalisation (Fernández-Fueyo et al., 2014). Certain metabolites of lichen, however, have better metal-chelating qualities than those of fungi, which may be advantageous for increasing the bioavailability of minerals in intricate food systems. Lichen-derived usnic acid, for instance, forms more stable compounds with zinc and iron at gut pH than do fungal siderophores. This could improve absorption and provide greater protection against precipitation (Manojlović et al., 2010).

9.2.7 Practical Implementation Considerations

Fungi have a number of benefits over lichens for biofortification applications from the standpoint of practical implementation:

1. *Cultivation efficiency:* In following known protocols, edible mushrooms and fermentative fungi can be produced on a large scale. Lichens, on the other hand, not suitable for cultivation in artificial environments and develop very slowly (1–2 mm annually for many species) (Lichen biology, 2008).
2. *Consumer acceptability:* While direct lichen intake is restricted to particular cultural contexts, mainly in Arctic and subarctic locations, mushrooms and fermented foods have established niches in various culinary traditions (Łuczaj et al., 2012).
3. *Regulatory status:* While most jurisdictions have less developed regulatory frameworks for substances derived from lichen, certain edible fungus have "generally recognised as safe" (GRAS) status for food applications (Hyde et al., 2019). However, lichens provide special metabolites and biochemical pathways that may enhance fungal-based methods, especially when it comes to improving the bioavailability of minerals in processed foods. In order to address micronutrient deficiencies in a variety of population groups, integrated approaches that combine lichen-derived bioactive chemicals with fungal cultivation may offer synergistic benefits.

9.3 Technological Platforms and Emerging Biotechnological Trends (Fig. 9.4)

9.3.1 Conventional Breeding Techniques

1. *Pedigree selection:* In this conventional breeding method, parent plants with the appropriate nutritional profiles are systematically chosen and crossed. Through meticulous selection and crossing with wild relatives, the technique has produced high-iron beans in Rwanda, increasing the iron content from 50 mg/kg to 100 mg/kg (Haas et al., 2016). The procedure entails keeping thorough lineage records and carrying out several field tests in various environments.
2. *Recurrent selection:* This method has been shown to be successful in building up advantageous alleles for increased nutritional content. The first step in this repetitive breeding process is to create various base populations, from which superior individuals are gradually chosen for intermating. The method has been especially effective in creating sweet potatoes high in beta-carotene, increasing levels from initially insignificant to over 100 mg/kg (Low et al., 2017).

9.3.2 Agronomic Biofortification Approaches

1. *Applications of Nano-fertilisers:* Using nanotechnology (1–100 nm) to improve nutrient delivery efficiency, nano-fertilisers are a state-of-the-art method in biofortification. In comparison to traditional fertilisers, studies have shown that zinc oxide nanoparticles can raise the grain zinc content in wheat by as much as 22%

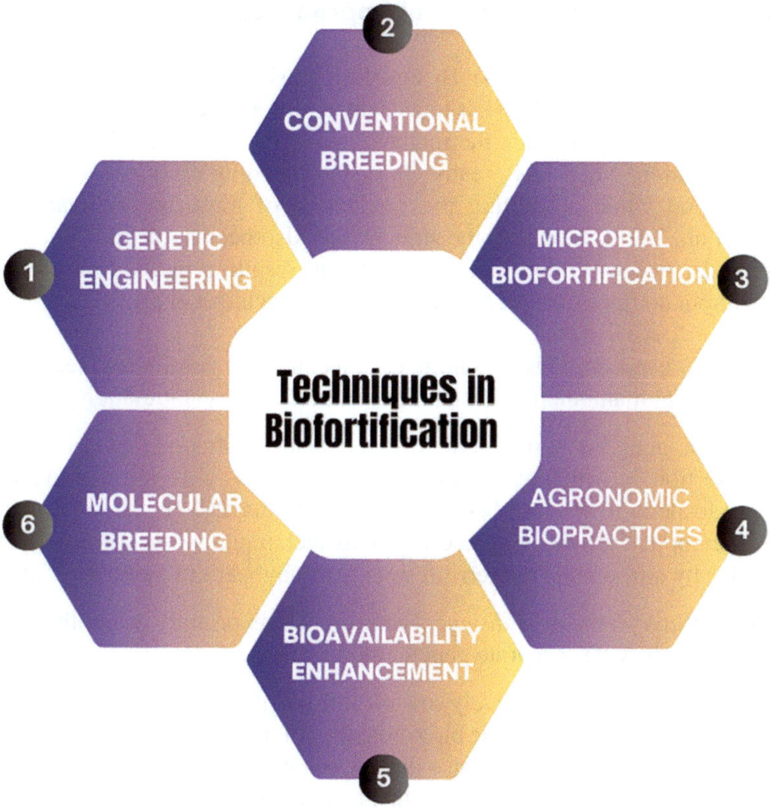

Fig. 9.4 Techniques in biofortification

(Sabir et al., 2014). They are more effective because of three key factors: smart-release systems that respond to environmental cues, particles with changed surfaces that facilitate absorption, and nutrition delivery inside a capsule. Iron oxide nanoparticles have demonstrated great promise in biofortifying rice, resulting in increased grain nutritional content and translocation rates (Elemike et al., 2019).

2. *Chelated fertiliser technology:* Organic molecule complexes protect critical minerals from soil interactions, addressing nutrient availability issues. Even in difficult alkaline soils, Fe-EDTA chelates have demonstrated exceptional efficacy in iron biofortification of cereals, preserving nutrient availability (Martinez-Navarrete et al., 2002). Three primary methods are included in the technology: mixed chelation systems, natural chelators that use amino acids, and EDTA chelation. Particularly effective are zinc-chelated fertilisers, which have been shown to raise grain zinc concentrations in wheat crops by as much as 30% (Lu et al., 2011).

9.3.3 Modern Biotechnological Approaches

Marker-assisted selection (MAS): By enabling it to be possible to precisely identify genes linked to nutrients, MAS has transformed biofortification breeding. For accelerating breeding programmes, the method uses QTL mapping, SNP markers, and marker validation. MAS has greatly shortened breeding cycle timeframes in rice biofortification while promoting the production of high-zinc cultivars (Garcia-Oliveira et al., 2018). In wheat, the strategy has also been successful in combining disease resistance with high grain zinc content, resulting in 20–40% greater zinc levels without compromising crop performance (Govindan et al., 2022).

1. *Genetic Engineering:* Using a number of techniques, genetic engineering allows for the direct manipulation of nutrient synthesis and accumulation pathways.

   1.1 Single-gene Insertion: Golden Rice is a successful example of single-gene insertion, in which bacterium and daffodil genes were inserted to allow rice endosperm to produce beta-carotene (Potrykus, 2017).

   1.2 Metabolic Pathway Engineering: This technique boosts the iron content of rice by introducing the ferritin gene from soybeans (Trijatmiko et al., 2016).

2. *CRISPR/Cas9 Genome Editing:* There are three primary ways that CRISPR technology provides accurate genetic modification:

   2.1 *Gene knockout*: Effectively lowers anti-nutritional elements such as phytic acid in rice, increasing the bioavailability of iron and zinc (Khan et al., 2019).

   2.2 *Gene insertion:* Used to raise the amount of vitamin C in tomatoes by up to 500% by altering the pathways involved in biosynthesis (Zhang et al., 2015).

   2.3 *Base editing:* AAllows for precise adjustments to particular nucleotides to maximise the function of genes linked to nutrients.

9.3.4 Advanced Integration Techniques

1. *Speed breeding:* By facilitating quick generation progression under controlled circumstances, speed breeding expedites biofortification initiatives. This method has shortened breeding cycles in wheat biofortification over 4–5 years to 2–3 generations annually (Watson et al., 2017). The technique combines optimised temperature regimes, prolonged photoperiods, and regulated environmental conditions.
2. *High-throughput phenotyping:* To analyse nutrients quickly, modern phenotyping systems use cutting-edge technologies. Mass spectrometry provides comprehensive nutrient profiling; near-infrared spectroscopy gives quick screening of organic nutrients; automated technologies enable large-scale screening programmes; and X-ray fluorescence spectroscopy allows for the non-destructive examination of thousands of samples every day Guild et al., 2017).

3. *Genomic selection:* To enhance breeding efficiency, this strategy makes use of whole-genome prediction models. Comprehensive phenotypic data collection, genome-wide marker data, advanced statistical modelling, selection based on modelling, and selection based on genomic predicted breeding values are important components (Bhat et al., 2016; Beebe et al., 2014).

References

Alleleyn, A. M. E., Van Avesaat, M., Ripken, D., Bleiel, S. B., Keszthelyi, D., Wilms, E., Troost, F. J., Hendriks, H. F. J., & Masclee, A. (2018). The effect of an encapsulated nutrient mixture on food intake and satiety: A double-blind randomized cross-over proof of concept study. *Nutrients, 10*(11), 1787. https://doi.org/10.3390/nu10111787

Andrino, A., Guggenberger, G., Kernchen, S., Mikutta, R., Sauheitl, L., & Boy, J. (2021). Production of organic acids by arbuscular mycorrhizal fungi and their contribution in the mobilization of phosphorus bound to iron oxides. *Frontiers in Plant Science, 12*. https://doi.org/10.3389/fpls.2021.661842

Astuti, M., Meliala, A., Dalais, F. S., & Wahlqvist, M. L. (2000). Tempe, a nutritious and healthy food from Indonesia. *Asia Pacific Journal of Clinical Nutrition, 9*(4), 322–325. https://doi.org/10.1046/j.1440-6047.2000.00176.x

Bačkor, M., & Loppi, S. (2009). Interactions of lichens with heavy metals. *Biologia Plantarum, 53*(2), 214–222. https://doi.org/10.1007/s10535-009-0042-y

Bačkor, M., Kováčik, J., Piovár, J., Pisani, T., & Loppi, S. (2009). Physiological aspects of cadmium and nickel toxicity in the lichens Peltigera rufescens and Cladina arbuscula Subsp. mitis. *Water Air & Soil Pollution, 207*(1–4), 253–262. https://doi.org/10.1007/s11270-009-0133-6

Bailey, R. L., West, K. P., Jr., & Black, R. E. (2015). The epidemiology of global micronutrient deficiencies. *Annals of Nutrition and Metabolism, 66*(Suppl. 2), 22–33. https://doi.org/10.1159/000371618

Becquer, A., Trap, J., Irshad, U., Ali, M. A., & Claude, P. (2014). From soil to plant, the journey of P through trophic relationships and ectomycorrhizal association. *Frontiers in Plant Science, 5*. https://doi.org/10.3389/fpls.2014.00548

Beebe, S. E., Rao, I. M., Devi, M. J., & Polania, J. (2014). Common beans, biodiversity, and multiple stresses: Challenges of drought resistance in tropical soils. *Crop and Pasture Science, 65*(7), 667. https://doi.org/10.1071/cp13303

Bellion, M., Courbot, M., Jacob, C., Blaudez, D., & Chalot, M. (2005). Extracellular and cellular mechanisms sustaining metal tolerance in ectomycorrhizal fungi. *FEMS Microbiology Letters, 254*(2), 173–181. https://doi.org/10.1111/j.1574-6968.2005.00044.x

Bergamaschi, L., Rizzio, E., Giaveri, G., Loppi, S., & Gallorini, M. (2007). Comparison between the accumulation capacity of four lichen species transplanted to a urban site. *Environmental Pollution, 148*(2), 468–476. https://doi.org/10.1016/j.envpol.2006.12.003

Bertéli, M. B. D., Barros, L., Reis, F. S., Ferreira, I. C. F. R., Glamočlija, J., Soković, M., Valle, J. S. D., Linde, G. A., Ruiz, S. P., & Colauto, N. B. (2021). Antimicrobial activity, chemical composition and cytotoxicity ofLentinus crinitusbasidiocarp. *Food & Function, 12*(15), 6780–6792. https://doi.org/10.1039/d1fo00656h

Bhardwaj, A. K., Chejara, S., Malik, K., Kumar, R., Kumar, A., & Yadav, R. K. (2022). Agronomic biofortification of food crops: An emerging opportunity for global food and nutritional security. *Frontiers in Plant Science, 13*. https://doi.org/10.3389/fpls.2022.1055278

Bhat, J. A., Ali, S., Salgotra, R. K., Mir, Z. A., Dutta, S., Jadon, V., Tyagi, A., Mushtaq, M., Jain, N., Singh, P. K., Singh, G. P., & Prabhu, K. V. (2016). Genomic selection in the era of next generation sequencing for complex traits in plant breeding. *Frontiers in Genetics, 7*. https://doi.org/10.3389/fgene.2016.00221

Bhatia, P., Prakash, R., & Prakash, N. T. (2014). Enhanced antioxidant properties as a function of selenium uptake by edible mushrooms cultivated on selenium-accumulated waste post-harvest wheat and paddy residues. *International Journal of Recycling of Organic Waste in Agriculture, 3*(4), 127–132. https://doi.org/10.1007/s40093-014-0074-y

Bjerke, J., Gwynnjones, D., & Callaghan, T. (2004). Effects of enhanced UV-B radiation in the field on the concentration of phenolics and chlorophyll fluorescence in two boreal and arctic? Alpine lichens. *Environmental and Experimental Botany, 53*(2), 139–149. https://doi.org/10.1016/j.envexpbot.2004.03.009

Bouis, H. E., & Saltzman, A. (2017). Improving nutrition through biofortification: A review of evidence from HarvestPlus, 2003 through 2016. *Global Food Security, 12*, 49–58. https://doi.org/10.1016/j.gfs.2017.01.009

Bouis, H. E., Hotz, C., McClafferty, B., Meenakshi, J. V., & Pfeiffer, W. H. (2011). Biofortification: A new tool to reduce micronutrient malnutrition. *Food and Nutrition Bulletin, 32*(1_suppl1), S31–S40. https://doi.org/10.1177/15648265110321s105

Budzyńska, S., Siwulski, M., Gąsecka, M., Magdziak, Z., Kalač, P., Niedzielski, P., & Mleczek, M. (2022). Biofortification of three cultivated mushroom species with three iron salts—Potential for a new iron-rich superfood. *Molecules, 27*(7), 2328. https://doi.org/10.3390/molecules27072328

Calcott, M. J., Ackerley, D. F., Knight, A., Keyzers, R. A., & Owen, J. G. (2017). Secondary metabolism in the lichen symbiosis. *Chemical Society Reviews, 47*(5), 1730–1760. https://doi.org/10.1039/c7cs00431a

Codex Alimentarius Commission:17-22 July 2017. (n.d.). Newsroom. Retrieved March 20, 2025, from https://www.fao.org/newsroom/detail/Codex-Alimentarius-Commission-17-22-July-2017/en#:~:text=The%20Codex%20Alimentarius%20Commission%20has%20adopted%20revisions%20updating%20its%20risk,producers%20through%20to%20final%20consumers

Conti, M., & Cecchetti, G. (2001). Biological monitoring: Lichens as bioindicators of air pollution assessment – A review. *Environmental Pollution, 114*(3), 471–492. https://doi.org/10.1016/s0269-7491(00)00224-4

Cuny, D., Davranche, L., Thomas, P., Kempa, M., & Van Haluwyn, C. (2004). Spatial and temporal variations of trace element contents in Xanthoria parietina thalli collected in a highly industrialized area in northern France as an element for a future epidemiological study. *Journal of Atmospheric Chemistry, 49*(1–3), 391–401. https://doi.org/10.1007/s10874-004-1254-3

Da Silva, M. C., Naozuka, J., Da Luz, J. M. R., De Assunção, L. S., Oliveira, P. V., Vanetti, M. C., Bazzolli, D. M., & Kasuya, M. C. (2011). Enrichment of Pleurotus ostreatus mushrooms with selenium in coffee husks. *Food Chemistry, 131*(2), 558–563. https://doi.org/10.1016/j.foodchem.2011.09.023

De Moura, F. F., Palmer, A. C., Finkelstein, J. L., Haas, J. D., Murray-Kolb, L. E., Wenger, M. J., Birol, E., Boy, E., & Peña-Rosas, J. P. (2014). Are biofortified staple food crops improving vitamin a and iron status in women and children? New evidence from efficacy trials. *Advances in Nutrition, 5*(5), 568–570. https://doi.org/10.3945/an.114.006627

Diagne, N., Ngom, M., Djighaly, P. I., Fall, D., Hocher, V., & Svistoonoff, S. (2020). Roles of arbuscular mycorrhizal fungi on plant growth and performance: Importance in biotic and abiotic stressed regulation. *Diversity, 12*(10), 370. https://doi.org/10.3390/d12100370

Dong, J. Z., Lei, C., Ai, X. R., & Wang, Y. (2012). Selenium enrichment on Cordyceps militaris link and analysis on its Main active components. *Applied Biochemistry and Biotechnology, 166*(5), 1215–1224. https://doi.org/10.1007/s12010-011-9506-6

Elemike, E. E., Uzoh, I. M., Onwudiwe, D. C., & Babalola, O. O. (2019). The role of nanotechnology in the fortification of plant nutrients and improvement of crop production. *Applied Sciences, 9*(3), 499. https://doi.org/10.3390/app9030499

Falandysz, J. (2008). Selenium in edible mushrooms. *Journal of Environmental Science and Health Part C, 26*(3), 256–299. https://doi.org/10.1080/10590500802350086

Falandysz, J., & Borovička, J. (2012). Macro and trace mineral constituents and radionuclides in mushrooms: Health benefits and risks. *Applied Microbiology and Biotechnology, 97*(2), 477–501. https://doi.org/10.1007/s00253-012-4552-8

Fernández-Fueyo, E., Castanera, R., Ruiz-Dueñas, F. J., López-Lucendo, M. F., Ramírez, L., Pisabarro, A. G., & Martínez, A. T. (2014). Ligninolytic peroxidase gene expression by Pleurotus ostreatus: Differential regulation in lignocellulose medium and effect of temperature and pH. *Fungal Genetics and Biology, 72*, 150–161. https://doi.org/10.1016/j.fgb.2014.02.003

Fernández-Moriano, C., Gómez-Serranillos, M. P., & Crespo, A. (2015). Antioxidant potential of lichen species and their secondary metabolites. A systematic review. *Pharmaceutical Biology, 54*(1), 1–17. https://doi.org/10.3109/13880209.2014.1003354

Field, M. S., Mithra, P., Estevez, D., & Peña-Rosas, J. P. (2020). Wheat flour fortification with iron for reducing anaemia and improving iron status in populations. *Cochrane Library, 7*. https://doi.org/10.1002/14651858.cd011302.pub2

Fróna, D., Szenderák, J., & Harangi-Rákos, M. (2019). The challenge of feeding the world. *Sustainability, 11*(20), 5816. https://doi.org/10.3390/su11205816

Gadd, G. M. (2007). Geomycology: Biogeochemical transformations of rocks, minerals, metals and radionuclides by fungi, bioweathering and bioremediation. *Mycological Research, 111*(1), 3–49. https://doi.org/10.1016/j.mycres.2006.12.001

Garcia-Oliveira, A. L., Chander, S., Ortiz, R., Menkir, A., & Gedil, M. (2018). Genetic basis and breeding perspectives of grain iron and zinc enrichment in cereals. *Frontiers in Plant Science, 9*. https://doi.org/10.3389/fpls.2018.00937

Garty, J. (2001). Biomonitoring atmospheric heavy metals with lichens: Theory and application. *Critical Reviews in Plant Sciences, 20*(4), 309–371. https://doi.org/10.1016/s0735-2689(01)80040-x

Garty, J., Tamir, O., Hassid, I., Eshel, A., Cohen, Y., Karnieli, A., & Orlovsky, L. (2001). Photosynthesis, chlorophyll integrity, and spectral reflectance in lichens exposed to air pollution. *Journal of Environmental Quality, 30*(3), 884–893. https://doi.org/10.2134/jeq2001.303884x

Gauslaa, Y., Yemets, O. A., Asplund, J., & Solhaug, K. A. (2015). Carbon based secondary compounds do not provide protection against heavy metal road pollutants in epiphytic macrolichens. *The Science of the Total Environment, 541*, 795–801. https://doi.org/10.1016/j.scitotenv.2015.09.114

Giordani, P., Brunialti, G., Bacaro, G., & Nascimbene, J. (2012). Functional traits of epiphytic lichens as potential indicators of environmental conditions in forest ecosystems. *Ecological Indicators, 18*, 413–420. https://doi.org/10.1016/j.ecolind.2011.12.006

Godinho, R., Verburg, T., Freitas, M., & Wolterbeek, H. (2008). Accumulation of trace elements in the peripheral and central parts of two species of epiphytic lichens transplanted to a polluted site in Portugal. *Environmental Pollution, 157*(1), 102–109. https://doi.org/10.1016/j.envpol.2008.07.021

González-Guerrero, M., Escudero, V., Saéz, Á., & Tejada-Jiménez, M. (2016). Transition metal transport in plants and associated endosymbionts: Arbuscular mycorrhizal fungi and rhizobia. *Frontiers in Plant Science, 7*. https://doi.org/10.3389/fpls.2016.01088

Govindan, V., Atanda, S., Singh, R. P., Huerta-Espino, J., Crespo-Herrera, L. A., Juliana, P., Mondal, S., Joshi, A. K., & Bentley, A. R. (2022). Breeding increases grain yield, zinc, and iron, supporting enhanced wheat biofortification. *Crop Science, 62*(5), 1912–1925. https://doi.org/10.1002/csc2.20759

Grube, M., Cernava, T., Soh, J., Fuchs, S., Aschenbrenner, I., Lassek, C., Wegner, U., Becher, D., Riedel, K., Sensen, C. W., & Berg, G. (2014). Exploring functional contexts of symbiotic sustain within lichen-associated bacteria by comparative omics. *The ISME Journal, 9*(2), 412–424. https://doi.org/10.1038/ismej.2014.138

Guild, G., Parkes, E., Nutti, M., Palacios-Rojas, N., & Stangoulis, J. (2017). High-throughput measurement methodologies for developing nutrient-dense crops. *African Journal of Food, Agriculture, Nutrition and Development, 17*(2), 11941–11954. https://doi.org/10.18697/ajfand.HarvestPlus06

Haas, J. D., Luna, S. V., et al. (2016). Consuming iron biofortified beans increases iron status in Rwandan women after 128 days in a randomized controlled feeding trial. *Journal of Nutrition, 146*(8), 1586–1592. https://doi.org/10.3945/jn.115.224741

Harastani, R., James, L. J., Walton, J., & Woolley, E. (2020). Tackling obesity: A knowledge-base to enable industrial food reformulation. *Innovative Food Science & Emerging Technologies, 64*, 102433. https://doi.org/10.1016/j.ifset.2020.102433

Haselwandter, K., Häninger, G., Ganzera, M., Haas, H., Nicholson, G., & Winkelmann, G. (2013). Linear fusigen as the major hydroxamate siderophore of the ectomycorrhizal Basidiomycota Laccaria laccata and Laccaria bicolor. *Biometals, 26*(6), 969–979. https://doi.org/10.1007/s10534-013-9673-8

Hawksworth, D. L., & Lücking, R. (2017). Fungal diversity revisited: 2.2 to 3.8 million species. *Microbiology Spectrum, 5*(4). https://doi.org/10.1128/microbiolspec.funk-0052-2016

Hložková, K., Matěnová, M., Žáčková, P., Strnad, H., Hršelová, H., Hroudová, M., & Kotrba, P. (2015). Characterization of three distinct metallothionein genes of the Ag-hyperaccumulating ectomycorrhizal fungus amanita strobiliformis. *Fungal Biology, 120*(3), 358–369. https://doi.org/10.1016/j.funbio.2015.11.007

Hotz, C., & Gibson, R. S. (2007). Traditional food-processing and preparation practices to enhance the bioavailability of micronutrients in plant-based diets1. *Journal of Nutrition, 137*(4), 1097–1100. https://doi.org/10.1093/jn/137.4.1097

Hu, T., Liang, Y., Zhao, G., Wu, W., Li, H., & Guo, Y. (2018). Selenium biofortification and antioxidant activity in cordyceps militaris supplied with selenate, selenite, or selenomethionine. *Biological Trace Element Research, 187*(2), 553–561. https://doi.org/10.1007/s12011-018-1386-y

Hurrell, R. F. (2020). The potential of iodine and iron double-fortified salt compared with iron-fortified staple foods to increase population iron status. *Journal of Nutrition, 151*, 47S–63S. https://doi.org/10.1093/jn/nxaa204

Hyde, K. D., Xu, J., Rapior, S., Jeewon, R., Lumyong, S., Niego, A. G. T., Abeywickrama, P. D., Aluthmuhandiram, J. V. S., Brahamanage, R. S., Brooks, S., Chaiyasen, A., Chethana, K. W. T., Chomnunti, P., Chepkirui, C., Chuankid, B., De Silva, N. I., Doilom, M., Faulds, C., Gentekaki, E., et al. (2019). The amazing potential of fungi: 50 ways we can exploit fungi industrially. *Fungal Diversity, 97*(1), 1–136. https://doi.org/10.1007/s13225-019-00430-9

Joner, E. J., Van Aarle, I. M., & Vosatka, M. (2000). Phosphatase activity of extra-radical arbuscular mycorrhizal hyphae: A review. *Plant and Soil, 226*(2), 199–210. https://doi.org/10.1023/a:1026582207192

Kalac, P. (2000). A review of trace element concentrations in edible mushrooms. *Food Chemistry, 69*(3), 273–281. https://doi.org/10.1016/s0308-8146(99)00264-2

Kalač, P. (2012). A review of chemical composition and nutritional value of wild-growing and cultivated mushrooms. *Journal of the Science of Food and Agriculture, 93*(2), 209–218. https://doi.org/10.1002/jsfa.5960

Kaur, T., Rana, K. L., Kour, D., Sheikh, I., Yadav, N., Kumar, V., Yadav, A. N., Dhaliwal, H. S., & Saxena, A. K. (2020). Microbe-mediated biofortification for micronutrients: Present status and future challenges. In *New and future developments in microbial biotechnology and bioengineering* (pp. 1–17). Elsevier.

Khan, M. S. S., Basnet, R., Islam, S. A., & Shu, Q. (2019). Mutational analysis of OSPLDA1 reveals its involvement in phytic acid biosynthesis in rice grains. *Journal of Agricultural and Food Chemistry, 67*(41), 11436–11443. https://doi.org/10.1021/acs.jafc.9b05052

Kiani, A. K., Dhuli, K., Donato, K., Aquilanti, B., Velluti, V., Matera, G., Iaconelli, A., Connelly, S. T., Bellinato, F., Gisondi, P., & Bertelli, M. (2022). Main nutritional deficiencies. *Journal of Preventive Medicine and Hygiene, 63*(2 Suppl 3), E93–E101. https://doi.org/10.15167/2421-4248/jpmh2022.63.2S3.2752. PMID: 36479498; PMCID: PMC9710417.

Kieliszek, M., Błażejak, S., Gientka, I., & Bzducha-Wróbel, A. (2015). Accumulation and metabolism of selenium by yeast cells. *Applied Microbiology and Biotechnology, 99*(13), 5373–5382. https://doi.org/10.1007/s00253-015-6650-x

Knapp, S., & van der Heijden, M. G. A. (2018). A global meta-analysis of yield stability in organic and conservation agriculture. *Nature Communications, 9*, 3632.

Kosanić, M., Ranković, B., Stanojković, T., Rančić, A., & Manojlović, N. (2014). Cladonia lichens and their major metabolites as possible natural antioxidant, antimicrobial and anticancer agents. *LWT, 59*(1), 518–525. https://doi.org/10.1016/j.lwt.2014.04.047

Lavu, R. V. S., Van De Wiele, T., Pratti, V. L., Tack, F., & Du Laing, G. (2015). Selenium bioaccessibility in stomach, small intestine and colon: Comparison between pure Se compounds, Se-enriched food crops and food supplements. *Food Chemistry, 197*, 382–387. https://doi.org/10.1016/j.foodchem.2015.08.001

Lawrey, J. D., & Rudolph, E. D. (1975). Lichen accumulation of some heavy metals from acids surface substrates of coal mine ecosystems insoutheastern Ohio. *The Ohio Journal of Science, 75*, 113–117.

Lehmann, A., & Rillig, M. C. (2014). Arbuscular mycorrhizal contribution to copper, manganese and iron nutrient concentrations in crops – A meta-analysis. *Soil Biology and Biochemistry, 81*, 147–158. https://doi.org/10.1016/j.soilbio.2014.11.013

Lehmann, A., Veresoglou, S. D., Leifheit, E. F., & Rillig, M. C. (2013). Arbuscular mycorrhizal influence on zinc nutrition in crop plants—A meta-analysis. *Soil Biology and Biochemistry, 69*, 123–131. https://doi.org/10.1016/j.soilbio.2013.11.001

Lichen Biology. (2008). Cambridge University Press eBooks. https://doi.org/10.1017/cbo9780511790478.

Lockyer, S., White, A., & Buttriss, J. L. (2018). Biofortified crops for tackling micronutrient deficiencies – What impact are these having in developing countries and could they be of relevance within Europe? *Nutrition Bulletin, 43*(4), 319–357. https://doi.org/10.1111/nbu.12347

Low, J. W., Mwanga, R. O., Andrade, M., Carey, E., & Ball, A. (2017). Tackling vitamin a deficiency with biofortified sweetpotato in sub-Saharan Africa. *Global Food Security, 14*, 23–30. https://doi.org/10.1016/j.gfs.2017.01.004

Lu, X., Tian, X., Cui, J., Zhao, A., Yang, X., & Mai, W. (2011). Effects of combined phosphorus-zinc fertilization on grain zinc nutritional quality of wheat grown on potentially zinc-deficient calcareous soil. *Soil Science, 176*(12), 684–690. https://doi.org/10.1097/ss.0b013e3182331635

Lücking, R., Hodkinson, B. P., & Leavitt, S. D. (2017). The 2016 classification of lichenized fungi in the Ascomycota and Basidiomycota – Approaching one thousand genera. *The Bryologist, 119*(4), 361. https://doi.org/10.1639/0007-2745-119.4.361

Łuczaj, Ł., Pieroni, A., Tardío, J., Pardo-De-Santayana, M., Sõukand, R., Svanberg, I., & Kalle, R. (2012). Wild food plant use in 21st century Europe: The disappearance of old traditions and the search for new cuisines involving wild edibles. *Acta Societatis Botanicorum Poloniae, 81*(4), 359–370. https://doi.org/10.5586/asbp.2012.031

Machida, M., Yamada, O., & Gomi, K. (2008). Genomics of aspergillus oryzae: Learning from the history of Koji Mold and exploration of its future. *DNA Research, 15*(4), 173–183. https://doi.org/10.1093/dnares/dsn020

Madaan, K., Sharma, S., & Kalia, A. (2023). Effect of selenium and zinc biofortification on the biochemical parameters of Pleurotus spp. under submerged and solid-state fermentation. *Journal of Trace Elements in Medicine and Biology, 82*, 127365. https://doi.org/10.1016/j.jtemb.2023.127365

Manojlović, N. T., Vasiljević, P., Jušković, M., Najman, S., Janković, S., & Milenković-Anđelković, A. (2010). HPLC analysis and cytotoxic potential of extracts from the lichen, Thamnolia vermicularis var. subuliformis. *Journal of Medicinal Plant Research, 4*(9):817–823

Martinez-Navarrete, N., et al. (2002). Iron deficiency and iron fortified foods – A review. *Food Research International, 35*(2–3), 225–231. https://doi.org/10.1016/s0963-9969(01)00189-2

Mason, J. B., Lotfi, M., Dalmiya, N., Sethuraman, K., & Deitchler, M. (2001). The micronutrient report: Current progress and trends in the control of vitamin A, iodine, and iron deficiencies. Ottawa, Canada: Micronutrient Initiative, Tulane University, UNICEF, & International Development Research Centre. ISBN 1-894217-18-7.

Mateos, A., & González, C. (2015). Physiological response and sulfur accumulation in the biomonitor Ramalina celastri in relation to the concentrations of SO2 and NO2 in urban environments. *Microchemical Journal, 125*, 116–123. https://doi.org/10.1016/j.microc.2015.11.025

Mitrović, T., Stamenković, S., Cvetković, V., Nikolić, M., Tošić, S., & Stojičić, D. (2011). Lichens as source of versatile bioactive compounds. *DOAJ (DOAJ: Directory of Open Access Journals)*. https://doaj.org/article/a4bd53dc480f40459dd3da8141bbbbaf

Mleczek, M., Siwulski, M., Rzymski, P., Budzyńska, S., Gąsecka, M., Kalač, P., & Niedzielski, P. (2016). Cultivation of mushrooms for production of food biofortified with lithium. *European Food Research and Technology, 243*(6), 1097–1104. https://doi.org/10.1007/s00217-016-2823-9

Naik, B., Kumar, V., Rizwanuddin, S., Mishra, S., Kumar, V., Saris, P. E. J., Khanduri, N., Kumar, A., Pandey, P., Gupta, A. K., Khan, J. M., & Rustagi, S. (2024). Biofortification as a solution for addressing nutrient deficiencies and malnutrition. *Heliyon, 10*(9), e30595. https://doi.org/10.1016/j.heliyon.2024.e30595

Nava, N. T., et al. (2010). HPLC analysis and cytotoxic potential of extracts from the lichen, Thamnolia vermicularis var. subuliformis. *Journal of Medicinal Plants Research, 4*(9), 817–823. https://doi.org/10.5897/jmpr10.104

Olson, R., Gavin-Smith, B., Ferraboschi, C., & Kraemer, K. (2021). Food fortification: The advantages, disadvantages and lessons from sight and life programs. *Nutrients, 13*(4), 1118. https://doi.org/10.3390/nu13041118

Onyeaka, H., Nwaiwu, O., Obileke, K., Miri, T., & Al-Sharify, Z. T. (2023). Global nutritional challenges of reformulated food: A review. *Food Science & Nutrition, 11*(6), 2483–2499. https://doi.org/10.1002/fsn3.3286

Paoli, L., Fačkovcová, Z., Guttová, A., Maccelli, C., Kresáňová, K., & Loppi, S. (2019). Evernia goes to school: Bioaccumulation of heavy metals and photosynthetic performance in lichen transplants exposed indoors and outdoors in public and private environments. *Plants, 8*(5), 125. https://doi.org/10.3390/plants8050125

Pateiro, M., Gómez, B., Munekata, P. E. S., Barba, F. J., Putnik, P., Kovačević, D. B., & Lorenzo, J. M. (2021). Nanoencapsulation of promising bioactive compounds to improve their absorption, stability, functionality and the appearance of the final food products. *Molecules, 26*(6), 1547. https://doi.org/10.3390/molecules26061547

Pawlik-Skowrońska, B., & Bačkor, M. (2010). Zn/Pb-tolerant lichens with higher content of secondary metabolites produce less phytochelatins than specimens living in unpolluted habitats. *Environmental and Experimental Botany, 72*(1), 64–70. https://doi.org/10.1016/j.envexpbot.2010.07.002

Pérez-Corona, M., Sánchez-Martínez, M., Valderrama, M., Rodríguez, M., Cámara, C., & Madrid, Y. (2010). Selenium biotransformation by Saccharomyces cerevisiae and Saccharomyces bayanus during white wine manufacture: Laboratory-scale experiments. *Food Chemistry, 124*(3), 1050–1055. https://doi.org/10.1016/j.foodchem.2010.07.073

Potrykus, I. (2017). The GMO-crop potential for more, and more nutritious food is blocked by unjustified regulation. *Journal of Innovation & Knowledge, 2*(2), 90–96. https://doi.org/10.1016/j.jik.2017.03.003

Sabir, S., Arshad, M., & Chaudhari, S. K. (2014). Zinc oxide nanoparticles for revolutionizing agriculture: Synthesis and applications. *The Scientific World Journal, 2014*, 1–8. https://doi.org/10.1155/2014/925494

Sajid, M., & Fessi, H. (2012). Nanometric encapsulation of food ingredients by emulsification–diffusion method. *International Journal of Pharmaceutics, 426*(1–2), 1–9.

Sarret, G., Manceau, A., Cuny, D., Van Haluwyn, C., Déruelle, S., Hazemann, J., Soldo, Y., Eybert-Bérard, L., & Menthonnex, J. (1998). Mechanisms of lichen resistance to metallic pollution. *Environmental Science & Technology, 32*(21), 3325–3330. https://doi.org/10.1021/es970718n

Scheid, S. S., Faria, M. G. I., Velasquez, L. G., Valle, J. S. D., Gonçalves, A. C., Dragunski, D. C., Colauto, N. B., & Linde, G. A. (2020). Iron biofortification and availability in the mycelial biomass of edible and medicinal basidiomycetes cultivated in sugarcane molasses. *Scientific Reports, 10*(1). https://doi.org/10.1038/s41598-020-69699-0

Srivastav, A., & Wadhawan, N. (2025). Innovations in food fortification technologies. In *Chemical formulation of fortified foods for optimal nutrition* (pp. 1–6). KD Publications.

Starzyńska-Janiszewska, A., Stodolak, B., & Wikiera, A. (2015). Proteolysis in tempeh-type products obtained with Rhizopus and aspergillus strains from grass pea (Lathyrus sativus) seeds. *Acta Scientiarum Polonorum Technologia Alimentaria, 14*(2), 125–132. https://doi.org/10.17306/j.afs.2015.2.14

Stocker-Wörgötter, E. (2007). Metabolic diversity of lichen-forming ascomycetous fungi: Culturing, polyketide and shikimatemetabolite production, and PKS genes. *Natural Product Reports, 25*(1), 188–200. https://doi.org/10.1039/b606983p

Sun, Z., Wu, Y., Long, S., Feng, S., Jia, X., Hu, Y., Ma, M., Liu, J., & Zeng, B. (2024). Aspergillus oryzae as a cell factory: Research and applications in industrial production. *Journal of Fungi, 10*(4), 248. https://doi.org/10.3390/jof10040248

Tamayo, E., et al. (2014). Genome-wide analysis of copper, iron and zinc transporters in the arbuscular mycorrhizal fungus Rhizophagus irregularis. *Frontiers in Plant Science, 5*. https://doi.org/10.3389/fpls.2014.00547

Trijatmiko, K. R., Dueñas, C., Tsakirpaloglou, N., Torrizo, L., Arines, F. M., Adeva, C., Balindong, J., Oliva, N., Sapasap, M. V., Borrero, J., Rey, J., Francisco, P., Nelson, A., Nakanishi, H., Lombi, E., Tako, E., Glahn, R. P., Stangoulis, J., Chadha-Mohanty, P., et al. (2016). Biofortified indica rice attains iron and zinc nutrition dietary targets in the field. *Scientific Reports, 6*(1). https://doi.org/10.1038/srep19792

Turło, J., Gutkowska, B., & Herold, F. (2010). Effect of selenium enrichment on antioxidant activities and chemical composition of Lentinula edodes (Berk.) Pegler. Mycelial extracts. *Food and Chemical Toxicology, 48*(4):1085–1091. https://doi.org/10.1016/j.fct.2010.01.030

Vetter, J. (2003). Chemical composition of fresh and conserved Agaricus bisporus mushroom. *European Food Research and Technology, 217*(1), 10–12. https://doi.org/10.1007/s00217-003-0707-2

Wang, H., Ng, T., & Liu, Q. (2002). Isolation of a new heterodimeric lectin with mitogenic activity from fruiting bodies of the mushroom Agrocybe cylindracea. *Life Sciences, 70*(8), 877–885. https://doi.org/10.1016/s0024-3205(01)01463-1

Wang, M., Yang, Y., & Chen, W. (2018). Manganese, zinc, and pH affect cadmium accumulation in Rice grain under Field conditions in southern China. *Journal of Environmental Quality, 47*(2), 306–311. https://doi.org/10.2134/jeq2017.06.0237

Watson, A., Ghosh, S., Williams, M. J., Cuddy, W. S., Simmonds, J., Rey, M., Hatta, M., et al. (2017). Speed breeding is a powerful tool to accelerate crop research and breeding. *Nature Plants, 4*(1), 23–29. https://doi.org/10.1038/s41477-017-0083-8

World Health Organization (WHO). (2020, February 21). *Obesity*. Retrieved March 20, 2025, from https://www.who.int/health-topics/obesity#tab=tab_1

World Health Organization (WHO). (2021). Food and Agricultural Organization of the United Nations. Vitamin and Mineral Requirements in Human Nutrition, 2nd ed.; World Health Organization: Geneva; Available online: https://apps.who.int/iris/bitstream/handle/10665/42716/9241546123.pdf?ua=1. Accessed on 3 May 2021.

Zambare, V. P., & Christopher, L. P. (2012). Biopharmaceutical potential of lichens. *Pharmaceutical Biology, 50*(6), 778–798. https://doi.org/10.3109/13880209.2011.633089

Zhang, Y., Butelli, E., Alseekh, S., Tohge, T., Rallapalli, G., Luo, J., Kawar, P. G., Hill, L., Santino, A., Fernie, A. R., & Martin, C. (2015). Multi-level engineering facilitates the production of phenylpropanoid compounds in tomato. *Nature Communications, 6*(1). https://doi.org/10.1038/ncomms9635

Zhao, L., Zhao, G., Zhao, Z., Chen, P., Tong, J., & Hu, X. (2004). Selenium distribution in a Se-enriched mushroom species of the GenusGanoderma. *Journal of Agricultural and Food Chemistry, 52*(12), 3954–3959. https://doi.org/10.1021/jf049965i

Zielińska-Dawidziak, M. (2015). Plant ferritin – A source of iron to prevent its deficiency. *Nutrients, 7*(2), 1184–1201.

Sustainable Food Innovation: Leveraging Food by-Products for Nutritional Enhancement and Food Security

10

Sarra Akermi, Boutheina Ben Hadj Hmida, Khaoula Elhadef, Moufida Chaari, Fakhreddine Ben Amara, Lotfi Mellouli, and Slim Smaoui

Abstract

In many developing countries, the abundance of agro-industrial by-products exacerbates food insecurity and malnutrition. These by-products, often discarded as waste, are rich in valuable nutrients and bioactive compounds that can enhance food products or create new food sources. Utilizing agro-industrial by-products as functional food ingredients offers a sustainable and innovative way to improve the nutritional profile and functionality of foods. These by-products are packed with dietary fibers, antioxidants, vitamins, and minerals, which can significantly benefit human health. Incorporating them into food formulations not only reduces environmental waste but also supports a circular economy by transforming waste into valuable resources. This chapter explores the potential of various agro-industrial by-products, such as fruit peels, seed husks, and vegetable residues, as functional ingredients. It highlights their nutritional benefits, functional properties, and the technological advancements that facilitate their integration into mainstream food products. Additionally, the chapter addresses consumer acceptance and regulatory considerations, emphasizing the need for further research and industrial applications to fully realize the benefits of these sustainable

The authors Sarra Akermi, Boutheina Ben Hadj Hmida, Khaoula Elhadef, Moufida Chaari, and Fakhreddine Ben Amara have equally contributed to this chapter.

S. Akermi · K. Elhadef · M. Chaari · F. B. Amara · L. Mellouli · S. Smaoui (✉)
Laboratory of Microbial Biotechnology, Enzymatic and Biomolecules, Centre of Biotechnology of Sfax, Sfax, Tunisia
e-mail: slim.smaoui@cbs.rnrt.tn

B. B. H. Hmida
Laboratory of Biochemistry and Enzymatic Engineering of Lipases, National Engineering School of Sfax (ENIS), Sfax, Tunisia

G. Sanghvi et al. (eds.), *Sustainable Food Fortification*,
https://doi.org/10.1007/978-981-95-1217-1_10

ingredients. By leveraging agro-industrial by-products, the food industry can move toward more sustainable and health-oriented food systems, ultimately contributing to global food security and environmental sustainability.

Keywords

Food waste · Agro-industrial by-products · Functional properties · Industrial applications · Food security · Environmental sustainability

10.1 Introduction

Sustainable food innovation is an urgent and transformative movement aimed at reshaping how we produce, distribute, and consume food to ensure environmental sustainability, economic viability, and social equity (Khan et al., 2021). At its core, sustainable food innovation integrates technological advancements, ecological stewardship, and consumer awareness to create a resilient food system that meets the needs of today without compromising future generations (Galanakis, 2024). This innovation spans multiple domains, including precision agriculture, plant-based and lab-grown alternatives, regenerative farming, and the reduction of food waste through smart logistics and packaging (Mešić et al., 2024). Companies and researchers are pioneering alternative proteins, such as insect-based foods, algae, and cultured meats, to reduce the environmental footprint associated with traditional livestock farming, which contributes significantly to deforestation, greenhouse gas emissions, and excessive water usage (Kumar et al., 2023a, b; Hadi & Brightwell, 2021). Similarly, regenerative agriculture aims to restore soil health, enhance biodiversity, and sequester carbon, mitigating climate change while ensuring productive farmland for years to come (Sher et al., 2024; McCauley & Barlow, 2023). Furthermore, digitalization and artificial intelligence are optimizing food supply chains, enabling predictive analytics to manage food distribution efficiently and reduce waste (Onyeaka et al., 2023).

An integral component of sustainable food innovation is the strategic utilization of food by-products, which plays a crucial role in minimizing waste and maximizing resource efficiency (Pal et al., 2024; De Barros & Gouvinhas, 2021). Rather than discarding agricultural and food processing remnants, industries and researchers are exploring ways to repurpose these by-products into valuable commodities, such as animal feed, biofuels, fertilizers, and even ingredients for new food products (Wibisono et al., 2025; Yang et al., 2021). By harnessing the nutritional potential of by-products like fruit and vegetable peels, whey from dairy processing, or brewer's spent grains, manufacturers can create nutrient-rich supplements, functional foods, and biodegradable packaging alternatives (Selvamuthukumaran, 2024; Ospina-Maldonado et al., 2024). Additionally, leveraging food by-products contributes to circular economies, where waste is repurposed into new inputs, significantly reducing environmental impact while promoting economic sustainability (Haque et al., 2023). The widespread adoption of this approach aligns with global sustainability

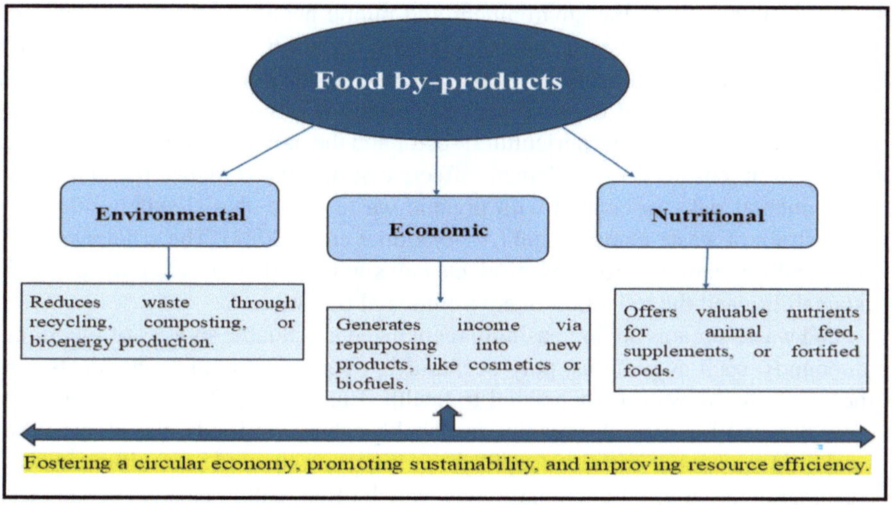

Fig. 10.1 Importance of food by-products

goals, enhancing food security, reducing landfills, and lowering greenhouse gas emissions associated with food waste (Fig. 10.1). As awareness grows around the importance of resource efficiency, integrating food by-products into sustainable food innovation will further strengthen efforts toward a responsible and regenerative food system (Brennan, 2024; Michel et al., 2024). Through collaboration between stakeholders, including farmers, businesses, policymakers, and consumers, sustainable food innovation promises to create a food system that supports both human health and ecological balance (Rosenthal et al., 2021).

10.2 Nutritional Enhancement Through Food by-Products

A substantial quantity of food waste and by-products is generated annually by the global food industry. Approximately 38% of this waste arises during food processing stages. These by-products originate from diverse sources (Sorrenti et al., 2023). Animal-based food processing yields waste materials, such as carcasses, hooves, heads, feathers, offal, bones, fat and meat trimmings, and blood. In the seafood sector, waste includes skin, bones, oils, and blood. Dairy industry by-products consist of whey, curd, and milk sludge produced during separation processes. Plant-based food processing also contributes significantly, with waste materials including peels, stems, seeds, shells, bran, and residues left after the extraction of oils, juices, starches, and sugars (Aït-Kaddour et al., 2024). The improper disposal of food industry waste poses serious environmental challenges due to its poor biological stability, high organic content, significant nutritional value, elevated water activity, susceptibility to oxidation, and active enzymatic properties (Kumar et al., 2022). The substantial volume of waste, along with its microbial degradation, can negatively impact both the environment and public health. Furthermore, the high costs associated with waste treatment represent an additional financial burden for food manufacturers, who often already operate under tight profit margins (Afif et al., 2022). This added expense not only affects the food industry but also has broader implications for the agricultural sector and the national economy. Therefore, there is an urgent need to implement effective strategies to reduce the health and environmental risks associated with organic waste while also alleviating the economic strain of waste management (Ashokkumar et al., 2022). The efficient valorization and utilization of food by-products can significantly contribute to economic sustainability and the reduction of environmental pollution.

Food waste streams are increasingly recognized as valuable sources of functional compounds, such as proteins, polysaccharides, fibers, lipids, phytochemicals, and other bioactive substances beneficial to health (Liu et al., 2023). With rising consumer demand for natural, organic, and health-enhancing foods, the global functional food market has expanded rapidly, reaching an estimated $100 billion in 2013 (Guenther et al., 2022). Functional foods, which offer health benefits beyond basic nutrition, play a crucial role in preventing chronic diseases like cardiovascular disorders, diabetes, and cancer (Sharma & Yadav, 2023). Bioactive compounds in fruits and plants, including phenolics and flavonoids, help combat oxidative stress and reduce disease risk, while dietary fibers contribute to cholesterol regulation and improved digestive health (Muscolo et al., 2024). Agricultural and food processing by-products, especially from fruits, vegetables, meat, seafood, and dairy, can be repurposed as raw materials for functional food development. This not only addresses environmental concerns but also strengthens industry competitiveness. Although much of this waste is still discarded, many food industries are now valorizing it to produce high-value ingredients, enhancing both economic sustainability and public health (Helkar et al., 2016).

10.2.1 Fruit and Vegetable by-Products

By-products from fruit processing industries, such as pomegranate peels and seeds, are rich in bioactive compounds with notable pharmaceutical and nutraceutical potential (Marra et al., 2022). Similarly, apple pomace, a major waste product from apple juice production, contains approximately 74% total dietary fiber and exhibits functional properties like high swelling capacity, water and oil retention, and a glucose dialysis retardation index of 36.91%. These characteristics make it particularly beneficial for gastrointestinal health (Rana et al., 2015). The majority of apple pomace (about 95%) consists of peel and pulp, rich in cell wall polysaccharides and phenolic compounds, and serves as a key raw material for pectin extraction (Costa et al., 2022).

De-oiled sunflower press cake is gaining attention as a high-quality plant-based protein source, offering a safe alternative to soy and egg proteins due to its low levels of antinutritional factors and absence of toxic compounds. Orange pomace also stands out for its high fiber content (40.47%), significant mineral presence, low fat

(2.14%), and impressive water-holding capacity (4.40 ml/g), making it suitable for functional food applications that require moisture retention and texture enhancement (Hadidi et al., 2024). Today, fibers and pectins derived from fruit waste, especially apples and citrus, are widely used as ingredients in both functional foods and pharmaceutical formulations. Pectin plays a role in drug delivery systems, while polyphenolic extracts from these by-products are marketed in Asia as nutritional supplements and health-promoting additives (Rațu et al., 2023). For more examples of functional ingredients derived from food by-products, refer to Table 10.1.

10.2.2 Grain by-Products

Rice bran, a by-product generated during rice milling, accounts for approximately 10% of the total weight of unpolished rice (Huang et al., 2020). It is a valuable source of essential nutrients, including dietary fiber, vitamins, minerals, essential fatty acids, sterols, and potent antioxidants, such as polyphenols, tocotrienols, vitamin E, and carotenoids (Sohail et al., 2017). These bioactive compounds contribute to its growing use in the food, nutraceutical, and pharmaceutical sectors. From the milling paddy, about 70% of the yield is edible rice, while the remaining 30% comprises by-products, 20% husk, 8% bran, and 2% germ. Due to its rich antioxidant profile, rice bran offers protective effects against oxidative stress and cellular damage, potentially lowering the risk of chronic illnesses. Its high fiber content and therapeutic properties make it an ideal ingredient for developing functional foods (Sohail et al., 2017). In fact, rice bran has been successfully incorporated into various food products, such as bread, pasta, noodles, cakes, and ice creams without significantly altering their texture or quality. Additionally, rice bran oil, known for its abundance of natural antioxidants, is being explored for its role in chronic disease prevention (Manzoor et al., 2023). For more examples of functional ingredients derived from food by-products, refer to Table 10.1.

Table 10.1 Functions of food by-products

Food by-product	Functions	References
Banana	Source of fiber, probiotic protection	Kumari et al. (2023)
Pineapple peel powder	Texturizing agent, probiotic protection	Sah et al. (2016)
Olive pomace	Source of phenols, probiotic protection	Mateos et al. (2020)
Grape marc flour	Source of phenols, probiotic protection	Marinelli et al. (2018)
Rice bran	Colorant	Tan et al. (2023)
Fish oil	Source of omega-3	Balwan and Saba (2021)
Fluid whey	Texturizing agent	Himashree et al. (2022)

10.2.3 Marine by-Products

Marine resources, owing to their immense biodiversity, serve as a rich reservoir of functional food ingredients and bioactive compounds. These include fish oils, proteins, peptides, macroalgae, microalgae, and other health-promoting marine substances. Despite their recognized health benefits, particularly in reducing the risk of cardiovascular diseases, marine-derived functional components remain underutilized in the food industry (Ghosh et al., 2022). Seafood is an excellent source of high-quality proteins and lipids, especially rich in omega-3 polyunsaturated fatty acids (PUFAs), essential vitamins, and minerals. Known as a "rich food for the poor," fish contributes significantly to global nutrition. The nutritional value obtained from fish consumption directly correlates with intake volume (Mohanty, 2020). Processing by-products from marine sources can constitute up to 80% of the total biomass, depending on the species. These by-products include valuable components, such as ω-3 PUFAs from white fish livers, fatty tissue from oily fish, marine mammal blubber, protein hydrolysates from fish viscera, and bioactive compounds from crustaceans like chitosan, glucosamine, and their derivatives. Such compounds can be extracted and incorporated into functional food formulations to enhance their health value (Shahidi et al., 2019). Additionally, fish-based ingredients are easily digestible and naturally tender, making them suitable for a variety of applications. Marine-derived bioactives, including omega-3 oils, chitin, fish protein hydrolysates, algal extracts, carotenoids, antioxidants, shark cartilage, taurine, and fish bone derivatives, are increasingly being recognized for their nutraceutical potential. Among these, omega-3 fatty acids remain the most widely used and researched marine-based bioactives, due to their established benefits in promoting cardiovascular and overall health (Shahidi et al., 2019).

10.2.4 Meat by-Products

As defined by the European Commission, "meat" encompasses the edible parts derived from the carcasses of domesticated animals, such as cattle, pigs, sheep, goats, poultry, and wild game (Lautenschlaeger & Upmann, 2017). In contrast, animal by-products refer to entire animal bodies or parts, along with products of animal origin not intended for direct human consumption. The classification of these by-products varies across cultures, traditions, and religions, but generally includes organs, blood, bones, skins, meat trimmings, fatty tissues, hooves, feet, and horns (Helkar et al., 2016). These by-products are nutritionally valuable, being rich in proteins, lipids, and carbohydrates. Through processes like enzymatic hydrolysis, cooking, or fermentation, meat proteins can be transformed into bioactive peptides. These peptides have demonstrated a wide array of health-promoting properties, including antimicrobial, antioxidant, antihypertensive, antithrombotic, anticancer, and immunomodulatory effects. They may influence several physiological systems, such as cardiovascular, digestive, nervous, and immune systems. Moreover, emerging research highlights their potential role in managing chronic conditions like cancer, diabetes, obesity, and even certain mental health disorders (Helkar et al., 2016). For more examples of functional ingredients derived from food by-products, refer to Table 10.1.

10.2.5 Dairy by-Products

The dairy processing sector is a significant component of the food industry in India. One of its major by-products, whey, is a rich source of valuable soluble proteins, such as β-lactoglobulin, α-lactalbumin, immunoglobulins, bovine serum albumin, lactoferrin, and lactoperoxidase. These proteins are widely utilized in food formulations, typically in dried form, due to their excellent nutritional qualities, particularly their high content of essential amino acids (Guo & Wang, 2019). Beyond nutrition, whey proteins exhibit important functional properties like emulsification, gelation, and foaming, as well as biological activities, including antimicrobial, anticancer, and immunomodulatory effects (Minj & Anand, 2020). Enriched whey-based products are increasingly being explored in the development of functional foods to meet growing health and nutritional demands. Additionally, curd, another dairy by-product, is recognized as a probiotic-rich functional food. A growing body of evidence supports the role of probiotics in preventing and managing various health conditions, including acute diarrhea, antibiotic-associated diarrhea, necrotizing enterocolitis, and certain allergic disorders (Kheto et al., 2025). For more examples of functional ingredients derived from food by-products, refer to Table 10.1.

10.3 Technological Innovations in Food by-Product Extraction

10.3.1 Conventional Technologies

Conventional techniques have been used to extract bioactive compounds from a variety of food wastes. These techniques mostly rely on the solvent selection, extraction effectiveness, and variables, such as temperature and agitation. The primary methods for removing bioactive substances from discarded peels are highlighted below:

10.3.1.1 Liquid-Liquid Extraction

This process involves two distinct phases: the organic phase and the aqueous phase. For a successful extraction, the analyte (the target biomolecule) must be dissolved in the organic phase. However, this method has numerous disadvantages, including the need for an extensive quantity of organic solvents, the creation of emulsions, automation challenges, and labor intensiveness (Usman et al., 2022).

10.3.1.2 Solid-Phase Extraction

The chemical composition is guaranteed to be completely removed during this extraction process from a flowing liquid. Eventually, the required constituent is retrieved from the sorbent through elusion, while the chemical component remains on the solid sorbent. The ability to fully elute the analytical solute is a crucial prerequisite for successful solid-phase extraction. Due to its economical and efficient use of solvent, solid-phase extraction is a more advantageous technique compared to liquid-liquid extraction (Murakami et al., 2020). Recently, molecularly imprinted polymers for the selective enrichment of different chemicals have been developed using this technology. Among these polymers, acrylamide and ethylene glycol dimethacrylate copolymers and ionic liquid-imprinted silica particles are utilized to create materials with a high affinity for the template structures (Srivastava et al., 2021a, b, c). The primary methods employed for extracting natural compounds through liquid-solid extraction include percolation, decoction, Soxhlet extraction, maceration, and infusion, as they effectively promote the transfer of bioactive compounds from solid materials into a liquid solvent.

- Maceration

In order to recover the desired compounds, the maceration involves three main steps: first, the sample is ground into fine particles to maximize solvent contact; second, the solid material is immersed in a solvent in a closed vessel, occasionally with heat to improve diffusion; and third, the liquid extract is separated from the solid residue by centrifugation, pressing, or sieving (Bitwell et al., 2023).

- Soxhle

In Soxhlet extraction, the appropriate solvent is heated in a round-bottomed flask, whereas the sample is set up in a thimble inside the extractor. The solvent vaporizes, passes through the thimble, and condenses in a different chamber when it reaches its boiling point. The thimble is gradually filled with condensed liquid, and once it exceeds the overflow level, the solution can return to the flask with the extracted compounds (Martins et al., 2023).

- Percolation

To ensure uniform fluid contact, the sample is first immersed in a menstruum for 4 h before being put in a percolator. For a 24 h soaking operation, more menstruum is introduced after a consistent percolation rate has been achieved. The wet mass is then compressed to remove any remaining fluid, and the solvent is added to reach the required concentration. This method's main drawbacks are its high usage of solvent and lengthy extraction times, which necessitate 4–5 extraction repetitions (1 to 2 days) and a high substrate-to-solvent proportion (Srivastava et al., 2021a, b, c).

- Decoction

This technique involves boiling plant material in water for 30 min to 2 h, then filtering it through multiple layers of muslin cloth. Centrifugation can be used to enhance extraction, after which the solvent is evaporated, following the collection of the supernatant. Compared to maceration, decoction may enhance the solubility of various phytochemicals, including alkaloids, flavonoids, polysaccharides, and tannins (Kumar et al., 2023a).

- Infusion

This method is used to generate diluted solutions of the biomolecules from crude drugs that are easily soluble. The first step in this preparation is to macerate the raw plant or drug ingredients for a brief amount of time using either hot or cold water (Srivastava et al., 2021a, b, c). Table 10.2 summarizes the conventional extraction applications for the extraction of biomolecules from agri-food by-products.

10.3.1.3 Solid-Phase Microextraction

This method involves exposing coated fibers or a sorbent to a target sample for a predetermined period, after which it is transferred to HPLC or gas chromatography. It is characterized by its simplicity, automation potential, and efficient detection of low-concentration bioactive chemicals while minimizing solvent-related issues (Usman et al., 2022).

Table 10.2 Classical extraction techniques for biomolecules from agricultural residues

Waste/ by-product	Extraction method	Main findings	References
Sweet orange peels	Percolation	Methanol proved to be the most effective extraction solvent, achieving the highest yield of total phenolic content at 28%	Thakur et al. (2020)
Pomegranate peels	Decoction	With water as the solvent, TPC ranged from 148 to 237 mg GAE/g DW, while RSA varied between 307 and 472 mg AAE/g DW	Turrini et al. (2020)
Capsicum annuum waste	Soxhlet	This extraction method uses 200 mL of hexane or ethanol, followed by evaporation of the aqueous solution to isolate alkaloids like capsaicin and dihydrocapsaicin	Bayil Oğuzkan (2019)
Tomato peels	Maceration	Extraction of lycopene from tomato peels using refined olive oil as a solvent, obtaining a 99.3% overall yield	Kehili et al. (2019)
Jaboticaba peels	Infusion	Steeping the powder in hot water (100 °C) for a 3 min using 2 g, to achieve the highest concentration of TPC (47.16 mg/L)	Marsiglia et al. (2021)

TPC total phenolic content, *GAE* gallic acid equivalent, *DW* dry weight, *RSA* radical scavenging ability, *AAE* ascorbic acid equivalent

10.3.2 Emerging Technologies

For over a hundred years, bioactive molecules from plants have been extracted using conventional extraction. However, low purity, lengthy extraction times, heat-sensitive molecular breakdown, and solvents toxicities hinder their efficiency. These restrictions have prompted research into economical and environmentally responsible extraction practices, encouraging the development of innovative strategies (Chemat et al., 2020).

10.3.2.1 Supercritical Fluid Extraction (SFE)

SFE is an environmentally friendly extraction method that is frequently used at both the pilot and industrial stages. Given its great purity, low critical temperature and pressure, and affordability, carbon dioxide is the most often utilized supercritical fluid. The method is especially useful for extracting thermally unstable chemicals that cannot be purified by steam distillation since it depends on the increased solvating power of gases above their critical point. To effectively extract bioactive compounds from discarded peels, the SFE apparatus is made up of several crucial parts, such as a CO_2 solvent reservoir, a temperature-controlled chamber, a pressure regulator, and a collection vessel (Srivastava et al., 2021a, b, c; Uwineza & Waśkiewicz, 2020).

10.3.2.2 Microwave-Assisted Extraction (MAE)

MAE is a method that fuses the conventional solvents extraction with nonionizing electromagnetic radiation at microwave frequency ranging from 0.3 to 300 GHz. This technique involves ionic conduction and dipole rotation to make compound extraction easier (Sarfarazi et al., 2020). The microwave-induced electrical field in ion conduction propels charge carriers through electrophoretic migration, producing heat and friction. Dipolar molecules oscillate in response to the alternating electric field during dipole rotation, causing heating and chemical collisions (Echegaray et al., 2022). The generation of heat directly causes a rise in both temperature and pressure of water vapor inside the cells, which results in the swelling and breaking of cell walls, subsequently releasing intracellular substances. The effectiveness of MAE is affected by various elements, such as the properties of the solvent, the composition of the matrix, temperature of heating, time of extraction, and the microwave frequencies power. Interestingly, polar solvents, including water, methanol, and ethanol, have strong microwave heating and absorption capacities. Furthermore, the temperature may be higher than the solvent's boiling point, improving the effectiveness of extraction. Additionally, the extraction period is a crucial factor. Increased power speeds up extraction (1–2 min at 150 W). But extended exposure can harm delicate chemicals (Bagade & Patil, 2021).

10.3.2.3 Ultrasound-Assisted Extraction (UAE)

UAE became a novel method that could enhance heat and mass transport by rupturing plant cell walls due to the cavitation effect. The development, expansion, and collapse of microbubbles within a liquid exposed to ultrasonic irradiation (frequencies >20 kHz) is known as acoustic cavitation, and it is linked to the benefits of high-power ultrasound on enhancing extraction. This collapse is characterized by turbulence, intense shear stress, ultrasonic streaming, microjets near rigid surfaces, and localized elevations in pressure and temperature (Carpentieri et al., 2021). An ultrasonic bath and ultrasonic probe can be used to administer US treatment. The primary distinction between them is related to the acoustic energy's source. But the foundation of both setups is a piezoelectric transducer that produces ultrasonic power. US-based extraction procedures provide for low-cost industrial production lines, primarily because they require less solvent and take less time to extract, which lowers energy costs (Kumar et al., 2021).

10.3.2.4 High-Pressure Homogenization (HPH)

One of the best mechanical techniques for large-scale cell disruption is HPH technology. For 3–5 min, it is carried out at room temperature under extremely high pressures between 300 and 100 MPa (Rifna et al., 2023). A liquid with suspended solids is forced through a narrow gap valve using a high-pressure intensifier, followed by a reduction in pressure that leads to the creation of high shear and elongational forces, along with cavitation. As a result, the cells/macromolecules dispersed in the liquid are subjected to significant mechanical stress, causing them to deform and twist. This technology has been successfully used to enhance the release of intracellular biomolecules from food by-products, due to the complete lysis of plant cells caused by HPH (Yong et al., 2021). The United States Food and Drug Administration recently recognized the HHP-aided extraction technique as an entirely ecologically friendly technology because it simply needs power to operate and produces no waste (Rifna et al., 2023).

10.3.2.5 Pulsed Electric Fields (PEF)

Researchers' attention has recently been drawn to electrotechnologies owing to their enormous promise for the rapid, economical, and environmentally sustainable recovery of bioactive molecules. High-voltage pulse generators, treatment chambers (parallel plate, coaxial, or collinear topologies), pumps, temperature control systems, oscilloscopes, and voltage are the standard components for the process maintenance. In PEF technique, a wet plant matrix is electrically connected to the metallic electrodes of a treatment chamber, run either continuously or in batches, and subjected to electrical field pulses (1–10 kV/cm), repetitive (Hz-kHz) very brief micro to milliseconds, and a relatively little energy input (1–20 kJ/kg) provided by the electrical pulse generator (Xi et al., 2021). The electric fields generated by the application of PEF break the biomolecules in the waste peel's cellular membranes according to their dipole characteristics. Once the membrane potential of the peel matrix reaches one-volt, repulsive forces arise between the charged components, leading to the formation of pores and enhancing the permeability of the peel matrix (Pattnaik et al., 2021).

10.3.2.6 Dielectric Barrier Discharge Plasma Extraction (DBDE)

Non-thermal plasma, also known as cold plasma, occurs when an electric field is applied to gas; this energy helps the released electrons accelerate and ionize the gas molecules, which in turn produces more free electrons that exacerbate new ionizations; additionally, an energized electron causes dissociations of molecules, leading to the formation of new radicals and atoms. When excited molecules return to the steady phase, they release excess energy in the form of UV and other electromagnetic radiation. As a result, energetic atoms, such as free radicals, nitrogen species, ozone, electrons, positive, and negative ions constitute most of the plasma. In recent years, there has been a noticeable increase in interest in the application of cold plasma extraction techniques in the food industry. By destroying the secondary configuration with released free radicals or by understanding the chemical variations of side chains throughout the action of the various chemically vigorous species encompassing the plasma, it can break down specific bonds for the extraction of bioactive substances from waste peels (Misra et al., 2016).

10.3.2.7 High-Voltage Electrical Discharges (HVED)

HVED is introduced directly into a water solution, resulting in ionization between two electrodes when a current is applied within the range of 10^3 to 10^4 A, a high voltage of 10^3 to 10^4 V, and frequencies ranging from 10^{-2} to 10^{-3} Hz through a plasma channel (Castro-Muñoz et al., 2022). HVED is a damp biomaterial cell breakdown process that relies on the electrical decomposition of water. A high-voltage pulse, a treatment chamber, an oscilloscope for data collection, a cooling system, a peristaltic pump, and instruments for monitoring voltage and current are all part of its system. A brief (2–5 μs) high-current/high-voltage electrical discharge (40–60 kV; 10 kA) creates a plasma channel that allows high energy to be released right away into the aqueous suspension that is set in a batch treatment chamber across a high-voltage needle electrode and a plated grounded electrode. A continuous HVED extraction system with an ongoing circular gap-type treatment chamber has been suggested to shorten the extraction time. This system has a much higher processing capacity and prevents clogging (Li et al., 2019).

10.3.2.8 Pressurized Liquid Extraction (PLE)

Pressure is used during PLE, which permits the utilization of temperatures higher than the solvents' boiling points. By improving its dissolution and mass transfer properties, it can more effectively enter solid materials, lowering the need for solvents and preventing the bioactive chemicals from degrading. The extraction procedure involves inserting the sample into the extractor and utilizing an HPLC pump to move the solvent into the extraction container. An electric heater jacket is used to continuously maintain the sample's temperature in the extraction cell until the necessary pressure is reached. The extraction process starts as soon as the ideal temperature and pressure are achieved. After the extraction has ended, the pump and heating system are turned off. To remove the pressured liquid extractor and eliminate any remaining solvent, inert gases like nitrogen can be used (Pimentel-Moral et al., 2020).

10.3.2.9 Enzyme-Assisted or Co-assisted Extraction (EAE)

Considering numerous benefits, the enzyme-assisted extraction approach has become more and more important in research. In general, enzymes are highly efficient and specialized substances. The following are the main benefits of using the EAE method to extract bioactive compounds from discarded peel: (i) The EAE approach is a higher-yielding method for bioactive molecules because it ruptures cell walls and produces preferred bioactive compounds; (ii) EAE increases system transparency by removing unnecessary cellular wall components; and (iii) it provides the benefits of increased catalytic competence and significantly preserves the peel samples' distinctive value. Under mild processing settings, EAE refers to the enzymes' capacity to disrupt the cellular components and alter the structure of food by-products, making it easier to extract and release the targeted bioactive substances from discarded peels. This extraction technique can be improved upon or used in conjunction with other techniques to extract bioactive compounds (Rifna et al., 2023). Various emerging technologies for the extraction of bioactive molecules from by-products are summarized in Table 10.3.

Table 10.3 Application of different novel techniques for biomolecule extraction from by-product peels

By-products	Extraction process	Major findings	References
Banana peels	Microwave-assisted extraction (MWAE)	Successfully extract phenolic compounds in just 6 min using MWAE (960 W, S/F ratio of 2:100 g/mL), demonstrating significantly greater efficiency compared to Soxhlet extraction, which requires 9 to 10 h	Jurić et al. (2019)
Opuntia ficus-indica peels		Effectively yield pure betalain extracts with MWAE at 70 °C for 1 hour, while maintaining their color and purity, ensuring stability	Xie et al. (2018)
Blueberry pomace	Pulsed electric fields (PEF	Improve anthocyanin extraction as the PEF process intensifies, using electric field intensities of 10–20 kV/cm, specific energy input up to 41.03 kJ/kg, and pulse numbers ranging from 10 to 100 with a 2-second width	Lončarić et al. (2020)
Tomato waste		Boost carotenoid extraction yields by 56.4% and double total phenolic concentration by applying PEF at 2 kV/cm with 700 pulses	Andreou et al. (2020)
Tomato peels	Supercritical fluid extraction (SFE)	Recover oleoresin from tomato peels with the highest lycopene content (5.28 mg/g dw) by performing SFE-CO2 for 155 min at a pressure of 400 bar and 74 °C	Mihalcea et al. (2021)
Wild thyme by-product		Improve meat quality by preserving color, reducing oxidation, and enhancing microbial safety with SFE at 350 bar and 50 °C	Šojić et al. (2020)

(continued)

Table 10.3 (continued)

By-products	Extraction process	Major findings	References
Jabuticaba by-products	Ultrasound-assisted extraction (UAE)	Achieve up to 350% higher anthocyanin recovery by applying UAE at 19 kHz for 3 min extractions at 33 °C with a mixture of 50 g water and 100 g solvent, compared to using pure ethanol or water	Gadioli Tarone et al. (2021)
Goji berry extracts and pomegranate fruit peels		Demonstrate the strongest antioxidant, antimicrobial, and antifungal activity with pomegranate peel extract, while goji berry extracts primarily exhibit antibacterial effects	Skenderidis et al. (2019)
Tomato peels	High-pressure homogenization (HPH)	Achieve greater extraction of polyphenols (+32.2%), proteins (+70.5%), and antioxidant activity (+23.3%) with HPH processing (10 passes) compared to high shear mixing, highlighting its superior effectiveness in enhancing bioactive compound recovery	Jurić et al. (2019)
Potato peels		Alter potato peel pectins with HPH by increasing galacturonic acid content while decreasing esterification degree and molecular weight. this modification enhances the viscosity and emulsifying properties of the pectin	Xie et al. (2018)
Grape peels	High-voltage electrical discharges (HVED)	Reduce HVED energy input by performing the subsequent solid-liquid extraction in 20% (w/v) aqueous glycerol	El Kantar et al. (2019)
Spent coffee grounds		Achieve the highest polyphenol content of 59.83 mg/g with HVED, outperforming solvent extraction by 20.03% and the converged electric field type method by 11.78%	Deng et al. (2018)
Broccoli by-products	Pressurized liquid extraction (PLE)	Gave the greatest content of bioactive compounds (6.37 mg GAE/g dw and 4.59 mg QE/g dw) and the highest antioxidant activity for DPPH and ABTS	Marinelli et al. (2020)
Blackberry peel		Achieve maximum TP (7.36 mg GAE/g) and AA (76.03 mmol TE/g) with purity of 91% and 90%, respectively, under the best PLE conditions (acidified water +140 °C)	Machado et al. (2015)

(continued)

Table 10.3 (continued)

By-products	Extraction process	Major findings	References
Tomato pomace	Dielectric barrier discharge plasma extraction (DBDE)	Enhanced the antioxidant capacity of the mentioned waste by changing their concentration profiles of phenolic compounds	Bao et al. (2020)
Haskap		Increased the anthocyanins extraction rate of (13.35–20.47%), promoted the number of anthocyanin monomers, and enhanced the antioxidant activity of the extracts	Li et al. (2023)
Broccoli by-products	Enzyme-assisted extraction (EAE)	Cellulase reduced the sugar content while increasing the uronic acid content	Rivas et al. (2022)
Radish root ground	Ultrasound-enzyme assisted extraction	Using cellulases, pectinases, amylases, glucanases, hemicellulases leads to higher total antioxidant capacity, with the highest extraction of total phenolic content	Rani et al. (2021)
Purple-heart radish	Microwave-enzyme assisted extraction	Papin combined with MWAE facilitated cell rupture and improved the extraction yields in short time	Lin et al. (2022)

S/F ratio solvent-to-feed ratio, *GAE* gallic acid equivalents, *QE* quercetin equivalents, *dw* dry weight, *AA* antioxidant activity, *TE* trolox equivalents, *TP* total polyphenols

10.4 Impact on Food Security

Food by-products play a crucial role in enhancing global food security by addressing resource efficiency and reducing waste (Singh et al., 2019). These by-products, derived from food processing industries, contain valuable nutrients that can be repurposed for human consumption, animal feed, or soil enrichment. For example, fruit peels, vegetable trimmings, or grains left after milling can be transformed into functional ingredients, supplements, or fortified foods, contributing to a sustainable and inclusive food system (Sarangi et al., 2024). Utilizing food by-products also alleviates pressure on agricultural production, minimizes environmental impact, and supports circular economies. By integrating food by-products into the supply chain, nations can reduce dependency on conventional resources, lower costs, and ensure a more equitable distribution of nutrients, ultimately addressing food insecurity on a global scale (Wijerathna-Yapa & Pathirana, 2022). This approach not only emphasizes innovation but also highlights the importance of responsible consumption and production patterns, paving the way for a resilient and sustainable future (Fig. 10.2).

For instance, whey from cheese production, which was historically seen as waste, is now widely used in the health and fitness industry for producing protein supplements (Buchanan et al., 2023). Similarly, the oil extracted from fruit seeds, such as avocado or mango can serve as a nutritious and sustainable alternative to conventional cooking oils (Srivastava et al., 2021a, b, c). Furthermore, by-products like

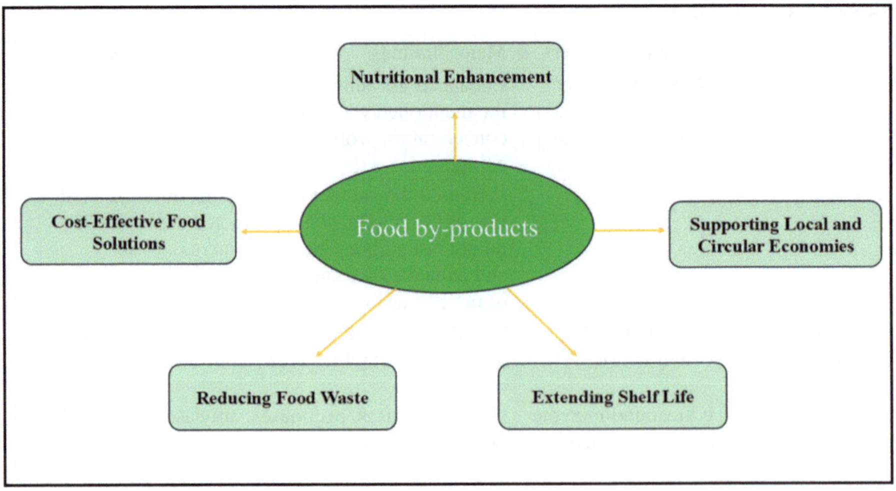

Fig. 10.2 The impact of food by-products on food security

coffee grounds or brewer's spent grain can be upcycled into high-quality ingredients for baked goods or even transformed into bioenergy, contributing to energy sustainability (Vasileiadou, 2024). The utilization of these by-products not only reduces food waste but also promotes economic opportunities, particularly for small and medium enterprises, by opening up new markets and creating value-added products (Minervini et al., 2022). In agriculture, by-products can be processed into organic fertilizers and animal feed, thereby enhancing soil health and livestock productivity while reducing reliance on synthetic alternatives (Xu & Geelen, 2018). This closed-loop approach aligns with global efforts to combat food insecurity by maximizing resource efficiency and minimizing environmental impact. Importantly, the integration of food by-products into the food system reflects a shift toward a circular economy, where the lifecycle of resources is extended, fostering both economic and environmental sustainability (Stillitano et al., 2021). By rethinking waste, humanity can address the dual challenges of feeding a growing population and preserving the planet's finite resources (Frόna et al., 2019).

10.5 Sustainability and Environmental Impact

A crucial factor in determining sustainable development is its environmental footprint. Repurposing food by-products helps mitigate this impact while also reducing waste treatment costs. Additionally, it provides financial benefits to companies by adding value to what would otherwise be discarded. In this context, the concept of a circular economy emerges as an efficient resource management approach, emphasizing minimal raw material consumption, extensive product and component reuse, and high-quality recycling within production systems (Leder et al., 2020). For by-product processing to align with sustainable practices, it is essential to assess the environmental impact associated with extracting valuable raw materials. Furthermore, evaluating the economic feasibility of these processes is crucial. This includes estimating the market price of the potential product, identifying its possible applications, and considering storage, transportation, volume, flexibility, and the geographical availability of the by-products.

The environmental benefits of repurposing food by-products extend across multiple sectors, from reducing greenhouse gas emissions and food waste to enhancing food security and creating sustainable alternatives for packaging and biofuel production (Awogbemi & Von Kallon, 2022). By leveraging advanced technologies and innovative processing techniques, food industry by-products can be transformed into valuable resources that contribute to both economic growth and environmental protection (Fig. 10.3). As global populations continue to rise and resource scarcity becomes a growing concern, maximizing the potential of food industry by-products represents a crucial step toward achieving long-term sustainability.

Emerging studies highlight the potential to transform agricultural and industrial residues into valuable products while addressing pressing global challenges, such as climate change, resource depletion, and food security (Ofori et al., 2021). Additionally, the growing recognition of food by-products as a sustainable resource has highlighted their potential to reduce waste, lower greenhouse gas emissions, and preserve natural resources. These by-products, which originate from fruit, vegetable, grain, and other food processing activities, offer a significant opportunity to enhance environmental sustainability (Romero et al., 2022; Afraz et al., 2023). Rather than being discarded as waste, they can be transformed into valuable materials, such as animal feed, food additives, biodegradable packaging, biofuels, and even pharmaceutical ingredients (Dilucia et al., 2020; Yukesh Kannah et al., 2020; Ofori et al., 2021).

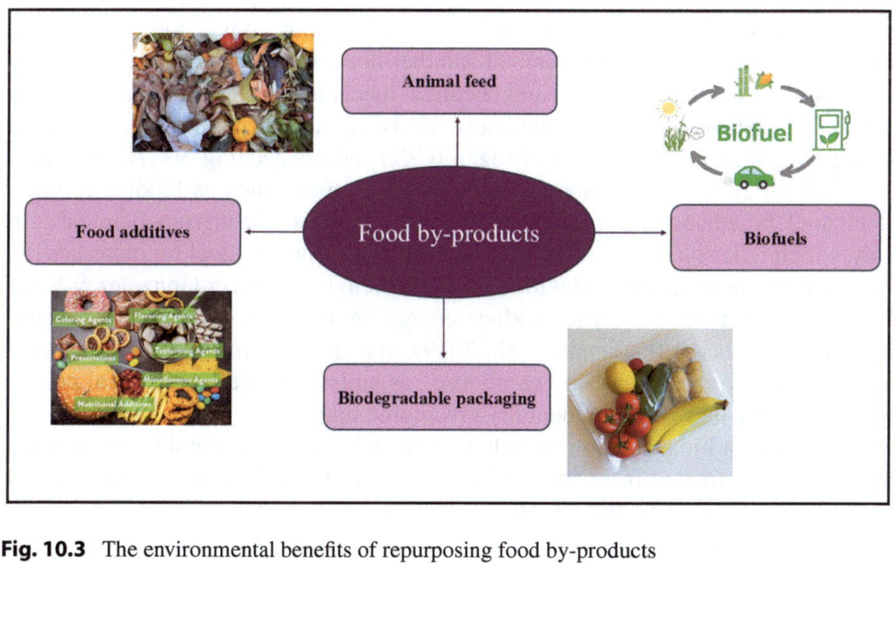

Fig. 10.3 The environmental benefits of repurposing food by-products

Another major benefit of food by-products is their role in the development of eco-friendly packaging alternatives (Dilucia et al., 2020). Residues from fruit and vegetable processing are rich in bioactive compounds, such as polyphenols and antioxidants, which can enhance the properties of biodegradable films and edible coatings (Betoret et al., 2023; Dilucia et al., 2020). These materials improve the flexibility and durability of packaging while offering antimicrobial and antioxidant benefits that extend the shelf life of perishable goods (Elhadef et al., 2023; Elhadef et al., 2021). For example, apple-processing by-products have been successfully integrated into renewable active packaging, incorporating bacterial cellulose-based nanomaterials coated with biodegradable polymers (Yadav et al., 2023). Such advancements not only help utilize food waste effectively but also address global concerns surrounding plastic pollution and synthetic additives in packaging (Dilucia et al., 2020). Apple pomace and citrus peels, which are naturally high in pectin and cellulose, are being repurposed into antioxidant-rich films that extend food freshness while reducing plastic waste (Chaari & Smaoui, 2024). Cutting-edge advancements include nano-enhanced films, where cellulose nanocrystals extracted from pineapple leaves boost tensile strength by 40%, as well as bio-based coatings like poly-hydroxy-alkanoate (PHA) produced from fermented lactic acid. These sustainable materials are aligned with the rapidly expanding global bioplastics market, currently valued at $300 billion and expected to grow by 15% annually by 2030 (Galanakis, 2024; Dilucia et al., 2020).

Beyond packaging and food applications, food by-products serve as a valuable resource for biofuel production and pharmaceutical applications (Osorio et al., 2021; Yukesh Kannah et al., 2020). Considering the vast number of agricultural by-products and their detrimental effects on human health, ecosystems, and the environment, transforming this biomass into biofuels offers a more sustainable solution. Utilizing agricultural residues for biofuel generation is a practical and eco-friendly approach to waste management that supports sanitation, conserves resources, and contributes to balancing carbon emissions. Two main conversion methods are employed in this context: biochemical and thermochemical processes. Biochemical conversion integrates biological and chemical mechanisms and is particularly well-suited for feedstocks with high moisture levels, significant cellulose and hemicellulose content, and a carbon-to-nitrogen (C/N) ratio exceeding 30 (Awogbemi & Von Kallon, 2022). This route typically yields biofuels, such as biodiesel, biogas (methane), bioethanol, and biobutanol. On the other hand, the thermochemical pathway, which relies on high-temperature chemical reactions, is ideal for materials rich in lignin, low in moisture, and with a C/N ratio below 30 (Ibarra-Gonzalez & Rong, 2019). This method is used to produce energy-dense outputs like bio-oil, syngas, biochar, and biocoal (Pattanaik et al., 2019). When evaluating both approaches in terms of energy efficiency and greenhouse gas emissions, the biochemical process tends to offer a more environmentally favourable option.

Additionally, bioactive compounds like phenolic acids, flavonoids, carotenoids, and vitamins extracted from fruit residues have been widely used in cosmetics and pharmaceuticals due to their antioxidant and therapeutic properties (Eliopoulos et al., 2022; Osorio et al., 2021). In addition, residues from the wine industry, such as grape marc, are being repurposed for bioethanol production, animal feed, and natural dyes within a single processing framework, decreasing landfill waste by 80% (Betoret et al., 2023; Osorio et al., 2021). Additionally, agricultural leftovers like maize stalks are being used in dual-purpose applications—cellulose extracts serve as sustainable packaging materials, while lignin is converted into biochar, fostering closed-loop agricultural systems (Sarker et al., 2024). This approach has the potential to recover an estimated 60 million tons of bioactive compounds annually from fruit and vegetable waste alone (Dilucia et al., 2020).

10.6 Consumer Acceptance and Market Trends

Historical consumers have looked down upon food by-products and have been inclined toward relating the foods with objectionable offal, rubbish, or poor quality (Ungureanu et al., 2023). This has arisen from post-industrial Western cultural values promoting cosmetic perfection and convenience over resourcefulness (Vrooman et al., 2024). Such stigmatization has constituted an overwhelming marketing challenge because such products have been linked with poverty, scarcity, or unclean places (Stuart & Sartorius, 2022).

The food industry has undergone change like no other in recent history (Béné, 2022). Companies today are leaders in sustainability and waste reduction (Zorpas et al., 2021). With consumers increasingly concerned with the environmental impact of their food, companies are compelled to think outside the box (Sandberg et al., 2023). Instead of letting what is left or leftovers go to waste or by-products, most companies have discovered creative ways of reusing these kinds of products (Ncube et al., 2023). What was once overlooked is now recognized as an innovation driver (Chaudhary et al., 2022). Food that might otherwise go to waste is now being reconsidered, prompting firms to develop innovative ways to avoid food loss without meeting the mounting demand for eco-friendly products. Changing consumer attitudes toward foods that utilize by-products is a time-consuming and educational process (Altintzoglou et al., 2021). As sustainably sourced food products become more prominent on shelves and menus, consumer resistance diminishes (McClements et al., 2021). Once considered a waste reduction effort, these efforts now change product development, marketing, and even how value and quality are defined in the food market. With time, efforts that initially seemed niche begin to become mainstream (Bauwens et al., 2024).

A prime example of this innovation is the upcycling of brewer's spent grains (BSG), the main by-product of beer production. In 2020 alone, global beer production generated approximately 36.4 million tons of wet BSG, representing about 85% of brewing waste (Kavalopoulos et al., 2021; Nyhan et al., 2023). Rather than being discarded or used solely as animal feed, BSG is increasingly recognized for its nutritional value, containing up to 30% protein and 40% dietary fiber, making it suitable for a range of human foods and beverages (Ikram et al., 2017; Lynch et al., 2016). Innovative companies are transforming BSG into high-protein, high-fiber flour for bread, snacks, and even plant-based burgers, as well as exploring its use in bioplastics and biofuels (Zhang et al., 2024). This shift not only reduces waste but also supports the development of sustainable, affordable, and nutritious food ingredients, addressing both environmental and health challenges (Herrero et al., 2023).

10.7 Conclusion

In conclusion, the strategic use of food by-products offers a transformative path toward a more sustainable and resilient food system. The repurposing of nutrient-rich residues from agricultural and food processing industries can reduce waste and enhance the nutritional profile of food products, contributing to food security worldwide. The integration of these innovations demands collaboration across sectors, including policymakers, food scientists, and industry leaders, to ensure scalable and impactful solutions. As we advance toward a future where sustainability is at the core of food production, embracing these approaches will play a pivotal role in addressing global hunger, resource conservation, and environmental protection. The journey ahead is one of innovation, responsibility, and collective action, where every by-product has the potential to create value for both humans and the planet.

References

Afif, K., Rebolledo, C., & Roy, J. (2022). Evaluating the effectiveness of the weight-based packaging tax on the reduction at source of product packaging: The case of food manufacturers and retailers. *International Journal of Production Economics, 245*(March), 108391. https://doi.org/10.1016/j.ijpe.2021.108391

Afraz, M. T., Xindong, X., Adil, M., Manzoor, M. F., Zeng, X.-A., Han, Z., & Aadil, R. M. (2023). Subcritical and supercritical fluids to valorize industrial fruit and vegetable waste. *Food, 12*(12), 2417. https://doi.org/10.3390/foods12122417

Aït-Kaddour, A., Hassoun, A., Tarchi, I., Loudiyi, M., Boukria, O., Cahyana, Y., Ozogul, F., & Khwaldia, K. (2024). Transforming plant-based waste and by-products into valuable products using various "food industry 4.0" enabling technologies: A literature review. *Science of the Total Environment, 955*(December), 176872. https://doi.org/10.1016/j.scitotenv.2024.176872

Altintzoglou, T., Honkanen, P., & Whitaker, R. D. (2021). Influence of the involvement in food waste reduction on attitudes towards sustainable products containing seafood by-products. *Journal of Cleaner Production, 285*(February), 125487. https://doi.org/10.1016/j.jclepro.2020.125487

Andreou, V., Dimopoulos, G., Dermesonlouoglou, E., & Taoukis, P. (2020). Application of pulsed electric fields to improve product yield and waste valorization in industrial tomato processing. *Journal of Food Engineering, 270*(April), 109778. https://doi.org/10.1016/j.jfoodeng.2019.109778

Ashokkumar, V., Flora, G., Radhakrishnan Venkatkarthick, K., SenthilKannan, C. K., Mary Stephy, G., Kamyab, H., Chen, W.-H., Thomas, J., & Ngamcharussrivichai, C. (2022). Advanced technologies on the sustainable approaches for conversion of organic waste to valuable bioproducts: Emerging circular bioeconomy perspective. *Fuel, 324*(September), 124313. https://doi.org/10.1016/j.fuel.2022.124313

Awogbemi, O., & Von Kallon, D. V. (2022). Valorization of agricultural wastes for biofuel applications. *Heliyon, 8*(10). https://doi.org/10.1016/j.heliyon.2022.e11117

Bagade, S. B., & Patil, M. (2021). Recent advances in microwave assisted extraction of bioactive compounds from complex herbal samples: A review. *Critical Reviews in Analytical Chemistry, 51*(2), 138–149. https://doi.org/10.1080/10408347.2019.1686966

Balwan, W. K., & Saba, N. (2021). Study of role of fish oil in human health. *Global Academic Journal of Medical Sciences, 3*(1), 14–18.

Bao, Y., Reddivari, L., & Huang, J.-Y. (2020). Development of cold plasma pretreatment for improving Phenolics extractability from tomato pomace. *Innovative Food Science & Emerging Technologies, 65*(October), 102445. https://doi.org/10.1016/j.ifset.2020.102445

Bauwens, T., Hartley, K., Hekkert, M., & Kirchherr, J. (2024). Building innovation ecosystems for circularity: Start-up business models in the food and construction sectors in The Netherlands. *Journal of Cleaner Production, 481*(November), 143970. https://doi.org/10.1016/j.jclepro.2024.143970

Bayil Oğuzkan, S. (2019). Extraction of Capsinoid and its analogs from pepper waste of different genotypes. *Natural Product Communications, 14*(7), 1934578X19865673. https://doi.org/10.1177/1934578X19865673

Béné, C. (2022). Why the great food transformation may not happen – A deep-dive into our food systems' political economy, controversies and politics of evidence. *World Development, 154*(June), 105881. https://doi.org/10.1016/j.worlddev.2022.105881

Betoret, N., Betoret, E., & Glicerina, V. T. (2023). Valorization and utilization of food wastes and by-products: Recent trends, innovative technologies and sustainability challenges. *Food, 13*(1), 9. https://doi.org/10.3390/foods13010009

Bitwell, C., Indra, S. S., Luke, C., & Kakoma, M. K. (2023). A review of modern and conventional extraction techniques and their applications for extracting phytochemicals from plants. *Scientific African, 19*(March), e01585. https://doi.org/10.1016/j.sciaf.2023.e01585

Brennan, C. S. (2024). Regenerative food innovation: The role of agro-food chain by-products and plant origin food to obtain high-value-added foods. *Food, 13*(3), 427. https://doi.org/10.3390/foods13030427

Buchanan, D., Martindale, W., Romeih, E., & Hebishy, E. (2023). Recent advances in whey processing and valorisation: Technological and environmental perspectives. *International Journal of Dairy Technology, 76*(2), 291–312. https://doi.org/10.1111/1471-0307.12935

Carpentieri, S., Soltanipour, F., Ferrari, G., Pataro, G., & Donsì, F. (2021). Emerging green techniques for the extraction of antioxidants from agri-food by-products as promising ingredients for the food industry. *Antioxidants, 10*(9), 1417. https://doi.org/10.3390/antiox10091417

Castro-Muñoz, R., Díaz-Montes, E., Gontarek-Castro, E., Boczkaj, G., & Galanakis, C. M. (2022). A comprehensive review on current and emerging technologies toward the valorization of bio-based wastes and by products from foods. *Comprehensive Reviews in Food Science and Food Safety, 21*(1), 46–105. https://doi.org/10.1111/1541-4337.12894

Chaari, M., & Smaoui, S. (2024). Physicochemical properties of citrus fruits: Analytical approach for physicochemical parameters of citrus fruit and juice. In A. K. Gupta, J. Kour, & P. Mishra (Eds.), *Citrus fruits and juice: Processing and quality profiling* (pp. 69–87). Springer Nature. https://doi.org/10.1007/978-981-99-8699-6_4

Chaudhary, S., Kaur, P., Talwar, S., Islam, N., & Dhir, A. (2022). Way off the mark? Open innovation failures: Decoding what really matters to chart the future course of action. *Journal of Business Research, 142*(March), 1010–1025. https://doi.org/10.1016/j.jbusres.2021.12.062

Chemat, F., Vian, M. A., Fabiano-Tixier, A.-S., Nutrizio, M., Jambrak, A. R., Munekata, P. E. S., Lorenzo, J. M., Barba, F. J., Binello, A., & Cravotto, G. (2020). A review of sustainable and intensified techniques for extraction of food and natural products. *Green Chemistry, 22*(8), 2325–2353. https://doi.org/10.1039/C9GC03878G

Costa, J. M., Ampese, L. C., Ziero, H. D. D., Sganzerla, W. G., & Forster-Carneiro, T. (2022). Apple pomace biorefinery: Integrated approaches for the production of bioenergy, biochemicals, and value-added products – An updated review. *Journal of Environmental Chemical Engineering, 10*(5), 108358. https://doi.org/10.1016/j.jece.2022.108358

De Barros, A. N., & Gouvinhas, I. (Eds.). (2021). *Innovation in the food sector through the valorization of food and agro-food by-products*. BOD–Books on Demand.

Deng, Y., Ting, J., & Xi, J. (2018). Circulating polyphenols extraction system with high-voltage electrical discharge: Design and performance evaluation. *ACS Sustainable Chemistry & Engineering, 6*(11), 15402–15410. https://doi.org/10.1021/acssuschemeng.8b03827

Dilucia, F., Lacivita, V., Conte, A., & Del Nobile, M. A. (2020). Sustainable use of fruit and vegetable by-products to enhance food packaging performance. *Food, 9*(7), 857. https://doi.org/10.3390/foods9070857

Echegaray, N., Munekata, P. E. S., Gullón, P., Dzuvor, C. K. O., Gullón, B., Kubi, F., & Lorenzo, J. M. (2022). Recent advances in food products fortification with anthocyanins. *Critical Reviews in Food Science and Nutrition, 62*, 1553.

Elhadef, K., Akermi, S., Hlima, H. B., Ennouri, K., Fourati, M., Braïek, O. B., Mellouli, L., & Smaoui, S. (2021). Tunisian pistachio Hull extracts: Phytochemical content, antioxidant activity, and foodborne pathogen inhibition. *Journal of Food Quality, 2021*(1), 9953545. https://doi.org/10.1155/2021/9953545

Elhadef, K., Chaari, M., Akermi, S., Nirmal, N. P., Khaneghah, A. M., Abdelkafi, S., Michaud, P., Ali, D. S., Mellouli, L., & Smaoui, S. (2023). Production of functional raw chicken meat by incorporation of date palm seed extract: An assessment of microbiological, chemical and sensory properties. *Journal of Food Measurement and Characterization, 17*(5), 5117–5133. https://doi.org/10.1007/s11694-023-02017-0

Eliopoulos, C., Markou, G., Langousi, I., & Arapoglou, D. (2022). Reintegration of food industry by-products: Potential applications. *Food, 11*(22), 3743. https://doi.org/10.3390/foods11223743

Fróna, D., Szenderák, J., & Harangi-Rákos, M. (2019). The challenge of feeding the world. *Sustainability, 11*(20), 5816. https://doi.org/10.3390/su11205816

Galanakis, C. M. (2024). The future of food. *Food, 13*(4), 506. https://doi.org/10.3390/foods13040506

Ghosh, S., Sarkar, T., Pati, S., Kari, Z. A., Edinur, H. A., & Chakraborty, R. (2022). Novel bioactive compounds from marine sources as a tool for functional food development. *Frontiers in Marine Science, 9*(February). https://doi.org/10.3389/fmars.2022.832957

Guenther, M., Driver, T., & Saunders, C. (2022). *The overall benefits of food safety–a literature review. AERU client report prepared for the New Zealand food safety science and research Centre (NZFSSRC)*. AERU: Lincoln University.

Guo, M., & Wang, C. (2019). Chemistry of whey proteins. In *Whey protein production, chemistry, functionality, and applications* (pp. 39–65). Wiley. https://doi.org/10.1002/9781119256052.ch3

Hadi, J., & Brightwell, G. (2021). Safety of alternative proteins: Technological, environmental and regulatory aspects of cultured meat, plant-based meat, insect protein and single-cell protein. *Food, 10*(6), 1226. https://doi.org/10.3390/foods10061226

Hadidi, M., Aghababaei, F., & McClements, D. J. (2024). Sunflower meal/cake as a sustainable protein source for global food demand: Towards a zero-hunger world. *Food Hydrocolloids, 147*(February), 109329. https://doi.org/10.1016/j.foodhyd.2023.109329

Haque, F., Fan, C., & Lee, Y.-Y. (2023). From waste to value: Addressing the relevance of waste recovery to agricultural sector in line with circular economy. *Journal of Cleaner Production, 415*(August), 137873. https://doi.org/10.1016/j.jclepro.2023.137873

Helkar, P. B., Sahoo, A. K., & Patil, N. J. (2016). Review: Food industry by-products used as a functional food ingredients. *International Journal of Waste Resources, 6*(3), 1–6.

Herrero, M., Hugas, M., Lele, U., Wirakartakusumah, A., & Torero, M. (2023). A shift to healthy and sustainable consumption patterns. In *Science and innovations for food systems transformation* (p. 59).

Himashree, P., Sengar, A. S., & Sunil, C. K. (2022). Food thickening agents: Sources, chemistry, properties and applications – A review. *International Journal of Gastronomy and Food Science, 27*(March), 100468. https://doi.org/10.1016/j.ijgfs.2022.100468

Huang, S., Benchamas, G., & Huang, G. (2020). Whole processing and use of Rice Polishings. *Innovative Food Science & Emerging Technologies, 63*(July), 102373. https://doi.org/10.1016/j.ifset.2020.102373

Ibarra-Gonzalez, P., & Rong, B.-G. (2019). A review of the current state of biofuels production from lignocellulosic biomass using thermochemical conversion routes. *Chinese Journal of Chemical Engineering, 27*(7), 1523–1535. https://doi.org/10.1016/j.cjche.2018.09.018

Ikram, S., Huang, L. Y., Zhang, H., Wang, J., & Yin, M. (2017). Composition and nutrient value proposition of brewers spent grain. *Journal of Food Science, 82*(10), 2232–2242. https://doi.org/10.1111/1750-3841.13794

Jurić, S., Ferrari, G., Velikov, K. P., & Donsì, F. (2019). High-pressure homogenization treatment to recover bioactive compounds from tomato peels. *Journal of Food Engineering, 262*(December), 170–180. https://doi.org/10.1016/j.jfoodeng.2019.06.011

Kantar, E., Sally, H. N., Rajha, N. B., Vorobiev, E., Maroun, R. G., & Louka, N. (2019). Green extraction of polyphenols from grapefruit peels using high voltage electrical discharges, deep eutectic solvents and aqueous glycerol. *Food Chemistry, 295*(October), 165–171. https://doi.org/10.1016/j.foodchem.2019.05.111

Kavalopoulos, M., Stoumpou, V., Christofi, A., Mai, S., Barampouti, E. M., Moustakas, K., Malamis, D., & Loizidou, M. (2021). Sustainable valorisation pathways mitigating environmental pollution from brewers' spent grains. *Environmental Pollution, 270*(February), 116069. https://doi.org/10.1016/j.envpol.2020.116069

Kehili, M., Sayadi, S., Frikha, F., Zammel, A., & Allouche, N. (2019). Optimization of lycopene extraction from tomato peels industrial by-product using maceration in refined olive oil. *Food and Bioproducts Processing, 117*(September), 321–328. https://doi.org/10.1016/j.fbp.2019.08.004

Khan, N., Ray, R. L., Kassem, H. S., Hussain, S., Zhang, S., Khayyam, M., Ihtisham, M., & Asongu, S. A. (2021). Potential role of technology innovation in transformation of sustainable food systems: A review. *Agriculture, 11*(10), 984. https://doi.org/10.3390/agriculture11100984

Kheto, A., Adhikary, U., Dhua, S., Sarkar, A., Kumar, Y., Vashishth, R., Shrestha, B. B., & Saxena, D. C. (2025). A review on advancements in emerging processing of whey protein: Enhancing functional and nutritional properties for functional food applications. *Food Safety and Health, 3*(1), 23–45. https://doi.org/10.1002/fsh3.12067

Kumar, K., Srivastav, S., & Sharanagat, V. S. (2021). Ultrasound assisted extraction (UAE) of bioactive compounds from fruit and vegetable processing by-products: A review. *Ultrasonics Sonochemistry, 70*(January), 105325. https://doi.org/10.1016/j.ultsonch.2020.105325

Kumar, V., Sharma, N., Umesh, M., Selvaraj, M., Al-Shehri, B. M., Chakraborty, P., Duhan, L., et al. (2022). RETRACTED: Emerging challenges for the agro-industrial food waste utilization: A review on food waste biorefinery. *Bioresource Technology, 362*(October), 127790. https://doi.org/10.1016/j.biortech.2022.127790

Kumar, A., Nirmal, P., Kumar, M., Jose, A., Tomer, V., Oz, E., Proestos, C., et al. (2023a). Major phytochemicals: Recent advances in health benefits and extraction method. *Molecules, 28*(2), 887. https://doi.org/10.3390/molecules28020887

Kumar, P., Mehta, N., Abubakar, A. A., Verma, A. K., Kaka, U., Sharma, N., Sazili, A. Q., Pateiro, M., Kumar, M., & Lorenzo, J. M. (2023b). Potential alternatives of animal proteins for sustainability in the food sector. *Food Reviews International, 39*(8), 5703–5728.

Kumari, P., Gaur, S. S., & Tiwari, R. K. (2023). Banana and its by-products: A comprehensive review on its nutritional composition and pharmacological benefits. *eFood, 4*(5), e110. https://doi.org/10.1002/efd2.110

Lautenschlaeger, R., & Upmann, M. (2017). How meat is defined in the European Union and in Germany. *Animal Frontiers, 7*(4), 57–59. https://doi.org/10.2527/af.2017.0446

Leder, N., Kumar, M., & Rodrigues, V. S. (2020). Influential factors for value creation within the circular economy: Framework for waste valorisation. *Resources, Conservation and Recycling, 158*(July), 104804. https://doi.org/10.1016/j.resconrec.2020.104804

Li, Z., Fan, Y., & Xi, J. (2019). Recent advances in high voltage electric discharge extraction of bioactive ingredients from plant materials. *Food Chemistry, 277*(March), 246–260. https://doi.org/10.1016/j.foodchem.2018.10.119

Li, J., Li, Z., Ma, Q., & Zhou, Y. (2023). Enhancement of anthocyanins extraction from Haskap by cold plasma pretreatment. *Innovative Food Science & Emerging Technologies, 84*(March), 103294. https://doi.org/10.1016/j.ifset.2023.103294

Lin, Y., Pi, J., Jin, P., Liu, Y., Mai, X., Li, P., & Fan, H. (2022). Enzyme and microwave co-assisted extraction, structural characterization and antioxidant activity of polysaccharides from *purple-heart radish. Food Chemistry, 372*(March), 131274. https://doi.org/10.1016/j.foodchem.2021.131274

Liu, Z., de Souza, T. S. P., Holland, B., Dunshea, F., Barrow, C., & Suleria, H. A. R. (2023). Valorization of food waste to produce value-added products based on its bioactive compounds. *PRO, 11*(3), 840. https://doi.org/10.3390/pr11030840

Lončarić, A., Celeiro, M., Jozinović, A., Jelinić, J., Kovač, T., Jokić, S., Babić, J., Moslavac, T., Zavadlav, S., & Lores, M. (2020). Green extraction methods for extraction of polyphenolic compounds from blueberry pomace. *Food, 9*(11), 1521. https://doi.org/10.3390/foods9111521

Lynch, K. M., Steffen, E. J., & Arendt, E. K. (2016). Brewers' spent grain: A review with an emphasis on food and health. *Journal of the Institute of Brewing, 122*(4), 553–568. https://doi.org/10.1002/jib.363

Machado, A. P., Fonseca, D., Pasquel-Reátegui, J. L., Barbero, G. F., & Martínez, J. (2015). Pressurized liquid extraction of bioactive compounds from blackberry (*Rubus Fruticosus* L.) residues: A comparison with conventional methods. *Food Research International, 77*(November), 675–683. https://doi.org/10.1016/j.foodres.2014.12.042

Manzoor, A., Pandey, V. K., Dar, A. H., Fayaz, U., Dash, K. K., Shams, R., Ahmad, S., et al. (2023). Rice bran: Nutritional, phytochemical, and pharmacological profile and its contribution to human health promotion. *Food Chemistry Advances, 2*(October), 100296. https://doi.org/10.1016/j.focha.2023.100296

Marinelli, V., Padalino, L., Conte, A., Del Nobile, M. A., & Briviba, K. (2018). Red grape Marc flour as food ingredient in durum wheat spaghetti: Nutritional evaluation and bioaccessibility of bioactive compounds. *Food Science and Technology Research, 24*(6), 1093–1100. https://doi.org/10.3136/fstr.24.1093

Marinelli, V., Spinelli, S., Angiolillo, L., Del Nobile, M. A., & Conte, A. (2020). Emerging techniques applied to by-products for food fortification. *Journal of Food Science and Technology, 57*(3), 905–914. https://doi.org/10.1007/s13197-019-04123-8

Marra, F., Petrovicova, B., Canino, F., Maffia, A., Mallamaci, C., & Muscolo, A. (2022). Pomegranate wastes are rich in bioactive compounds with potential benefit on human health. *Molecules, 27*(17), 5555. https://doi.org/10.3390/molecules27175555

Marsiglia, W. I., de Lima, M., de Sousa, L., Oliveira, C., Santiago, Â. M., Araújo, V. S., Amorim, B. C., de Melo, R., Almeida, L. J., & Santos, N. C. (2021). Evaluation of bioactive compounds in an infused drink prepared from the powder of Jaboticaba (Myrciaria Cauliflora) peels. *Journal of Food Processing and Preservation, 45*(9), e15731. https://doi.org/10.1111/jfpp.15731

Martins, R., Barbosa, A., Advinha, B., Sales, H., Pontes, R., & Nunes, J. (2023). Green extraction techniques of bioactive compounds: A state-of-the-art review. *PRO, 11*(8), 2255. https://doi.org/10.3390/pr11082255

Mateos, R., Sarria, B., & Bravo, L. (2020). Nutritional and other health properties of olive pomace oil. *Critical Reviews in Food Science and Nutrition, 60*(20), 3506–3521.

McCauley, K., & Barlow, K. (2023). Regenerative agriculture: Increasing plant diversity and soil carbon sequestration on agricultural landscapes. *SURG Journal, 15*(1). https://doi.org/10.21083/surg.v15i1.7196

McClements, D. J., Barrangou, R., Hill, C., Kokini, J. L., Lila, M. A., Meyer, A. S., & Liangli, Y. (2021). Building a resilient, sustainable, and healthier food supply through innovation and technology. *Annual Review of Food Science and Technology, 12*(1), 1–28. https://doi.org/10.1146/annurev-food-092220-030824

Mešić, A., Jurić, M., Donsì, F., Bandić, L. M., & Jurić, S. (2024). Advancing climate resilience: Technological innovations in plant-based, alternative and sustainable food production systems. *Discover Sustainability, 5*(1), 423. https://doi.org/10.1007/s43621-024-00581-z

Michel, M., Eldridge, A. L., Hartmann, C., Klassen, P., Ingram, J., & Meijer, G. W. (2024). Benefits and challenges of food processing in the context of food systems, value chains and sustainable development goals. *Trends in Food Science & Technology, 153*(November), 104703. https://doi.org/10.1016/j.tifs.2024.104703

Mihalcea, L., et al. (2021). Supercritical CO2 extraction and microencapsulation of lycopene-enriched oleoresins from tomato peels: Evidence on Antiproliferative and Cytocompatibility activities. *Antioxidants, 10*(2), 222. https://doi.org/10.3390/antiox10020222

Minervini, F., Comitini, F., De Boni, A., Fiorino, G. M., Rodrigues, F., Tlais, A. Z. A., Carafa, I., & De Angelis, M. (2022). Sustainable and health-protecting food ingredients from bio-processed food by-products and wastes. *Sustainability, 14*(22), 15283. https://doi.org/10.3390/su142215283

Minj, S., & Anand, S. (2020). Whey proteins and its derivatives: Bioactivity, functionality, and current applications. *Dairy, 1*(3), 233–258. https://doi.org/10.3390/dairy1030016

Misra, N. N., Pankaj, S. K., Segat, A., & Ishikawa, K. (2016). Cold plasma interactions with enzymes in foods and model systems. *Trends in Food Science & Technology, 55*(September), 39–47. https://doi.org/10.1016/j.tifs.2016.07.001

Mohanty, B. (2020). Contribution of fish in human nutrition.

Murakami, H., Omiya, M., Miki, Y., Umemura, T., Esaka, Y., Inoue, Y., & Teshima, N. (2020). Evaluation of the adsorption properties of nucleobase-modified sorbents for a solid-phase extraction of water-soluble compounds. *Talanta, 217*(September), 121052. https://doi.org/10.1016/j.talanta.2020.121052

Muscolo, A., Mariateresa, O., Giulio, T., & Mariateresa, R. (2024). Oxidative stress: The role of antioxidant phytochemicals in the prevention and treatment of diseases. *International Journal of Molecular Sciences, 25*(6), 3264. https://doi.org/10.3390/ijms25063264

Ncube, A., Mtetwa, S., Bukhari, M., Fiorentino, G., & Passaro, R. (2023). Circular economy and green chemistry: The need for radical innovative approaches in the design for new products. *Energies, 16*(4), 1752. https://doi.org/10.3390/en16041752

Nyhan, L., Sahin, A. W., Schmitz, H. H., Siegel, J. B., & Arendt, E. K. (2023). Brewers' spent grain: An unprecedented opportunity to develop sustainable plant-based nutrition ingredients addressing global malnutrition challenges. *Journal of Agricultural and Food Chemistry, 71*(28), 10543–10564. https://doi.org/10.1021/acs.jafc.3c02489

Ofori, S. A., Cobbina, S. J., & Obiri, S. (2021). Climate change, land, water, and food security: Perspectives from sub-Saharan Africa. *Frontiers in Sustainable Food Systems, 5*(July). https://doi.org/10.3389/fsufs.2021.680924

Onyeaka, H., Tamasiga, P., Nwauzoma, U. M., Miri, T., Juliet, U. C., Nwaiwu, O., & Akinsemolu, A. A. (2023). Using artificial intelligence to tackle food waste and enhance the circular economy: Maximising resource efficiency and minimising environmental impact: A review. *Sustainability, 15*(13), 10482. https://doi.org/10.3390/su151310482

Osorio, L. L. D. R., Flórez-López, E., & Grande-Tovar, C. D. (2021). The potential of selected agri-food loss and waste to contribute to a circular economy: Applications in the food, cosmetic and pharmaceutical industries. *Molecules, 26*(2), 515.

Ospina-Maldonado, S., Martin-Gómez, H., & Cardoso-Ugarte, G. A. (2024). From waste to wellness: A review on the harness of food industry by-products for sustainable functional food production. *International Journal of Food Science and Technology, 59*(11), 8680–8692. https://doi.org/10.1111/ijfs.17571

Pal, P., Singh, A. K., Srivastava, R. K., Rathore, S. S., Sahoo, U. K., Subudhi, S., Sarangi, P. K., & Prus, P. (2024). Circular bioeconomy in action: Transforming food wastes into renewable food resources. *Food, 13*(18), 3007. https://doi.org/10.3390/foods13183007

Pattanaik, L., Pattnaik, F., Saxena, D. K., & Naik, S. N. (2019). Chapter 5 – Biofuels from agricultural wastes. In A. Basile & F. Dalena (Eds.), *Second and third generation of feedstocks* (pp. 103–142). Elsevier. https://doi.org/10.1016/B978-0-12-815162-4.00005-7

Pattnaik, M., Pandey, P., Martin, G. J. O., Mishra, H. N., & Ashokkumar, M. (2021). Innovative technologies for extraction and microencapsulation of bioactives from plant-based food waste and their applications in functional food development. *Food, 10*(2), 279. https://doi.org/10.3390/foods10020279

Pimentel-Moral, S., de la Luz Cádiz-Gurrea, M., Rodríguez-Pérez, C., & Segura-Carretero, A. (2020). Chapter 7 – Recent advances in extraction technologies of phytochemicals applied for the revaluation of agri-food by-products. In B. Prakash (Ed.), *Functional and preservative properties of phytochemicals* (pp. 209–239). Academic Press. https://doi.org/10.1016/B978-0-12-818593-3.00007-5

Rana, S., Gupta, S., Rana, A., & Bhushan, S. (2015). Functional properties, phenolic constituents and antioxidant potential of industrial apple pomace for utilization as active food ingredient. *Food Science and Human Wellness, 4*(4), 180–187. https://doi.org/10.1016/j.fshw.2015.10.001

Rani, A., Arfat, Y., Aziz, R. S., Ali, L., Ahmed, H., Asim, S., Rashid, M., & Hocart, C. H. (2021). Enzymatically assisted extraction of antioxidant and anti-mutagenic compounds from radish (*Raphanus Sativus). Environmental Technology & Innovation, 23*(August), 101620. https://doi.org/10.1016/j.eti.2021.101620

Rațu, R. N., Veleșcu, I. D., Stoica, F., Usturoi, A., Arsenoaia, V. N., Crivei, I. C., Postolache, A. N., et al. (2023). Application of agri-food by-products in the food industry. *Agriculture, 13*(8), 1559. https://doi.org/10.3390/agriculture13081559

Rifna, E. J., Misra, N. N., & Dwivedi, M. (2023). Recent advances in extraction technologies for recovery of bioactive compounds derived from fruit and vegetable waste peels: A review. *Critical Reviews in Food Science and Nutrition, 63*(6), 719–752.

Rivas, M. Á., Benito, M. J., Martín, A., de Guía, M., Córdoba, S. R.-M., & Casquete, R. (2022). Improve the functional properties of dietary fibre isolated from broccoli by-products by using different technologies. *Innovative Food Science & Emerging Technologies, 80*(August), 103075. https://doi.org/10.1016/j.ifset.2022.103075

Romero, J., Cruz, R. M. S., Díez-Méndez, A., & Albertos, I. (2022). Valorization of berries' agro-industrial waste in the development of biodegradable pectin-based films for fresh Salmon (Salmo Salar) shelf-life monitoring. *International Journal of Molecular Sciences, 23*(16), 8970. https://doi.org/10.3390/ijms23168970

Rosenthal, A., et al. (2021). Healthy food innovation in sustainable food system 4.0: Integration of entrepreneurship, research, and education. *Current Opinion in Food Science, 42*(December), 215–223. https://doi.org/10.1016/j.cofs.2021.07.002

Sah, B. N. P., Vasiljevic, T., McKechnie, S., & Donkor, O. N. (2016). Physicochemical, textural and rheological properties of probiotic yogurt fortified with fibre-rich pineapple Peel powder during refrigerated storage. *LWT, 65*(January), 978–986. https://doi.org/10.1016/j.lwt.2015.09.027

Sandberg, H., Alnoor, A., & Tiberius, V. (2023). Environmental, social, and governance ratings and financial performance: Evidence from the European food industry. *Business Strategy and the Environment, 32*(4), 2471–2489. https://doi.org/10.1002/bse.3259

Sarangi, P. K., Pal, P., Singh, A. K., Sahoo, U. K., & Prus, P. (2024). Food waste to food security: Transition from bioresources to sustainability. *Resources, 13*(12), 164. https://doi.org/10.3390/resources13120164

Sarfarazi, M., Jafari, S. M., Rajabzadeh, G., & Galanakis, C. M. (2020). Evaluation of microwave-assisted extraction Technology for Separation of bioactive components of saffron (*Crocus Sativus* L.). *Industrial Crops and Products, 145*(March), 111978. https://doi.org/10.1016/j.indcrop.2019.111978

Sarker, A., Raju Ahmmed, S. M., Ahsan, J. R., Ghosh, M. K., & Nandi, R. (2024). A comprehensive review of food waste valorization for the sustainable management of global food waste. *Sustainable Food Technology, 2*(1), 48–69. https://doi.org/10.1039/D3FB00156C

Selvamuthukumaran, M. (2024). *Wealth out of food processing waste: Ingredient recovery and valorization*. CRC Press.

Shahidi, F., Varatharajan, V., Peng, H., & Senadheera, R. (2019). Utilization of marine by-products for the recovery of value-added products. *Journal of Food Bioactives, 6*(June). https://doi.org/10.31665/JFB.2019.6184

Sharma, L., & Yadav, A. (2023). Role of functional foods in human health and disease prevention. In M. Thakur & T. Belwal (Eds.), *Bioactive components: A sustainable system for good health and well-being* (pp. 225–243). Springer Nature. https://doi.org/10.1007/978-981-19-2366-1_14

Sher, A., et al. (2024). Importance of regenerative agriculture: Climate, soil health, biodiversity and its socioecological impact. *Discover Sustainability, 5*(1), 462. https://doi.org/10.1007/s43621-024-00662-z

Singh, K., Kumar, T., Prince, V. K., Sharma, S., & Rani, J. (2019). A review on conversion of food wastes and by-products into value added products. *International Journal of Chemical Studies, 7*(2), 2068–2073.

Skenderidis, P., Mitsagga, C., Giavasis, I., Petrotos, K., Lampakis, D., Leontopoulos, S., Hadjichristodoulou, C., & Tsakalof, A. (2019). The in vitro antimicrobial activity assessment of ultrasound assisted Lycium Barbarum fruit extracts and pomegranate fruit peels. *Journal of Food Measurement and Characterization, 13*(3), 2017–2031. https://doi.org/10.1007/s11694-019-00123-6

Sohail, M., Rakha, A., Butt, M. S., Iqbal, M. J., & Rashid, S. (2017). Rice bran nutraceutics: A comprehensive review. *Critical Reviews in Food Science and Nutrition, 57*(17), 3771–3780.

Šojić, B., Tomović, V., Kocić-Tanackov, S., Kovačević, D. B., Putnik, P., Mrkonjić, Ž., Đurović, S., et al. (2020). Supercritical extracts of wild thyme (*thymus Serpyllum* L.) by-product as natural antioxidants in ground pork patties. *LWT, 130*(August), 109661. https://doi.org/10.1016/j.lwt.2020.109661

Sorrenti, V., Burò, I., Consoli, V., & Vanella, L. (2023). Recent advances in health benefits of bioactive compounds from food wastes and by-products: Biochemical aspects. *International Journal of Molecular Sciences, 24*(3), 2019. https://doi.org/10.3390/ijms24032019

Srivastava, N., Singh, A., Kumari, P., et al. (2021a). Chapter 21 – Advances in extraction technologies: Isolation and purification of bioactive compounds from biological materials. In R. P. Sinha & D.-P. Häder (Eds.), *Natural bioactive compounds* (pp. 409–433). Academic Press. https://doi.org/10.1016/B978-0-12-820655-3.00021-5

Srivastava, P., Bej, S., Yordanova, K., & Wolkenhauer, O. (2021b). Self-attention-based models for the extraction of molecular interactions from biological texts. *Biomolecules, 11*(11), 1591. https://doi.org/10.3390/biom11111591

Srivastava, Y., Semwal, A. D., & Dhiman, A. (2021c). A comprehensive review on processing, therapeutic benefits, challenges, and economic scenario of unconventional oils. *Journal of Food Processing and Preservation, 45*(2), e15152. https://doi.org/10.1111/jfpp.15152

Stillitano, T., Spada, E., Iofrida, N., Falcone, G., & De Luca, A. I. (2021). Sustainable agri-food processes and circular economy pathways in a life cycle perspective: State of the art of applicative research. *Sustainability, 13*(5), 2472. https://doi.org/10.3390/su13052472

Stuart, H., & Sartorius, N. (2022). *Paradigms lost, paradigms found: Lessons learned in the fight against the stigma of mental illness*. Oxford University Press.

Tan, B. L., Norhaizan, M. E., & Chan, L. C. (2023). Rice bran: From waste to nutritious food ingredients. *Nutrients, 15*(11), 2503. https://doi.org/10.3390/nu15112503

Tarone, A. G., Silva, E. K., Barros, H. D. D. F. Q., Cazarin, C. B. B., & Junior, M. R. M. (2021). High-intensity ultrasound-assisted recovery of anthocyanins from jabuticaba by-products using green solvents: Effects of ultrasound intensity and solvent composition on the extraction of phenolic compounds. *Food Research International, 140*, 110048.

Thakur, P., George, N., & Chakraborty, M. (2020). Qualitative phytochemical screening of sweet orange (citrus sinensis) peel extract-a preliminary study. *Plant Cell Biotechnology and Molecular Biology*, 29–34.

Turrini, F., Malaspina, P., Giordani, P., Catena, S., Zunin, P., & Boggia, R. (2020). Traditional decoction and PUAE aqueous extracts of pomegranate peels as potential low-cost anti-Tyrosinase ingredients. *Applied Sciences, 10*(8), 2795. https://doi.org/10.3390/app10082795

Ungureanu, N., et al. (2023). Management of by-products and waste from poultry meat industry. Annals of the Faculty of Engineering Hunedoara-International. *Journal of Engineering, 1*.

Usman, I., Hussain, M., Imran, A., Afzaal, M., Saeed, F., Javed, M., Afzal, A., Ashfaq, I., Al Jbawi, E., & Saewan, S. A. (2022). Traditional and innovative approaches for the extraction of bioactive compounds. *International Journal of Food Properties, 25*(1), 1215–1233.

Uwineza, P. A., & Waśkiewicz, A. (2020). Recent advances in supercritical fluid extraction of natural bioactive compounds from natural plant materials. *Molecules, 25*(17), 3847. https://doi.org/10.3390/molecules25173847

Vasileiadou, A. (2024). From organic wastes to bioenergy, biofuels, and value-added products for urban sustainability and circular economy: A review. *Urban Science, 8*(3), 121. https://doi.org/10.3390/urbansci8030121

Vrooman, J. C., Boelhouwer, J., & Gijsberts, M. (2024). A contemporary class structure: Capital disparities in The Netherlands. *PLoS One, 19*(1), e0296443. https://doi.org/10.1371/journal.pone.0296443

Wibisono, D. A., Setiawan, C.-Y. S., Tsung-Yen, W., & Chau, C.-F. (2025). Advancing industrial food byproduct management: Strategies, technologies, and progress in waste reduction. *PRO, 13*(1), 84. https://doi.org/10.3390/pr13010084

Wijerathna-Yapa, A., & Pathirana, R. (2022). Sustainable agro-food systems for addressing climate change and food security. *Agriculture, 12*(10), 1554. https://doi.org/10.3390/agriculture12101554

Xi, J., Li, Z., & Fan, Y. (2021). Recent advances in continuous extraction of bioactive ingredients from food-processing wastes by pulsed electric fields. *Critical Reviews in Food Science and Nutrition, 61*(10), 1738–1750.

Xie, F., Zhang, W., Lan, X., Gong, S., Jinhong, W., & Wang, Z. (2018). Effects of high hydrostatic pressure and high pressure homogenization processing on characteristics of potato Peel waste pectin. *Carbohydrate Polymers, 196*(September), 474–482. https://doi.org/10.1016/j.carbpol.2018.05.061

Xu, L., & Geelen, D. (2018). Developing biostimulants from agro-food and industrial by-products. *Frontiers in Plant Science, 9*(October). https://doi.org/10.3389/fpls.2018.01567

Yadav, A., Kumar, N., Upadhyay, A., & Pratibha, & Anurag, R. K. (2023). Edible packaging from fruit processing waste: A comprehensive review. *Food Reviews International, 39*(4), 2075–2106.

Yang, K., Qing, Y., Qifang, Y., Tang, X., Chen, G., Fang, R., & Liu, H. (2021). By-product feeds: Current understanding and future perspectives. *Agriculture, 11*(3), 207. https://doi.org/10.3390/agriculture11030207

Yong, S. X. M., Song, C. P., & Choo, W. S. (2021). Impact of high-pressure homogenization on the extractability and stability of phytochemicals. *Frontiers in Sustainable Food Systems, 4*, 593259.

Yukesh Kannah, R., Merrylin, J., Poornima Devi, T., Kavitha, S., Sivashanmugam, P., Kumar, G., & Rajesh Banu, J. (2020). Food waste valorization: Biofuels and value added product recovery. *Bioresource Technology Reports, 11*(September), 100524. https://doi.org/10.1016/j.biteb.2020.100524

Zhang, J., et al. (2024). Development of functional food ingredients from byproducts of the brewing industry. https://doi.org/10.13140/RG.2.2.25566.96322.

Zorpas, A. A., Navarro-Pedreño, J., Jeguirim, M., Dimitriou, G., Candel, M. B. A., Argirusis, C., Vardopoulos, I., Loizia, P., Chatziparaskeva, G., & Papamichael, I. (2021). Crisis in leadership vs waste management. *Euro-Mediterranean Journal for Environmental Integration, 6*(3), 80. https://doi.org/10.1007/s41207-021-00284-1

Nano-based Approaches for the Development of Sustainable Nutritive Feeds

11

VijayaDurga V. V. Lekkala, Dakshayani Lomada, Madhava C. Reddy, T. Sri Ranjani, Nentia Chib, and L. V. Ravishankar

Abstract

Malnutrition is a serious concern in developing countries due to the nonavailability of nutritious food. Crops are major sources of human energy because they are sources of bioactive compounds, minerals, and vitamins as natural sources. In underprivileged communities, nutrient deficiency is a major threat to children, pregnant women, and younger people due to lack of nutritious food, leading to a compromise in human health. The key issue influencing human health in developing countries is nutrient-deficient foods as a source of consumption from field and horticultural crops. Fortification of food with bio-stimulants and fertilizers is a potential way to enrich food naturally with nutrients. Overexposure to the present chemical fertilizers increses the risk of diseases, unfit for human consumption, and environmental pollution, low nutrient uptake without inadequate assimilation capacity. Nanotechnology applications may help improve the nutritive value of food crops through biofortification as plants readily acquire nano-based nutrients due to their small size and higher entrapment efficiency. Nanofertilizers offer several advantages over traditional methods, including maximum bioavailability, controlled release, and a high half-life that can be

V. V. V. Lekkala (✉) · D. Lomada
Department of Genetics and Genomics, Yogi Vemana University, Kadapa, Andhra Pradesh, India

M. C. Reddy
Department of Biotechnology and Bioinformatics, Yogi Vemana University, Kadapa, Andhra Pradesh, India

T. Sri Ranjani
Department of Zoology, D.K. Govt College for Women(A), Nellore, Andhra Pradesh, India

N. Chib · L. V. Ravishankar
Division of Plant Pathology, FoA, Sher-E-Kashmir University of Agriculture Science and Technology, Chatha Jammu, Jammu and Kashmir, India

G. Sanghvi et al. (eds.), *Sustainable Food Fortification*,
https://doi.org/10.1007/978-981-95-1217-1_11

retained for long periods, and prevent it from leaching in soil. Combining biofortification with nanotechnology can increase nutrient efficiency with a great impact. Further, this chapter highlights the potential use of biosynthesized nanobiofertilizers and nanonutrients, enhancing agronomic biofortification to achieve better nutritive food with sustainable practices.

Keywords

Nanotechnology · Biofortification · Micronutrients · Nanofertilizers · Sustainable agriculture

11.1 Introduction

Biofortification is an important concept in food security to provide the biological availability of nutrients to humans through food crops. It is a process of enhancing the quality of food to prevent malnutrition by different modern technological applications in agriculture, with the principles of biotechnology like plant breeding, agronomic practices, and nanotechnology. Mineral nutrition deficiency is one of the important factors in the global health index among different nations due to poor food quality. Developing countries are more vulnerable to this situation because of the technology gap in agriculture. Many developing countries, including African nations and South Asian nations, are still under similar situations of food deficit due to aberrant climatic conditions and challenges to society in agriculture and food security. The Indian subcontinent ranks 105th in the Global Hunger Index, compared to 127th, indicating a persistent issue of malnutrition; the circumstances in numerous African nations are even more severe, according to the Global Huger Index (GHI) study (GHI, 2024). Around 2 billion of the world population across the globe suffer from hidden nutrition known as "hidden hunger" (Hodge, 2016), lack of micronutrients in their food, but with the constant rise of food production, which targets only quantity but not quality (Muthayya et al., 2013).

Besides the growing surplus food due to nutrition rather than a healthy diet, it became a matter of concern in human health because the agricultural systems have not been designed for it (Khush, 2008). Despite increasing surplus in food production, overnutrition has become a concern for human health, as agricultural systems were not originally designed to ensure a balanced and healthy diet. The shift of agriculture from the point of quantity of food production accompanied by quality will rapidly enhance human nutrition and decrease the burden of malnutrition (Gilani & Nasim, 2007). Fighting malnutrition could be possible only if we prepare our agricultural systems with fortification, which will help eliminate "hidden hunger" or "micronutrient malnutrition" especially in underdeveloped countries, where diets are dominated by micronutrient-poor staple food crops. Nutrients are most important to every form of life, from minute single-celled to multicellular organisms. Forty different types of nutrients are required by human beings to live a

Table 11.1 Importance of mineral nutrition in human health and their role (Garg et al., 2018)

Microminerals	Vitamins	Amino acids (essential)	Fatty acids (essential)	Macro-minerals
Fe	A (retinol)	Histidine	Linoleic acid	K
Zn	D (calciferol)	Isoleucine	Linolenic acid	Ca
Cu	E (α-tocopherol)	Leucine		Mg
Mn	K (phylloquinone)	Lysine		S
I	C (ascorbic acid)	Methionine		P
Se	B_1 (thiamine)	Phenylalanine		Na
Mo	B_2 (riboflavin)	Threonine		Cl
Co	B_3 (niacin)	Tryptophan		
Ni	B_5 (pantothenic acid) B_6 (pyridoxine) B_7 (biotin) B_9 (folic acid, folacin), B_{12} (cobalamin)	Valine		

healthy life. The role of many essential micro- and macroelements in human health and their corresponding role of different bioenzymes in human health are listed in Table 11.1. The collective role of these micro- and macroelements is important in humans for the proper physical and mental development of human beings (White & Broadley, 2005).The majority of the global population depends on a major source of food grains as cereals, such as rice, wheat, maize, and root crops like potato and cassava, which contain an adequate amount of nutrients to meet their daily requirements to maintain a healthy body routine. The enhancement of these nutrient availabilities in major staple food resources will increase the socioeconomic development of human beings.

11.2 Malnutrition

Consumption of unbalanced diet in daily routine leads to malnutrition and its different forms such as (1) stunting due to low weight as per chronological age, mostly in children due to low access to food. (2)Wasting children are thin and have less height for their age due to acute food shortages. (3) Micronutrient deficiencies due to suboptimal nutrient intake due to lack of intake and absorption, or mostly a single source of nutrition. (4) Lower body mass index as per healthy body height (Fanzo et al., 2019).

11.2.1 Global Scenario of Malnutrition Crisis

Around 2.37 billion people globally do not have access to good food, 786 million people are undernourished, 118 million people are under a hunger crisis, 20.5 billion newborn children have low weight, and finally, 149.2 million children and 12 million adult deaths happen around the globe, which is unavoidable (Initiatives, 2021). Around 0.570.8 million women between 18 and 49 years face anemia, and 538.7 million adults are diabetic (FOOD and AFFORDABLE, 2022).

11.3 Importance of Biofortification in Crop Production

Different programs to overcome malnutrition are followed, such as (1) food fortification, (2) dietary diversification, and (3) crop biofortification procedure of enhancing the quality of crops.

11.3.1 Biofortification and Its Importance in Enhancing Cropping Systems

Agronomic biofortification is an important concept of biofortification when compared to conventional plant breeding and transgenic methods as it is easily adaptable and welcomed by all scientific communities. In agronomic biofortification, the usage of bioagents and modifications of fertilizer consortium will alter the habitats to utilize and unlock the potential nutrient resources (Table 11.2).

Table 11.2 Biofortification in different crops

Crop	Fortified nutrient	Status	References
Cereals			
Rice			
	Iron	Research	He et al. (2013), Yuan et al. (2013)
	Zinc	Research	Mabesa et al. (2013), Wei et al. (2012)
	Se	Research	Chen et al. (2002), Fang et al. (2008)
Wheat			
	Iron	Research	Aciksoz et al. (2011)
	Zinc	Research	Cakmak (2010), Yang et al. (2011)
	Se	Research	Aro et al. (1995)
	Pi-fertilizer + mycorrhiza	Research	Noori et al. (2014)
	Organic + chemical fertilizers (iron)	Research	Ramzani et al. (2016)
Pea			
	Zn^{2+}	Research	Poblaciones and Rengel (2016)
Common bean			
	Zinc	Research	Ibrahim and Ramadan (2015)
	N, P, K, Cu, Mn, zinc (organic + chemicalfertilizers)	Research	Westermann et al. (2011)
Oil seed			
Mustard			
	Se, rhizosphere bacteria	Research	Yasin et al. (2015)

11.3.2 Nanoparticles in Biofortification

Nutrient availability can be positively approached through the application of engineered nanoparticles due to progressive controlled release and simultaneously reducing the mitigating and leaching losses, besides enhancing the bioavailability of nutrients in plant tissues. Cellular homeostasis is an important factor for maintaining the proper physiological conditions in a plant cell. The cell will die if any intervention involving the presence of oxidizing nanoparticles, such as oxide nanoparticles (NPs), disrupts the natural equilibrium condition of the cell and causes the creation of reactive oxygen species (ROS). The majority of in vitro research showed that ROS production may occur when the cell is in equilibrium. Therefore, the mechanism of the nanoparticles should be precisely predicted to reduce oxidative stress of the plant (Fallah et al., 2024; Zia-ur-Rehman et al., 2023).

11.3.3 Different Forms of Nanonutrients and Their Application Benefits

Engineered nanoparticles are synthesized with desirable physio-chemical properties and considered the best safe alternative in biofortification of food crops (Elemike et al., 2019). Biosynthesis of nanomaterials through microbial- and plant-based strategies reduces elements and/or oxides or ionic salts to form nanoscale materials. Green nanotechnology involves microbial and plant extracts; the presence of phytochemicals in these extracts is responsible for the preparation of nanoparticles. These bioactive compounds act as capping agents for the synthesis of nanoparticles. Green synthesis approaches are sustainable and less toxic. However, certain limitations are enumerated in green synthesis; batch-to-batch variability in properties is a major shortcoming (Jeevanandam et al., 2022; Kumar et al., 2021; Purohit et al., 2019).

11.4 Nanobiotechnology

Nanotechnology is defined as one of the fastest-growing technologies in the twenty-first century with the combination of chemistry, physics, and biology. The design,synthesis and characterization of these nanomaterials are depends upon the controlling size,shape at nanometer scale range(1-100nm) (Bheemidi et al., 2011). It has emerged that interdisciplinary biomedical research is quickly advancing in diagnostics, tissue engineering, and drug delivery. The main application of nanomaterials is to manipulate, build, and improve biological components at the nanoscale range (Singh et al. 2021). Nanoparticles with exact size targets for specific ligands are used as drugs for site-specific diseases. Nanostructured drugs selectively target and protect the host against diseases. Nanotechnology allows researchers to manipulate matter at the atomic level and interact with cells at the molecular level to understand the mechanism of molecules. Finally, it generates highly efficient and therapeutic drugs for the diagnosis of diseases in early stages. Nanostructured drugs offer several advantages over traditional methods, including maximum bioavailability of drugs with minimal side effects. Nanomedicine also provides high half-life while encapsulated with peptide drugs. Control the release of drugs with a minimal amount of usage. The main advantages of using nanomedicine as a drug-carrying capacity are their unique properties like carrier capacity, stability, and hydrophilic and hydrophobic encapsulation efficacies (Lekkala et al., 2024).

11.4.1 Importance of Nanotechnology

Agriculture is a major source of food production and acts as an interface between the environment and humans. Excessive use of fertilizers changes the natural pools of nutrients in soil, which ultimately decreases crop productivity(Roy-Bolduc & Hijri, 2011). Overdose of pesticides in agricultural lands leads to the production of free radicals, which scavenge enzymes. Synthetic modes of pesticides used for plants usually bring resistance against pests (El-Shenawy et al., 2011). Extensive use of these pesticides causes several drastic effects on human health, livestock, and pollinating insects in the ecosystem. Nanoformulations are an appropriate solution to overcome these problems (Pandit et al., 2022).

11.5 Synthesis of Nanoelements for Biofortification

Malnutrition deficits, referred to as hidden hunger, proliferate worldwide, impacting health, productivity, and development (Welch, 2002). Dietary inadequacies impact a significant portion of the population. The World Health Organization estimates that 2 billion individuals globally experience iron deficiency, whereas 750 million are affected by iodine deficiency (Organization and UNICEF, 2006; Ramakrishnan, 2002). Zinc insufficiency is recognized as a major public health concern (Black, 2003). The primary cause of malnutrition is the emphasis on agricultural practices aimed at increasing crop yield rather than improving quality. These tactics result in the depletion of essential elements in grains. Altered food chain dynamics at the consumer level, along with the incorporation of nutritional supplements such as fruits, legumes, vegetables, and dairy products into their diets, result in significant issues stemming from nutritional inadequacies, adversely affecting their health owing to malnutrition. These highlights pertain to the advancement and incorporation of nutritional enhancements for the crops. Consequently, agricultural scientists concentrated on the use of nutrients in crops by a natural process known as biofortification (Khan et al., 2020; Pandey et al., 2016). Biofortification pertains to the bioavailability of vital nutrients during plant growth using agronomic and genetic methods. This is a sustainable and economical strategy for nutritional supplements. Biological fortification to improve dietary quality for a growing population is the most effective alternative strategy. Various micronutrients have been incorporated into fortification schemes; for instance, zinc is added to maize flour and wheat for the preparation of tortillas and bread, which are two primary staple foods in the country (Bouis et al., 2011; Joint, 1998). Primarily, there are two categories of biofortification: agronomic and genetic biofortification (Khan et al., 2020). Genetic biofortification encompasses both traditional and transgenic methodologies. Conventional breeding is the paramount method of genetic biofortification due to its acceptability and sustainability in nature. More advanced genotypes are crossed than those yielding lower nutritional content (Pérez-Massot et al., 2013). In commercial farms, genetic variation manifests through gene linkage and pool introgression for traits (Lyons et al., 2005). Transgenic methods enable the enhancement of nutrient-related genes from diverse species and their expression in crops with lower nutritional value (Pérez-Massot et al., 2013). Notwithstanding their advantages, these methodologies need substantial investment of time and resources during the study process. In adittion to reduced acceptance of transgenic crops by scientific community and farmers regarding its commercialization and consumption benefits for the human population (Garg et al., 2018; Hefferon, 2016). The aforementioned solutions are constrained by agronomic approaches within the biofortification program. Agronomic biofortification primarily relies on the use of nutrient-dense fertilizers, including nitrogen, potassium, phosphorus, and micronutrients such as zinc, copper, and manganese, either via foliar spray or soil amendment to enhance yield. The efficacy of biofortification is contingent upon various aspects, including the transport, accumulation within plant tissues, and absorption of nutrient intake from the soil (Athar et al., 2020; Erisman et al., 2008; Garg et al., 2018). Nanotechnology is one of the rapidly advancing technologies applied in agriculture to strengthen crops at the genetic level, hence improving nutrient uptake for sustainable nutritional feeds. In contrast to conventional methods, nanofertilizers offer several advantages, including enhanced seedling growth, improved nitrogen metabolism, higher biomolecule synthesis (proteins, carbohydrates), and augmented nutrient translocation by mitigating various stressors. Nonetheless, a comparatively reduced application of nanofertilizers significantly enhances nutrient content in food crops while minimizing environmental harm (Al-Mamun et al., 2021; Thakur et al., 2018; Zulfiqar et al., 2019).

11.5.1 Types of Nanoelements

Micronutrient deficiencies in plants are prevalent in agricultural lands, ultimately influencing nutrient quality and crop yield, hence impacting human health (Assunção et al., 2022; Dhaliwal et al., 2022). The overexploitation of soil resources and intensive agricultural methods result in the depletion of water and soil nutrients (Saraiva et al., 2023). Micronutrients significantly contribute alongside the primary macronutrients: nitrogen, potassium, and phosphorus. A deficiency in micronutrients within plants results in a significant prevalence of micronutrient deficiencies among the human population (Azam et al., 2022; Ciurli et al., 2022; Guardiola-Márquez et al., 2022). A significant portion of the global population is deficient in micronutrients such as zinc (Zn), copper (Cu), and iron (Fe). These vital micronutrients significantly contribute to the physiological and metabolic processes in both plants

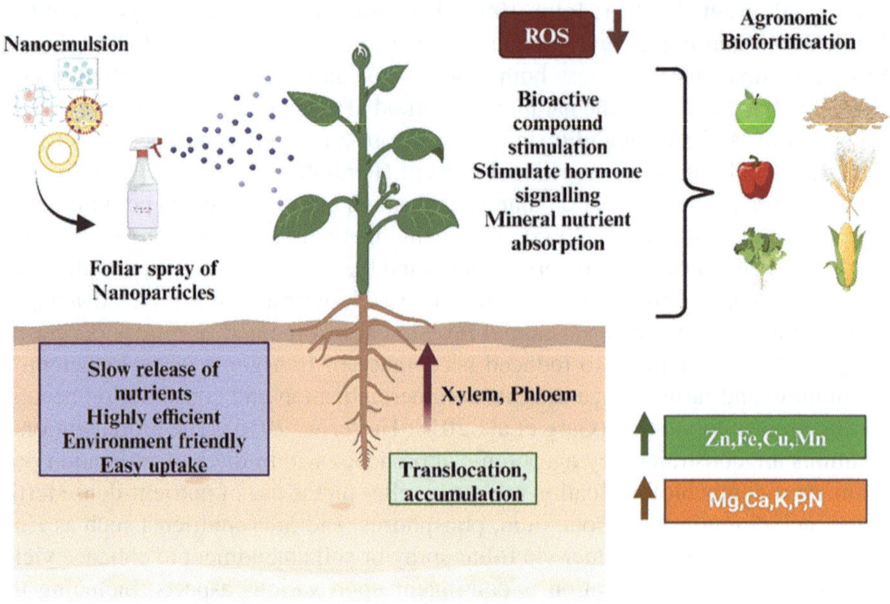

Fig. 11.1 Application of nanonutrients for biofortification

and humans (Guardiola-Márquez et al., 2022). Traditional methods of utilizing inorganic nutrient forms have low absorption efficiency, with a significant portion lost through leaching. When soluble nutrient forms are introduced by fertilizers, they are swiftly absorbed by plants and converted into soluble forms as nanofertilizers. Nanofertilizers present an innovative strategy and promising solutions to enhance crop productivity under abiotic stress and improve soil nutrient balance (El-Bialy et al., 2023; Shalaby et al., 2022; Verma et al., 2022). Several significant nanoparticles are addressed in the subsequent sections (Fig. 11.1).

11.5.2 Zinc Nanoparticles

Zinc is an essential vitamin that is significantly deficient in most developing countries. It serves as a crucial structural component, catalyst, and regulatory ion and is considered a primary micronutrient in the human diet (Cakmak, 2008; Hacisalihoglu, 2020). Chasapis et al. (2012) indicate that insufficient zinc intake might result in delayed wound healing, impaired growth, dermatological problems, frequent infections, and diminished taste and smell. The reduced solubility of zinc oxide renders it more economical than zinc sulfate. Consequently, the method of biofortification can be employed to enhance zinc efficacy. Various strategies for the application of Zn have been employed to enhance natural Zn biofortification (Montalvo et al., 2016). Maize is the primary grain crop providing sustenance for both animals and humans (Canton, 2021). It serves as a crucial dietary staple for about 200 million individuals, providing substantial protein and caloric content (Nuss and Tanumihardjo, 2010). Foliar treatment of ZnO nanoparticles results in superior yield and increased grain concentration compared to soil application in maize (Umar et al., 2021). The foliar spraying of cocktail solutions including Fe, Zn, and N has enhanced the concentrations in grain maize growers (Xue et al., 2023). The combined application of Si+, Zn+, and AgNPs by foliar treatment enhanced several yield and growth parameters, including leaf area index, chlorophyll content, plant height, protein grain content, and harvest index over a designated time period (Kandil et al., 2023). Wheat is a fundamental food source in several countries and supplies essential proteins for health. The presence of phytate diminishes the micronutrient zinc concentration. Zn-complexed chitosan-Thiamine Pyrophosphate (TPP) nanoparticles as a nanofertilizer show significant impacts at comparatively low concentrations in a field-scale investigation (Dapkekar et al., 2018). Under field experimental conditions, ZnO nanoparticles have less inhibitory effects on plant development compared to zinc sulfate. ZnO nanoparticles enhanced the Zn content in grains and exhibited reduced toxicity compared to ZnSO4 in terms of Zn uptake (Du et al., 2019).

The application of 0.5% ZnO nanoparticles and 0.5% Fe_2O_3 nanoparticles (ZFN) in chickpeas enhances yield, nutrient absorption, and content during the preflowering stage (Dhaliwal et al., 2021). During the panicle stage, the application of ZnO nanoparticles results in a greater concentration of zinc in rice grains compared to zinc during the tillering and basal stages (Yang et al., 2021), outperforming zinc salts. The application of foliar sprays containing a minimal concentration of 40 ppm green-manufactured zinc nanoparticles enhanced the physiological and agronomic characteristics of rice crops (Elshayb et al., 2021). Under drought conditions in sorghum, the application of ZnO nanoparticles as a soil fertilizer enhanced the efficacy of nitrogen, hence promoting plant development (Dimkpa et al., 2019). In tomato plants, the foliar application of nano-Zn suppressed the in vitro growth of *Pseudomonas syringae* bacterial speck by enhancing the assimilation of Zn, Ca, and P in a nanobiofortified form (Parra-Torrejón et al., 2023).

11.5.3 Copper Nanoparticles

Copper nanoparticles have attracted considerable attention owing to their cost-effectiveness, accessibility, high conductivity, and non-toxicity (Rawat et al., 2018). Copper-based nanofertilizers are frequently employed in crop protection due to their position as an essential micronutrient in the metabolism of healthy organisms. Copper plays a role in biological and physiological functions including protein trafficking, mitochondrial respiration, cellular respiration, hormone signaling, and antioxidant activity in plants (Vishveshvar et al., 2018). The foliar application of copper oxide nanoparticles in lettuce enhances the concentration of flavonoids and phenols. Under hydroponic conditions, it also augments the leaf count in lettuce plants (Fortis-Hernández et al., 2024). These nanoparticles also influence the growth and germination of *Vigna radiata* seedlings. The majority of plants at lower doses remain unaffected, whereas inhibition occurs at elevated concentrations of CuONPs in comparison to control samples (Singh et al., 2017). The foliar spray of CuNPs enhanced the nutraceutical and physical quality, as well as the concentration of nutrients in melon fruits (Hernández et al., 2022). The amalgamation of CuNPs with spirulina extract in French beans enhanced yield and growth relative to controls of 1000 and 1500 mg/L, with a lesser effect than 500 mg/L (Elbanna et al., 2024).

11.5.4 Iron Nanoparticles

Iron nanoparticles are essential minerals for human growth. Iron deficiency due to malnutrition is a significant factor that precipitates anemia, particularly in youngsters (Olivares et al., 1999). In humans, dietary iron is a crucial micronutrient that is integral to the synthesis of hemoglobin and cytochromes (Claessens et al., 2013). Iron oxide nanoparticles produced from *Glycyrrhiza glabra* enhanced the lengths of both roots and shoots at concentrations of 100 and 50 ppm (Ahmad et al., 2024). Foliar fertilizers applied to soybean as nanostructured $FePO_4$ potentially enhanced the iron content in the evaluated samples compared to the control (Knijnenburg et al., 2018). Iron biofortification in the wheat genotypes WL711 and IITR26 enhances the germination rate and physiological growth parameters in crops (Sundaria et al., 2019). The integration of nanobiofertilizers with ZnO and Fe_2O_3 promotes plant growth and increases biofortified glucosinolates and micronutrients in broccoli and microgreens (Guardiola-Márquez et al., 2023) (Fig. 11.1).

11.5.5 Movement of Nanoparticle-Foliar Application

Foliar application of nanoparticles acts as a promising strategy for plant protection and productivity. Intense precipitation and soil intricacy diminish the efficiency of fertilizer utilization in plants. The low solubility of fertilizers results in diminished plant growth due to the inability to absorb trace elements (Sun et al., 2020). Compared to traditional fertilizer application, foliar methods offer various advantages, including cost-effectiveness, low soil impact, and rapid absorption. They frequently supply essential nutrients and vitamins for plant development, and research indicates that slow-release nanofertilizers improve the absorption of phosphorus, potassium, and nitrogen (Meier et al., 2020).

The foliar application and fortification of nanoparticles enhance efficiency via stomatal and cuticular channels (Lv et al., 2019). Lipophilic chemicals enter foliage via the cuticular channel, whereas polar and nonpolar substances traverse cuticular paths with diameters ranging from 0.6 to 4.8 nm. Nanoparticles smaller than 4.8 nm traverse the cuticular route. The stomatal entry of the nanoparticle channel has a hydrophilic diameter of at least 40 nm. Reduced cuticular pore size facilitates entry via the stomatal channel in leaves (Ali et al., 2021; Eichert & Goldbach, 2008; Schreiber, 2005).

The root surface possesses a negative charge due to the excretion of organic acids from root hairs, facilitating the adsorption and deposition of positively charged nanoparticles (Huang et al., 2022). Nanoparticles traverse the root hair cuticle, aiming for and ultimately arriving at the epidermis (Schwab et al., 2016). Apoplastic and symplastic pathways were identified; apoplastic movement occurs through intercellular gaps, whereas the symplastic pathway transpires via plasmodesmata. The apoplastic mobility of nanoparticles (NPs) is generally acknowledged, as evidenced by their presence in the intercellular spaces of root tissues across numerous studies (Li et al., 2016; Tombuloglu et al., 2019). (Wang et al., 2012; Zhu et al., 2008). Casparian strips serve as a barrier to apoplastic transport. Nutrients penetrate via the lateral root zones at the root tip (Schymura et al., 2017). ZnO nanoparticles have been observed to infiltrate through lateral roots into the xylem of maize (Lv et al., 2015).

11.6 Future Prospects

Nanotechnology is an emerging scientific discipline, and its application across several sectors has demonstrated significant efficacy, particularly in agriculture and related industries, by embracing innovative scientific advancements such as nanotechnology. It facilitates sustainable methodologies and environmentally friendly applications to augment production.

11.7 Conclusion

The era of nanotechnology is immensely emerging and successful in becoming widely applicable in sustainable agriculture. New agricultural technologies and farming practices must be established and executed. This chapter has tried to discuss the nanobiofortification presently under development in agricultural sciences. One way to achieve and reduce global hunger is to increase the nutrients through food. An attempt must be made with the upcoming fortification to enhance the nutrients through nanoparticles via green synthesis of various nanoparticles. Though nanoapplications are validated and patented, their stability and integrity remain one of the most important questions. The prospects should be concerned about how different nutrients and combinations will help promote this technology to make agriculture profitable and sustainable. Green nanosynthesis is an ideal and important way to green technology, which uses opportunistic and sustainable resources in agriculture. The maindraw back of nanonutrients applications in agriculture is the stability and efficiency must be studied for the further upcoming research.

References

Aciksoz, S. B., Yazici, A., Ozturk, L., & Cakmak, I. (2011). Biofortification of wheat with iron through soil and foliar application of nitrogen and iron fertilizers. *Plant and Soil, 349*, 215–225.

Ahmad, S., Ahmad, N., Islam, M. S., Ahmad, M. A., Ercisli, S., Ullah, R., Bari, A., & Munir, I. (2024). Rice seeds biofortification using biogenic ıron oxide nanoparticles synthesized by using *Glycyrrhiza glabra*: A study on growth and yield ımprovement. *Scientific Reports, 14*, 12368.

Ali, S., Mehmood, A., & Khan, N. (2021). Uptake, translocation, and consequences of nanomaterials on plant growth and stress adaptation. *Journal of Nanomaterials, 2021*, 6677616.

Al-Mamun, M. R., Hasan, M. R., Ahommed, M. S., Bacchu, M. S., Ali, M. R., & Khan, M. Z. H. (2021). Nanofertilizers towards sustainable agriculture and environment. *Environmental Technology & Innovation, 23*, 101658.

Aro, A., Alfthan, G., & Varo, P. (1995). Effects of supplementation of fertilizers on human selenium status in Finland. *Analyst, 120*, 841–843.

Assunção, A. G., Cakmak, I., Clemens, S., González-Guerrero, M., Nawrocki, A., & Thomine, S. (2022). Micronutrient homeostasis in plants for more sustainable agriculture and healthier human nutrition. *Journal of Experimental Botany, 73*, 1789–1799.

Athar, T., Khan, M. K., Pandey, A., Yilmaz, F. G., Hamurcu, M., Hakki, E. E., & Gezgin, S. (2020). Biofortification and the involved modern approaches. *Journal of Elementology, 25*.

Azam, M., Bhatti, H. N., Khan, A., Zafar, L., & Iqbal, M. (2022). Zinc oxide nano-fertilizer application (foliar and soil) effect on the growth, photosynthetic pigments and antioxidant system of maize cultivar. *Biocatalysis and Agricultural Biotechnology, 42*, 102343.

Bheemidi, V. S., Tiruckovela, M., Chettipalli, N. D., & Yanamadala, S. V. (2011). Novel applications of nanotechnology in life sciences. *Journal of Bioanalysis & Biomedicine, 3*.

Black, R. E. (2003). Zinc deficiency, infectious disease and mortality in the developing world. *The Journal of Nutrition, 133*, 1485S–1489S.

Bouis, H. E., Eozenou, P., & Rahman, A. (2011). Food prices, household income, and resource allocation: Socioeconomic perspectives on their effects on dietary quality and nutritional status. *Food and Nutrition Bulletin, 32*, S14–S23.

Cakmak, I. (2008). Enrichment of cereal grains with zinc: Agronomic or genetic biofortification? *Plant and Soil, 302*, 1–17.

Cakmak, I. (2010). Biofortification of cereals with zinc and iron through fertilization strategy. In *19th world congress of soil science* (pp. 1–6). Citeseer.

Canton, H. (2021). Food and agriculture organization of the United Nations – FAO. In *The Europa directory of international organizations 2021* (pp. 297–305). Routledge.

Chasapis, C. T., Loutsidou, A. C., Spiliopoulou, C. A., & Stefanidou, M. E. (2012). Zinc and human health: An update. *Archives of Toxicology, 86*, 521–534.

Chen, L., Yang, F., Xu, J., Hu, Y., Hu, Q., Zhang, Y., & Pan, G. (2002). Determination of selenium concentration of rice in China and effect of fertilization of selenite and selenate on selenium content of rice. *Journal of Agricultural and Food Chemistry, 50*, 5128–5130.

Ciurli, A., Giagnoni, L., Pastorelli, R., Sega, D., Zamboni, A., Renella, G., & Varanini, Z. (2022). A novel P nanofertilizer has no impacts on soil microbial communities and soil microbial activity. *Applied Soil Ecology, 178*, 104570.

Claessens, M., Contor, L., Dhonukshe-Rutten, R., De Groot, L. C., Fairweather-Tait, S. J., Gurinovic, M., Koletzko, B., Van Ommen, B., Raats, M. M., & Van't Veer, P. (2013). EURRECA – Principles and future for deriving micronutrient recommendations. *Critical Reviews in Food Science and Nutrition, 53*, 1135–1146.

Dapkekar, A., Deshpande, P., Oak, M. D., Paknikar, K. M., & Rajwade, J. M. (2018). Zinc use efficiency is enhanced in wheat through nanofertilization. *Scientific Reports, 8*, 6832.

Dhaliwal, S. S., Sharma, V., Shukla, A. K., Verma, V., Behera, S. K., Singh, P., Alotaibi, S. S., Gaber, A., & Hossain, A. (2021). Comparative efficiency of mineral, chelated and nano forms of zinc and iron for improvement of zinc and iron in chickpea (*Cicer arietinum* L.) through biofortification. *Agronomy, 11*, 2436.

Dhaliwal, S. S., Sharma, V., Shukla, A. K., Verma, V., Kaur, M., Shivay, Y. S., Nisar, S., Gaber, A., Brestic, M., & Barek, V. (2022). Biofortification – A frontier novel approach to enrich micronutrients in field crops to encounter the nutritional security. *Molecules, 27*, 1340.

Dimkpa, C. O., Singh, U., Bindraban, P. S., Elmer, W. H., Gardea-Torresdey, J. L., & White, J. C. (2019). Zinc oxide nanoparticles alleviate drought-induced alterations in sorghum performance, nutrient acquisition, and grain fortification. *Science of the Total Environment, 688*, 926–934.

Du, W., Yang, J., Peng, Q., Liang, X., & Mao, H. (2019). Comparison study of zinc nanoparticles and zinc sulphate on wheat growth: From toxicity and zinc biofortification. *Chemosphere, 227*, 109–116.

Eichert, T., & Goldbach, H. E. (2008). Equivalent pore radii of hydrophilic foliar uptake routes in stomatous and astomatous leaf surfaces–further evidence for a stomatal pathway. *Physiologia Plantarum, 132*, 491–502.

Elbanna, H. M., Ahmed, O. K., Fayed, S. A.-K., Hammam, K. A.-M., & Yousef, R. S. (2024). Enhancing French basil growth through synergistic foliar treatment with copper nanoparticles and spirulina sp. *BMC Plant Biology, 24*, 512.

El-Bialy, S. M., El-Mahrouk, M. E., Elesawy, T., Omara, A. E.-D., Elbehiry, F., El-Ramady, H., Áron, B., Prokisch, J., Brevik, E. C., & Solberg, S. Ø. (2023). Biological nanofertilizers to enhance growth potential of strawberry seedlings by boosting photosynthetic pigments, plant enzymatic antioxidants, and nutritional status. *Plants, 12*, 302.

Elemike, E. E., Uzoh, I. M., Onwudiwe, D. C., & Babalola, O. O. (2019). The role of nanotechnology in the fortification of plant nutrients and improvement of crop production. *Applied Sciences, 9*, 499.

Elshayb, O. M., Farroh, K. Y., Amin, H. E., & Atta, A. M. (2021). Green synthesis of zinc oxide nanoparticles: Fortification for rice grain yield and nutrients uptake enhancement. *Molecules, 26*, 584.

El-Shenawy, N., El-Ahmary, B., & Al-Eisa, R. (2011). Mitigating effect of ginger against oxidative stress induced by atrazine herbicides in mice liver and kidney. *Journal of Biofertilizers & Biopesticides, 2*.

Erisman, J. W., Sutton, M. A., Galloway, J., Klimont, Z., & Winiwarter, W. (2008). How a century of ammonia synthesis changed the world. *Nature Geoscience, 1*, 636–639.

Fallah, S., Yusefi-Tanha, E., & Peralta-Videa, J. R. (2024). Interaction of nanoparticles and reactive oxygen species and their impact on macromolecules and plant production. *Plant Nano Biology*, 100105.

Fang, Y., Wang, L., Xin, Z., Zhao, L., An, X., & Hu, Q. (2008). Effect of foliar application of zinc, selenium, and iron fertilizers on nutrients concentration and yield of rice grain in China. *Journal of Agricultural and Food Chemistry, 56*, 2079–2084.

Fanzo, J., Hawkes, C., Udomkesmalee, E., Afshin, A., Allemandi, L., Assery, O., Baker, P., Battersby, J., Bhutta, Z., & Chen, K. (2019). 2018 global nutrition report.

Food, R., and Affordable, H. D. M. (2022). Food security and nutrition in the World.

Fortis-Hernández, M., González-Rodríguez, T., Espinosa-Palomeque, B., Preciado-Rangel, P., Gallegos-Robles, M. A., & Rueda-Puente, E. O. (2024). Foliar biofortification with copper nanoparticles and its effect on phytochemical quality and enzymatic activity in lettuce. *Horticultura Brasileira, 42*, e2617.

Garg, M., Sharma, N., Sharma, S., Kapoor, P., Kumar, A., Chunduri, V., & Arora, P. (2018). Biofortified crops generated by breeding, agronomy, and transgenic approaches are improving lives of millions of people around the world. *Frontiers in Nutrition, 5*, 12.

Gilani, G. S., & Nasim, A. (2007). Impact of foods nutritionally enhanced through biotechnology in alleviating malnutrition in developing countries. *Journal of AOAC International, 90*, 1440–1444.

Guardiola-Márquez, C. E., Santos-Ramírez, M. T., Segura-Jiménez, M. E., Figueroa-Montes, M. L., & Jacobo-Velázquez, D. A. (2022). Fighting obesity-related micronutrient deficien-

cies through biofortification of agri-food crops with sustainable fertilization practices. *Plants, 11*, 3477.
Guardiola-Márquez, C. E., García-Sánchez, C. V., Sánchez-Arellano, Ó. A., Bojorquez-Rodríguez, E. M., & Jacobo-Velázquez, D. A. (2023). Biofortification of broccoli microgreens (*Brassica oleracea* var. italica) with Glucosinolates, zinc, and iron through the combined application of bio-and Nanofertilizers. *Food, 12*, 3826.
Hacisalihoglu, G. (2020). Zinc (Zn): The last nutrient in the alphabet and shedding light on Zn efficiency for the future of crop production under suboptimal Zn. *Plants, 9*, 1471.
He, W., Shohag, M., Wei, Y., Feng, Y., & Yang, X. (2013). Iron concentration, bioavailability, and nutritional quality of polished rice affected by different forms of foliar iron fertilizer. *Food Chemistry, 141*, 4122–4126.
Hefferon, K. L. (2016). Can biofortified crops help attain food security? *Current Molecular Biology Reports, 2*, 180–185.
HernáNdez, M. F., Lopez, J. O., Rangel, P. P., Valencia, R. T., Fortiz, E. L., Andrade-Sifuentes, A., et al. (2022). Biofortification with copper nanoparticles (Nps Cu) and its effect on the physical and nutraceutical quality of hydroponic melon fruits. *Notulae Botanicae Horti Agrobotanici Cluj-Napoca, 50*, 12568–12568.
Hodge, J. (2016). Hidden hunger: approaches to tackling micronutrient deficiencies.
Huang, D., Dang, F., Huang, Y., Chen, N., & Zhou, D. (2022). Uptake, translocation, and transformation of silver nanoparticles in plants. *Environmental Science: Nano, 9*, 12–39.
Ibrahim, E. A., & Ramadan, W. A. (2015). Effect of zinc foliar spray alone and combined with humic acid or/and chitosan on growth, nutrient elements content and yield of dry bean (*Phaseolus vulgaris* L.) plants sown at different dates. *Scientia Horticulturae, 184*, 101–105.
Initiatives, D. (2021). *The state of global nutrition*. Development Initiatives Bristol.
Jeevanandam, J., Kiew, S. F., Boakye-Ansah, S., Lau, S. Y., Barhoum, A., Danquah, M. K., & Rodrigues, J. (2022). Green approaches for the synthesis of metal and metal oxide nanoparticles using microbial and plant extracts. *Nanoscale, 14*, 2534–2571.
Joint, F. (1998). WHO Expert Consultation on Human vitamin and mineral requirements. *Vitamin and mineral requirements in human nutrition: report of a joint FAO/WHO expert consultation, Bangkok, Thailand* 2.
Kandil, E. E., Lamlom, S. F., Gheith, E.-S. M., Javed, T., Ghareeb, R. Y., Abdelsalam, N. R., & Hussain, S. (2023). Biofortification of maize growth, productivity and quality using nano-silver, silicon and zinc particles with different irrigation intervals. *The Journal of Agricultural Science, 161*, 339–355.
Khan, M. K., Pandey, A., Hamurcu, M., Hakki, E. E., & Gezgin, S. (2020). Role of molecular approaches in improving genetic variability of micronutrients and their utilization in breeding programs. In *Wheat and barley grain biofortification* (pp. 27–52). Elsevier.
Khush, G. (2008). Biofortification of crops for reducing malnutrition.
Knijnenburg, J. T., Hilty, F. M., Oelofse, J., Buitendag, R., Zimmermann, M. B., Cakmak, I., & Grobler, A. F. (2018). Nano-and Pheroid technologies for development of foliar iron fertilizers and iron biofortification of soybean grown in South Africa. *Chemical and Biological Technologies in Agriculture, 5*, 1–10.
Kumar, U., Priyanka, M., et al. (2021). Fe chelation and zinc solubilization: A promising approach for cereals biofortification. In *Soil microbiomes for sustainable agriculture: Functional annotation* (pp. 149–174).
Lekkala, V. V., Reddy, M. C., Reddy, V. C., Kanthirigala, S. K., Chitta, S., Reddy, K. R., & Lomada, D. (2024). Advancements in nanoparticles-based therapies for biomedical applications. *Nano-Structures & Nano-Objects, 40*, 101365.
Li, J., Hu, J., Ma, C., Wang, Y., Wu, C., Huang, J., & Xing, B. (2016). Uptake, translocation and physiological effects of magnetic iron oxide (γ-Fe2O3) nanoparticles in corn (*Zea mays* L.). *Chemosphere, 159*, 326–334.
Lv, J., Zhang, S., Luo, L., Zhang, J., Yang, K., & Christie, P. (2015). Accumulation, speciation and uptake pathway of ZnO nanoparticles in maize. *Environmental Science: Nano, 2*, 68–77.

Lv, J., Christie, P., & Zhang, S. (2019). Uptake, translocation, and transformation of metal-based nanoparticles in plants: Recent advances and methodological challenges. *Environmental Science: Nano, 6*, 41–59.

Lyons, G., Ortiz-Monasterio, I., Stangoulis, J., & Graham, R. (2005). Selenium concentration in wheat grain: Is there sufficient genotypic variation to use in breeding? *Plant and Soil, 269*, 369–380.

Mabesa, R., et al. (2013). Contrasting grain-Zn response of biofortification rice (*Oryza sativa* L.) breeding lines to foliar Zn application. *Field Crops Research, 149*, 223–233.

Meier, S., Moore, F., Morales, A., González, M.-E., Seguel, A., Meriño-Gergichevich, C., Rubilar, O., Cumming, J., Aponte, H., & Alarcón, D. (2020). Synthesis of calcium borate nanoparticles and its use as a potential foliar fertilizer in lettuce (*Lactuca sativa*) and zucchini (*Cucurbita pepo*). *Plant Physiology and Biochemistry, 151*, 673–680.

Montalvo, D., Degryse, F., Da Silva, R., Baird, R., & McLaughlin, M. (2016). Agronomic effectiveness of zinc sources as micronutrient fertilizer. *Advances in Agronomy, 139*, 215–267.

Muthayya, S., Rah, J. H., Sugimoto, J. D., Roos, F. F., Kraemer, K., & Black, R. E. (2013). The global hidden hunger indices and maps: An advocacy tool for action. *PLoS One, 8*, e67860.

Noori, M., Adibian, M., Sobhkhizi, A., & Eyidozehi, K. (2014). Effect of phosphorus fertilizer and mycorrhiza on protein percent, dry weight, weight of 1000 grain in wheat.

Nuss, E. T., & Tanumihardjo, S. A. (2010). Maize: A paramount staple crop in the context of global nutrition. *Comprehensive Reviews in Food Science and Food Safety, 9*, 417–436.

Olivares, M., Walter, T., Hertrampf, E., & Pizarro, F. (1999). Anaemia and iron deficiency disease in children. *British Medical Bulletin, 55*, 534–543.

Organization, W. H., and UNICEF. (2006). Preventing and controlling micronutrient deficiencies in populations affected by an emergency. In *Preventing and controlling micronutrient deficiencies in populations affected by an emergency* (p. 2).

Pandey, A., Khan, M. K., Hakki, E. E., Thomas, G., Hamurcu, M., Gezgin, S., Gizlenci, O., & Akkaya, M. S. (2016). Assessment of genetic variability for grain nutrients from diverse regions: Potential for wheat improvement. *Springerplus, 5*, 1–11.

Pandit, C., Roy, A., Ghotekar, S., Khusro, A., Islam, M. N., Emran, T. B., Lam, S. E., Khandaker, M. U., & Bradley, D. A. (2022). Biological agents for synthesis of nanoparticles and their applications. *Journal of King Saud University-Science, 34*, 101869.

Parra-Torrejón, B., Cáceres, A., Sánchez, M., Sainz, L., Guzmán, M., Bermúdez-Perez, F. J., Ramírez-Rodríguez, G. B., & Delgado-López, J. M. (2023). Multifunctional nanomaterials for biofortification and protection of tomato plants. *Environmental Science & Technology, 57*, 14950–14960.

Pérez-Massot, E., Banakar, R., Gómez-Galera, S., Zorrilla-López, U., Sanahuja, G., Arjó, G., Miralpeix, B., Vamvaka, E., Farré, G., & Rivera, S. M. (2013). The contribution of transgenic plants to better health through improved nutrition: Opportunities and constraints. *Genes & Nutrition, 8*, 29–41.

Poblaciones, M., & Rengel, Z. (2016). Soil and foliar zinc biofortification in field pea (*Pisum sativum* L.): Grain accumulation and bioavailability in raw and cooked grains. *Food Chemistry, 212*, 427–433.

Purohit, J., Chattopadhyay, A., & Singh, N. K. (2019). Green synthesis of microbial nanoparticle: Approaches to application. In *Microbial nanobionics: Volume 2, basic research and applications* (pp. 35–60).

Ramakrishnan, U. (2002). Prevalence of micronutrient malnutrition worldwide. *Nutrition Reviews, 60*, S46–S52.

Ramzani, P. M. A., Khalid, M., Naveed, M., Ahmad, R., & Shahid, M. (2016). Iron biofortification of wheat grains through integrated use of organic and chemical fertilizers in pH affected calcareous soil. *Plant Physiology and Biochemistry, 104*, 284–293.

Rawat, S., Pullagurala, V. L., Hernandez-Molina, M., Sun, Y., Niu, G., Hernandez-Viezcas, J. A., Peralta-Videa, J. R., & Gardea-Torresdey, J. L. (2018). Impacts of copper oxide nanoparticles on bell pepper (Capsicum annum L.) plants: A full life cycle study. *Environmental Science: Nano, 5*, 83–95.

Roy-Bolduc, A., & Hijri, M. (2011). The use of mycorrhizae to enhance phosphorus uptake: A way out the phosphorus crisis. *Journal of Biofertilizers & Biopesticides, 2*, 1–5.

Saraiva, R., Ferreira, Q., Rodrigues, G. C., & Oliveira, M. (2023). Nanofertilizer use for adaptation and mitigation of the agriculture/climate change dichotomy effects. *Climate, 11*, 129.

Schreiber, L. (2005). Polar paths of diffusion across plant cuticles: New evidence for an old hypothesis. *Annals of Botany, 95*, 1069–1073.

Schwab, F., Zhai, G., Kern, M., Turner, A., Schnoor, J. L., & Wiesner, M. R. (2016). Barriers, pathways and processes for uptake, translocation and accumulation of nanomaterials in plants–critical review. *Nanotoxicology, 10*, 257–278.

Schymura, S., Fricke, T., Hildebrand, H., & Franke, K. (2017). Elucidating the role of dissolution in CeO2 nanoparticle plant uptake by smart radiolabeling. *Angewandte Chemie International Edition, 56*, 7411–7414.

Shalaby, T. A., El-Bialy, S. M., El-Mahrouk, M. E., Omara, A. E.-D., El-Beltagi, H. S., & El-Ramady, H. (2022). Acclimatization of in vitro banana seedlings using root-applied bio-nanofertilizer of copper and selenium. *Agronomy, 12*, 539.

Singh, A., Singh, N., Hussain, I., Singh, H., & Yadav, V. (2017). Synthesis and characterization of copper oxide nanoparticles and its impact on germination of *Vigna radiata* (L.) R Wilczek. *Tropical Plant Research, 4*, 246–253.

Singh, K., Kumar, L., Prasad, D., Sharma, S., & Gupta, G. (2021). Research and Reviews: Journal of Pharmaceutics and Nanotechnology.

Sun, H., Du, W., Peng, Q., Lv, Z., Mao, H., & Kopittke, P. M. (2020). Development of ZnO nanoparticles as an efficient Zn fertilizer: Using synchrotron-based techniques and laser ablation to examine elemental distribution in wheat grain. *Journal of Agricultural and Food Chemistry, 68*, 5068–5075.

Sundaria, N., Singh, M., Upreti, P., Chauhan, R. P., Jaiswal, J., & Kumar, A. (2019). Seed priming with iron oxide nanoparticles triggers iron acquisition and biofortification in wheat (*Triticum aestivum* L.) grains. *Journal of Plant Growth Regulation, 38*, 122–131.

Thakur, S., Thakur, S., & Kumar, R. (2018). Bio-nanotechnology and its role in agriculture and food industry. *Journal of Molecular and Genetic Medicine, 12*, 1747–0862.

Tombuloglu, H., Slimani, Y., Tombuloglu, G., Almessiere, M., & Baykal, A. (2019). Uptake and translocation of magnetite (Fe3O4) nanoparticles and its impact on photosynthetic genes in barley (*Hordeum vulgare* L.). *Chemosphere, 226*, 110–122.

Umar, W., Hameed, M. K., Aziz, T., Maqsood, M. A., Bilal, H. M., & Rasheed, N. (2021). Synthesis, characterization and application of ZnO nanoparticles for improved growth and Zn biofortification in maize. *Archives of Agronomy and Soil Science, 67*, 1164–1176.

Verma, K. K., Song, X.-P., Joshi, A., Rajput, V. D., Singh, M., Sharma, A., Singh, R. K., Li, D.-M., Arora, J., & Minkina, T. (2022). Nanofertilizer possibilities for healthy soil, water, and food in future: An overview. *Frontiers in Plant Science, 13*, 865048.

Vishveshvar, K., Aravind Krishnan, M., Haribabu, K., & Vishnuprasad, S. (2018). Green synthesis of copper oxide nanoparticles using *Ixiro coccinea* plant leaves and its characterization. *BioNanoScience, 8*, 554–558.

Wang, Z., Xie, X., Zhao, J., Liu, X., Feng, W., White, J. C., & Xing, B. (2012). Xylem-and phloem-based transport of CuO nanoparticles in maize (Zea mays L.). *Environmental Science & Technology, 46*, 4434–4441.

Wei, Y., Shohag, M., & Yang, X. (2012). Biofortification and bioavailability of rice grain zinc as affected by different forms of foliar zinc fertilization.

Welch, R. M. (2002). The impact of mineral nutrients in food crops on global human health. *Plant and Soil, 247*, 83–90.

Westermann, D., Terán, H., Muñoz-Perea, C., & Singh, S. (2011). Plant and seed nutrient uptake in common bean in seven organic and conventional production systems. *Canadian Journal of Plant Science, 91*, 1089–1099.

White, P. J., & Broadley, M. R. (2005). Biofortifying crops with essential mineral elements. *Trends in Plant Science, 10*, 586–593.

Xue, Y.-F., Li, X.-J., Yan, W., Miao, Q., Zhang, C.-Y., Huang, M., Sun, J.-B., Qi, S.-J., Ding, Z.-H., & Cui, Z.-L. (2023). Biofortification of different maize cultivars with zinc, iron and selenium by foliar fertilizer applications. *Frontiers in Plant Science, 14*, 1144514.

Yang, X., et al. (2011). Impacts of phosphorus and zinc levels on phosphorus and zinc nutrition and phytic acid concentration in wheat (Triticum aestivum L.). *Journal of the Science of Food and Agriculture, 91*, 2322–2328.

Yang, G., Yuan, H., Ji, H., Liu, H., Zhang, Y., Wang, G., Chen, L., & Guo, Z. (2021). Effect of ZnO nanoparticles on the productivity, Zn biofortification, and nutritional quality of rice in a life cycle study. *Plant Physiology and Biochemistry, 163*, 87–94.

Yasin, M., El Mehdawi, A. F., Jahn, C. E., Anwar, A., Turner, M. F., Faisal, M., & Pilon-Smits, E. A. (2015). Seleniferous soils as a source for production of selenium-enriched foods and potential of bacteria to enhance plant selenium uptake. *Plant and Soil, 386*, 385–394.

Yuan, L., Wu, L., Yang, C., & Lv, Q. (2013). Effects of iron and zinc foliar applications on rice plants and their grain accumulation and grain nutritional quality. *Journal of the Science of Food and Agriculture, 93*, 254–261.

Zhu, H., Han, J., Xiao, J. Q., & Jin, Y. (2008). Uptake, translocation, and accumulation of manufactured iron oxide nanoparticles by pumpkin plants. *Journal of Environmental Monitoring, 10*, 713–717.

Zia-ur-Rehman, M., Anayatullah, S., Irfan, E., Hussain, S. M., Rizwan, M., Sohail, M. I., Jafir, M., Ahmad, T., Usman, M., & Alharby, H. F. (2023). Nanoparticles assisted regulation of oxidative stress and antioxidant enzyme system in plants under salt stress: A review. *Chemosphere, 314*, 137649.

Zulfiqar, F., Navarro, M., Ashraf, M., Akram, N. A., & Munné-Bosch, S. (2019). Nanofertilizer use for sustainable agriculture: Advantages and limitations. *Plant Science, 289*, 110270.

[illegible]

Public Health Aspects of Food Fortification: Comprehensive Strategies, Impact Analysis, and Policy Recommendations for Enhancing Nutritional Outcomes

12

Dhara Zaveri

Abstract

It has become a significant public health intervention that attempts to combat micronutrient deficiencies across a wide variety of communities all over the world. Fortification of food has developed as a result of this crucial intervention. This comprehensive review of the most recent data from research carried out in a range of countries all over the world evaluates the efficacy of a number of different fortification strategies and includes the most recent guidelines that have been released by the World Health Organization (WHO). The objective of this chapter is to give an in-depth analysis of state-of-the-art technological breakthroughs, implementation issues that are encountered in practice, and key performance metrics that are specific to effective fortification programs. A comprehensive impact evaluation, cost-effectiveness, nutrient stability, and bioavailability are some of the aspects that are the primary emphasis of this chapter. The purpose of this chapter is to provide public health practitioners and policymakers with specific policy recommendations that are intended to assist them in effectively scaling up sustainable food fortification activities. The book chapter also provides an analysis of quantitative impact measurements. Additionally, the chapter pulls from more than 50 sources that have been published during the last 5 years and have been subjected to peer review.

Keywords

Food fortification · Public health aspects · Micronutrient deficiencies · World Health Organization · Cost-effectiveness

D. Zaveri (✉)
Department of Botany, Faculty of Science, The Maharaja Sayajirao University of Baroda, Vadodara, Gujarat, India

G. Sanghvi et al. (eds.), *Sustainable Food Fortification*,
https://doi.org/10.1007/978-981-95-1217-1_12

12.1 Introduction

Micronutrient malnutrition represents a pervasive global challenge affecting billions of individuals, particularly in low- and middle-income countries. Food fortification, defined as the deliberate addition of essential micronutrients to food during processing, has proven to be one of the most cost-effective strategies for addressing these nutrient deficiencies. The worldwide dissemination of food fortification programs, strongly endorsed by international authorities such as the World Health Organization (2014), has been credited with significant improvements in population health outcomes, notably in reducing common deficiencies like iron deficiency anemia and vitamin A deficiency. This chapter critically examines the public health aspects of food fortification, focusing on compelling evidence derived from recent studies and WHO guidelines.

Food fortification specifically targets several key micronutrients, including vitamin A, iron, iodine, and zinc, among others. This intervention not only enhances individual nutritional status and well-being but also contributes to broader socioeconomic development by reducing the disease burden, lowering healthcare costs, and promoting productivity within communities. Public health practitioners and policymakers face the critical challenge of ensuring that fortification strategies are scientifically robust, equitable, and sustainable over time. In particular, issues such as nutrient stability, bioavailability, and the development of appropriate monitoring frameworks are central to the ongoing debate surrounding effective fortification.

Moreover, as the landscape of global nutrition continues to evolve, it has become increasingly evident that food fortification must be approached with flexibility and adaptability. The rise in prevalence of noncommunicable diseases necessitates that fortification strategies not only address deficiencies but also take into account the overall dietary habits of populations (WHO, 2020). The integration of fortification with existing food production systems can foster long-term sustainability and enhance community engagement in health initiatives. Implementing participatory approaches, whereby community stakeholders are involved in the design and rollout of fortification programs, can ensure that these interventions meet local needs and preferences (Ruel & Alderman, 2013).

Addressing logistical challenges remains a cornerstone of effective fortification. This includes ensuring the consistent supply of fortified foods, as well as monitoring and evaluating the impact of such initiatives on population health. The development of strong partnerships between governments, nongovernmental organizations, and private sector players can facilitate resource sharing and foster innovative solutions to overcome barriers to access (Hoddinott et al., 2015). Importantly, a transparent system for ingredient sourcing and fortification standards will bolster consumer trust and promote greater uptake of fortified products.

Lastly, continual education and awareness campaigns are vital in promoting acceptance and understanding of food fortification among the public. Public health campaigns that effectively communicate the benefits of micronutrient-rich foods can help dispel myths and encourage healthier eating practices. As the body of evidence for the effectiveness of food fortification continues to grow, stakeholders must remain engaged and responsive to emerging research and population needs. In conclusion, food fortification represents a powerful strategy in the fight against malnutrition and health disparities, necessitating a multi-faceted approach that prioritizes scientific integrity, community involvement, and systematic evaluation. The following sections outline the methodologies applied in current research, implementation challenges encountered in diverse real-world settings, and essential policy recommendations necessary for enhancing the effectiveness of fortification programs.

12.2 Methodologies

12.2.1 Framework for Evaluating Food Fortification Efficacy

The methodologies employed in assessing food fortification programs typically encompass a mixture of quantitative and qualitative research. Quantitative approaches include the measurement of serum micronutrient levels, consumption surveys, and health impact assessments. Qualitative studies often supplement these findings by exploring stakeholder attitudes and the sociodemographic factors influencing program success. Recent studies (Mahboob et al., 2020; Ye et al., 2023) have emphasized the integration of these approaches to fully capture the multifaceted impacts of fortification interventions.

Standardized evaluation frameworks, such as logic models and impact pathway analyses, have been developed to guide the thorough assessment of fortification programs. These frameworks facilitate the identification of critical inputs, processes, outputs, and outcomes directly related to fortification interventions. The comprehensive framework recommended by the WHO (2020) integrates cost-effectiveness analyses with health outcome metrics to provide a holistic understanding of intervention impact, emphasizing the linkage between investment and public health improvements.

12.2.2 Data Collection and Statistical Analysis

Robust data collection methodologies are essential for producing reliable assessments of food fortification programs. Researchers employ diverse study designs, both cross-sectional and longitudinal, to measure intervention outcomes across different time periods. Data collection instruments include 24-hour dietary recalls, food frequency questionnaires, and biochemical assessments aimed at accurately measuring micronutrient levels. Advanced statistical methods such as regression analyses, randomized controlled trials (RCTs), and meta-analyses (Adebayo et al., 2021; Athe et al., 2014; Gupta, 2014) are employed to establish causal relationships between fortification and resultant health outcomes, thus enhancing the validity of findings.

Recently published meta-analyses have underscored the importance of accounting for confounding variables, such as socioeconomic status and regional dietary patterns, when assessing fortification programs. The integration of big data analytics and machine learning techniques has further refined the evaluation of public health interventions by enabling the prediction of long-term health benefits across diverse populations.

12.2.3 Nutrient Stability and Bioavailability Assessment

A critical aspect of food fortification is ensuring that added nutrients remain stable and bioavailable from production to consumption. Researchers assess nutrient stability using laboratory-based shelf-life studies across different storage conditions. Bioavailability studies, often conducted using controlled feeding trials and in vitro digestion models, evaluate the extent to which nutrients are released, absorbed, and utilized by the human body. Recent technological advances have enhanced the accuracy of these assessments (Allen, 2006; Baldelli et al., 2023; Fairweather-Tait & Teucher, 2002).

The application of encapsulation and microencapsulation techniques has improved the stability of heat-sensitive vitamins and minerals during food processing. Laboratory experiments coupled with pilot-scale studies have provided robust data on the kinetics of nutrient degradation, thereby guiding manufacturers in optimizing fortification formulations. These research efforts are crucial in ensuring that fortified foods deliver the intended public health benefits.

12.3 Implementation Challenges: An In-depth Look into Barriers

12.3.1 Technological Challenges in Fortification

The implementation of food fortification programs is frequently hindered by various technical challenges that need to be addressed systematically. These challenges include variations in fortification technology, nutrient stability issues during processing, and the maintenance of nutrient bioavailability through storage and distribution. Moreover, complex chemical interactions between added nutrients and the food matrix may hinder the efficacy of fortification interventions. For instance, iron fortification is complicated by the risk of oxidation, which can lead to nutrient loss, and the potential for taste alterations in the final product, ultimately affecting consumer acceptance and uptake (Hamad & Singh, 2025; Hurrell, 2021; Vatandoust et al., 2021).

In addition to these hurdles, careful consideration of local dietary habits is critical for the success of food fortification initiatives. Community acceptance of fortified products often hinges on cultural preferences and existing nutritional deficiencies within the population. Research indicates that engaging community members in the design and implementation of fortification programs can significantly enhance acceptance rates and compliance (Saran et al., 2025). Furthermore, public awareness campaigns to educate consumers about the benefits of fortified foods can bolster demand and facilitate a greater understanding of nutrition.

Another layer of complexity arises from regulatory frameworks that govern fortification practices. Different countries have established varying standards, often resulting in inconsistencies that can confuse manufacturers and consumers alike. Streamlining regulations while ensuring safety and efficacy may create a more conducive environment for fortification efforts across different regions (Marks et al., 2018). Additionally, collaboration between government entities, industry stakeholders, and academic institutions can foster innovation in fortification techniques, potentially leading to improved nutrient formulations that are both effective and palatable.

To further advance these programs, ongoing monitoring and evaluation are essential. This includes assessing both the immediate impact of fortification on nutrient intake and the long-term health outcomes of the target populations. Utilizing technologies such as mobile health applications can facilitate the collection of real-time data, enabling stakeholders to make necessary adjustments to fortification strategies swiftly. Ultimately, a comprehensive approach that addresses both technical and social challenges is crucial for improving the success rates of food fortification efforts globally.

Advances in fortification technologies such as encapsulated iron and stabilized vitamin formulations have been developed to address these issues. Despite these improvements, the translation of laboratory successes to large-scale industrial production remains problematic. The consistency of micronutrient dosage throughout the supply chain, from production to consumption, is critical. Quality control measures and real-time monitoring systems are increasingly being integrated into production processes to ensure that products meet prescribed fortification standards.

12.3.2 Socioeconomic and Cultural Barriers

Socioeconomic disparities and cultural resistance can impede the successful implementation of food fortification policies. In some regions, traditional dietary habits and locally produced foods are resistant to industrial fortification techniques. Policymakers must therefore balance the introduction of fortified foods with the preservation of cultural dietary practices to avoid unintended consequences such as consumer rejection or market disruptions.

Interventions such as community-based education programs and stakeholder consultations are paramount in building public trust in fortification initiatives. Recent studies (Abdisa et al., 2023) have shown that effective communication strategies that explain the benefits and safety of fortified foods can mitigate resistance and improve uptake.

Moreover, as public awareness grows, it becomes increasingly important to engage local leaders and influencers in these education initiatives. By collaborating with community figures, who already have established relationships and credibility, programs can create a more persuasive argument for the benefits of fortified foods. Various methods, such as workshops, cooking demonstrations, and taste tests, can be employed to not only educate but also encourage participation and feedback from the community, ensuring that their voices are heard in shaping the programs' effectiveness. Monitoring and evaluating these interventions should be a continuous process to assess their impact on public perception and health outcomes. Data collection mechanisms, such as surveys and interviews, can help gauge the community's understanding and acceptance of fortified foods over time. This approach will provide invaluable insights, allowing for necessary adjustments to be made to the educational content and methods used, enhancing the overall effectiveness of fortification efforts.

Furthermore, integrating fortification initiatives into broader public health campaigns can maximize their visibility and acceptance. For instance, aligning fortified food messaging with existing nutrition programs can lead to a unified front in educating the public about dietary choices and health benefits. This holistic approach not only fosters a supportive environment for fortification but also reinforces the importance of nutrition in overall well-being.

As these collaborative and adaptive strategies unfold, it is critical to maintain transparency about the processes involved in fortification, including sourcing ingredients, nutritional standards, and quality assurance. This transparency will fortify consumer confidence and encourage more widespread acceptance and participation. In conclusion, combining sound research with innovative outreach methods is essential in promoting fortified foods as a beneficial and trusted element of public health nutrition (Fig. 12.1).

12.3.3 Regulatory and Supply Chain Constraints

Regulatory frameworks governing food fortification vary considerably between regions and can present challenges in standardizing fortification levels. A lack of harmonized international standards complicates trade and implementation, particularly for multi-national food producers. Apart from this, ensuring the consistent supply of high-quality fortified premixes is a challenge, particularly in regions with underdeveloped infrastructure and limited regulatory oversight.

The WHO (2020) underscores the importance of establishing robust legal and regulatory frameworks to guide fortification efforts. Emerging research (Karki et al., 2024; Rodas-Moya et al., 2023) recommends the establishment of regional centers of excellence to standardize testing procedures and facilitate the exchange of best practices among countries. These centers could also spearhead initiatives aimed at training local personnel in quality assurance and monitoring methodologies.

Furthermore, collaborations between governmental organizations and private sector entities can play a crucial role in resource mobilization and technology transfer. As highlighted by Raile et al. (2025) public-private partnerships create a

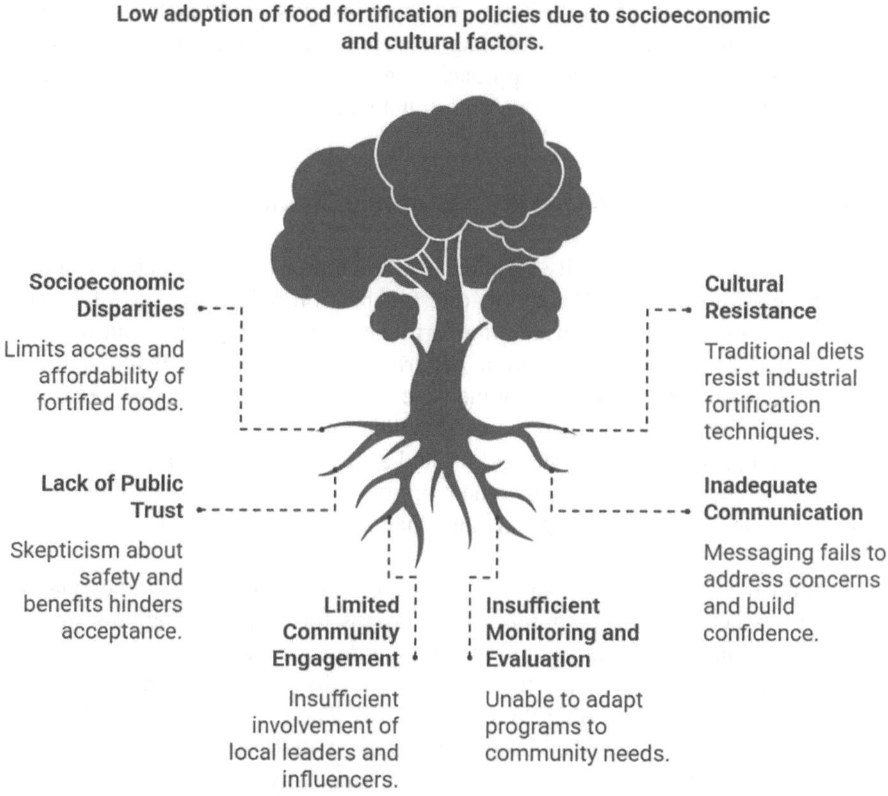

Fig. 12.1 Details on low adaptation of food fortification policies

dynamic platform for innovation, enabling access to cutting-edge technology and research that traditional government funding structures might overlook.

In addition to enhancing testing and monitoring capabilities, these centers of excellence should also focus on public education campaigns. Engaging communities through awareness programs about the benefits of food fortification can lead to improved acceptance and increased consumption of fortified products. This grassroots approach, as suggested by Dhanjani et al. (2024), is essential not only for the success of fortification initiatives but also for the empowerment of individuals to make informed dietary choices.

To further this endeavor, it is crucial to establish clear metrics for success, as outlined in the framework proposed by Damschroder et al. (2022). Evaluating the impact of fortification programs on public health outcomes will provide valuable insights and encourage continuous improvement. Regular assessments and feedback loops can help adapt strategies to meet evolving needs, ensuring that the health benefits of fortification are maximized.

Finally, fostering international cooperation and knowledge sharing among countries could reinforce these efforts. Regional summits and conferences, where stakeholders can present their findings and share insights, could create a collaborative environment that drives the global agenda for food fortification. By drawing on diverse experiences and strategic approaches, nations can unite their efforts against malnutrition and improve health outcomes on a larger scale.

12.3.4 Monitoring and Quality Assurance Challenges

Effective monitoring and quality assurance are not merely beneficial but essential to maintain and improve the efficacy of food fortification programs. The key performance indicators (KPIs) for evaluating fortification include product nutrient content, verified consumer consumption patterns, and quantifiable health outcome metrics. However, designing comprehensive monitoring systems that effectively capture these diverse elements remains a challenging endeavor. Variability in analytical techniques across laboratories, limited access to advanced testing facilities, and resource constraints in low-income countries exacerbate these challenges significantly, often leading to gaps in data that are crucial for assessing program success.

A multi-tiered monitoring strategy, incorporating both process and outcome indicators, is necessary to comprehensively assess fortification programs successfully. Recent innovations in portable analytical devices and remote sensing technologies have paved the way for real-time quality assurance and monitoring capabilities (Waller et al., 2022; Watt et al., 2019). The integration of mobile applications for data collection and cloud-based analytical platforms has further enhanced the ability to monitor fortification performance, especially in remote or underserved areas, thereby maintaining product effectiveness and safety.

12.4 Cost-Effectiveness and Public Health Impact Metrics

Numerous studies highlight the cost-effectiveness of food fortification as a public health intervention. Cost-effectiveness analyses compare the costs associated with fortification programs against the benefits realized in terms of reduced disease burden and improved nutritional status. Fortification programs typically require modest investments relative to health care savings. For example, iron and folic acid fortification programs have demonstrated a cost-to-benefit ratio that is among the most favorable in nutrition interventions (Osendarp et al., 2018; Rodrigues et al., 2021).

Public health impact metrics incorporate quantitative measures such as Disability-Adjusted Life Years (DALYs) averted, changes in nutritional deficiencies, and measurable improvements in cognitive and physical development among populations. In addition to these traditional measures, newer studies have started incorporating quality-adjusted life outcomes and return on investment (ROI) analyses that consider both direct and indirect economic benefits. This integration of various types of metrics into program evaluations not only enhances the ability of policymakers to make data-driven decisions but also provides a clearer narrative to stakeholders about the value and effectiveness of fortification initiatives.

An extensive body of literature (Makkar et al., 2022) underscores that fortification strategies yield high returns on investment, particularly when implemented at scale. These studies suggest that even modest improvements in population micronutrient intake can translate into significant public health benefits and economic productivity gains.

Furthermore, recent research has highlighted the vital role of community engagement in the success of these fortification initiatives. When local populations are actively involved in the planning and implementation processes, the likelihood of sustained uptake and adherence to fortified products increases (Dalaba et al., 2022). Educational campaigns that inform communities about the benefits of micronutrients not only enhance awareness but also foster a sense of ownership among stakeholders.

The ongoing COVID-19 pandemic has shed light on the interconnectedness of nutrition and health, reinforcing the argument for robust fortification programs. As the prevalence of micronutrient deficiencies became more pronounced during this period, policymakers have recognized the urgent need to integrate nutrition strategies into public health frameworks (WHO, 2020). This integration can lead to more comprehensive responses to both chronic and acute health challenges, thereby promoting healthier populations.

As countries strive to meet their dietary goals, innovative delivery mechanisms such as mobile fortification units and the use of digital platforms for distribution are gaining traction. These advancements offer new avenues for ensuring that fortified foods reach underserved populations, including remote and rural communities where conventional supply chains may falter (Mildon et al., 2015). In conclusion, while the economic aspects of fortification strategies are compelling, the broader social, educational, and technological dimensions play a significant role in shaping their success. Continued investment in research and evaluation is essential to refine these strategies further and to adapt them to diverse contexts, thereby maximizing their impact on global health outcomes (Fig. 12.2).

12.5 Detailed Analysis of Fortification Technology: Innovations, Its Efficacy, and Implications for Public Health

12.5.1 Advances in Fortification Methods and Ingredient Formulation

Modern fortification technology is characterized by sophisticated advances in ingredient formulation and delivery systems, which specifically enhance nutrient stability and bioavailability. Microencapsulation techniques have allowed for the controlled release of vitamins and minerals, thereby mitigating degradation during high-temperature processing and prolonged storage periods, which cause loss of

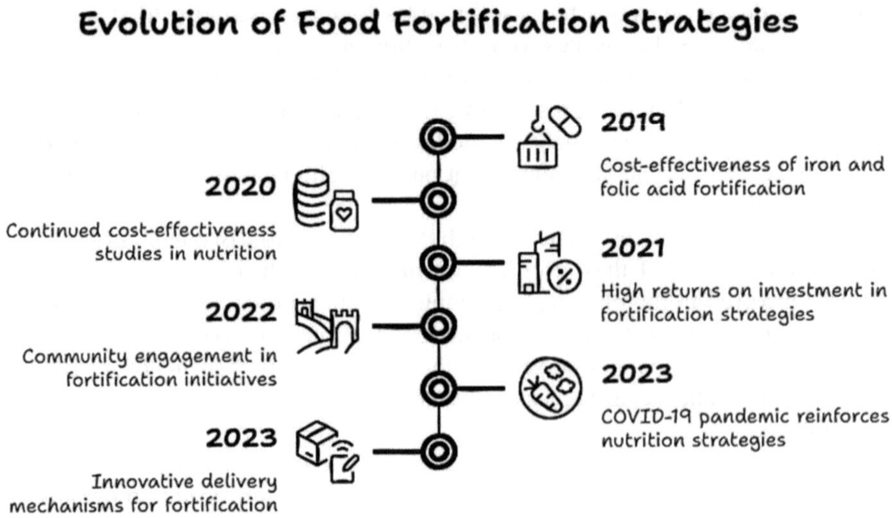

Fig. 12.2 Details on evolution of food fortification strategies

potency. For instance, research by Tang et al. (2022) demonstrated that encapsulated vitamin A exhibited enhanced stability under extreme environmental conditions compared to nonencapsulated versions, showing promise in maintaining its nutritional value after extended shelf life and heat exposure.

The recent developments in nano-encapsulation have shown promise for improving the absorption of micronutrients even at low dosages. These technologies employ nanostructured carriers that enhance solubility and gastrointestinal uptake, thereby increasing the overall efficacy of fortified products (Subramanian, 2021; Thuy et al., 2022). Given that many micronutrients are susceptible to oxidative degradation, such technological innovations represent significant strides in the field.

The ongoing research into biocompatible materials is expanding the potential applications of nano-encapsulation in various food products. For instance, recent studies have highlighted the feasibility of using plant-based polymers to create more sustainable and environmentally friendly nano-carriers. This could significantly reduce reliance on synthetic materials, which often raise concerns regarding health and safety. By employing these natural alternatives, manufacturers can not only enhance nutrient delivery but also appeal to a growing consumer preference for organic and sustainable options.

Furthermore, synergistic effects between micronutrients and bioactive compounds are being explored, which may further enhance health outcomes. Preliminary findings suggest that encapsulated vitamins, when combined with certain phytochemicals, can yield greater benefits than when consumed in isolation. This phase of research not only aims to improve nutrient bioavailability but also seeks to leverage the combined properties of these compounds to improve overall wellness.

Despite these advancements, challenges remain in the regulatory landscape. Current guidelines for encapsulated supplements often lag behind innovations in technology, leading to discrepancies in how products are marketed and perceived by consumers. As such, stakeholders in the industry are calling for updated regulations that can adapt to the pace of technological advancements while ensuring safety and efficacy.

Looking ahead, the integration of artificial intelligence and machine learning into the development process is expected to streamline the creation of more effective nano-encapsulation systems. By analyzing vast amounts of data, these tools could optimize formulations for targeted delivery and better absorption rates tailored to specific populations. This approach could prove invaluable in addressing public health issues related to micronutrient deficiencies, particularly in vulnerable groups such as children and pregnant women.

Ultimately, as these technologies mature, they hold promise not only for enhancing the nutritional quality of foods but also for contributing to global health initiatives aimed at reducing malnutrition. By prioritizing innovation and collaboration across disciplines, we can build a more resilient food system capable of meeting the nutritional needs of diverse populations while promoting sustainability.

Nutrient stability is a critical determinant of the effectiveness of a fortification program. Various extrinsic and intrinsic factors influence nutrient stability, including pH levels, temperature, humidity, and interactions with other food components. It is well documented that certain micronutrients are sensitive to environmental conditions; for instance, vitamin C degrades rapidly in the presence of oxygen and heat (Yin et al., 2022).

To address these challenges, food technologists have developed stabilization strategies designed thoughtfully to ensure nutrient integrity throughout the food manufacturing process. These strategies include the incorporation of antioxidants to minimize oxidative damage and the use of protective coatings that shield sensitive nutrients from deleterious conditions during food processing. Further innovative research (Bamigbade et al., 2025; Edo et al., 2025) has highlighted the potential of novel polysaccharide-based carriers, which offer enhanced stabilization capabilities without compromising functional properties, thereby ensuring the bioavailability of nutrients post-consumption (Fig. 12.3).

12.5.2 Bioavailability Factors and Their Public Health Implications

Bioavailability, defined as the proportion of a nutrient that is absorbed and utilized by the body, is a central concern in food fortification. Numerous factors influence bioavailability, including the chemical form of the nutrient, the presence of inhibitors or enhancers in the meal, and individual metabolic differences. The use of chelated minerals, for example, has been shown to enhance bioavailability by protecting minerals from antagonistic interactions in the gastrointestinal tract (Biabani et al., 2024; Mohammadizad et al., 2024).

Additionally, research on fortification has increasingly focused on understanding how different dietary matrices influence nutrient absorption. Studies (Devarshi

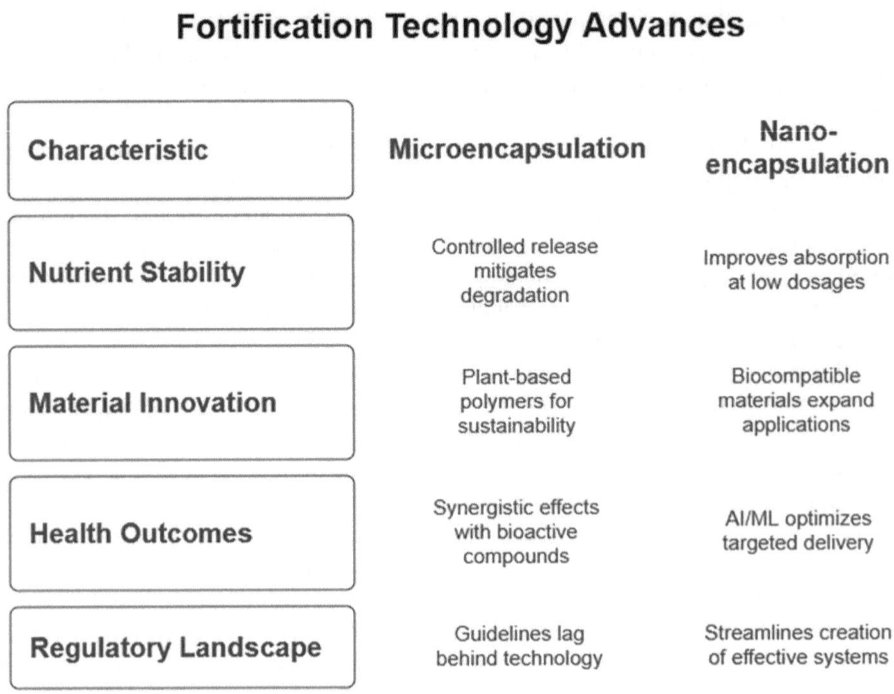

Fig. 12.3 Details on fortification technology advancements

et al., 2024; Prynne et al., 2010) emphasize the significance of meal composition, noting that certain dietary fibers and phytates can inhibit the uptake of iron and zinc. As such, public health practitioners must consider both the nutrient formulation and the typical dietary patterns of target populations to maximize the benefits of fortification interventions.

Furthermore, the interaction between various nutrients in fortified foods also warrants attention. For instance, the presence of fat-soluble vitamins can be affected by the presence of polyunsaturated fatty acids, which might alter their bioavailability in different meal contexts. Research has also shown that synergistic effects can enhance absorption rates; for example, vitamin C is known to facilitate iron absorption when consumed together (Pan et al., 2024). Therefore, strategic combinations of fortificants might yield superior outcomes in nutrient retention and absorption. The longitudinal studies tracking the long-term impacts of fortification programs reveal the necessity of ongoing evaluation and adaptation. Data collected from various demographics over time indicate that preferences and dietary behaviors can shift, influenced by cultural trends, economic changes, and accessibility to food sources. This evolving landscape requires that public health initiatives remain flexible, allowing for adjustments to fortification strategies that align with emerging dietary habits.

Collaboration between researchers, policymakers, and local communities is crucial in this context. By engaging with stakeholders, health practitioners can better understand localized dietary practices and incorporate traditional foods in fortification schemes. This approach not only respects cultural contexts but also enhances the likelihood of acceptance and sustained consumption of fortified products.

Finally, the integration of technology in monitoring and evaluating the effectiveness of fortification programs holds promise. Advances in nutritional epidemiology and data analytics can provide real-time insights into nutrient status within populations, enabling more targeted and effective interventions. Ultimately, addressing the multifaceted nature of nutrient absorption and understanding the broader dietary implications will bolster the effectiveness of fortification as a public health strategy.

12.6 Quantitative Impact Assessment of Food Fortification

The quantitative assessment of food fortification programs involves measuring both direct and indirect health outcomes. Direct outputs include measurable changes in micronutrient levels, reductions in deficiency-related morbidity, and improvements in child growth indicators. Indirect outputs encompass broader socioeconomic benefits, such as enhanced work capacity and reduced healthcare costs.

Further, the application of longitudinal studies across various demographics and settings has enabled researchers to track the sustained benefits of fortification programs over multiple years, revealing important insights about long-term public health impacts. Data from large-scale fortification initiatives in Latin America, for instance, have demonstrated significant improvements in cognitive and motor development in children, with concomitant reductions in anemia incidence, providing compelling evidence for the effectiveness of these interventions (Medise, 2021; Tam et al., 2020). The application of longitudinal studies has enabled researchers to track the sustained benefits of fortification programs over multiple years. Data from large-scale fortification initiatives in Latin America demonstrated significant improvements in cognitive and motor development in children, with concomitant reductions in anemia incidence (Darnton-Hill et al., 1999; Roberts et al., 2022).

12.7 Key Performance Indicators and Monitoring Tools

Effective monitoring and evaluation (M&E) strategies are critical to ensuring that food fortification initiatives achieve targeted public health outcomes. The development of key performance indicators (KPIs) for fortification programs enables both policymakers and public health professionals to assess process efficiency, product quality, and health impacts systematically.

KPIs in food fortification include:

- Product nutrient content consistency, as measured by laboratory assays.
- Consumer consumption patterns and market penetration metrics.

- Biochemical indicators of nutritional status within the target population.
- Cost-effectiveness metrics, including cost savings per DALY averted.
- Quality assurance scores derived from internal audits and third-party reviews.

Recently, several innovative monitoring tools have been deployed to refine data collection and analysis. For instance, mobile-based platforms now allow field workers to conduct rapid nutrient assessments using portable devices. Cloud-based data systems further facilitate the aggregation and analysis of data from disparate regions (Gupta & Sharma, 2023; Xu et al., 2022). These approaches are instrumental in ensuring that implementation deviations are promptly identified and corrected.

Impact assessment tools such as randomized community trials and spatial epidemiological models have been used to predict and measure the reach of fortification programs over time. The integration of these monitoring approaches with cost-effectiveness analyses enables a comprehensive understanding of both program performance and its long-term sustainability. Moreover, the evolving landscape of public health interventions necessitates the continuous refinement of these tools. As technology advances, new data collection methods such as mobile health applications and big data analytics are increasingly used to enhance the precision of impact assessments. For instance, location-based data can uncover geographic disparities in nutrient deficiencies, thereby enabling targeted fortification efforts in at-risk communities.

The stakeholder engagement also plays a crucial role in the success of these programs. By actively involving local communities, health professionals, and policymakers in the planning and implementation phases, programs can be tailored to suit cultural preferences and local dietary habits. This collaborative approach not only fosters a sense of ownership among the community members but also enhances the credibility and acceptance of the fortification initiatives.

The ongoing training for local health workers is essential to ensure that they are well equipped to monitor and educate the community about the benefits of fortified products. Regular updates to training modules that reflect new scientific evidence and community feedback can help maintain the momentum of fortification programs. This cyclical process of evaluation and adaptation forms the backbone of a resilient public health strategy.

At last, the long-term sustainability of these programs relies on securing funding and support from both governmental and nongovernmental organizations. Establishing public-private partnerships can provide additional resources and encourage innovation in product development and distribution channels. By leveraging a multi-sectoral approach, the overall impact of fortification programs can be amplified, paving the way for a healthier future for vulnerable populations.

12.8 Global Research Trends and WHO Guidelines in Food Fortification: An Overview of Current Insights and Future Directions

In recent years, global research has increasingly focused on the optimization of food fortification strategies, with a strong emphasis on technological innovation, cost-effectiveness, and environmental sustainability. The latest WHO Food Fortification Guidelines (WHO, 2020) serve as an authoritative reference that consolidates evidence from various regions and contexts, guiding appropriate nutrient levels, safety margins, and monitoring mechanisms.

Current global research trends have underscored the following themes:

- The optimization of nutrient formulations through novel encapsulation techniques.
- The integration of real-time monitoring and evaluation frameworks.
- The assessment of sociocultural factors that influence program adoption.
- The development of international regulatory standards and harmonized guidelines.
- The quantification of direct and indirect economic benefits of fortification programs.

These collaborative research initiatives, supported by advanced statistical and analytical methodologies, underscore the critical role of food fortification in achieving global nutrition security. In aligning with the WHO guidelines, these efforts not only emphasize the importance of adopting evidence-based interventions tailored specifically to the nutritional needs and dietary practices of the population in question but also promote shared learning among diverse stakeholders to optimize fortification effectiveness.

These research initiatives are supported by advanced statistical and analytical methodologies, which highlight the critical role of food fortification in global nutrition security. The WHO guidelines emphasize the importance of adopting evidence-based interventions that are tailored to the nutritional needs and dietary practices of the population in question.

12.9 Policy Recommendations

Based on the extensive review of current research and practice, several policy recommendations emerge for the successful implementation and scaling up of food fortification programs:

12.9.1 Strengthening Regulatory Frameworks

National governments and international bodies should work collaboratively to develop and enforce harmonized regulatory standards for food fortification. Establishing clear benchmarks for nutrient addition, stability, and bioavailability is critical to ensuring product quality across all sectors. Policymakers should invest in the creation of regional centers of excellence that can support training, laboratory testing, and quality assurance initiatives.

The WHO 2020 guidelines recommend the establishment of mandatory fortification standards, supported by legal frameworks that ensure consistent implementation. Enhanced regulatory oversight, coupled with robust monitoring systems, will build consumer confidence and ensure that fortified foods meet population health needs.

12.9.2 Enhancing Technological Innovation and Quality Assurance

Federal and local governments should prioritize investments in research and development to foster technological innovations in food fortification. Adoption of newer technologies such as nano-encapsulation and advanced monitoring devices will significantly improve nutrient stability and absorption. Policymakers should encourage public-private partnerships aimed at bridging the gap between laboratory research and industrial production.

The implementation of rigorous quality assurance protocols is paramount for ensuring program effectiveness. Investing in the development of continuous quality improvement programs, which employ standardized KPIs and real-time data collection tools, will ensure that fortification programs consistently achieve the desired public health outcomes and meet the nutritional needs of target populations. Cross-sector collaborations that involve industry stakeholders, academic institutions, and nongovernmental organizations can further enhance the processes of quality assurance, ultimately contributing to improved health outcomes.

Furthermore, the integration of innovative technologies in these quality assurance frameworks cannot be overlooked. Utilizing mobile applications for data collection, as well as blockchain for traceability of nutrient sources, represents a forward-thinking approach to ensure transparency and accountability within the supply chain. This technological advancement not only improves data accuracy but also facilitates immediate responses to potential deficiencies or contamination issues, ensuring that the nutritional programs remain effective and safe for consumption (Saran et al., 2025).

The ongoing training and capacity building for program implementers are essential elements that must be prioritized. By equipping local health workers and program managers with the latest knowledge and skills related to nutritional standards and quality assurance practices, we can foster a culture of excellence and responsiveness within communities (Sultan et al., 2025). This empowerment leads to sustainability as community stakeholders become more invested in the health outcomes of their populations.

In conclusion, the synergy between quality assurance protocols, technological innovation, and human capacity development holds the key to advancing nutritional fortification programs. With dedicated efforts and collaboration among all stakeholders, it is possible to achieve significant improvements in public health and ensure that vulnerable populations receive the nutrients they need to thrive (WHO, 2020).

12.9.3 Addressing Socioeconomic and Cultural Barriers

To improve the uptake of fortified products, policies must be rooted in an understanding of local dietary practices and cultural norms. Community engagement initiatives and educational campaigns are critical to dispelling myths and building trust in fortification interventions. Policymakers should work with local leaders to customize public messaging that resonates with the community's values and addresses concerns regarding taste, safety, and efficacy.

Designed the specific strategies to subsidize or otherwise incentivize the use of fortified foods should be implemented effectively and inclusively, especially in regions with high levels of poverty and food insecurity. Evidence suggests that integrating conditional cash transfer programs with fortification initiatives can further improve adoption rates and overall nutritional outcomes, as these approaches not only enhance access but also address economic barriers that hinder the consumption of fortified products (Kronebusch & Damon, 2019; Ramos et al., 2021).

12.9.4 Integrating Cost-Effectiveness into Public Health Planning

Policymakers should integrate comprehensive cost-effectiveness analyses into the planning and evaluation of fortification programs systematically. By clearly demonstrating the long-term health and economic benefits of these interventions to stakeholders, public health advocates can secure greater political and financial commitment. Tools such as return on investment (ROI) assessments and disability adjusted life year (DALY) based cost effectiveness analyses should be routinely employed as standardized measures to track the performance of fortification initiatives and support resource allocation toward high-impact interventions.

Investment in fortification programs should be framed not only as a nutritional intervention but also as a crucial economic strategy. The use of transparent and standardized financial metrics will support policy decisions and facilitate the allocation of resources toward high-impact interventions. Moreover, international funding agencies and development banks should prioritize funding for programs with proven cost-effectiveness and broad public health impact.

In addition to these financial considerations, public-private partnerships can play a pivotal role in scaling up fortification initiatives. By leveraging the expertise and resources of the private sector, governments can enhance the reach and effectiveness of their programs. Strategic collaborations could include joint ventures with local food manufacturers to ensure that fortified products are widely available and affordable, thus increasing access among vulnerable populations.

Furthermore, it is vital to establish robust monitoring and evaluation frameworks. These systems not only track the progress of fortification programs but also provide data necessary for refining strategies to maximize impact. Regular assessments can reveal insights into dietary patterns, public acceptance of fortified foods, and areas where awareness needs to be improved. This ongoing evaluation helps sustain momentum and justify future investments in health interventions. For example, a study published by the Global Alliance for Improved Nutrition emphasizes the need for continuous monitoring to adapt to changing demographics and nutritional landscapes (Global Alliance for Improved Nutrition, 2021).

Lastly, community engagement is critical to the success of fortification programs. Stakeholder involvement, particularly from the populations that benefit from these interventions, ensures that the initiatives are culturally appropriate and meet the actual needs of the community. Educational campaigns can raise awareness about the importance of micronutrients and the benefits of consuming fortified foods, ultimately empowering individuals to make informed dietary choices.

In conclusion, fortification programs represent a vital intersection of nutrition and economic development. By investing in these initiatives, utilizing standardized metrics for transparency, fostering public-private partnerships, establishing solid monitoring frameworks, and engaging communities, governments can navigate the complexities of malnutrition and promote sustainable health improvements. Only through a concerted and collaborative approach can we expect to tackle the challenges posed by nutrient deficiencies on a global scale.

12.9.5 Enhancing Monitoring, Evaluation, and Impact Assessment

A robust and resilient monitoring and evaluation system is indispensable for operational success. Policymakers should mandate the use of innovative tools that integrate real-time data collection, cloud-based analytics, and remote sensing technologies to continuously track program performance. The development of an international fortification monitoring framework, modeled on the WHO guidelines, would enable cross-country comparisons and the sharing of best practices.

Multisectoral collaboration is essential in designing impact assessment protocols that incorporate both process and outcome indicators. This approach must include periodic audits, community feedback mechanisms, and the use of contemporary tools such as mobile health (mHealth) platforms. Enhanced M&E systems are critical not only for verifying the efficacy of fortification programs but also for making timely adjustments to ensure sustainable public health outcomes.

12.10 Conclusion

In summary, the public health dimensions of food fortification reflect a complex interplay of technological innovation, rigorous regulatory oversight, socioeconomic factors, and robust evaluation mechanisms. The extensive body of research reviewed herein confirms that food fortification is among the most effective interventions for reducing micronutrient deficiencies and enhancing public health outcomes globally. Drawing from evidence spanning biochemical studies, comprehensive economic evaluations, and community-based research, food fortification emerges as a multi-dimensional strategy with significant potential for long-term impact and success.

As global health challenges continue to evolve in complexity, further investments in research, innovation, and systematic evaluation of fortification programs will be essential to ensure their effectiveness and sustainability. It is incumbent upon both national governments and international organizations to actively collaborate in the pursuit of nutrition security, particularly for vulnerable populations who bear the brunt of micronutrient deficiencies. With a clear evidence base and comprehensive policy guidance, food fortification stands as a critical tool in the global fight against micronutrient deficiencies and serves as a key contributor to improved health outcomes worldwide.

As global health challenges continue to evolve, further investments in research, innovation, and systematic evaluation of fortification programs will be essential. It is incumbent upon both national governments and international organizations to collaborate in the pursuit of nutrition security, particularly for vulnerable populations. With a clear evidence base and comprehensive policy guidance, food fortification stands as a critical tool in the fight against micronutrient deficiencies and a key contributor to improved global health outcomes.

Ultimately, the path forward requires concerted efforts, coordinated policies, and persistent innovation. The integration of quantitative impact assessments, advanced fortification technologies, and community-focused strategies will ensure that food fortification remains a responsive and adaptable component of public health practice well into the future.

References

Abdisa, D. K., Jaleta, D. D., Tsegaye, D., Jarso, M. H., Jaleta, G. D., Tolesa, G. F., & Kitila, K. M. (2023). Effect of community based nutritional education on knowledge, attitude and compliance to IFA supplementation among pregnant women in rural areas of Southwest Ethiopia: A quasi experimental study. *BMC Public Health, 23*(1), 1923.

Adebayo, F. A., Itkonen, S. T., Öhman, T., Kiely, M., Cashman, K. D., Lamberg-Allardt, C., & Odin Consortium. (2021). Safety of vitamin D food fortification and supplementation: Evidence from randomized controlled trials and observational studies. *Food, 10*(12), 3065.

Allen, L. (2006). *Guidelines on food fortification with micronutrients*. World Health Organization.

Athe, R., Rao, M. V. V., & Nair, K. M. (2014). Impact of iron-fortified foods on Hb concentration in children (<10 years): A systematic review and meta-analysis of randomized controlled trials. *Public Health Nutrition, 17*(3), 579–586.

Baldelli, A., Ren, M., Liang, D. Y., Lai, S., Hartono, B., Sum, K., & Pratap-Singh, A. (2023). Sprayed microcapsules of minerals for fortified food. *Journal of Functional Foods, 101*, 105401.

Bamigbade, G. B., Abdin, M., Subhash, A., Arachchi, M. P., Ullah, N., Gan, R. Y., et al. (2025). Plant polysaccharide-capped nanoparticles: A sustainable approach to modulate gut microbiota and advance functional food applications. *Comprehensive Reviews in Food Science and Food Safety, 24*(2), e70156.

Biabani, N., Taherpour, K., Ghasemi, H. A., Akbari Gharaei, M., Hafizi, M., & Nazaran, M. H. (2024). Advanced chelate technology-based trace minerals reduce inflammation and oxidative stress in Eimeria-infected broilers by modulating NF-kB and Nrf2 pathways. *Scientific Reports, 14*(1), 24227.

Dalaba, M. A., Nonterah, E. A., Chatio, S. T., Adoctor, J. K., Dambayi, E., Nonterah, E. W., et al. (2022). Engaging community members in setting priorities for nutrition interventions in rural northern Ghana. *PLOS Global Public Health, 2*(9), e0000447.

Damschroder, L. J., Reardon, C. M., Widerquist, M. A. O., & Lowery, J. (2022). The updated consolidated framework for implementation research based on user feedback. *Implementation Science, 17*(1), 75.

Darnton-Hill, I., Mora, J. O., Weinstein, H., Wilbur, S., & Nalubola, P. R. (1999). Iron and folate fortification in the Americas to prevent and control micronutrient malnutrition: An analysis. *Nutrition Reviews, 57*(1), 25–31.

Devarshi, P. P., Mao, Q., Grant, R. W., & Hazels Mitmesser, S. (2024). Comparative absorption and bioavailability of various chemical forms of zinc in humans: A narrative review. *Nutrients, 16*(24), 4269.

Dhanjani, S., Allen, H., Varman, B., Callender, C., Dave, J. M., & Thompson, D. (2024). Community-based participatory obesity prevention interventions in rural communities: A scoping review. *Nutrients, 16*(14), 2201.

Edo, G. I., Mafe, A. N., Akpoghelie, P. O., Gaaz, T. S., Yousif, E., Yusuf, O. S., et al. (2025). The utilization of biopolymer hydrogels to encapsulate and protect probiotics in foods. *Process Biochemistry, 153*, 66–91.

Fairweather-Tait, S. J., & Teucher, B. (2002). Iron and calcium bioavailability of fortified foods and dietary supplements. *Nutrition Reviews, 60*(11), 360–367.

Global Alliance for Improved Nutrition. (2021). *Annual report 2020–2021*. https://www.gainhealth.org/sites/default/files/annual_reports/GAIN-Annual%20Report%2020-21_0.pdf

Gupta, A. (2014). Fortification of foods with vitamin D in India. *Nutrients, 6*(9), 3601–3623.

Gupta, U., & Sharma, R. (2023). A study of cloud-based solution for data analytics. In *Data analytics for internet of things infrastructure* (pp. 145–161). Cham.

Hamad, A., & Singh, P. (2025). Boosting nutritional value: The role of iron fortification in meat and meat products. *Biometals, 38*, 1–19.

Hoddinott, J. F., Gillespie, S., & Yosef, S. (2015). *Public-private partnerships and the reduction of undernutrition in developing countries* (Vol. 1487). The International Food Policy Research Institute.

Hurrell, R. F. (2021). Iron fortification practices and implications for iron addition to salt. *The Journal of Nutrition, 151*, 3S–14S.

Karki, R., Ojha, P., Dongol, D. M. S., Maharjan, S., Manandhar, U., & Maharjan, S. (2024). Food fortification: Global experience, importance, challenges and potential in Nepal. *Journal of Food Science and Technology Nepal*. https://doi.org/10.3126/jfstn.v14i14.71713

Kronebusch, N., & Damon, A. (2019). The impact of conditional cash transfers on nutrition outcomes: Experimental evidence from Mexico. *Economics & Human Biology, 33*, 169–180.

Mahboob, U., Ohly, H., Joy, E. J., Moran, V., Zaman, M., & Lowe, N. M. (2020). Exploring community perceptions in preparation for a randomised controlled trial of biofortified flour in Pakistan. *Pilot and Feasibility Studies, 6*, 1–11.

Makkar, S., Minocha, S., Bhat, K. G., John, A. T., Swaminathan, S., Thomas, T., et al. (2022). Iron fortification through universal distribution of double-fortified salt can increase wages and be cost-effective: An ex-ante modeling study in India. *The Journal of Nutrition, 152*(2), 597–611.

Marks, K. J., Luthringer, C. L., Ruth, L. J., Rowe, L. A., Khan, N. A., De-Regil, L. M., et al. (2018). Review of grain fortification legislation, standards, and monitoring documents. *Global Health: Science and Practice, 6*(2), 356–371.

Medise, B. E. (2021). The role of iron for supporting children's growth and development. *World Nutrition Journal, 5*(S1), 16–24.

Mildon, A., Klaas, N., O'Leary, M., & Yiannakis, M. (2015). Can fortification be implemented in rural African communities where micronutrient deficiencies are greatest? Lessons from projects in Malawi, Tanzania, and Senegal. *Food and Nutrition Bulletin, 36*(1), 3–13.

Mohammadizad, T., Taherpour, K., Ghasemi, H. A., Shirzadi, H., Tavakolinasab, F., & Nazaran, M. H. (2024). Potential benefits of advanced chelate-based trace minerals in improving bone mineralization, antioxidant status, immunity, and gene expression modulation in heat-stressed broilers. *PLoS One, 19*(10), e0311083.

Osendarp, S. J., Martinez, H., Garrett, G. S., Neufeld, L. M., De-Regil, L. M., Vossenaar, M., & Darnton-Hill, I. (2018). Large-scale food fortification and biofortification in low-and middle-income countries: A review of programs, trends, challenges, and evidence gaps. *Food and Nutrition Bulletin, 39*(2), 315–331.

Pan, X., Köberle, M., & Ghashghaeinia, M. (2024). Vitamin C-dependent uptake of non-heme iron by enterocytes, its impact on erythropoiesis and redox capacity of human erythrocytes. *Antioxidants, 13*(8), 968.

Prynne, C. J., McCarron, A., Wadsworth, M. E., & Stephen, A. M. (2010). Dietary fibre and phytate – a balancing act: Results from three time points in a British birth cohort. *British Journal of Nutrition, 103*(2), 274–280.

Raile, E. D., Austin, E. K., Wallner, M. P., Peterson, J., Lewandowski, B., Kapps, D., et al. (2025). Technology license agreements as public–private partnerships for economic development: Evaluations using surveys, input–output modeling, and regression analysis. *Journal of Innovation and Entrepreneurship, 14*(1), 14.

Ramos, D., da Silva, N. B., Ichihara, M. Y., Fiaccone, R. L., Almeida, D., Sena, S., et al. (2021). Conditional cash transfer program and child mortality: A cross-sectional analysis nested within the 100 million Brazilian cohort. *PLoS Medicine, 18*(9), e1003509.

Roberts, M., Tolar-Peterson, T., Reynolds, A., Wall, C., Reeder, N., & Rico Mendez, G. (2022). The effects of nutritional interventions on the cognitive development of preschool-age children: A systematic review. *Nutrients, 14*(3), 532.

Rodas-Moya, S., Giudici, F. M., Owolabi, A., Samuel, F., Kodish, S. R., Lachat, C., et al. (2023). A generic theory of change-based framework with core indicators for monitoring the effectiveness of large-scale food fortification programs in low-and middle-income countries. *Frontiers in Nutrition, 10*, 1163273.

Rodrigues, V. B., Silva, E. N. D., & Santos, M. L. P. (2021). Cost-effectiveness of mandatory folic acid fortification of flours in prevention of neural tube defects: A systematic review. *PLoS One, 16*(10), e0258488.

Ruel, M. T., & Alderman, H. (2013). Nutrition-sensitive interventions and programmes: How can they help to accelerate progress in improving maternal and child nutrition? *The Lancet, 382*(9891), 536–551.

Saran, R., Kasar, J., Jha, M., Sahu, J. K., Babu, S. S., & Mutreja, A. (2025). Building blocks of life: Improving nutrition and health outcomes through fortification and breastfeeding in the first 1000 days of a child's life. *Frontiers in Nutrition, 12*, 1517247.

Subramanian, P. (2021). Lipid-based nanocarrier system for the effective delivery of nutraceuticals. *Molecules, 26*(18), 5510.

Sultan, M. A., Miller, E., Tikkanen, R. S., Singh, S., Kullu, A., Cometto, G., et al. (2025). Competency-based education and training for community health workers: A scoping review. *BMC Health Services Research, 25*(1), 263.

Tam, E., Keats, E. C., Rind, F., Das, J. K., & Bhutta, Z. A. (2020). Micronutrient supplementation and fortification interventions on health and development outcomes among children under-five in low-and middle-income countries: A systematic review and meta-analysis. *Nutrients, 12*(2), 289.

Tang, W., Zhuang, J., Anselmo, A. C., Xu, X., Duan, A., Zhang, R., et al. (2022). Enhanced stability and clinical absorption of a form of encapsulated vitamin a for food fortification. *Proceedings of the National Academy of Sciences, 119*(51), e2211534119.

Thuy, V. N., Van, T. V., Dao, A. H., & Lee, B. J. (2022). Nanostructured lipid carriers and their potential applications for versatile drug delivery via oral administration. *OpenNano, 8*, 100064.

Vatandoust, A., Mannar, M. V., & Diosady, L. L. (2021). Organoleptic effects of salt fortification with iron and iodine: A review. *The Journal of Nutrition, 151*(7), 1690–1702.

Waller, A. W., Gaytán-Martínez, M., & Andrade Laborde, J. E. (2022). In-country method validation of a paper-based, smartphone-assisted iron sensor for corn flour fortification programs. *Foods, 11*(3), 276.

Watt, M. S., Pearse, G. D., Dash, J. P., Melia, N., & Leonardo, E. M. C. (2019). Application of remote sensing technologies to identify impacts of nutritional deficiencies on forests. *ISPRS Journal of Photogrammetry and Remote Sensing, 149*, 226–241.

World Health Organization. (2014). *Guideline: Fortification of food-grade salt with iodine for the prevention and control of iodine deficiency disorders*. WHO.

World Health Organization. (2020). *WHO guideline development group meeting: Fortification of wheat flour with vitamins and minerals in public health*. World Health Organization.

Xu, C., Du, X., Fan, X., Giuliani, G., Hu, Z., Wang, W., et al. (2022). Cloud-based storage and computing for remote sensing big data: A technical review. *International Journal of Digital Earth, 15*(1), 1417–1445.

Ye, R., Wu, Y., Sun, C., Wang, Q., Mao, Y., Zhou, H., & Raat, H. (2023). Diffusion of a micronutrient home fortification program for infants and toddlers in a multi-ethnic population in rural western China. *BMC Public Health, 23*(1), 889.

Yin, X., Chen, K., Cheng, H., Chen, X., Feng, S., Song, Y., & Liang, L. (2022). Chemical stability of ascorbic acid integrated into commercial products: A review on bioactivity and delivery technology. *Antioxidants, 11*(1), 153.

Neuro-Nutrition: Fortifying Foods for Cognitive Well-Being

13

Prachi Patel, Ashok Kumar Bishoyi, and Gaurav Sanghvi

Abstract

The brain needs daily nutrition to function, repair, and maintain behavior, cognition, and physical regulation. Neuro-nutrition studies how vitamins, minerals, fats, and antioxidants affect brain growth and function. This chapter reviews these interactions' scientific understanding using peer-reviewed studies from the past 5 years. Brain chemistry, nutritional benefits, and cognitive-optimizing diets are detailed. Sample meal plans and daily strengthening items are included. The information presented here shows how diet can boost brain health and cognitive performance. The brain needs daily nutrition to function, repair, and maintain behavior, cognition, and physical regulation. Neuro-nutrition studies how vitamins, minerals, fats, and antioxidants affect brain growth and function. This chapter reviews these interactions' scientific understanding using peer-reviewed studies from the past 5 years. Brain chemistry, nutritional benefits, and cognitive-optimizing diets are detailed. Sample meal plans and daily strengthening items are included. The information presented here shows how diet can boost brain health and cognitive performance. Omega-3 fatty acids, whole grains, and fresh fruits and vegetables in the Mediterranean diet may reduce cognitive decline, according to new studies. These nutrition choices and water are vital since even mild dehydration can impair concentration and cognition. Water-rich foods like cucumbers and oranges may help one assess their hydration and make better decisions. Neuro-nutrition research will uncover deeper linkages between diet

P. Patel
Department of Botany, The Maharaja Sayajirao University of Baroda, Vadodara, Gujarat, India

A. K. Bishoyi · G. Sanghvi (✉)
Department of Microbiology, Marwadi University, Rajkot, Gujarat, India

G. Sanghvi et al. (eds.), *Sustainable Food Fortification*,
https://doi.org/10.1007/978-981-95-1217-1_13

and cognition, enabling cognitively tailored nutrition planning. Brain-healthy diets and a balanced lifestyle can extend cognitive life.

Keywords

Neuro-nutrition · Cognitive performance · Brain health · Diet and cognition · Micronutrients · Mediterranean diet · Nutritional neuroscience

13.1 Introduction

As modern society continues to face challenges associated with aging, neurodegenerative disorders, and cognitive decline, the quest for interventions that might delay or mitigate these processes has never been more critical. Diet plays a crucial role in modulating several aspects of brain function, including neurotransmission, synaptic plasticity, neurogenesis, and energy metabolism. Importantly, scientific advances over the last decade have begun to elucidate how specific nutrients interact with brain systems to enhance cognitive performance.

Neuro-nutrition merges the fields of neuroscience and nutrition, creating a transformative paradigm for understanding how what we eat influences mental faculties, mood, and overall brain health. From the role of omega-3 fatty acids in optimizing neuronal membrane fluidity to the importance of antioxidants in curbing neuroinflammation, the evidence is mounting that diet is a modifiable factor in brain function. The purpose of this chapter is to provide a comprehensive, scientifically rigorous overview of these relationships, complete with specific nutrient recommendations and practical dietary strategies.

Recent studies have highlighted the impact of micronutrients on cognitive performance and emotional well-being. For instance, vitamin D has been linked to neuroprotection and has shown potential in reducing the risk of neurodegenerative conditions such as Alzheimer's disease (Llewellyn et al., 2010). Similarly, B vitamins—particularly B6, B12, and folate—play a critical role in the synthesis of neurotransmitters and the maintenance of myelin, both essential for optimal cognitive function (Kennedy, 2016). A deficiency in these vitamins can lead to mood disorders and cognitive decline, underscoring the need for dietary sources of these nutrients.

Moreover, the emerging field of the gut-brain axis has illuminated the significant role of the microbiome in mental health. Fermented foods, rich in probiotics, can positively influence gut bacteria, promoting the production of neurotransmitters like serotonin—about 90% of which is produced in the gut (Anderson et al., 2017). This connection reinforces the idea that a diet high in fiber and fermented products may not only aid digestion but also enhance mental health.

Nutritional interventions, such as the Mediterranean diet, have been associated with reduced cognitive decline and improved mood. This diet emphasizes high consumption of fruits, vegetables, whole grains, legumes, fish, and healthy fats, with evidence suggesting its rich antioxidant and anti-inflammatory properties contribute to enhanced brain health (Martínez-Lapiscina et al., 2013). Incorporating specific foods like blueberries and dark chocolate, both known for their high levels of flavonoids, has also been shown to improve cognitive function (Wang, 2021).

In conclusion, the connection between nutrition and brain health is becoming increasingly clear. The integration of neuro-nutrition into everyday dietary practices and public health policies could serve as a powerful tool to promote mental well-being and cognitive resilience. As researchers continue to unravel these intricate relationships, it becomes evident that prioritizing nutrition is essential for nurturing both our bodies and minds.

The guide is organized into three primary sections: first, an examination of brain chemistry and the specific biochemical pathways that underlie cognitive performance; second, a detailed look at essential nutrients and the fortifying foods that support them; and third, actionable strategies for implementing a neuro-nutrition based diet in daily life. By combining these elements, readers will acquire a robust understanding of the interplay between diet and brain function and gain practical insights into how nutritional choices can lead to measurable improvements in cognitive performance.

To further enhance understanding, the guide will delve into the role of specific nutrients like omega-3 fatty acids, antioxidants, and B vitamins in enhancing neuroplasticity—the brain's ability to adapt and grow. Research indicates that omega-3 fatty acids found in fish such as salmon and walnuts not only contribute to brain structure but also support synaptic function, which is crucial for learning and memory (Innis, 2008). Additionally, antioxidants, predominantly found in berries and green leafy vegetables, combat oxidative stress, a factor linked to cognitive decline. Thus, understanding these nutrients' effects can empower readers to make better dietary choices tailored to enhance their cognitive resilience.

The second section will comprehensively explore nutrient-dense foods that serve as the building blocks for improved brain health. Foods such as avocados, blueberries, and quinoa are rich in compounds that promote cognitive function. For example, berries are packed with flavonoids that have been shown to enhance communication between brain cells, thereby improving memory (Joseph et al., 2005). Moreover, incorporating whole grains into the diet can aid in steady glucose delivery to the brain, which is essential for maintaining focus and attention throughout the day (Wirström et al., 2013). This balanced diet can ultimately create a more optimistic environment for cognitive engagement.

Finally, the guide will address the practical aspect of adopting a neuro-nutrition-based diet through everyday habits and meal planning strategies. Meal prepping, mindful eating practices, and the integration of superfoods into familiar recipes can make the transition to a brain-healthy diet achievable and sustainable. Simple tactics, such as creating a weekly menu centered on neuro-nutrition principles or exploring new recipes that incorporate a variety of brain-boosting ingredients, can greatly simplify the process. The guide aims to inspire individuals to take actionable steps towards enhancing their cognitive capabilities through informed dietary choices while dispelling the myth that healthy eating must come at the cost of enjoyment (Jiwani et al., 2022).

13.2 Brain Chemistry and Neuro-Nutrition

13.2.1 The Brain's Metabolic Landscape

The human brain, accounting for just about 2% of total body weight, consumes roughly 20% of the body's energy—a testament to its high metabolic demand. Glucose is the primary energy substrate for neurons; however, an increasing body of literature suggests that ketone bodies, lactate, and specific micronutrients are also critical for proper brain function. Energy metabolism in the brain is a complex interplay of oxidative phosphorylation, glycolysis, and alternative metabolic pathways that ensure a continuous supply of energy even during periods of dietary fluctuation.

The intricacies of this metabolic landscape highlight the brain's incredible adaptability and resilience. For instance, during periods of fasting or prolonged exercise, the brain demonstrates a remarkable ability to utilize ketone bodies as an alternative fuel source. These ketones, derived from fatty acids, not only provide energy but may also produce neuroprotective benefits, assisting in the maintenance of cognitive function over time. This adaptability underscores the importance of dietary choices; a diet rich in healthy fats may enhance cognitive performance, particularly in individuals facing metabolic challenges.

Moreover, various micronutrients—such as B vitamins, antioxidants, and omega-3 fatty acids—play significant roles in optimizing brain metabolism. The B vitamins, for example, are integral in various enzymatic reactions, facilitating the conversion of glucose into usable energy. Antioxidants help mitigate oxidative stress, which can impair neuronal function and lead to neurodegenerative diseases. Omega-3 fatty acids, particularly DHA, are critical for maintaining membrane fluidity and supporting synaptic plasticity, both of which are essential for learning and memory.

As research in neuroenergetics progresses, interest in metabolic flexibility is gaining momentum. Metabolic flexibility refers to the brain's ability to switch between different energy substrates based on availability and demand. Understanding this phenomenon could open new avenues for therapeutic interventions aimed at preventing cognitive decline and promoting brain health throughout the lifespan. Furthermore, lifestyle factors—including physical activity, sleep quality, and social engagement—are emerging as crucial elements that influence metabolic health and cognitive resilience.

Ultimately, maintaining a balanced diet, engaging in regular physical activity, and prioritizing mental health may not only bolster brain energy metabolism but also pave the way for a healthier, more vibrant life. The synergy between nutrition, metabolism, and brain function serves as a reminder of the interconnectedness of our physiological systems—a reminder that what we consume has profound effects far beyond mere physical energy, extending to our very capacity for thought, creativity, and emotional well-being.

Recent studies, including those published by Harrison et al. (2021), Walters and Usachev (2023), have highlighted the importance of mitochondrial efficiency in brain cells. Mitochondria are not only responsible for the generation of adenosine triphosphate (ATP) but also crucial for calcium signaling, reactive oxygen species (ROS) regulation, and programmed cell death. The health of these organelles is dependent upon an array of nutritional inputs, underscoring a direct correlation between dietary choices and neuronal energy metabolism.

13.2.2 Neurotransmitters and Synaptic Plasticity

Neurotransmitters are chemical messengers that facilitate communication between neurons. They have distinct roles in regulating mood, memory, and motor function. For instance, serotonin modulates mood and emotional stability, while dopamine plays a pivotal role in reward and motivation. The synthesis and regulation of these neurotransmitters require the presence of specific nutrients. Tryptophan, a precursor to serotonin, and tyrosine, a precursor to dopamine, are both derived from dietary proteins. Additionally, enzymes involved in neurotransmitter synthesis require cofactors and vitamins (e.g., vitamin B6, folate, and vitamin C) to function properly.

Synaptic plasticity—the ability of synapses to strengthen or weaken over time in response to increases or decreases in activity—is a hallmark of learning and memory. Numerous studies (e.g. Vauzour, 2012; Davinelli et al., 2023; Melgar-Locatelli et al., 2023) have underscored the impact of dietary compounds on the modulation of synaptic plasticity. Polyphenols, omega-3 fatty acids, and flavonoids have each been shown to influence signaling pathways that enhance long-term potentiation (LTP), a process critical for memory formation.

Moreover, the consumption of these dietary compounds is not only beneficial during formative years but also plays a crucial role in cognitive health throughout the lifespan. Recent research has indicated that a diet rich in these nutrients may mitigate age-related cognitive decline and assist in the prevention of neurodegenerative diseases (Fekete et al., 2023). For instance, studies have demonstrated that individuals with higher omega-3 fatty acid intake exhibit improved synaptic function and a lower risk of developing Alzheimer's disease, suggesting that dietary interventions can be a viable strategy for promoting brain health.

In addition to dietary considerations, lifestyle factors such as physical activity and cognitive engagement also significantly contribute to synaptic plasticity. Exercise has been shown to stimulate the release of neurotrophic factors like brain-derived neurotrophic factor (BDNF), which is essential for synaptic growth and the maintenance of neuronal health (Marin Bosch et al., 2021). Engaging in mentally stimulating activities—ranging from puzzles to social interactions—further enhances cognitive flexibility and resilience, thereby making the brain more adaptable to changes and challenges.

The interplay between diet, lifestyle, and synaptic plasticity underscores the importance of a holistic approach to cognitive health. Future research will be crucial in elucidating the mechanisms by which these factors interact, providing insights that could lead to personalized dietary and lifestyle recommendations for optimizing brain function across different life stages.

Ultimately, fostering an environment that supports synaptic plasticity not only enhances learning and memory but also contributes to overall mental well-being, illustrating the intricate connection between nutrition, activity, and cognitive health.

13.2.3 Inflammation, Oxidative Stress, and Neurodegeneration

Inflammatory processes and oxidative stress are among the leading contributors to neurodegeneration, contributing to conditions such as Alzheimer's disease, Parkinson's disease, and other forms of cognitive decline. Chronic systemic inflammation, often a consequence of poor dietary habits and sedentary lifestyles, leads to the production of pro-inflammatory cytokines that can cross the blood-brain barrier. Once in the brain, these cytokines trigger an inflammatory cascade that contributes to neuronal death.

Similarly, oxidative stress—resulting from an imbalance between ROS production and the body's ability to detoxify these reactive intermediates—can lead to cellular damage and impaired neuronal communication. Fortunately, a variety of dietary antioxidants, including vitamins E and C, flavonoids, and carotenoids, have been shown to significantly reduce oxidative stress within the central nervous system (CNS). Research spearheaded by le emphasizes the reciprocity between dietary antioxidants and the mitigation of neuroinflammation, thereby enhancing overall cognitive resilience.

Moreover, the synergistic effects of multiple antioxidants may hold promise for further reducing neuroinflammation and oxidative stress. For instance, a study by Heo and Lee (2004) indicates that combined supplementation of vitamins C and E, along with the flavonoid quercetin, not only lowers ROS levels but also enhances neurotransmitter synthesis and release. This suggests that a multifaceted approach to antioxidant intake could yield greater benefits for cognitive health.

In addition to dietary sources, lifestyle choices play a crucial role in modulating oxidative stress. Regular physical activity has been shown to enhance the body's endogenous antioxidant defenses, promoting neuroplasticity and cognitive function. According to Ashcroft et al. (2022), exercise stimulates the production of brain-derived neurotrophic factor (BDNF), which is vital for neuronal survival and growth, thereby counteracting the damaging effects of oxidative stress on the CNS.

Furthermore, the emerging field of nutrigenomics explores how individual genetic makeup can influence the effectiveness of antioxidant-rich foods on brain health. Personalized nutrition, tailored to one's genetic predisposition, could optimize the benefits of specific antioxidants. This paradigm shift underscores the importance of custom dietary interventions based on genetic profiles to maximize neuroprotection and cognitive function.

Finally, the interaction between the gut microbiome and the central nervous system is gaining attention for its role in regulating oxidative stress. Recent findings suggest that certain gut bacteria can produce metabolites that exhibit antioxidant properties, thereby contributing to brain health. As such, fostering a healthy gut microbiome through a diverse, fiber-rich diet may enhance the body's ability to combat neuroinflammation and oxidative stress.

In conclusion, a holistic approach—incorporating antioxidant-rich foods, regular exercise, personalized nutrition, and gut health—holds significant potential for mitigating oxidative stress and promoting cognitive resilience. Continued research is essential to unravel the complexities of these interactions and develop effective strategies for overall brain health.

13.3 Essential Nutrients for Cognitive Function

13.3.1 Macronutrients and Brain Function

Macronutrients—carbohydrates, proteins, and fats—are essential not only for maintaining overall health but also for optimal brain function.

Carbohydrates play a crucial role in brain function by providing glucose, the primary energy substrate for neurons. This is particularly important as neurons are heavily reliant on glucose to perform their various tasks, including processing information and maintaining cognitive functions such as memory and attention. However, the quality of the carbohydrates consumed is paramount to achieving optimal brain health. While simple sugars can lead to rapid spikes in energy, they often result in subsequent crashes that can impair cognitive performance. In contrast, complex carbohydrates, such as those found in whole grains, legumes, and certain vegetables, are digested more slowly. This slow digestion allows for a gradual and steady release of glucose into the bloodstream, ensuring a consistent supply of energy that is crucial for sustained cognitive engagement throughout the day. Moreover, the fiber present in complex carbohydrates aids in maintaining overall digestive health, which can further influence cognitive function by promoting a healthy gut microbiome. A well-balanced intake of high-quality carbohydrates can thus provide the necessary fuel for not just maintaining focus but also enhancing cognitive resilience against fatigue and stress.

Proteins serve as essential building blocks for neurotransmitters and enzymes, which are indispensable in regulating numerous brain functions. The amino acids derived from dietary proteins play a direct role in the synthesis of neurotransmitters such as serotonin and dopamine, which significantly influence mood, behavior, and cognitive performance. For instance, serotonin, often dubbed the "feel-good" neurotransmitter, is primarily synthesized from the amino acid tryptophan. This connection highlights the importance of a balanced diet rich in proteins for emotional well-being. Furthermore, branched-chain amino acids (BCAAs) like leucine, isoleucine, and valine are critical not only for energy production but also for muscle recovery and growth, which can indirectly impact mental health through improved physical vitality. On the other hand, amino acids like tryptophan and tyrosine are directly linked to the synthesis of vital brain chemicals; tryptophan leads to the production of serotonin, while tyrosine is a precursor for dopamine and norepinephrine, neurotransmitters associated with focus and alertness. This intricate interplay between amino acids and neurotransmitter production underscores the necessity of consuming a diverse range of protein sources, including animal and plant-based options, to support optimal brain health and cognitive function.

Fats, especially polyunsaturated fatty acids (PUFAs), are vital components of cell membranes within the brain, playing a crucial role in maintaining their integrity and functionality. Among these, omega-3 fatty acids, including eicosapentaenoic acid (EPA) and docosahexaenoic acid (DHA), are not only predominant in neuronal membranes but also critical in modulating brain processes. DHA, in particular, is recognized for its unique ability to enhance neuronal membrane fluidity, which is essential for optimal synaptic function and communication between neurons. This fluidity facilitates the efficient transmission of electrical signals, ultimately affecting cognitive processes such as learning and memory. Recent studies, including those conducted by, have provided compelling evidence that links DHA with improved memory retention and overall brain health. Their research sheds light on how increased levels of DHA are associated with better cognitive outcomes in both children and older adults, suggesting that dietary sources rich in this fatty acid, such as fatty fish and algae, could be invaluable for maintaining cognitive performance across the lifespan. Moreover, the anti-inflammatory properties of omega-3 fatty acids add another layer of benefit, as inflammation is a known contributor to neurodegenerative diseases. Thus, incorporating these essential fats into one's diet not only supports immediate cognitive functionality but also may offer protective effects against cognitive decline in the long term.

13.3.2 Micronutrients That Enhance Cognitive Performance

Micronutrients, including vitamins and minerals, are essential co-factors in numerous physiological reactions that underpin cognitive function. The following are some of the key micronutrients with documented benefits for the brain:

13.3.2.1 B-Vitamins

B-vitamins—such as B6, B9 (folate), and B12—are imperative for maintaining neurological function. They are deeply involved in homocysteine metabolism, where elevated homocysteine levels have been linked to an increased risk of neurodegenerative diseases. A recent randomized controlled trial by Smith et al. (2010) concluded that supplementation with B-vitamins not only lowered homocysteine levels but also improved working memory in older adults. Dietary sources include leafy greens, legumes, eggs, and fortified cereals.

Additionally, the role of B-vitamins extends beyond memory enhancement, as they contribute significantly to the synthesis of neurotransmitters, which are necessary for effective communication between neurons. For instance, vitamin B6 is crucial for the production of serotonin, a neurotransmitter that regulates mood and emotional well-being. Inadequate levels of this vitamin could potentially lead to mood disorders, underscoring the importance of a balanced diet rich in these essential nutrients.

Moreover, emerging research suggests a synergistic effect between B-vitamins and other nutrients in promoting cognitive health. Studies have indicated that omega-3 fatty acids, particularly those found in fish, can work in tandem with B-vitamins to enhance brain function and protect against cognitive decline. This interplay highlights the necessity of a holistic approach to nutrition, aiming for a diverse diet that not only includes B-vitamins but also a variety of other vitamins, minerals, and healthy fats.

Furthermore, as our understanding of the gut-brain axis grows, a healthy gut microbiome is recognized as playing a pivotal role in the absorption and metabolism of B-vitamins. Probiotics and fermented foods can support gut health, potentially enhancing the body's utilization of these vital nutrients. A balanced diet combined with adequate hydration and regular exercise could thus serve as a multifaceted strategy to bolster neurological health.

In summary, while B-vitamins are undeniably crucial for maintaining optimal brain function, their benefits are magnified when consumed as part of a broader dietary context. As research continues to uncover the intricate relationships between nutrition and cognitive health, individuals are encouraged to prioritize a nutrient-rich diet, ensuring they receive a comprehensive array of vitamins and minerals necessary for their long-term neurological well-being.

13.3.2.2 Vitamin D

Traditionally associated with bone health, vitamin D plays emerging roles in cognitive health as well. It is involved in the regulation of neurotrophic factors, which support neuron survival and synaptic plasticity, and modulates immune function within the CNS. Recent epidemiological studies (e.g., Wang et al., 2021; Chen et al., 2025) have noted a correlation between optimal serum levels of vitamin D and improved cognitive performance, particularly in aging populations. Dietary sources include fatty fish, fortified dairy products, and exposure to sunlight.

Furthermore, the implications of vitamin D deficiency extend beyond cognitive decline, as inadequate levels have been linked to various neurodegenerative disorders, including Alzheimer's disease and multiple sclerosis. Research by Yang et al. (2020) indicates that lower vitamin D levels may accelerate the progression of these conditions, emphasizing the need for adequate vitamin D status as a preventive measure.

In addition to dietary sources, many individuals may benefit from supplementation, especially those living in regions with limited sunlight exposure or those with dietary restrictions. However, it is essential to approach supplementation under medical supervision, as excessive intake of vitamin D can lead to toxicity and adverse health effects.

Awareness campaigns and public health initiatives aimed at educating populations about the importance of maintaining appropriate vitamin D levels are crucial. These programs can focus on dietary education, the benefits of safe sun exposure, and the potential cognitive benefits of vitamin D. Moreover, future research is warranted to explore the mechanisms by which vitamin D influences brain health, potentially paving the way for new therapeutic strategies in the context of aging-related cognitive decline.

Ultimately, fostering a holistic approach towards nutrition and lifestyle may prove vital for cognitive health maintenance throughout the lifespan. Integrating vitamin D into a broad spectrum of brain health strategies—including regular physical activity, mental stimulation, and social engagement—could enhance not just cognitive function but overall well-being in aging populations.

13.3.2.3 Antioxidant Vitamins (C and E)

Vitamins C and E serve as primary defenders against oxidative stress by neutralizing free radicals. Their roles in scavenging ROS protect neuronal integrity. A double-blind study conducted by Nazzi et al. (2024) demonstrated that antioxidants could mitigate age-related cognitive decline when administered as part of a comprehensive dietary program. Rich sources include citrus fruits, berries, nuts, seeds, and green leafy vegetables.

Moreover, the impact of these vitamins extends beyond mere cognitive function, as they are crucial for overall health maintenance. Recent research has highlighted the synergistic effects of antioxidants when combined with other dietary compounds such as omega-3 fatty acids and flavonoids. For instance, a study by Power et al. (2022) revealed that a combination of vitamin E with omega-3 supplements significantly improves not only memory performance but also mood regulation in older adults.

Additionally, it is important to recognize the role of lifestyle factors in enhancing the efficacy of these vitamins. Regular physical activity, adequate sleep, and stress management techniques further bolster the brain's resilience against age-related decline. A holistic approach that integrates these elements along with antioxidant-rich foods creates a synergistic effect, promoting long-term cognitive health.

On the other hand, it is essential to remain cautious about excessive supplementation. While some vitamins are beneficial in moderate amounts, high doses can lead to toxicity and adverse health effects. Thus, maintaining a balanced diet remains the most effective strategy for ensuring adequate nutrient intake without the risks associated with over-supplementation.

In conclusion, the inclusion of vitamins C and E in a diversified diet is a proactive measure in the fight against oxidative stress and cognitive decline. Future research should continue to explore the complex interactions between various dietary components and their collective impact on brain health, with an emphasis on developing strategies to implement these findings into practical dietary guidelines for the aging population.

13.3.2.4 Minerals: Magnesium, Zinc, and Selenium

Minerals play crucial roles in neuronal signaling and enzyme activation. Magnesium is vital for synaptic plasticity and supports the function of NMDA receptors, which are involved in memory formation. Zinc is essential in neurotransmission and has been linked to enhanced cognitive performance, particularly in tasks involving attention and memory. Selenium, on the other hand, works synergistically with antioxidant enzymes to reduce oxidative stress. Emerging research, such as that by Zhang et al. (2023), has underscored the importance of these minerals in supporting optimal brain function.

Additionally, copper, another trace mineral, plays a significant role in neurological health by participating in the synthesis of neurotransmitters and contributing to the maintenance of myelin, the protective sheath surrounding neurons. Copper deficiencies have been associated with neurodegenerative diseases, highlighting its necessity for neuronal integrity and function. Furthermore, iron is critical for oxygen transport and is involved in energy metabolism within the brain. A lack of iron can lead to cognitive deficits, particularly in young children, thereby underscoring the importance of adequate nutrition during developmental stages.

Recent studies, including those by Rahi et al. (2024), suggest that a balanced intake of these minerals can enhance neurogenesis—the process of generating new neurons—which is vital for learning and memory. Dietary sources rich in these micronutrients, such as leafy greens, nuts, legumes, and whole grains, should be emphasized to improve brain health across the lifespan. In contrast, excessive supplementation can lead to toxicity, emphasizing the need for careful regulation of mineral intake.

Moreover, the interplay between diet and lifestyle factors, such as physical exercise, cannot be overlooked. Regular aerobic activity has been shown to increase the availability of these essential minerals in the brain, thereby enhancing their positive effects on cognitive functions. As more evidence surfaces regarding the neuroprotective benefits of these minerals combined with a healthy lifestyle, strategies focusing on nutrition and physical activity are becoming pivotal in preventing cognitive decline and promoting mental agility as we age.

13.3.3 Omega-3 Fatty Acids: The Brain's Building Blocks

Omega-3 fatty acids occupy a place of preeminent importance in neuro-nutrition. DHA, in particular, is a critical component of neuronal membranes and contributes significantly to membrane fluidity—a factor that is directly related to the efficacy of neurotransmission. Multiple clinical trials, including one by Yurko-Mauro et al. (2010), have demonstrated that regular consumption of omega-3-rich foods can enhance cognitive function, improve mood, and even attenuate the progression of neurodegenerative disease in at-risk populations.

Omega-3 fatty acids also exhibit anti-inflammatory properties, reducing the production of pro-inflammatory cytokines and playing a preventative role against chronic brain inflammation. Natural dietary sources include fatty fish (such as salmon, mackerel, and sardines), walnuts, flaxseeds, and chia seeds. Incorporation of these foods into the diet has been associated with both short-term cognitive improvements and long-term neuroprotection.

Furthermore, studies suggest that the benefits of omega-3 fatty acids extend beyond mere cognitive support, influencing mood regulation and mental health. Regular consumption of these essential fats has been linked to lower rates of depression and anxiety, highlighting their holistic contribution to overall mental well-being. This connection can be particularly crucial in our fast-paced, modern lifestyles, where stress and mental strain can undermine brain function.

In addition to dietary sources, omega-3 supplementation has become increasingly popular as individuals seek to enhance their cognitive capacities and emotional health. Fish oil capsules and algae-based supplements offer alternative avenues for those who may not frequently consume fish or plant-based sources of omega-3. Moreover, emerging research continues to explore the interaction between omega-3 fatty acids and gut microbiota, revealing how these fats may influence gut health, which, in turn, impacts brain function through the gut-brain axis.

Despite the numerous benefits, it is essential to approach omega-3 intake with balance and moderation. Overconsumption and reliance on supplements without a comprehensive dietary plan can lead to potential side effects, including blood thinning and excessive vitamin A intake from fish sources. Therefore, harnessing the full potential of omega-3 fatty acids involves not only appropriate consumption but also a holistic approach to nutrition, emphasizing a balanced diet rich in a variety of nutrients.

Future research is expected to delve deeper into understanding the optimal ratios of omega-3 to omega-6 fatty acids, which are often abundant in typical Western diets. By decreasing omega-6 intake while enhancing omega-3 consumption, individuals may cultivate a more favorable inflammatory environment within their bodies, further supporting cognitive health and mitigating risks associated with neurodegenerative diseases.

Ultimately, integrating omega-3 fatty acids into the daily diet presents a simple yet powerful strategy for improving mental acuity and emotional resilience, fostering a healthier brain for years to come.

13.3.4 Polyphenols and Flavonoids: Nature's Cognitive Protectors

Polyphenols and flavonoids present in a diverse array of plant-based foods have been shown to encourage cerebral blood flow, enhance synaptic plasticity, and reduce the risk of cognitive decline. Specifically, research reviewed (Ramakrishna Pillai et al., 2022) suggests that regular consumption of foods high in these bioactive compounds can bolster memory performance and delay the progression of cognitive impairments. Blueberries, green tea, dark chocolate, and red grapes are exemplary sources of these compounds.

It is believed that polyphenols exert their benefits by modulating cell signaling pathways that regulate inflammation and oxidative stress. In addition, these compounds may influence neuronal survival by upregulating brain-derived neurotrophic factor (BDNF), an essential protein for neurogenesis and synaptic plasticity.

Moreover, recent studies have highlighted the potential of polyphenols to improve cognitive function and contribute to neuroprotection in age-related neurodegenerative diseases such as Alzheimer's and Parkinson's. By enhancing BDNF signaling, polyphenols may promote the survival of neurons and foster the formation of new connections, which are vital for maintaining cognitive flexibility.

Additionally, polyphenols are thought to exert their effects through interactions with gut microbiota, which play a crucial role in the metabolism of these compounds. The fermentation of polyphenols by gut bacteria produces various metabolites that can cross the blood-brain barrier, thereby influencing neuroinflammation and neuronal health directly. This gut-brain axis has emerged as a significant area of research, demonstrating that the health of the gastrointestinal system can impact mental health and cognitive function.

Furthermore, dietary sources rich in polyphenols, such as berries, dark chocolate, green tea, and leafy greens, are encouraged as part of a balanced diet aimed at enhancing cognitive resilience. Longitudinal studies suggest that regular consumption of such foods correlates with lower rates of cognitive decline and improved overall brain health.

In conclusion, while the mechanisms behind the protective effects of polyphenols are still being explored, the evidence supporting their role in promoting neuronal health and combating cognitive decline is compelling. Future research will undoubtedly expand our understanding of how these dietary compounds can be utilized not only for their health benefits but also for their potential therapeutic applications in neurodegenerative conditions (Fig. 13.1).

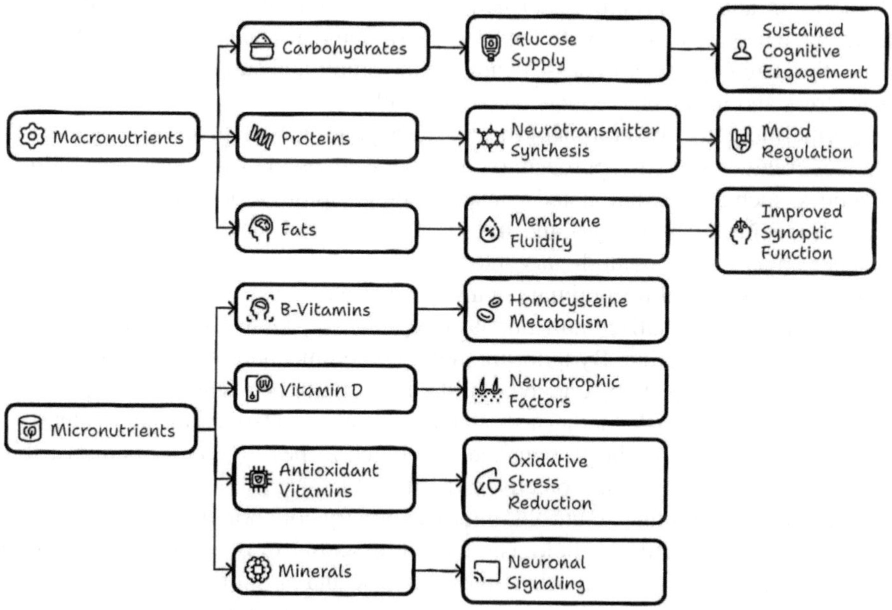

Fig. 13.1 Essential nutrients for cognitive function

13.4 The Impact of Dietary Patterns on Cognitive Performance

13.4.1 Mediterranean and MIND Diets

The Mediterranean diet, characterized by high intakes of fruits, vegetables, whole grains, legumes, nuts, and olive oil, combined with moderate consumption of fish and poultry, has long been associated with reduced risks of cardiovascular and neurodegenerative diseases. The MIND (Mediterranean-DASH Intervention for Neurodegenerative Delay) diet is a hybrid dietary pattern that incorporates similar principles with the explicit goal of reducing cognitive decline.

Recent prospective cohort studies (e.g., Morris et al., 2015) have provided robust evidence linking adherence to these dietary patterns with a lower incidence of Alzheimer's disease and other forms of dementia. Specifically, the anti-inflammatory and antioxidant components of these diets work synergistically to protect against neural inflammation and oxidative damage, thus promoting healthier cognitive aging.

Moreover, a growing body of research indicates that lifestyle factors, including physical activity and social engagement, also play critical roles in cognitive health. Studies suggest that individuals who regularly engage in aerobic exercise exhibit a marked improvement in cognitive function, potentially due to enhanced blood flow to the brain and the promotion of neurogenesis. Social interactions, on the other hand, have been associated with lower levels of dementia incidence, likely because they stimulate cognitive processes and provide emotional support, which can mitigate stress and its adverse effects on brain health.

Furthermore, the integration of these dietary patterns with other healthy lifestyle choices creates a holistic approach to cognitive wellness. For instance, mindfulness practices, such as meditation and yoga, are being explored for their potential benefits in reducing mental stress and improving focus and memory retention. These practices may complement dietary interventions by fostering a balanced emotional state, which is crucial for optimal cognitive function.

In addition, personalized nutrition is emerging as a promising avenue in the fight against cognitive decline. By tailoring dietary recommendations to the individual's genetic makeup, lifestyle, and health status, practitioners can more effectively address the unique risk factors associated with each person. This individualized approach has the potential to enhance adherence to healthy eating patterns, thus maximizing their cognitive protective benefits.

As research continues to unfold, interdisciplinary collaboration among neurologists, nutritionists, and psychologists will be essential to develop comprehensive strategies aimed at preventing Alzheimer's disease and promoting cognitive longevity. Ultimately, fostering a deeper understanding of how dietary habits interweave with other lifestyle factors will empower individuals to make informed choices and take active steps towards preserving their brain health throughout the aging process.

13.4.2 Ketogenic and Low-Carbohydrate Diets

Amid contemporary discussions on diet and brain function, ketogenic and low-carbohydrate diets have gained considerable attention. These dietary strategies shift the primary energy source from glucose to ketone bodies—a process that may offer benefits to brain health, particularly in populations with impaired glucose metabolism. Ketones are derived from fatty acid oxidation and can provide a more efficient fuel for the brain during periods of metabolic stress.

Clinical research led by Morris et al. (2015) has demonstrated that ketogenic diets can improve cognitive performance in individuals with mild cognitive impairment by reducing neuroinflammation and enhancing mitochondrial function. Nonetheless, it is critical to maintain a balance, as extreme carbohydrate restriction might lead to deficits in essential micronutrients if not carefully managed.

Furthermore, the study highlights the importance of incorporating a variety of nutrient-dense foods within the ketogenic framework to mitigate the risks associated with micronutrient deficiencies. For instance, including leafy greens, nuts, seeds, and fatty fish can provide essential vitamins and minerals while adhering to the low-carbohydrate paradigm. Additionally, researchers have suggested that periodic reintroduction of carbohydrates—known as targeted ketogenic dieting—may enhance sustained cognitive benefits without compromising overall nutritional status.

As interest in ketogenic diets continues to grow, it is also essential to consider individual variability in response to such dietary changes. Genetic factors, existing health conditions, and even microbiome composition can influence how well a person adapts to a ketogenic approach. Future research should aim to identify these variances to personalize dietary recommendations further. This may pave the way for more specialized protocols tailored to individual needs, ultimately maximizing the cognitive and health benefits while minimizing potential risks.

Moreover, the psychological aspects of adopting a ketogenic lifestyle should not be overlooked. For many, the restrictive nature of the diet can lead to feelings of deprivation, which may trigger binge-eating episodes or an unhealthy relationship with food. Incorporating behavioral strategies alongside nutritional education remains crucial, empowering individuals to make sustainable dietary choices without exacerbating mental health concerns.

In summary, while the findings of present a promising avenue for improving cognitive function through ketogenic diets, it is vital to approach this dietary strategy holistically. By promoting a balanced intake of nutrients, accommodating individual differences, and addressing behavioral health, practitioners and individuals alike can optimize the advantages of ketogenic diets in the pursuit of enhanced cognitive performance and overall wellness.

13.4.3 Anti-inflammatory Dietary Strategies

Inflammation is a key driver of cognitive decline, and dietary choices that reduce chronic inflammation are vital in maintaining neuronal health. Key anti-inflammatory nutrients include omega-3 fatty acids, antioxidants, and fiber-rich foods. The consumption of whole, minimally processed foods is linked to reduced levels of systemic inflammation. In addition, bioactive compounds such as curcumin found in turmeric have been shown to inhibit inflammatory cascades within the CNS.

A comprehensive review by James and Garza (2012) highlights how dietary polyphenols, in combination with an overall anti-inflammatory eating pattern, can reduce microglial activation—a process implicated in neurodegeneration. The regular inclusion of cruciferous vegetables, fermented foods, and spices may contribute significantly to lowering inflammation and preserving neural integrity.

Furthermore, recent studies have indicated that the synergistic effects of these dietary components could extend beyond just reducing inflammation. For example, ingredients rich in omega-3 fatty acids, such as fatty fish and flaxseeds, are known to modulate neurotransmitter function and enhance synaptic plasticity, thereby promoting cognitive resilience. The incorporation of such foods into a balanced diet may bolster brain health alongside the anti-inflammatory properties of polyphenols.

In addition to dietary interventions, lifestyle factors such as regular physical activity and adequate sleep play crucial roles in maintaining neurological health. Engaging in moderate exercise has been shown to increase the production of brain-derived neurotrophic factor (BDNF), which is vital for neurogenesis and the survival of neurons. Coupled with a nutrient-rich diet, these lifestyle choices create a holistic approach to mitigating the risks associated with neurodegenerative diseases.

Emerging research also suggests that certain mindfulness practices and stress-reduction techniques, like yoga and meditation, can have a favorable impact on brain function. By lowering cortisol levels and promoting relaxation, these practices may further enhance the protective effects of diet and exercise on neural health.

As we continue to explore the multifaceted relationships between diet, lifestyle, and neurodegenerative conditions, it becomes increasingly clear that adopting a comprehensive, integrative approach may offer the best defense against cognitive decline. Future research should aim to elucidate the specific mechanisms through which these various factors interact, ultimately contributing to tailored recommendations for preventing neurodegeneration (Fig. 13.2).

13.5 Scientific Evidence and Recent Developments

13.5.1 Recent Peer-Reviewed Studies

Over the past 5 years, numerous studies have documented the effects of specific nutrients and dietary patterns on brain function. For instance, a meta-analysis by Benwell et al. (2020) confirmed the association between enhanced omega-3 fatty acid intake and improved cognitive performance in both healthy individuals and patients with mild cognitive impairment. Another study by Hernandez et al. (2020) demonstrated that regular consumption of a diet rich in polyphenols led to measurable increases in cerebral blood flow and improvements in executive functioning.

Similarly, randomized controlled trials (RCTs) conducted by Smith et al. (2010) and Zhang et al. (2023) reported that interventions involving B-vitamin

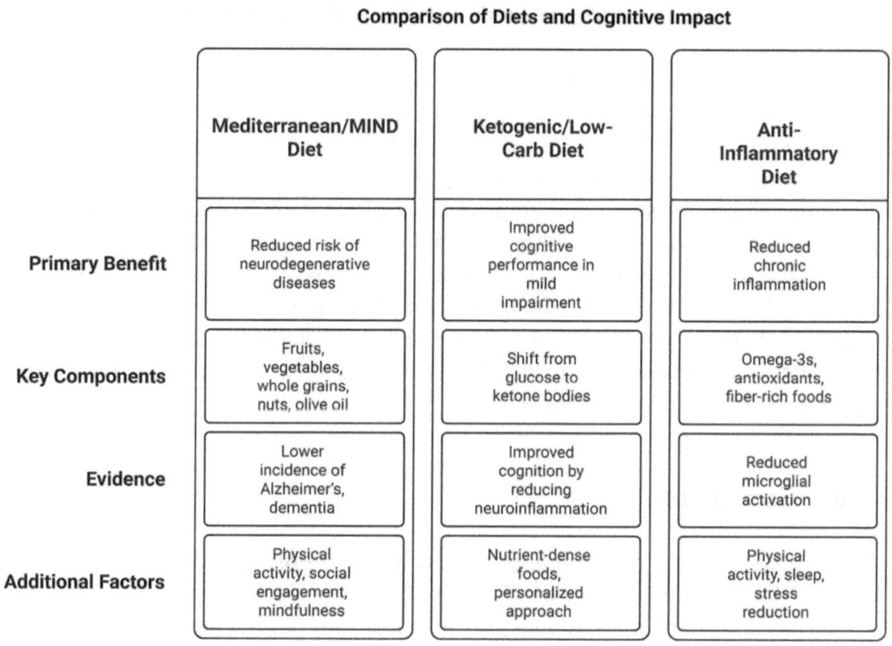

Fig. 13.2 Comparison of diets and cognitive impact

supplementation and mineral fortification not only reduced biomarkers of inflammation but also translated into statistically significant enhancements in working memory and attentional control. Additionally, emerging research from James and Garza (2012) has identified promising links between ketogenic diets and an improvement in neuroplasticity, particularly in older adults.

13.5.2 Mechanistic Insights from Recent Research

At the molecular and cellular levels, a plethora of recent studies have shed light on the biological mechanisms by which nutrients confer cognitive benefits. Key pathways include the upregulation of BDNF, improvements in mitochondrial biogenesis, modulation of the PI3K/Akt pathway, and suppression of pro-inflammatory transcription factors such as NF-κB. These mechanisms illustrate how dietary compounds can actively influence synaptic plasticity and neuronal survival.

For example, a study by Baur and Sinclair (2006) detailed how the polyphenol resveratrol, found in red grapes and berries, activated SIRT1 pathways, which in turn enhanced the expression of neuroprotective genes. Similarly, the beneficial effects of vitamin D on nerve growth factor (NGF) have been documented, suggesting that vitamin D supplementation can help ameliorate cognitive decline in aging populations.

13.5.3 Integration of Multi-omics Data in Neuro-Nutrition

Advances in multi-omics technologies—integrating genomics, metabolomics, and proteomics—are further elucidating the intricate relationships between diet and brain function. These integrative approaches allow researchers to map the influence of nutrients on gene expression profiles and metabolic pathways on a comprehensive scale. Such investigations help tailor nutritional interventions to individual genetic profiles, paving the way for personalized neuro-nutrition strategies.

The potential of personalized nutrition is underscored in a recent review by Walters and Usachev (2023), which discussed how dietary modifications based on genetic predispositions can yield substantial improvements in cognitive function. As the field continues to evolve, future studies are anticipated to refine neuro-nutrition guidelines even further.

13.6 Conclusion

The intricate interplay between nourishment and neurobiology is now better understood thanks to advances in scientific research over the past 5 years. It is clear that the nutritional components found in a balanced, nutrient-dense diet play crucial roles in supporting mitochondrial function, reducing inflammation, enhancing synaptic plasticity, and ultimately improving cognitive performance. By focusing on whole foods rich in omega-3 fatty acids, antioxidants, B-vitamins, and essential minerals, individuals can actively fortify their brain health and ward off cognitive decline.

This chapter has outlined the scientific foundation of neuro-nutrition and provided actionable dietary strategies that can be tailored to individual needs. The evidence is compelling enough to suggest that nutritional choices are modifiable and that even modest improvements in dietary habits can translate into meaningful enhancements in cognitive performance. It is imperative for both healthcare practitioners and patients to integrate neuro-nutrition principles into daily routines, thereby embracing a proactive approach to long-term brain health.

As research continues to evolve, future guidelines will likely become more refined, paving the way for a new era in personalized nutrition therapies that specifically target brain health. In the meantime, adopting a consistent, science-backed dietary plan is one of the most practical and effective tools available for ensuring cognitive vitality into the later years of life.

References

Anderson, S. C., Cryan, J. F., & Dinan, T. (2017). *The psychobiotic revolution: Mood, food, and the new science of the gut-brain connection*. National Geographic Books.

Ashcroft, S. K., Ironside, D. D., Johnson, L., Kuys, S. S., & Thompson-Butel, A. G. (2022). Effect of exercise on brain-derived neurotrophic factor in stroke survivors: A systematic review and meta-analysis. *Stroke, 53*(12), 3706–3716.

Baur, J. A., & Sinclair, D. A. (2006). Therapeutic potential of resveratrol: The in vivo evidence. *Nature Reviews Drug Discovery, 5*(6), 493–506. https://doi.org/10.1038/nrd2060

Benwell, C. S., Davila-Pérez, P., Fried, P. J., Jones, R. N., Travison, T. G., Santarnecchi, E., et al. (2020). EEG spectral power abnormalities and their relationship with cognitive dysfunction in patients with Alzheimer's disease and type 2 diabetes. *Neurobiology of Aging, 85*, 83–95.

Chen, H., Pang, X., & Huang, Y. (2025). Higher dietary vitamin D intake influences brain and mental function in elderly Americans: A cross-sectional analysis. *Frontiers in Nutrition, 12*, 1564568.

Davinelli, S., Medoro, A., Ali, S., Passarella, D., Intrieri, M., & Scapagnini, G. (2023). Dietary flavonoids and adult neurogenesis: Potential implications for brain aging. *Current Neuropharmacology, 21*(3), 651–668.

Fekete, M., Lehoczki, A., Tarantini, S., Fazekas-Pongor, V., Csípő, T., Csizmadia, Z., & Varga, J. T. (2023). Improving cognitive function with nutritional supplements in aging: A comprehensive narrative review of clinical studies investigating the effects of vitamins, minerals, antioxidants, and other dietary supplements. *Nutrients, 15*(24), 5116.

Harrison, P. W., Ahamed, A., Aslam, R., Alako, B. T., Burgin, J., Buso, N., et al. (2021). The european nucleotide archive in 2020. *Nucleic Acids Research, 49*(D1), D82–D85.

Heo, H. J., & Lee, C. Y. (2004). Protective effects of quercetin and vitamin C against oxidative stress-induced neurodegeneration. *Journal of Agricultural and Food Chemistry, 52*(25), 7514–7517.

Hernandez, A. R., Winesett, S. P., Federico, Q. P., Williams, S. A., Burke, S. N., & Clark, D. J. (2020). A cross-species model of dual-task walking in young and older humans and rats. *Frontiers in Aging Neuroscience, 12*, 276.

Innis, S. M. (2008). Dietary omega 3 fatty acids and the developing brain. *Brain Research, 1237*, 35–43.

James, W. P. T., & Garza, C. (2012). Nutrition and human microbiome: 24 th marabou symposium, Stockholm, Sweden. Foreword. *Nutrition Reviews, 70*, S1.

Jiwani, R., Robbins, R., Neri, A., Renero, J., Lopez, E., & Serra, M. C. (2022). Effect of dietary intake through whole foods on cognitive function: Review of randomized controlled trials. *Current Nutrition Reports, 11*(2), 146–160.

Joseph, J. A., Shukitt-Hale, B., & Casadesus, G. (2005). Reversing the deleterious effects of aging on neuronal communication and behavior: Beneficial properties of fruit polyphenolic compounds. *The American Journal of Clinical Nutrition, 81*(1), 313S–316S.

Kennedy, D. O. (2016). B vitamins and the brain: Mechanisms, dose and efficacy – A review. *Nutrients, 8*(2), 68.

Llewellyn, D. J., Lang, I. A., Langa, K. M., Muniz-Terrera, G., Phillips, C. L., Cherubini, A., et al. (2010). Vitamin D and risk of cognitive decline in elderly persons. *Archives of Internal Medicine, 170*(13), 1135–1141.

Marin Bosch, B., Bringard, A., Logrieco, M. G., Lauer, E., Imobersteg, N., Thomas, A., et al. (2021). A single session of moderate intensity exercise influences memory, endocannabinoids and brain derived neurotrophic factor levels in men. *Scientific Reports, 11*(1), 14371.

Martínez-Lapiscina, E. H., Clavero, P., Toledo, E., Estruch, R., Salas-Salvadó, J., San Julián, B., & Martinez-Gonzalez, M. Á. (2013). Mediterranean diet improves cognition: The PREDIMED-NAVARRA randomised trial. *Journal of Neurology, Neurosurgery & Psychiatry, 84*(12), 1318–1325.

Melgar-Locatelli, S., de Ceglia, M., Mañas-Padilla, M. C., Rodriguez-Pérez, C., Castilla-Ortega, E., Castro-Zavala, A., & Rivera, P. (2023). Nutrition and adult neurogenesis in the hippocampus: Does what you eat help you remember. *Frontiers in Neuroscience, 17*, 1147269.

Morris, M. C., Tangney, C. C., Wang, Y., Sacks, F. M., Bennett, D. A., & Aggarwal, N. T. (2015). MIND diet associated with reduced incidence of Alzheimer's disease. *Alzheimer's & Dementia, 11*(9), 1007–1014.

Nazzi, C., Avenanti, A., & Battaglia, S. (2024). The involvement of antioxidants in cognitive decline and neurodegeneration: Mens sana in corpore Sano. *Antioxidants, 13*(6), 701.

Power, R., Nolan, J. M., Prado-Cabrero, A., Roche, W., Coen, R., Power, T., & Mulcahy, R. (2022). Omega-3 fatty acid, carotenoid and vitamin E supplementation improves working memory in older adults: A randomised clinical trial. *Clinical Nutrition, 41*(2), 405–414.

Rahi, B., Rashid, F., Sultana, R., Benoit, J., Parvez, F., & Khan, K. (2024). Impact of nutritional minerals biomarkers on cognitive performance among Bangladeshi rural adolescents—A pilot study. *Nutrients, 16*(22), 3865.

Ramakrishna Pillai, J., Wali, A. F., Menezes, G. A., Rehman, M. U., Wani, T. A., Arafah, A., et al. (2022). Chemical composition analysis, cytotoxic, antimicrobial and antioxidant activities of Physalis angulata L.: A comparative study of leaves and fruit. *Molecules, 27*(5), 1480.

Smith, A. D., Smith, S. M., De Jager, C. A., Whitbread, P., Johnston, C., Agacinski, G., et al. (2010). Homocysteine-lowering by B vitamins slows the rate of accelerated brain atrophy in mild cognitive impairment: A randomized controlled trial. *PLoS One, 5*(9), e12244.

Vauzour, D. (2012). Dietary polyphenols as modulators of brain functions: Biological actions and molecular mechanisms underpinning their beneficial effects. *Oxidative Medicine and Cellular Longevity, 2012*(1), 914273.

Walters, G. C., & Usachev, Y. M. (2023). Mitochondrial calcium cycling in neuronal function and neurodegeneration. *Frontiers in Cell and Developmental Biology, 11*, 1094356.

Wang, Y. (2021). *Characterising the response to blueberry dietary interventions aimed at improving cognition and vascular function*. University of Northumbria at Newcastle.

Wang, R., Wang, W., Hu, P., Zhang, R., Dong, X., & Zhang, D. (2021). Association of dietary vitamin D intake, serum 25 (OH) D3, 25 (OH) D2 with cognitive performance in the elderly. *Nutrients, 13*(9), 3089.

Wirström, T., Hilding, A., Gu, H. F., Östenson, C. G., & Björklund, A. (2013). Consumption of whole grain reduces risk of deteriorating glucose tolerance, including progression to prediabetes. *The American Journal of Clinical Nutrition, 97*(1), 179–187.

Yang, T., Wang, H., Xiong, Y., Chen, C., Duan, K., Jia, J., & Ma, F. (2020). Vitamin D supplementation improves cognitive function through reducing oxidative stress regulated by telomere length in older adults with mild cognitive impairment: A 12-month randomized controlled trial. *Journal of Alzheimer's Disease, 78*(4), 1509–1518.

Yurko-Mauro, K., McCarthy, D., Rom, D., Nelson, E. B., Ryan, A. S., Blackwell, A., et al. (2010). Beneficial effects of docosahexaenoic acid on cognition in age-related cognitive decline. *Alzheimer's & Dementia, 6*(6), 456–464.

Zhang, F., Li, X., & Wei, Y. (2023). Selenium and selenoproteins in health. *Biomolecules, 13*(5), 799.

Agro-industrial By-Products: Alternative Solution for Functional Food Ingredients

14

Martha Monzerrath Orozco-Sifuentes, Jesús Andrés Salas-Tovar, Soriano-Melgar Lluvia de Abril Alexandra, Campos-Múzquiz Lizeth Guadalupe, Nery-Flores Sendar Daniel, Palomo-Ligas Lissethe, Esparza-González Sandra Cecilia, Salazar-Villa Edith, and Rodríguez-Herrera Raúl

Abstract

The food industry has evolved to create products that meet consumer needs. Factors such as population growth, economic development, and accelerated lifestyle have influenced the search for more nutritious and functional foods. Functional foods are defined as those whose chemical composition provides health benefits, helping in the treatment or prevention of diseases. One of the alternatives for the food industry to respond to these demands is the use of agro-industrial wastes, which are generated and quantified in the order of tons worldwide, and if they are not properly processed, can become highly polluting, generators of greenhouse gases, undesirable odors, and pest incidence. Due to their biological nature, these residues have a high enzymatic activity and are a potential source of metabolites with functional activity, whose extraction and incorporation into raw materials or finished products represent a range of possibilities not only in the food industry but also in those related to pharmacology, materials, and agronomy itself. The use of phytochemicals, such as anthocyanins, flavonols, catechins, tannins, and triterpenes, among others, offers benefits including antioxidant, anti-inflammatory, antihypertensive, and hypoglycemic activities. The adoption of new techniques focused on the extraction, purification, characterization, and application of compounds of interest can not only benefit the increase in the production of functional foods but also favor the circular economy and environmental conservation.

M. M. Orozco-Sifuentes · J. A. Salas-Tovar · S.-M. L. de Abril Alexandra ·
C.-M. L. Guadalupe · N.-F. S. Daniel · P.-L. Lissethe · E.-G. S. Cecilia · R.-H. Raúl (✉)
School of Chemistry, Autonomous University of Coahuila, Saltillo, Coahuila, Mexico
e-mail: raul.rodriguez@uadec.edu.mx

S.-V. Edith
School of Biology, Autonomous University of Sinaloa, Culiacán, Sinaloa, Mexico

G. Sanghvi et al. (eds.), *Sustainable Food Fortification*,
https://doi.org/10.1007/978-981-95-1217-1_14

Keywords

Bioactive compounds · Polyphenols · Waste sources · Plants · Aliments

14.1 Introduction

According to the Food and Agriculture Organization of the United Nations (FAO), about 1.3 billion tons of food are wasted worldwide, which is equivalent to about a quarter of global production. About half of this waste comes mainly from fruits, seeds, vegetables and roots, followed by products such as fish (35%) and cereals (30%) and to a lesser extent meat and dairy products (20%) (Benítez, 2015).

The loss of these products occurs at different stages of the agri-food chain, such as production, post-harvest, storage, transportation and processing (Fig. 14.1). In most cases, the final destination of these wastes is landfills and some are even left in open fields or incinerated, and because of their organic nature, are decomposed by the action of bacteria, generating a high enzymatic activity. Under these conditions, agro-industrial wastes represent a source of contamination that can affect soil and water, as well as the generation of greenhouse gases that pollute the atmosphere (Romero-Sáez, 2022).

One way to counteract the problem of high agroindustrial waste generation is to implement circular economy strategies, which consist of integrating waste into the generation of new products. Biological transformation into composts, fertilizers, bioplastics, recycled paper, and even the extraction of their compounds to add value to pharmaceuticals, food, and cosmetics, are alternatives to close their usefulness cycle and contribute to a culture of responsible consumption (Salazar-López et al., 2023).

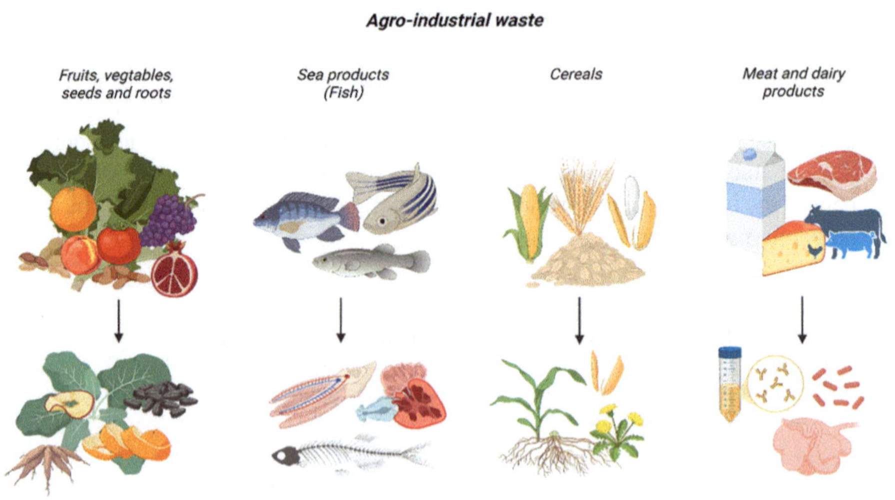

Fig. 14.1 Main sources of agro-industrial waste. (Created in https://BioRender.com)

Today's society is more aware that a healthy and varied diet is essential for a better quality of life. Agro-industrial residues are a source of compounds of interest as nutrients (amino acids, proteins, fatty acids, carbohydrates, fiber, minerals, vitamins) and secondary metabolites, such as polyphenols, pigments, essential oils, and saponins, among others. These biomolecules are characterized by their functional or bioactive activity, such as antioxidants, antimicrobial, anti-inflammatory, anticancer, hypoglycemic, antihypertensive, and others (Schiebel et al., 2024).

In the following chapter, we will discuss how agroindustrial wastes and their phytochemical potential have been exploited for their application in different health fields, such as food, pharmaceuticals, and biomaterials. We will also address aspects to be considered in obtaining and characterizing phytochemicals from residues from plant species, such as fruits, vegetables, and desert plants.

14.2 Phytochemical Characterization of Agro-industrial Wastes

The utilization of agro-industrial wastes worldwide is an activity that is bearing fruit in a wide variety of applications. The nature of the residues and the physicochemical characterization of their components are an essential part of defining the potential use of these residues. In food products, molecules with bioactive potential have generated great interest. The development of healthier foods with additional functional value is highly demanded by a society increasingly conscious of acquiring a better quality of life through food.

14.2.1 Obtaining Bioactive Compounds

An important step in obtaining bioactive compounds is the extraction technique and medium. Today, the extraction assisted by hybrid technologies, such as ultrasound-microwave (EAUM) and the use of environmentally friendly solvents (water, ethanol, and ethanol), have been widely accepted in the food field. The application of physical phenomena, such as the emission of sound waves, cavitation and microwave radiation, facilitates the breaking and permeability of vegetable membranes, shortening processing times. Other factors such as the power used (96–700 Watts), dehydration of the sample (powder), and particle size help standardize and make the extraction process more efficient (Madera-Santana et al., 2022).

14.2.2 Quantification of Polyphenolic Compounds by Folin-Ciocalteu

Among the techniques for the characterization of bioactive molecules with potential use in food processing is the determination of polyphenolic compounds. These organic compounds make up a large family (about 8000) of secondary metabolites, widely distributed among plant species. They are characterized by having phenolic rings in their molecular structure, accompanied by hydroxyl groups (-OH), and are classified according to the number of rings they possess. Some of its main groups are flavonoids, phenolic acids, stilbenes, phenolic alcohols, lignans, tannins, catechins, glucosinolates, and anthocyanins (Robles-Botero et al., 2020). These groups are characterized by their beneficial activities for health, such as their antioxidant, antiparasitic, antibiotic, anti-inflammatory, hypoglycemic, anticancer, anti-allergic, vasodilator, and detoxifying actions, among others (Chensom et al., 2019).

The polyphenol content is determined by the Folin-Ciocalteu (molybdate and sodium tungtate) assay, a spectrophotometric technique based on the ability of phenols to react with an oxidizing agent (Folin-Ciocalteu reagent), forming a blue complex (phosphomolybdic-phosphotungstic) with maximum absorbance at a wavelength of 725 to 765 nm. A standard curve based on gallic acid (GA) is used for quantification and can be reported as mg GAE/g sample (mg gallic acid equivalents per g sample) (Sokolovska & Jankulovska, 2022).

14.2.3 Assays to Measure Antioxidant Activity

One of the bioactive properties of polyphenolic groups is their antioxidant activity, i.e., the capacity of a substance or compound to delay the oxidation of molecules. Free radicals are oxidative molecules, with an unpaired electron, highly reactive, which are produced naturally by the metabolism of organisms, and despite having short lifetimes, their excess causes damage to cellular structures. There are different types of free radicals; reactive oxygen species (ROS) are an example. These radicals react with biomolecules, such as DNA, RNA, proteins, and lipids, among others, and damage their structure, causing disease (Mejía-Reyes et al., 2022).

Compounds with antioxidant action protect cells against oxidative damage, halting aging and the progression of diseases, such as cancer, arthritis, diabetes, cardiovascular, and neurodegenerative, among others). There are enzymatic and nonenzymatic antioxidants; the former are part of the main defense machinery of organisms, while the latter comprises those compounds of low molecular weight and plant origin, such as tocopherols, vitamins A, E, C, and polyphenols (Tikoo et al., 2022).

Tests to detect antioxidant activity include assays based on the transfer of a hydrogen atom (TAH), such as oxygen radical absorbance capacity (ORAC), hydroxyl radical absorbance capacity (HORAC), and total oxyradical scavenging capacity (TOSC). One-electron transfer-based assays include ferric ion reducing antioxidant power (FRAP) and cupric ion reducing antioxidant capacity (CUPRAC). Mixed assays include both one-electron and one-hydrogen atom transfer are ABTS (2,2′-azino-bis-(3-ethylbenzothiazoline-6-sulfonic acid) and DPPH (1,1-diphenyl-2-picrylhydrazyl) (Munteanu & Apetrei, 2021).

Among the most commonly used techniques to measure antioxidant activity in foods are the ABTS, DPPH, and FRAP assays. ABTS is a colorimetric test and consists of generating the ABTS+ radical (absorbs at a wavelength of 743 nm) from the oxidation of the ABTS reagent with ammonium persulfate. When an antioxidant agent reacts with the ABTS+ radical (green-blue color), the solution discolors, and the degree of discoloration corresponds to the degree of radical inhibition. Trolox (6-hydroxy-2,5,7,8-tetramethylchroman-2-carboxylic acid) acts as an inhibitor of the radical and is used as a standard to construct a calibration curve; the results can be reported in mg ET/g sample (milligram equivalents of Trolox per g of sample) (Maqsoudlou et al., 2020).

Similarly, the DPPH radical elimination assay is one of the most widely used. This radical, stable in solution with methanol, is violet in color, but when it receives electrons or hydrogens (H+) from an antioxidant, it is reduced and changes to a pale yellow color, presenting a maximum absorbance at 517 nm. As a standard, antioxidant (Trolox, catechin epicatechin, etc.) is used, and the results can be expressed in mg ET/g sample (milligram equivalents of Trolox per g of sample) and in percent age of inhibition (Mejía-Reyes et al., 2022).

The FRAP assay is based on the ability of an antioxidant to reduce ferric ion (Fe3+) to ferrous ion (Fe2+). The ferrous ion forms a complex with 2,4,6-tripyridyl-s-triazine (TPTZ), in an acid medium (acetic acid-sodium acetate buffer, pH 3.6), of blue color with intensity proportional to the reducing capacity of the sample, with maximum absorption at 593 nm. Trolox can be used as a standard reagent (Robles-Botero et al., 2020).

The ORAC assay is related to the ability of antioxidants to trap peroxyl radicals, which cause lipid oxidation in foods. It measures the inhibition of chemiluminescence of a fluorescent probe induced by an oxidant (Benítez-Estrada et al., 2020). The cupric ion reducing antioxidant capacity assay (CUPRAC), developed as a variant of the FRAP assay, uses copper (Cu) instead of iron (Fe) (Mejía-Reyes et al., 2022). The HORAC assay is based on the oxidation of a fluorescent probe by hydroxyl radicals, which are generated through the Fenton system. The TOSC assay consists of inhibiting the formation of ethylene in the presence of antioxidant compounds that compete with α-keto-γ-betolbutyric acid (KMBA) for ROS (Munteanu & Apetrei, 2021).

14.2.4 Chromatographic Techniques

Among the techniques used for identification of bioactive compounds are high-performance liquid chromatography (HPLC) and ultra-high performance liquid chromatography (UHPLC), which are based on the separation of a mixture of compounds, that is distributed between a solid stationary phase and a liquid mobile phase with different polarity, under controlled flow and pressure; in UHPLC, the particle sizes analyzed can be less than 2 μm, while in HPLC, they are 5 to 3 μm (Grande-Tovar et al., 2021). This type of equipment can be associated with mass spectrometry (MS), a technique based on the mass/charge ratio (*m*/*z*) of the ions in a sample for their identification. Gas chromatography (GC), on the other hand, separates compounds by partitioning between a liquid stationary phase and a gaseous mobile phase (Munteanu & Apetrei, 2021). These techniques generate chromatograms, whose signals in the form of peaks are displayed at specific retention times for the different types of compounds. The data are compared with databases, and the compound present in the sample is deduced.

14.3 Functionalization of Food Products Employing Fruit and Vegetable Residues

Functional foods are a current trend focused on improving consumers' health. These are expected to provide benefits beyond their inherent nutritional contribution. Because of bioactive compounds, such as antioxidants, phytochemicals, and dietary fiber, functional foods are a dietary approach intended to improve outputs or prevent the development of non-transmissible chronic diseases (Vignesh et al., 2024). Although a functional food could be natural owing to its nutritional charge, such foods may be obtained via processing through fortification or enrichment. While enrichment implies adding micronutrients to increase their concentration in foods, fortification signifies supplementation of nutrients not constitutively present in a product. Nonetheless, Frumuzachi et al. (2025) emphasized the relevance of an appropriate differentiation between a functional and a functionalized food. Even though a functionalized food that has been either enriched or fortified has the potential to be considered a functional food, the lack of clinical validation studies is the characteristic that may differentiate both.

By-products obtained from the food agroindustry are described as having high nutritional value. These may be used during functionalized food designing once valuable bioactive compounds are recovered (i.e., dietary fibers, antioxidants, biopolymers, and substances potentially acting as food additives) (Ţiţa et al., 2024). In some cases, roughly one-half of biomass is discarded after food processing, and the main discarded segments of fruits and vegetables are peels, seeds, and pomace. Because of their relative abundance and ease of accessibility, wasted peels from fruits and vegetable processing are extensively used as bioactive molecule sources for food enrichment or fortification. Commonly isolated compounds from the peel are carotenoids, tannins, flavonoids, alkaloids, glycosides, terpenoids, sterols, and anthocyanins (Durán-Aranguren et al., 2023; Zaini et al., 2020). These compounds may display a wide range of functions, such as anti-inflammatory, antibacterial, oxidative stress reductants, coronary heart disease reducing agents, antidiabetic, and antihypertensive, among others (Backes et al., 2020; Ţiţa et al., 2024).

Residues derived from vegetables and fruits have been harnessed in food products ranging from sausages, dairy, beverages, confectionery, bakery, etc. The data collected from multiple studies dealing with this topic are summarized in Table 14.1. The works aimed to employ residues such as a fibrous source waste from a industrial processing is almost limited to drying and milling. Meanwhile, investigations attempting to incorporate phenolics and pigments generally require a solvent extraction procedure accompanied by the assistance of another technology (ultrasound, microwave, pulsed electric field, etc.).

Table 14.1 Fruit and vegetable residues used for food functionalization

Functionalized food	Residue used as bioactive source	Residue processing	Improved properties	Additional observations	References
Chicken sausage	Banana peel and Carrot pomace	Drying and milling to obtain powder	Increase in dietary fiber and reduced fat and cholesterol content	Only the lower addition level studied was acceptable at the sensorial level. Banana peel powder reduced the rate of lipid oxidation of sausage. The displeasure of higher substitution values was related to changes in color and texture (hardness and cohesiveness)	Yadav et al. (2018) and Zaini et al. (2020)
Yogurt	Carrot peel and pulp	Drying, milling, solvent extraction, solvent evaporation, and freeze drying	Higher total phenolic content and antioxidant activity	Carotenoid pigments were beneficial for color and appearance. However, taste and aroma were affected by carrot extract addition	Sharifi et al. (2023)
Juice	Tomato peel	Drying, grounding, sieving, solvent extraction (assisted by various emerging technologies), and emulsification	Increased antioxidant activity	High lycopene yield was achieved with combined ultrasound-pulsed electric field-assisted extraction. The lemon beverage showed significant color changes after supplementation	Kumar et al. (2023)

(continued)

Table 14.1 (continued)

Functionalized food	Residue used as bioactive source	Residue processing	Improved properties	Additional observations	References
Tagliatelle pasta	Grape and olive pomace	Drying (freeze-drying and oven drying) and milling	Fortification with phenolic compounds and greater fiber content	The fortified tagliatelle displayed a similar behavior during cooking to control samples. Phenols were retained after cooking	Balli et al. (2021)
Brownie	Biomass from peach palm, lime, and orange	Drying and milling. Ultrasound-assisted solvent extraction of phytochemicals and encapsulation	Reduced fat and increased fiber content. Supplementation with phytochemicals	Most tested formulations altered sensorial parameters significantly. Only one formulation mostly mimicked the sensorial profile obtained by the control brownie	Durán-Aranguren et al. (2023)
Doughnut glaze and Brazilian pastry called "beijinho"	Fig peel (*Ficus carica* L.) and blackthorn fruit epicarp (*Prunus spinosa* L.)	Freeze-drying, milling, sieving, ultrasound-assisted solvent extraction, and lyophilization	Increased antioxidant activity	Extracts provided a purple color once applied to doughnuts in glaze, besides displaying antimicrobial activity	Backes et al. (2020)
Gummy	Peel of purple passion fruit (*Passiflora edulis* f. *edulis*)	Drying, grounding, sieving, and ultrasound-assisted solvent extraction	More content of anthocyanins and increased antioxidant activity	The simulated digestion process stimulated the bioavailability of anthocyanins	Sakulrang et al. (2025)

It is worth mentioning that adding extracts or derivatives of food residues, in some cases, could lead to undesirable sensorial and functional changes. Therefore, formulations must be conceived taking into account consumers' acceptance. Additionally, during investigations focused on functionalized food development, analytical methods that guarantee the effective transference and bioavailability of bioactive compounds to the matrix must be implemented. Likewise, clinical proofs are required to back up health-benefit claims regarding the developed product so it can acquire the functional food designation.

14.4 Functional Coatings and Bioactive Compounds Derived from Horticultural Waste for Postharvest Preservation

The increasing generation of agro-industrial waste has driven the development of innovative strategies for its revalorization in food preservation. In particular, functional coatings based on biopolymers extracted from vegetable waste have proven to be a sustainable and effective alternative for extending the shelf life of fresh fruits and vegetables. These coatings, in addition to acting as a protective barrier against moisture loss and gas exchange, can incorporate bioactive compounds with antioxidant and antimicrobial properties, thereby improving postharvest quality and reducing dependence on synthetic and/or chemical preservatives (Basri et al., 2021; Carnaval et al., 2024).

14.4.1 Biopolymers and Their Functionality in Coatings

The biopolymers used in the formulation of functional coatings come from various agro-industrial sources, including polysaccharides, such as cellulose, starch, and pectin, as well as proteins and lipids, among others. Pectin, extracted from fruit by-products, such as apples, has demonstrated excellent gelling and barrier properties when combined with green solvents, enhancing its functionality in sustainable packaging applications (Gómez Vargas et al., 2024). Likewise, coatings based on modified starch, such as those obtained from mango seeds, have shown a significant reduction in weight loss and microbiological deterioration in fresh fruits, such as kiwi and tomato (Singh et al., 2024). Polysaccharides, such as arabinoxylan and β-glucans, have been studied for their ability to form strong and flexible films in products, such as apples, reducing fruit respiration and preserving firmness during storage (Ali et al., 2019). The incorporation of proteins and lipid compounds in these coatings not only improves their mechanical and barrier properties but also allows the inclusion of bioactive agents that can contribute to food preservation without altering its organoleptic or nutraceutical quality (Jorge et al., 2023). Additionally, recent research has shown that combining these biopolymers with plant extracts can improve thermal stability and moisture resistance, increasing their potential application in the food industry (Tapia-Blácido et al., 2023).

14.4.2 Incorporation of Bioactive Compounds in Functional Coatings

The addition of bioactive compounds in biodegradable coatings is a key strategy for improving the stability and safety of fresh products, which has gained great momentum in recent years. Phenolic and flavonoid extracts obtained from horticultural by-products have shown high efficacy in inhibiting pathogenic microorganisms and reducing oxidative stress in fruits and vegetables (Sorrenti et al., 2023). Moreover, essential oils and secondary metabolites derived from agro-industrial waste, such as those obtained from citrus peels and grape pomace, have shown a positive effect on the postharvest preservation of fruits by acting as natural barriers against microbiological contamination (Sun et al., 2024). The combination of these extracts with biopolymer nanoparticles enables the controlled release of antimicrobial compounds, enhancing coating stability and extending the freshness of perishable products (Martins et al., 2024). Therefore, the incorporation of extracts from horticultural waste into biofilms has been shown to be effective in preserving strawberries and blueberries, reducing weight loss and prolonging firmness during storage (Ferreira et al., 2016).

14.4.3 Utilization of Agro-industrial Waste and Extraction of Bioactive Compounds from Vegetables

The utilization of agro-industrial waste for the extraction of bioactive compounds is one of the most promising strategies in food preservation. Transforming these by-products into functional ingredients not only helps reduce waste but also enables the development of materials with direct applications in the food industry (Carnaval et al., 2024). In particular, the extraction of polyphenols, flavonoids, and glucosinolates from vegetable waste has proven to be effective for formulating bioactive coatings with antioxidant and antimicrobial properties (Damas-Job et al., 2023; Seguí & Barrera, 2025). Recent studies have highlighted the feasibility of using broccoli and kale waste as rich sources of glucosinolates and sulfur compounds that enhance fruit and vegetable preservation by reducing microbial growth (Quizhpe et al., 2024). The use of broccoli extracts in cherry tomatoes not only enhances antioxidant retention but also reduces postharvest fungal proliferation (Damas-Job et al., 2023). The application of grape pomace extracts as postharvest treatments effectively extended the shelf life of strawberries by reducing weight loss and mold growth compared to untreated fruits (Martínez-Vázquez et al., 2022). Additionally, the extraction of polysaccharides, such as pectin and cellulose from vegetable peels and stems has enabled the development of biodegradable materials with improved barrier properties, ideal for sustainable packaging (Basri et al., 2021).

The use of green technologies during the extraction of these compounds, such as ultrasound-assisted extraction or the application of supercritical fluids, has optimized the obtainment of bioactive ingredients with greater stability and functionality. These techniques have enabled the development of advanced formulations of biodegradable coatings that extend food shelf life without compromising its sensory quality or food safety (Sun et al., 2024). The integration of these compounds into functional coatings and active packaging is an innovative approach for developing more efficient and sustainable preservation systems. With the growing demand for eco-friendly products and increasing interest in reducing food waste, the utilization of agro-industrial waste is emerging as a key solution for the food and packaging industry in the coming years.

14.5 Desert Plant Residues and Their Potential Applications in the Food Industry

Desert plants are characterized by the production of multiple chemical compounds, development of structures, and morphological changes that help them survive under adverse conditions of temperature, water availability, and osmotic pressures. These harsh environments often experience high temperatures during the day and low temperatures at night, in addition to long periods of drought. Although several of these plants have been used as human food for more than 2000 years (e.g., *Yucca* spp., *Ferocactus* spp., *Opuntia* spp., *Prosapis* spp., *Agave* spp., etc.), few are exploited industrially or semi-industrially, thereby generating waste or by-products. These agro-industrial residues can be transformed into other foods or used to obtain compounds that can be used in the food industry. An example of this is the production of cheaper proteins from agro-industrial waste (Sá et al., 2020). In this section, we will focus on the residues from the industrialization of five plant species (Maguey, Xoconoxthe, Cactus, Mesquite, and Jatropha).

14.5.1 Maguey

Most of Agave (*Agave* spp) parts have been used in human activities, in food its sap (nonalcoholic beverage, and syrup) and alcoholic beverage (pulque), flowers, and inflorescence are used as food, its fibers are used in wickerwork, its sugars from its leaves (pencas) and pseudostems (piña) for the production of syrup and alcoholic beverages (tequila, mezcal, bacanora etc.), its "pencas" for obtaining membranes for food coverage, needle-like thorns in wickerwork, even its pests (worms) are used as food. During the industrial production of *Agave* (Fig. 14.2), some by-products (leaves, bagasse, and fibers) are generated (Álvarez-Chávez et al., 2021). These residues are composed of cellulose, lignin, hemicellulose, simple carbohydrates, pectin, fructans, as well as phytochemicals like fructans, saponins, and flavonoids. The prebiotic activity of *Agave* fructans has been reported, which have the potential to lower blood glucose and enhance lipid homeostasis. The addition of these and other agave polysaccharides to different foods would improve their nutritional value and technological properties (Márquez-Rangel et al., 2023). Meanwhile, saponins and flavonoids from *Agave* spp. have been associated with anti-inflammatory, antimicrobial,

Fig. 14.2 Maguey (*Agave* spp) plant during inflorescence period. (Picture taken by the authors)

antioxidant, and anticancer properties (Álvarez-Chávez et al., 2021). A drawback to directly using *Agave* bagasse in animal feed is its low digestibility due to its high lignin content. One way to increase this digestibility is a treatment with calcium hydroxide (Ramírez-Cortina et al., 2012).

14.5.2 Xoconoxtle

There are different species known by this name, even from different genera. They are characterized because they are grouped together with other cacti; they have acids and not very fleshy fruits with high nutritional value and antioxidants, due to their content of dietary fiber (DF) and polyphenols (Fig. 14.3). These fruits can be consumed fresh without the shell, or added as ingredients in broths (with or without shell). The fruit shells contain phenolic acids, such as catechin, caffeoyl tartaric acid, 5–5′-dehydrodiferulic acid, sinensetine, 1,2,2′-trisinepoylgentiobiose, dihydroquercetin, delphinidin, quercetin, and 5,3′, 4′ -trihydroxy3-methoxy-6:7-methyl enedioxyflavone4'-*O*-glucuronide. The highest antioxidant activity was found in the shells (Coutino-Laguna et al., 2022). Other compounds reported in xoconoxtle shells are tocopherol and isorhamnetin *O*-hexoside di-rhamnoside with high antioxidant properties (Morales et al., 2014). These compounds can be isolated and purified and used in various industries, such as food, cosmetics, and

Fig. 14.3 *Xoconoxtle* plant during fructification period. (Picture taken by the authors)

pharmaceuticals. It has been reported that the addition of xoconoxtle flour to animal feed increases the content of ether extract, protein, total phenols, antioxidant activity, and propionic acid. In addition, higher digestion rate and lower methane production were observed (Espino-García et al., 2023). Compounds such as betalains and vulgaxanthins have been reported in xoconoxtle shells, which are important in human nutrition and the food industry (Hernández-Fuentes et al., 2015).

14.5.3 Cactus Pear Fruits

The peel of the cactus pear fruit (*Opuntia ficus-indica* L. Mill) is discarded during fresh consumption of the fruit, or for the production of cheeses, juices, or liqueurs. Manzur-Valdespino et al. (2020) added green and red fruit peel powders during the production of tablets, finding higher total phenolic content (TPC) in formulations with green cactus pear residue, while antioxidant powder was higher in formulations with red cactus pear residue. The phytochemical compounds in this peel are attributed with hypoglycemic, anticancer, cholesterol control and body weight reducing properties (Tesoriere et al., 2005). During the production of nopalitos (small pieces of young, fresh cladodes), the external parts of the cladode are removed. This mucilage is composed of xylose, glucose, rhamnose, arabinose, galacturonic acid, fucose, and galactose, and it has the capacity to absorb huge amounts of water, forming a viscous colloid with interesting rheological properties, in addition to high calcium content, total phenolic content, flavonols, and flavonoids (Elshewy et al., 2023). The phytochemical composition of this fruit can be affected by the degree of maturity, cultivar, and their interaction (Cardador-Martínez et al., 2011). The chemical and physical properties of the peel of this fruit give it potential for multiple food applications.

14.5.4 Mesquite

The fruits of this legume are consumed for their juice, where the fruit is chewed, and the solid part (peel and seed) is discarded. The flour from the ripe fruit is also used in baking or to obtain phytochemical compounds. Different species of this plant are distributed in various countries of the American continent. In Argentina, mesquite (*Prosopis nigra*) fruits are used to obtain fine flours, discarding residues of the highest granulometry. These residues have high fiber and fructose content, which can vary according to the *Prosopis* genotype evaluated (Correa Uriburu et al., 2022). In addition, rich potassium, iron, fibers, free and bound phenolic compounds, C-glycosides derivatives from apigenin (Pérez et al., 2020). Other C-glycosyl flavones reported from *Prosopis alba* are isoschaftoside, vitexin, schaftoside, vicenin II, and isovitexin (Correa Uriburu et al., 2022). These phenolic compounds have high antioxidant activity with the potential to inhibit enzymes related to carbohydrate metabolism (α-glucosidase and α-amylase) (Pérez et al., 2020). Other species of this plant (*Prosopis cineraria*) have been reported to contain higher contents of moisture and protein (24%) than other plants (Iqbal et al., 2022). These properties give this residue the potential to be used as a novel renewable, and sustainable material for the production of bioactive food formulations (Correa Uriburu et al., 2022).

14.5.5 Jatropha

The fruits of this plant are not consumed directly as food, but, they have medicinal properties and can also be used to obtain oil or biofuels, where waste is produced, that is, generally not used. However, proteins isolated from the residues of this plant have high nutritional quality and are rich in essential amino acids (except lysine), where the main protein fraction comprises glutelin (42.03%), followed by globulins (20.17%) (León-Villanueva et al., 2018).

14.6 Applications of Agro-industrial Waste Compounds in Metabolic Diseases

The agricultural industry annually generates large volumes of waste, which represents an environmental and health challenge if it is not properly managed (Mudalip et al., 2025). However, these wastes are a rich source of bioactive compounds with antioxidant, anti-inflammatory, and antimicrobial properties that can be used in the prevention and treatment of metabolic diseases, such as obesity, diabetes, dyslipidemia, gut dysbiosis, atherosclerotic cardiovascular diseases, and metabolic syndrome (Enciso-Martínez et al., 2024; Malenica et al., 2025). The valorization of these agro-industrial by-products in the formulation of nutraceuticals and functional foods has gained relevance in medical research (Comi et al., 2025).

14.6.1 Bioactive Compounds of Agro-industrial Waste

Agro-industrial waste contains a variety of bioactive compounds, including polyphenols, flavonoids, carotenoids, dietary fiber, and essential fatty acids, among other compounds with potential benefits for metabolic health (Vicente-Zurdo et al., 2025). Citrus residues are rich in polyphenols with antioxidant properties that can improve insulin sensitivity and reduce systemic inflammation, two key factors in the development of metabolic diseases (Maqbool et al., 2023). Likewise, cereal and legume residues contain dietary fiber, which contributes to the modulation of the gut microbiome and the regulation of blood glucose (Buljeta et al., 2023). Extracts from fruit peels and seed residues have been studied for their capacity to influence lipid metabolism and cardiovascular function, making them ideal candidates for nutraceutical applications (Can-Cauich et al., 2017). Residues from mango, passion fruit, tomato, carrot, and other plant sources contain carotenoids with antioxidant, anti-apoptotic, and anti-inflammatory effects, which contribute to the suppression of various effects of diabetes (Das et al., 2024).

14.6.2 Mechanisms of Action in Metabolic Diseases

The bioactive compounds present in agro-industrial waste have shown a positive impact on the management of metabolic diseases, such as obesity, diabetes mellitus, and cardiovascular pathologies, acting through various mechanisms of action (Yadav et al., 2024). Peptides derived from seeds and husks inhibit digestive enzymes, such as α-glucosidase and α-amylase, reducing carbohydrate absorption and promoting glycemic regulation (Colombo et al., 2024). In addition, the anthocyanin-rich phenolic compounds found in grape pomace, which consists of skins, seeds, and residual pulp, activate signaling pathways, such as AMPK, improving glucose uptake in tissues and increasing insulin sensitivity (Arruda et al., 2021). On the other hand, antioxidants, such as ferulic acid, present in rice bran, contribute to pancreatic protection by reducing oxidative stress in β-cells of the pancreas, thus preserving their function and capacity for insulin secretion (Pérez-Ternero et al., 2017).

14.6.3 Extraction Methods of Bioactive Compounds Isolated from Agro-industrial Waste

The extraction of bioactive compounds is a critical stage for their use, where methodological selection has a direct impact on the yield, purity, and sustainability of the process. Conventional methods such as maceration, Soxhlet extraction, and hydrodistillation stand out for their operational simplicity, although they have significant limitations: high consumption of organic solvents, long processing times, and risk of thermal degradation of thermolabile molecules (Fierascu et al., 2019). Despite this, these techniques have made it possible to obtain anthocyanins from red dragon peel peels, polyphenols from pomegranate peels, and pectins from citrus and passion fruit waste, demonstrating their usefulness in specific matrices (Floridas-Galán & Paredes-Concepción, 2024). In contrast, unconventional methods emerge optimizing efficiency and minimizing environmental impacts. Ultrasonically assisted extraction takes advantage of cavitation phenomena for selective cell breakdown, while microwaves induce volumetric heating that accelerates the release of metabolites. Technologies such as supercritical fluids (supercritical CO_2) offer selectivity adjustable by pressure and temperature, ideal for nonpolar molecules, such as carotenoids. In addition, enzymatic methods use cellulases or pectinases to degrade plant walls, preserving sensitive bioactive structures (Amran et al., 2021). The transition to these alternatives responds to the need for scalable processes with a smaller ecological footprint, where factors such as the polarity of the target compound, the complexity of the residual matrix (porosity, lignocellulosic content) and the requirements of the final product (food grade vs. pharmaceutical) determine the technological choice (Li et al., 2022). This differential approach makes it possible to valorize agricultural waste under circular economy principles, transforming by-products into high-value-added ingredients (Arun et al., 2020).

14.6.4 Challenges and Future Prospects

Despite the promising potential of bioactive compounds derived from agro-industrial waste, their application in metabolic health is challenged by factors, such as the standardization of extraction processes, the bioavailability of the compounds, and consumer acceptance. However, the growing demand for functional foods and nutraceuticals is an opportunity for research and development of innovative products based on the revalorization of this waste and promotes a more sustainable and circular production model (Khan et al., 2024).

14.7 Agro-industrial Waste Compounds in Neurodegenerative Diseases

Neurodegenerative diseases are a heterogeneous group of disorders characterized by the progressive loss of neuronal count, deterioration of neuronal integrity, and impairments in neuronal function. These changes ultimately lead to cognitive decline as well as sensory and motor dysfunction(Chiu et al., 2020). The pathophysiology of these diseases involves various processes, including oxidative stress, neuroinflammation, protein misfolding, mitochondrial dysfunction, endoplasmic reticulum stress, calcium imbalance, impaired autophagy, and neuronal apoptosis (Angeloni et al., 2022; Rahman et al., 2021).

The brain is particularly vulnerable to oxidative stress due to its high lipid content, the presence of metals that catalyze oxidative reactions, elevated oxygen consumption, and limited antioxidant defenses (John et al., 2020). Consequently, one strategy for preventing and treating neurodegenerative diseases involves the intake of exogenous antioxidants. In recent years, natural products, and their bioactive compounds, particularly antioxidants, have been extensively investigated. These compounds are being studied for their biological effects, nutritional properties, and potential therapeutic benefits. These compounds offer promising alternatives for combating oxidative stress in neurodegenerative conditions (Atlante et al., 2020; Rojas-García et al., 2023). An alternative and valuable source of antioxidants is agro-industrial waste. Harnessing these by-products not only reduces environmental impact but also adds value by transforming waste materials into functional ingredients with potential applications in the pharmaceutical and food industries.

14.7.1 Alzheimer

Alzheimer's disease (AD) is a progressive neurodegenerative disorder characterized by the extracellular accumulation of β-amyloid (Aβ) peptides and intracellular neurofibrillary tangles (NFTs) composed of hyperphosphorylated tau proteins. These pathological hallmarks contribute to a cascade of neuroinflammatory responses, driven by microglial activation, oxidative stress, and cholinergic dysfunction. These pathological features are particularly evident in the hippocampus and cerebral cortex, regions critical for memory formation and cognitive processing. The progressive degeneration of these areas contributes significantly to the characteristic learning and memory impairments observed in Alzheimer's disease.

The interplay between these mechanisms is central to AD pathogenesis and highlights potential therapeutic targets (Fig. 14.2) (Ciccone et al., 2023). Compounds present in pomegranate peel (PP) have shown beneficial effects on this disease. An in vitro study showed that PP extract reduced oxidative stress, improved cell viability, and decreased apoptosis in human H4 neurons co-cultured with $A\beta_{1-42}$. Additionally, the extract inhibited the transition of microglia from the M1 to the M2 phenotype, a crucial process in the induction of neuroinflammation (Alami et al., 2024). Similarly, another in vitro study found that polyphenols present in PP inhibited acetylcholinesterase activity, thereby increasing acetylcholine levels (Khokar et al., 2021).

In vivo studies have demonstrated that PP extracts exert neuroprotective effects in experimental models of AD (Fig. 14.4). These effects include a reduction in lipid peroxidation and nitric oxide levels, as well as decreased expression and activity of acetylcholinesterase. Furthermore, the extract enhanced the antioxidant defense

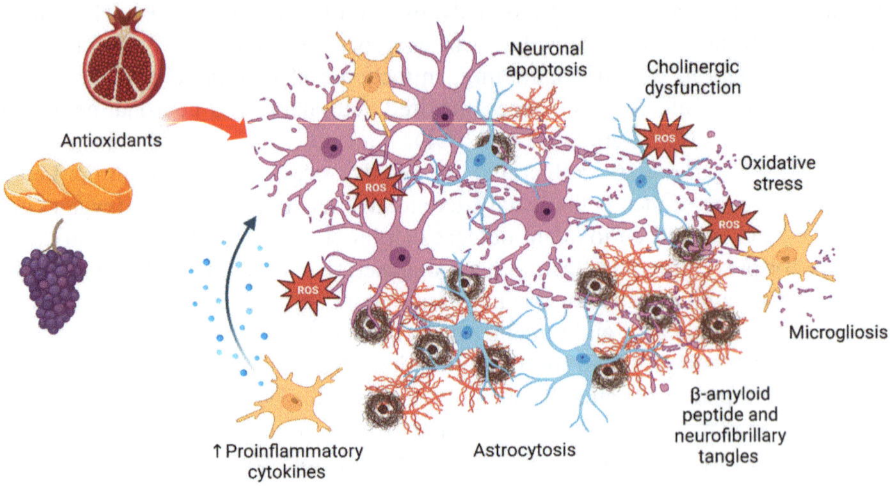

Fig. 14.4 Neuroprotective effects of antioxidants derived from agro-industrial waste Alzheimer's disease. Extracts from pomegranate, orange peel, and grape have been shown to mitigate neuronal damage associated with oxidative stress, neuronal apoptosis, and cholinergic dysfunction. These antioxidants reduce the production of reactive oxygen species (ROS), the expression of proinflammatory cytokines, and the formation of neurofibrillary tangles, thus contributing to neuronal protection

system by increasing glutathione levels and upregulating the activity of key antioxidant enzymes, including glutathione peroxidase and superoxide dismutase (Abd El-Aziz et al., 2023). Punicalagin, one of the major bioactive compounds in pomegranate peel, has also been shown to mitigate learning and memory deficits and exert anti-neuroinflammatory effects, as evidenced by reduced microglial activation and astrocytosis (Chen et al., 2023).

Grape pomace, an agro-industrial by-product, contains bioactive compounds with potential therapeutic effects against AD. Use of a red grape seed extract rich in proanthocyanidins has showed a reduction in amyloid plaque deposition and microgliosis in an experimental model of AD induced by D-galactose in rats (Kiran et al., 2024). Another in vivo study utilizing grape skin supplementation demonstrated improved cognitive performance in transgenic mice modeling AD. This effect was associated with increased expression of ADAM10, an α-secretase involved in non-amyloidogenic APP processing, in the hippocampus and cerebral cortex. Additionally, grape skin supplementation reduced the levels of pro-inflammatory cytokines (TNF-α and IL-1β) in both serum and hippocampus, decreased malondialdehyde (MDA) concentrations, and enhanced superoxide dismutase (SOD) activity, suggesting a potential neuroprotective role (Sheng et al., 2018).

Orange peel extracts have demonstrated potential neuroprotective effects against AD. In vitro, the extract exhibited significant antioxidant and anticholinesterase activities. In vivo, orange peel extract reduced acetylcholinesterase activity, lipid peroxidation, and nitric oxide levels. Additionally, the extract downregulated the expression of genes encoding amyloid precursor protein (APP) and beta-secretase enzyme (BACE1), key factors involved in amyloid plaque formation. These findings suggest that orange peel extracts may mitigate oxidative stress and cholinergic dysfunction, contributing to their potential as therapeutic agents in AD (Abd El-Aziz et al., 2023).

14.7.2 Parkinson's

Parkinson's disease (PD) is a progressive neurodegenerative disorder characterized by the selective loss of dopaminergic neurons in the substantia nigra pars compacta and the accumulation of Lewy bodies, which are intraneuronal protein aggregates primarily composed of misfolded α-synuclein. This neuronal degeneration leads to a significant reduction in striatal dopamine levels, resulting in hallmark motor symptoms, such as bradykinesia, resting tremor, muscular rigidity, and postural instability. Non-motor manifestations, including cognitive impairment, mood disturbances, and autonomic dysfunction, are also common and contribute substantially to disease burden (Saleem et al., 2021).

Extracts derived from agroindustrial waste have also demonstrated neuroprotective potential in the context of Parkinson's disease. In vitro studies have shown that pomegranate seed juice can mitigate MPTP-induced neurotoxicity in primary human neurons. This protective effect was evidenced by the attenuation of MPTP-induced reductions in intracellular NAD^+ and ATP levels, as well as a decrease in extracellular lactate dehydrogenase (LDH) activity, indicative of improved neuronal viability. Furthermore, these extracts counteracted MPTP-induced oxidative stress by preserving the activity of endogenous antioxidant enzymes, such as SOD and glutathione peroxidase (GPx) while maintaining reduced glutathione (GSH) levels (Braidy et al., 2013). In an in vivo model, pretreatment with pomegranate seed extract or juice demonstrated neuroprotective effects against paraquat-induced neurotoxicity in mice. The study revealed that treatment increased tyrosine hydroxylase levels in the substantia nigra, suggesting preservation of dopaminergic neurons. Additionally, there was an increase in striatal dopamine and its metabolite DOPAC, indicating enhanced dopaminergic function.

Oxidative stress markers were ameliorated, as evidenced by reduced MDA levels and enhanced activities of antioxidant enzymes. Inflammatory responses were modulated, with decreased pro-inflammatory cytokines and increased anti-inflammatory markers. Furthermore, pomegranate extract led to a reduction in apoptosis and downregulation of NF-κB gene expression in the striatum (Fathy et al., 2021). These findings suggest that pomegranate seed extract and juice may offer neuroprotective benefits by mitigating oxidative stress, modulating inflammatory responses, and preserving dopaminergic neuronal integrity in PD models. Grape pomace-derived compounds have shown significant neuroprotective effects in Parkinson's disease models.

In vitro studies demonstrated that oligomeric stilbenes extracted from grape stems protected dopaminergic neurons by reducing mitochondrial dysfunction and preventing apoptosis. These compounds also modulated inflammatory responses by decreasing the production of pro-inflammatory cytokines, suggesting their potential in mitigating neuroinflammatory processes (Pislyagin et al., 2025). In an in vivo model using a subacute MPTP mouse model of Parkinson's disease, grape pomace extract formulated in nutriosomes exhibited robust protective effects. The treatment effectively preserved tyrosine hydroxylase (TH)-positive neurons in the substantia nigra and striatum, indicative of reduced neuronal damage (Parekh et al., 2022). Moreover, in another study, red grape seed and skin extract administration in a 6-OHDA-induced Parkinson's model efficiently protected dopaminergic neurons, reduced neuronal degeneration, and improved motor function (Ben Youssef et al., 2021). These findings highlight the potential of grape pomace-derived compounds in protecting neuronal integrity, modulating oxidative stress, and reducing inflammation in Parkinson's disease models.

14.8 Applications of Compounds from Agro-industrial Waste in Infectious Diseases

The increasing prevalence of infectious diseases, worsened by the global rise in antimicrobial resistance, needs the exploration of sustainable and innovative solutions. Agro-industrial waste, rich in bioactive compounds, presents a promising possibility for developing functional food ingredients that may aid in infection prevention and management.

14.8.1 Food Preservation and Antimicrobial Properties

One of the most significant applications of bioactive compounds derived from agro-industrial waste is food preservation. Effective preservation methods play a crucial role in reducing microbial contamination and prolonging the shelf life of food products, thereby minimizing the risk of foodborne diseases (Teshome et al., 2022). Natural antimicrobial agents extracted from agricultural residues offer a sustainable and safer alternative to synthetic preservatives.

For instance, ethanolic extracts from wheat, sunflower, and maize crop residues have demonstrated bacteriostatic effects against foodborne pathogens, such as *Salmonella* Typhimurium and *Staphylococcus aureus*, highlighting their potential as natural food additives (Milica et al., 2023). Additionally, grape pomace extracts have been effectively utilized to extend the shelf life of meat products by inhibiting the growth of *Staphylococcus aureus* and *Escherichia coli* (Sagini et al., 2024).

Beyond preservation, agro-industrial waste serves as a valuable substrate to produce antimicrobial peptides. For example, *Lactococcus lactis* has been employed to produce bacteriocins from various agro-industrial by-products, including whey, mussel processing waste, octopus peptone, cull potatoes, condensed distillers' solubles, skimmed milk, fermented barley extract, and sweet whey (Bali et al., 2016). These bacteriocins exhibit potent antimicrobial properties, making them suitable candidates for functional food applications.

Another innovative approach involves the synthesis of nanoparticles using biowaste. These nanoparticles can be incorporated into edible coatings made from materials, such as cellulose, chitosan, and polyvinyl alcohol, enhancing their antimicrobial and preservation properties (Flores-Contreras et al., 2024). This method offers a novel way to extend food shelf life while simultaneously contributing to infection control by reducing microbial load.

14.8.2 Immunostimulation

Enhancing immune responses is a key strategy against infections. Agro-industrial waste has been repurposed in diets in aquaculture, promoting growth and disease resistance in aquatic organisms. A meta-analysis using a random-effects model suggested a slight positive effect on IL-1β expression, indicating its potential immunostimulatory role (Bertocci & Mannino, 2022).

Additionally, the antibacterial activity of *Origanum majorana* essential oil (EOM) and its nanocapsulated form (NOM) has been evaluated in silver catfish (*Rhamdia quelen*) infected with *Aeromonas hydrophila*. Fish treated with EOM and NOM exhibited higher lymphocyte levels, indicating stimulation of the immune system (da Cunha et al., 2018).

14.8.3 Modulation of Gut Microbiota

Dietary fibers and prebiotic compounds from agro-industrial waste can selectively promote the growth of beneficial gut bacteria, enhancing immune function and overall health. For example, pectic oligosaccharides derived from citrus peel waste have been shown to stimulate the proliferation of *Lactobacillus* and *Bifidobacterium* species, improving gut health (Foti & Romeo, 2024).

Brewer's spent grain (BSG), a by-product of the brewing industry, is rich in arabinoxylans and polyphenols, which have been identified as potential prebiotics. Studies indicate that BSG can modulate gut microbiota by promoting the growth of beneficial bacteria, such as *Bifidobacterium* and *Lactobacillus* while reducing the abundance of harmful bacteria like *Bacteroides, Escherichia,* and *Shigella.* Additionally, BSG increases the production of short-chain fatty acids (SCFAs) and branched short-chain fatty acids (BCFAs), both beneficial for gut health (Calvete-Torre et al., 2023; Kruk et al., 2025).

Cashew apple by-product (CAB) is another agro-industrial waste with prebiotic properties. CAB enhances the relative abundance of *Bifidobacterium* and *Lactobacillus/Enterococcus* while reducing harmful bacteria, such as *Bacteroides/Prevotella* and *Clostridium histolyticum*. Additionally, CAB increases SCFA production and lowers the pH of colonic fermentation media, further supporting its prebiotic effects (Menezes et al., 2021).

14.9 Anti-inflammatory Potential of Agro-industrial Waste: A Source of Bioactive Compounds for Functional Food Development

Inflammation is a fundamental biological response of the immune system, as a defense against pathogens, damaged cells, or toxic compounds. While acute inflammation is a protective response, chronic inflammation has been implicated in the pathogenesis of numerous diseases (Chen et al., 2015; Libby et al., 2024).

The inflammatory process is often mediated by pro-inflammatory cytokines like IL-1β, IL-6, and TNF-α, or others implicated in the inflammasome activation (Zhu et al., 2024). These mediators recruit immune cells for pathogen clearance or tissue repair. However, chronic inflammation leads to tissue damage and disease progression (Yu et al., 2024). This process is frequently accompanied by oxidative stress, which causes cellular damage and inflammatory cycle perpetuation (Guan, 2024).

In this section, several examples of bioactive compounds from agro-industrial waste with anti-inflammatory properties were described.

14.9.1 Anti-inflammatory Properties with Non-intestinal Targets

Recently, the saponin-rich plants, *Agave sisalana* and *Punica granatum* waste, were used in the development of mucoadhesive gels for dermal applications. The individual effect of both extracts diminished the phagocytosis in RAW 264 macrophages, effect that were superior in the combination of these extracts. The inhibition of phagocytosis was comparable to that of a gel containing dexamethasone, a commercial pharmaceutical product. Furthermore, the extracts inhibited macrophage spreading, demonstrating effects similar to dexamethasone. Additionally, these extracts have antihemolytic and protective membrane properties, because the extracts inhibit the hypotonicity and lysis of the erythrocyte membrane (Fracasso et al., 2023).

A comprehensive review documented the anti-inflammatory properties of Agavaceae species in experimental models, attributable to their flavonoid and saponin content. *A. angustifolia* and *A. tequilana* reduced pro-inflammatory cytokines and inflammation in edema and lupus models. *A. marmorata* inhibited NF-κB, while *A. intermixta* and *A. americana* showed potent oedema reduction. Cantalasaponin-1, from *A. americana, A. barbadensis*, and *Furcraea selloa var. marginata*, had strong anti-inflammatory and anti-neuroinflammatory effects, lowering IL-6 and TNF-α while increasing IL-10 (Herrera-Ruiz et al., 2022).

The effects of olive mill wastewater were studied in rabbits fed a diet supplemented with this residue. As a result, the residue caused a downregulation of genes *SIRT1*, *TNFA*, *AGER*, *BAX*, and *PPARA* in the liver, suggesting a preventive against the oxidative stress and inflammation. This second effect could be attributable to the modulation of certain phenolic compounds that could regulate the hepatic nuclear factor (NF-κB) pathway, a key transcription factor in the pro-inflammatory cascade (Cappelli et al., 2021).

The potential use of aromatic medicinal plants was also studied as anti-inflammatory. The plant *Bunium ferulaceum* utilization is limited to food flavoring and decoration. A study identified a rich phenolic profile in the plant's aerial part. These compounds exhibited a significant antioxidant activity in vitro and in vivo; the extract improved the mice liver redox state, increasing catalase and reduced glutathione and lipid peroxidation levels. Particularly, the extract reduced ear inflammation induced in the animals, suggesting potential therapeutic applications (Deghima et al., 2022).

14.9.2 Gastrointestinal Anti-inflammatory Effects

Other interesting findings are those describing the anti-inflammatory effects in the intestine. For example, the use of *Perilla frutescens* seed residue from oil production in the reduction of the inflammatory process in vitro and in vivo. Using a carcinogenesis colon model in rats, the authors observed that a crude extract of perilla, administered orally, reduces the number of aberrant crypt foci and decreases pro-inflammatory cytokines (IL-6, IL-1β, and TNF-α) in rat serum. The treatment also suppresses mRNA expression of these cytokines in colonic epithelial cells. Additionally, extract increases the gut microbiota diversity in the rat model, particularly favoring the abundance of beneficial bacteria, such as *Muribaculaceae*, *Lactobacillus*, and *Oscillospiraceae,* and reducing pathogenic bacteria (*Escherichia/Shigella* and *Paraprevotella).* Regarding the effects of perilla in vitro, the stimulated cell lines HCT-116 and HT-29 in the presence of IL-16 were significantly inhibited in their growth by extract exposure, confirming their antiproliferative effect in an inflammatory status. The diminishing in proinflammatory cytokines was also confirmed in RAW 264.7 cells induced by lipopolysaccharides (LPS) (Chantana et al., 2023).

The modulatory capacity of another residue, the pomace of grape (*Vitis vinifera* L.), was also studied in cancer cell line SW480 and in isolated mouse colon exposed to *Escherichia coli* LPS. An aqueous extract from the winemaking process reduced cell viability in the cell line, and in the mouse, cells inhibited mRNA levels NF-kB, COX-2, TNF-α, IL-6, IL-1β, IL-10, iNOS, and IFN-γ, in isolated colon. In addition, the extract improves the antioxidant defense by the increased genic expression of CAT and SOD (Recinella et al., 2022).

The waste form of lentils (*Lens culinaris)* was studied in Caco-2 cells stimulated by LPS. The most effective extract downregulated TLR expression and reduced iNOS and IL-1 production while upregulating IL-10. Additionally, the extracts demonstrated spasmolytic effects on ileum and colon tissues, suggesting a dual role in alleviating both inflammation and muscle spasms associated with inflammatory bowel diseases (Panaro et al., 2024).

14.9.3 Future Perspectives

Agro-industrial waste represents a promising source of bioactive compounds with anti-inflammatory properties. Future research should focus on conducting clinical studies to validate the efficacy and safety of these compounds, as well as developing innovative food formulations that incorporate them into diets to help manage various inflammation-related conditions.

14.10 Anticancer Activity of Agro-industrial Waste Compounds

Cancer is a disease resulting from the disruption of the normal cellular process known as cell division. There are different types of cancer depending on the cell type where the alteration began. Molecular analyses reveal highly described proteins involved in this process, such as p53, which is responsible for causing the cell to undergo apoptosis when there is damage to the DNA of the replicating cell. However, in cancer, this protein is altered without recognizing the DNA damage and allowing the uncontrolled replication of the altered cells (de Visser & Joyce, 2023).

Currently, different alternatives have been developed for cancer, including radiation, chemotherapy, surgery, and others. However, these are not specific, but generally attack the destruction of cells, including unaltered cells, causing general discomfort in the patient. Therefore, the scientific community has focused on finding new alternatives for this type of disease (Lee et al., 2024).

We have once again turned to plants as an alternative source of treatment, as in the early history of humankind. Hence, leaves, stems, and parts of different plants have been studied to obtain metabolites with diverse biological activities. Recently, with the increase in population, there has been a need to extract and add value to those considered waste, such as peels of some agro-industrial products like pomegranate peels, grape marc, and coffee pulp. Although these are residues, they possess many molecules with biological activities since they come from a living being, which, like humans, must defend itself against pathogens (Teniente et al., 2023).

As a first step, these bioactives are recovered through different technologies and subsequently characterized to identify the type of isolated molecule. Currently, libraries of bioactive compounds exist. Some of the most abundant bioactive compounds extracted are polyphenols. These polyphenols are characterized by being secondary metabolites of plants, presenting one or more aromatic rings, hydroxyl groups, esters, methyl esters, and glycosides, which are highly polymerized molecules. Polyphenols are classified as phenolic acids, flavonoids, tannins, lignans, and stilbenes (Rudra et al., 2021).

Some of the mechanisms described as the anticancer effect of polyphenols are the inhibition of the replication of cancer cells. This has been seen in polyphenols extracted from Tea. The presence of epigallocatechin gallate (EGCG) and theaflavin (TF) involved in cell replication has been identified. (Sur & Panda, 2017). The inhibition of cancer cell migration is another mechanism that has been described to be affected by the presence of polyphenols, such as resveratrol, which prevents the metastasis of MDA-MB 231 breast cancer cells in vitro by increasing the expression of proteins for the formation of junctions, such as E-cadherins (Sun et al., 2019). Likewise, resveratrol has shown efficacy in inducing cellular arrest and apoptosis. This was observed by analyzing in vitro the transcriptome of liver cancer cells (HepG2) exposed to resveratrol and it was found that the modification in gene expression is mainly in the arrest in the S phase of the cell cycle, concluding that it is associated with cellular arrest (Wu et al., 2019). Another mechanism described in cancer cells in the presence of polyphenols is epigenetics, with curcumin as an example of epigenetic modification through the inhibition of DNA methyltransferases, modifications on histones, regulation of microRNAs and actions on the binding sites of transcription factors to DNA, all of these mechanisms are involved in the expression of genes associated with the maintenance of cancer cells (Hassan et al., 2019).

Some polyphenols have also been reported to exhibit genotoxicity, which also leads to cell death through DNA damage; this could be due to degradation. This effect has been reported for phenolic compounds isolated from *Phyllanthus tenellus* Roxb, and for polyphenols extracted with ethanol from *Moringa olifera*, which also show genotoxicity to cancer cells (Senna-Pereira et al., 2022; Esparza et al., 2024). However, just as polyphenols have been observed to be beneficial against cancer cells, their effects have also been studied in vivo in mice, showing that they can be toxic, generating acute toxicity at doses above 500 mg/kg. Therefore, it is advisable to follow established guidelines for this type of product (de Barros et al., 2022).

14.11 Conclusions

The nature of agro-industrial wastes allows them to be widely explored globally. The benefits of utilizing these resources are promising. The wide range of applications in areas, such as medicine, the functional food industry, pharmaceuticals, and biomaterials, opens a possibility to reduce them as sources of pollution and instead generate products with high added value. These actions reflect the need to become aware of the importance of the culture of recycling and the responsible use of natural resources.

References

Abd El-Aziz, N. M., Shehata, M. G., Alsulami, T., Badr, A. N., Elbakatoshy, M. R., Ali, H. S., & El-Sohaimy, S. A. (2023). Characterization of orange peel extract and its potential protective effect against aluminum chloride-induced Alzheimer's disease. *Pharmaceuticals, 16*(1), 12. Retrieved from https://www.mdpi.com/1424-8247/16/1/12

Alami, M., Boumezough, K., Zerif, E., Zoubdane, N., Khalil, A., Bunt, T., et al. (2024). In vitro assessment of the neuroprotective effects of pomegranate (Punica granatum L.) polyphenols against tau phosphorylation, neuroinflammation, and oxidative stress. *Nutrients, 16*(21), 3667. Retrieved from https://www.mdpi.com/2072-6643/16/21/3667

Ali, U., Kanwar, S., Yadav, K., Basu, S., & Mazumder, K. (2019). Effect of arabinoxylan and β-glucan stearic acid ester coatings on post-harvest quality of apple (*Royal delicious*). *Carbohydrate Polymers, 209*, 338–349. https://doi.org/10.1016/j.carbpol.2019.01.002

Álvarez-Chávez, J., Villamiel, M., Santos-Zea, L., & Ramírez-Jiménez, A. K. (2021). Agave by-products: An overview of their nutraceutical value, current applications, and processing methods. *Polysaccharides, 2*(3), 720–743. https://doi.org/10.3390/polysaccharides2030044

Amran, M. A., Palaniveloo, K., Fauzi, R., Satar, N. M., Mohidin, T. B. M., Mohan, G., et al. (2021). Value-added metabolites from agricultural waste and application of green extraction techniques. *Sustainability, 13*(20), 11432. https://doi.org/10.3390/su132011432

Angeloni, C., Malaguti, M., Prata, C., Freschi, M., Barbalace, M. C., & Hrelia, S. (2022). Mechanisms underlying neurodegenerative disorders and potential neuroprotective activity of agrifood by-products. *Antioxidants, 12*(1), 94. https://doi.org/10.3390/antiox12010094

Arruda, H. S., Silva, E. K., Araujo, N. M. P., Pereira, G. A., Pastore, G. M., & Marostica Junior, M. R. (2021). Anthocyanins recovered from agri-food by-products using innovative processes: Trends, challenges, and perspectives for their application in food systems. *Molecules, 26*(9), 2632. https://doi.org/10.3390/molecules26092632

Arun, K. B., Madhavan, A., Sindhu, R., Binod, P., Pandey, A., & Sirohi, R. (2020). Remodeling agro-industrial and food wastes into value-added bioactives and biopolymers. *Industrial Crops and Products, 154*, 112621. https://doi.org/10.1016/j.indcrop.2020.112621

Atlante, A., Amadoro, G., Bobba, A., & Latina, V. (2020). Functional foods: An approach to modulate molecular mechanisms of Alzheimer's disease. *Cells, 9*(11), 2347. https://doi.org/10.3390/cells9112347

Backes, E., Leichtweis, M. G., Pereira, C., Carocho, M., Barreira, J. C. M., Kamal Genena, A., et al. (2020). Ficus carica L. and Prunus spinosa L. extracts as new anthocyanin-based food colorants: A thorough study in confectionery products. *Food Chemistry, 333*, 127457. https://doi.org/10.1016/j.foodchem.2020.127457

Bali, V., Panesar, P. S., & Bera, M. B. (2016). Trends in utilization of agro-industrial byproducts for production of bacteriocins and their biopreservative applications. *Critical Reviews in Biotechnology, 36*(2), 204–214. https://doi.org/10.3109/07388551.2014.947916

Balli, D., Cecchi, L., Innocenti, M., Bellumori, M., & Mulinacci, N. (2021). Food by products valorisation: Grape pomace and olive pomace (pâté) as sources of phenolic compounds and fiber for enrichment of tagliatelle pasta. *Food Chemistry, 355*, 129642. https://doi.org/10.1016/j.foodchem.2021.129642

Basri, M. S. M., Shah, N. N. A. K., Sulaiman, A., Tawakkal, I. S. M. A., Nor, M. Z. M., Ariffin, S. H., Ghani, N. H. A., & Salleh, F. S. M. (2021). Progress in the valorization of fruit and vegetable wastes: Active packaging, biocomposites, by-products, and innovative technologies used for bioactive compound extraction. In *Polymers* (Vol. 13, issue 20). MDPI. https://doi.org/10.3390/polym13203503

Ben Youssef, S., Brisson, G., Doucet-Beaupré, H., Castonguay, A. M., Gora, C., Amri, M., & Lévesque, M. (2021). Neuroprotective benefits of grape seed and skin extract in a mouse model of Parkinson's disease. *Nutritional Neuroscience, 24*(3), 197–211. https://doi.org/10.1080/1028415x.2019.1616435

Benítez, R. (2015). Pérdidas y desperdicios de alimentos en América Latina y el Caribe. *Oficina Regional de la FAO para América Latina y el Caribe*. https://www.fao.org/americas/news/news-detail/pda-benitez/es

Benítez-Estrada, A., Villanueva-Sánchez, J., González-Rosendo, G., Alcántar-Rodríguez, V. E., Puga-Díaz, R., & Quintero-Gutiérrez, A. G. (2020). Determination of the total antioxidant capacity of food and human plasma by photochemiluminescence: Correlation with spectrophotometric (FRAP) and fluorometric (ORAC) assays. *TIP Revista Especializada en Ciencias Químico-Biológicas, 23*(1), 1–9. https://doi.org/10.22201/fesz.23958723e.2020.0.233

Bertocci, F., & Mannino, G. (2022). Can agri-food waste be a sustainable alternative in aquaculture? A bibliometric and meta-analytic study on growth performance, innate immune system, and antioxidant defenses. *Food, 11*(13), 1861. https://doi.org/10.3390/foods11131861

Braidy, N., Selvaraju, S., Essa, M. M., Vaishnav, R., Al-Adawi, S., Al-Asmi, A., et al. (2013). Neuroprotective effects of a variety of pomegranate juice extracts against MPTP-induced cytotoxicity and oxidative stress in human primary neurons. *Oxidative Medicine and Cellular Longevity, 2013*(1), 685909. https://doi.org/10.1155/2013/685909

Buljeta, I., Šubarić, D., Babić, J., Pichler, A., Šimunović, J., & Kopjar, M. (2023). Extraction of dietary fibers from plant-based industry waste: A comprehensive review. *Applied Sciences, 13*(16), 9309. https://doi.org/10.3390/app13169309

Calvete-Torre, I., Sabater, C., Montilla, A., Moreno, F. J., Riestra, S., Margolles, A., & Ruiz, L. (2023). Physicochemical characterization and microbiota modulatory potential of brewer's spent grain and arabinoxylan-derived fractions: A valorization study. *LWT, 185*, 115107. https://doi.org/10.1016/j.lwt.2023.115107

Can-Cauich, C. A., Sauri-Duch, E., Betancur-Ancona, D., Chel-Guerrero, L., González-Aguilar, G. A., Cuevas-Glory, L. F., et al. (2017). Tropical fruit peel powders as functional ingredients: Evaluation of their bioactive compounds and antioxidant activity. *Journal of Functional Foods, 37*, 501–506. https://doi.org/10.1016/j.jff.2017.08.028

Cappelli, K., Ferlisi, F., Mecocci, S., Maranesi, M., Trabalza-Marinucci, M., Zerani, M., Dal Bosco, A., & Acuti, G. (2021). Dietary supplementation of olive mill waste water polyphenols in rabbits: Evaluation of the potential effects on hepatic apoptosis, inflammation and metabolism through RT-qPCR approach. *Animals : An Open Access Journal from MDPI, 11*(10), 2932. https://doi.org/10.3390/ani11102932

Cardador-Martínez, A., Jiménez-Martínez, C., & Sandoval, G. (2011). Revalorization of cactus pear (Opuntia spp.) wastes as a source of antioxidants. *Food Science and Technology, 31*, 782–788. https://doi.org/10.1590/S0101-20612011000300036

Carnaval, L. d. S. C., Jaiswal, A. K., & Jaiswal, S. (2024). Agro-food waste valorization for sustainable bio-based packaging. *Journal of Composites Science, 8*(2), 41. https://doi.org/10.3390/jcs8020041

Chantana, W., Hu, R., Buddhasiri, S., Thiennimitr, P., Tantipaiboonwong, P., & Chewonarin, T. (2023). The extract of *Perilla frutescens* seed residue attenuated the progression of aberrant crypt foci in rat colon by reducing inflammatory processes and altered gut microbiota. *Foods (Basel, Switzerland), 12*(5), 988. https://doi.org/10.3390/foods12050988

Chen, L., Deng, H., Cui, H., Fang, J., Zuo, Z., Deng, J., Li, Y., Wang, X., & Zhao, L. (2015). Inflammatory responses and inflammation-associated diseases in organs. *Oncotarget, 9*(6), 7204–7218. https://doi.org/10.18632/ONCOTARGET.23208

Chen, P., Chen, F., Lei, J., & Zhou, B. (2023). Pomegranate polyphenol punicalagin improves learning memory deficits, redox homeostasis, and neuroinflammation in aging mice. *Phytotherapy Research, 37*(9), 3655–3674. https://doi.org/10.1002/ptr.7848

Chensom, S., Okumura, H., & Mishima, T. (2019). Primary screening of antioxidant activity, total polyphenol content, carotenoid content, and nutritional composition of 13 edible flowers from Japan. *Preventive Nutrition and Food Science, 24*(2), 171–178. https://doi.org/10.3746/pnf.2019.24.2.171

Chiu, H.-F., Venkatakrishnan, K., & Wang, C.-K. (2020). The role of nutraceuticals as a complementary therapy against various neurodegenerative diseases: A mini-review. *Journal of Traditional and Complementary Medicine, 10*(5), 434–439. https://doi.org/10.1016/j.jtcme.2020.03.008

Ciccone, L., Nencetti, S., Rossello, A., & Orlandini, E. (2023). Pomegranate: A source of multifunctional bioactive compounds potentially beneficial in Alzheimer's disease. *Pharmaceuticals, 16*(7), 1036. https://doi.org/10.3390/ph16071036

Colombo, R., Pellicorio, V., Barberis, M., Frosi, I., & Papetti, A. (2024). Recent advances in the valorization of seed wastes as source of bioactive peptides with multifunctional properties. *Trends in Food Science & Technology, 144*, 104322. https://doi.org/10.1016/j.tifs.2023.104322

Comi, L., Giglione, C., Klinaku, F. T., Pialorsi, F., Tollemeto, V., Zurlo, M., et al. (2025). Valorizing agro-food waste for nutraceutical development: Sustainable approaches for managing metabolic dysfunction-associated steatotic liver disease and related co-morbidities. *Food Frontiers, 6*(2), 670. https://doi.org/10.1002/fft2.535

Correa Uriburu, F. M., Cattaneo, F., Maldonado, L. M., Zampini, I. C., Alberto, M. R., & Isla, M. I. (2022). Prosopis alba seed as a functional food waste for food formulation enrichment. *Foods, 11*(18), 2857. https://doi.org/10.3390/foods11182857

Coutino-Laguna, B. C., Flores-Gallegos, A. C., Ascacio-Valdes, J. A., Ilina, A., Saenz-Galindo, S., Castaneda-Facio, A. O., Esparza-Gonzalez, S. C., & Rodríguez-Herrera, R. (2022). Physicochemical and functional properties of the undervalued fruits of cactus *Cylindropuntia imbricate* ("xoconostle") and antioxidant potential. *Biocatalysis and Agricultural Biotechnology, 39*, 102245. https://doi.org/10.1016/j.bcab.2021.102245

da Cunha, J. A., de Ávila Scheeren, C., Fausto, V. P., de Melo, L. D. W., Henneman, B., Frizzo, C. P., de Almeida Vaucher, R., Castagna de Vargas, A., & Baldisserotto, B. (2018). The antibacterial and physiological effects of pure and nanoencapsulated Origanum majorana essential oil on fish infected with Aeromonas hydrophila. *Microbial Pathogenesis, 124*, 116–121. https://doi.org/10.1016/j.micpath.2018.08.040

Damas-Job, M. d. C., Soriano-Melgar, L. A., Rodríguez-Herrera, R., Peralta-Rodríguez, R. D., Rivera-Cabrera, F., & Martínez-Vazquez, D. G. (2023). Effect of broccoli fresh residues-based extracts on the postharvest quality of cherry tomato (*Solanum lycopersicum* L.) fruits. *Scientia Horticulturae, 317*, 112076. https://doi.org/10.1016/j.scienta.2023.112076

Das, J. N., Basak, S., Sablania, V., Negi, R., Yadav, N., Rustagi, S., Singh, S., Rai, A. K., Shreaz, S., & Yadav, A. N. (2024). Food waste as potential bioresources for extraction of carotenoid of nutraceutical importance: Current research and future challenges. *Journal of Applied Biology & Biotechnology, 12*(6), 65–79. https://doi.org/10.7324/JABB.2024.162743

de Barros, M. C., Silva, A. G. B., dos Santos Souza, T. G., Chagas, C. A., Machado, J. C. B., Ferreira, M. R. A., et al. (2022). Evaluation of acute toxicity, 28-day repeated dose toxicity, and genotoxicity of Moringa oleifera leaves infusion and powder. *Journal of Ethnopharmacology, 296*, 115504. https://doi.org/10.1016/j.jep.2022.115504

de Visser, K. E., & Joyce, J. A. (2023). The evolving tumor microenvironment: From cancer initiation to metastatic outgrowth. *Cancer Cell, 41*(3), 374–403. https://doi.org/10.1016/j.ccell.2023.02.016

Deghima, A., Righi, N., Rosales-Conrado, N., Eugenia León-González, M., Baali, F., Gómez-Mejía, E., Madrid, Y., & Bedjou, F. (2022). Phenolic profile, safety, antioxidant and anti-inflammatory activities of wasted Bunium ferulaceum Sm. aerial parts. *Food research international (Ottawa, Ont.), 160*, 111714. https://doi.org/10.1016/j.foodres.2022.111714

Durán-Aranguren, D. D., Muñoz-Daza, L. F., Castillo-Hurtado, L. J., Posada, J. A., Mussatto, S. I., Sierra, R., & Hernández-Carrión, M. (2023). Design of a baked good using food ingredients recovered from agro-industrial by-products of fruits. *LWT, 185*, 115174. https://doi.org/10.1016/j.lwt.2023.115174

Elshewy, A., Blando, F., Bahlol, H., El-Desouky, A., De Bellis, P., & Khalifa, I. (2023). Egyptian *Opuntia ficus-indica* (OFI) residues: Recovery and characterization of fresh mucilage from cladodes. *Horticulturae, 9*(7), 736. https://doi.org/10.3390/horticulturae9070736

Enciso-Martínez, Y., Zuñiga-Martínez, B. S., Ayala-Zavala, J. F., Domínguez-Avila, J. A., González-Aguilar, G. A., & Viuda-Martos, M. (2024). Agro-industrial by-products of plant origin: therapeutic uses as well as antimicrobial and antioxidant activity. *Biomolecules, 14*(7), 762. https://doi.org/10.3390/biom14070762

Esparza, C., Estrada, R., Sanchez, D. S., Saenz-Galindo, A., Valdes, J. A. A., Flores-Gallegos, A. C., & Rodríguez-Herrera, R. (2024). Ultrasound and microwave extraction from Moringa oleifera Lam.: Characterization and antiproliferative effect. *International Journal of Secondary Metabolite, 11*(2), 292–304. https://doi.org/10.21448/ijsm.1363300

Espino-García, J. J., Almaraz-Buendía, I., Peralta-Ortiz, J., Reyes-Munguía, A., Hernández-Soto, I., González-Montiel, L., & Campos-Montiel, R. G. (2023). Efecto de Xoconostle (Opuntia matudae Scheinvar) sobre la concentración de metano y las variables ruminales durante una fermentación in vitro de rastrojo de maíz. *Revista mexicana de ciencias pecuarias, 14*(2), 309–325. https://doi.org/10.22319/rmcp.v14i2.6093

Fathy, S. M., El-Dash, H. A., & Said, N. I. (2021). Neuroprotective effects of pomegranate (Punica granatum L.) juice and seed extract in paraquat-induced mouse model of Parkinson's disease. *BMC Complementary Medicine and Therapies, 21*(1), 130. https://doi.org/10.1186/s12906-021-03298-y

Ferreira, M. S., Linhares, R., & Martelli, M. (2016). Films and coatings from agro-industrial residues. In M. P. Montero-Garcia (Ed.), *Edible films and coatings* (pp. 193–214). CRC Press. https://doi.org/10.1201/9781315373713-20

Fierascu, R. C., Fierascu, I., Avramescu, S. M., & Sieniawska, E. (2019). Recovery of natural antioxidants from agro-industrial side streams through advanced extraction techniques. *Molecules, 24*(23), 4212. https://doi.org/10.3390/molecules24234212

Flores-Contreras, E. A., González-González, R. B., Pablo Pizaña-Aranda, J. J., Parra-Arroyo, L., Rodríguez-Aguayo, A. A., Iñiguez-Moreno, M., et al. (2024). Agricultural waste as a sustainable source for nanoparticle synthesis and their antimicrobial properties for food preservation. *Frontiers in Nanotechnology, 6*, 1346069. https://doi.org/10.3389/fnano.2024.1346069

Floridas-Galán, J. P., & Paredes-Concepción, P. (2024). Bioactive compounds from agroindustrial byproducts and their applications: From traditional methods to green technologies. *Scientia Agropecuaria, 15*(3), 461–476. https://doi.org/10.17268/sci.agropecu.2024.034

Foti, P., & Romeo, F. V. (2024). Valorization of olive mill wastewater and citrus peel waste as source of nutraceuticals. *Future Postharvest and Food, 1*(4), 441–443. https://doi.org/10.1002/fpf2.12041

Fracasso, J. A. R., Sikina, I. Y. G., da Costa, L. T. S., Guarnier, L. P., Ribeiro-Paes, J. T., de Ferreira, F. Y., de Almeida, L. V. C., de Castro Silva, B., de Barros Barbosa, D., Ximenes, V. F., Venkli, D., Viel, A. M., & Dos Santos, L. (2023). Toxicological profile and anti-inflammatory effect of mucoadhesive gel from residues of *Agave sisalana* and *Punica granatum*. *Gels (Basel, Switzerland), 9*(12), 942. https://doi.org/10.3390/gels9120942

Frumuzachi, O., Flanagan, A., Rohn, S., & Mocan, A. (2025). The dichotomy between functional and functionalized foods—A critical characterization of concepts. *Food Research International, 208*, 116173. https://doi.org/10.1016/j.foodres.2025.116173

Gómez Vargas, C., Ponce, N. M. A., Stortz, C. A., Fissore, E. N., Bonelli, P., Otálora González, C. M., & Gerschenson, L. N. (2024). Pectin obtention from agroindustrial wastes of *Malus domestica* using green solvents (citric acid and natural deep eutectic solvents). Chemical, thermal, and rheological characterization. *Frontiers in Chemistry, 12*. https://doi.org/10.3389/fchem.2024.1504582

Grande-Tovar, C., Araujo-Pabón, L., Flórez-López, E., & Aranaga-Arias, C. (2021). Determinación de la actividad antioxidante y antimicrobiana de residuos de mora (Rubus glaucus Benth). *Informador Técnico, 85*(1), 64–82. https://doi.org/10.23850/22565035.2932

Guan, J. (2024). The interplay between aging and inflammation and its role in chronic disease development. *Theoretical and Natural Science, 48*(1), 45–54. https://doi.org/10.54254/2753-8818/48/2024pj0121

Hassan, F. U., Rehman, M. S. U., Khan, M. S., Ali, M. A., Javed, A., Nawaz, A., & Yang, C. (2019). Curcumin as an alternative epigenetic modulator: Mechanism of action and potential effects. *Frontiers in Genetics, 10*, 514. https://doi.org/10.3389/fgene.2019.00514

Hernández-Fuentes, A. D., Bañuelos, A. F., Gallegos-Vazquez, C., Campos Montiel, R. G., & Pinedo-Espinoza, J. M. (2015). Antioxidant activity xoconostle (Opuntia joconostle) genotypes of the state of Zacatecas, Mexico. http://www.redalyc.org/articulo.oa?id=81339864012

Herrera-Ruiz, M., Jiménez-Ferrer, E., González-Cortazar, M., Zamilpa, A., Cardoso-Taketa, A., Arenas-Ocampo, M. L., Jiménez-Aparicio, A. R., & Monterrosas-Brisson, N. (2022). Potential use of *Agave* genus in neuroinflammation management. *Plants (Basel, Switzerland), 11*(17), 2208. https://doi.org/10.3390/plants11172208

Iqbal, J., Shaheen, A., Ghani, M. A., Iqbal, Q., Azam, M., Ali, B., et al. (2022). Nutritional and mineral composition analysis of desert plants: A case study. *Emirates Journal of Food and Agriculture, 34*(10), 882–887. https://doi.org/10.9755/ejfa.2022.v34.i10.2926

John, T., Samuel, B., Abolaji, O., Folashade, O., Oyetooke, A., & Oluwatosin, F. (2020). Functional foods and bioactive compounds: Roles in the prevention, treatment and management of neurodegenerative diseases. *GSC Biological and Pharmaceutical Sciences, 11*(2), 297–313. https://doi.org/10.30574/gscbps.2020.11.2.0143

Jorge, A. M. S., Gaspar, M. C., Henriques, M. H. F., & Braga, M. E. M. (2023). Edible films produced from agrifood by-products and wastes. *Innovative Food Science and Emerging Technologies, 88*, 103442. https://doi.org/10.1016/j.ifset.2023.103442

Khan, R., Anwar, F., Ghazali, F. M., & Mahyudin, N. A. (2024). Valorization of waste: Innovative techniques for extracting bioactive compounds from fruit and vegetable peels-A comprehensive review. *Innovative Food Science & Emerging Technologies, 97*, 103828. https://doi.org/10.1016/j.ifset.2024.10382

Khokar, R., Hachani, K., Hasan, M., Othmani, F., Essam, M., Al Mamari, A., et al. (2021). Anti-Alzheimer potential of a waste by-product (peel) of Omani pomegranate fruits: Quantification of phenolic compounds, in-vitro antioxidant, anti-cholinesterase and in-silico studies. *Biocatalysis and Agricultural Biotechnology, 38*, 102223. https://doi.org/10.1016/j.bcab.2021.102223

Kiran, V., Reddy, H., Kumar, P., Naik, M., Narasimha Rao, C., Jayasankar, J., et al. (2024). Histological analysis reveals that red grape seed extract (Vitis vinifera L.) rejuvenates antioxidant enzymes in male albino rats (Wistar strain) with D-galactose-induced Alzheimer's disease (AD) brain regions. *Pharmacognosy Research, 16*, 828–836. https://doi.org/10.5530/pres.16.4.95

Kruk, M., Lalowski, P., Płecha, M., Ponder, A., Rudzka, A., Zielińska, D., & Trząskowska, M. (2025). Prebiotic potential of spent brewery grain—In vitro study. *Food Chemistry, 463*, 141254. https://doi.org/10.1016/j.foodchem.2024.141254

Kumar, S., Rawson, A., Kumar, A., Ck, S., Vignesh, S., & Venkatachalapathy, N. (2023). Lycopene extraction from industrial tomato processing waste using emerging technologies, and its application in enriched beverage development. *International Journal of Food Science and Technology, 58*(4), 2141–2150. https://doi.org/10.1111/ijfs.16156

Lee, S. F., Kennedy, S. K., Caini, S., Wong, H. C., Yip, P. L., Poortmans, P. M., et al. (2024). Randomised controlled trials on radiation dose fractionation in breast cancer: Systematic review and meta-analysis with emphasis on side effects and cosmesis. *BMJ, 386*. https://doi.org/10.1136/bmj-2023-079089

León-Villanueva, A., Huerta-Ocampo, J. A., Barrera-Pacheco, A., Medina-Godoy, S., & de la Rosa, A. P. B. (2018). Proteomic analysis of non-toxic Jatropha curcas byproduct cake: fractionation and identification of the major components. *Industrial Crops and Products, 111*, 694–704. https://doi.org/10.1016/j.indcrop.2017.11.046

Li, J., Pettinato, M., Campardelli, R., De Marco, I., & Perego, P. (2022). High-pressure technologies for the recovery of bioactive molecules from agro-industrial waste. *Applied Sciences, 12*(7), 3642. https://doi.org/10.3390/app12073642

Libby, P., Smith, R. D., Rubin, E. J., Glassberg, M. K., Farkouh, M. E., & Rosenson, R. S. (2024). Inflammation unites diverse acute and chronic diseases. *European Journal of Clinical Investigation*. https://doi.org/10.1111/eci.14280

Madera-Santana, T. J., Rodríguez-Núñez, J. R., Barreras-Urbina, C. G., López-Martínez, L. X., & Fortiz-Hernández, J. (2022). Extracción de compuestos bioactivos de residuos agroindustriales empleando extracción asistida por ultrasonido y microondas. In L. C. Montoya-Ballesteros, M. E. Tiznado-Hernández, T. J. Madera-Santana, J. F. Ayala-Zavala, & G. A. González-Aguilar (Eds.), *Tecnología, ingeniería y biotecnología de alimentos de origen vegetal: aprovecha miento de sus subproductos* (Primera edición) (pp. 49–66). LEBERMEX.

Malenica, D., Kass, M., & Bhat, R. (2025). Optimization of antioxidant extraction from agro-industrial wastes using ultrasound & response surface methodology. *Food Chemistry Advances, 7*, 100966. https://doi.org/10.1016/j.focha.2025.100966

Manzur-Valdespino, S., Ramírez-Moreno, E., Arias-Rico, J., Jaramillo-Morales, O. A., Calderón-Ramos, Z. G., Delgado-Olivares, L., et al. (2020). Opuntia ficus-indica L. Mill residues—properties and application possibilities in food supplements. *Applied Sciences, 10*(9), 3260. https://doi.org/10.3390/app10093260

Maqbool, Z., Khalid, W., Atiq, H. T., Koraqi, H., Javaid, Z., Alhag, S. K., et al. (2023). Citrus waste as source of bioactive compounds: Extraction and utilization in health and food industry. *Molecules, 28*(4), 1636. https://doi.org/10.3390/molecules28041636

Maqsoudlou, A., Mohebodini, H., & Jafari, S. M. (2020). Antioxidant activity analysis of nanoencapsulated food ingredients. In *Characterization of Nanoencapsulated Food Ingredients* (pp. 617–664). Academic. https://doi.org/10.1016/B978-0-12-815667-4.00018-3

Márquez-Rangel, I., Cruz, M., Ruiz, H. A., Rodríguez-Jasso, R. M., Loredo, A., & Belmares, R. (2023). Agave waste as a source of prebiotic polymers: Technological applications in food and their beneficial health effect. *Food Bioscience, 56*, 103102. https://doi.org/10.1016/j.fbio.2023.103102

Martínez-Vázquez, D., Rodríguez Herrera, R., Del, S., Rangel Ortega, C., Verónica, A., Rodríguez, C., & Soriano-Melgar, L. A. (2022). Effect of grape waste extracts on the postharvest life of strawberry (*Fragaria vesca* L.). *Revista SNIQBA., 1*(2), 1–7. http://sniqba.com.mx/revsniqba-2022-02-10/

Martins, V. F. R., Pintado, M. E., Morais, R. M. S. C., & Morais, A. M. M. B. (2024). Recent highlights in sustainable bio-based edible films and coatings for fruit and vegetable applications. *Foods, 13*(2), 318. https://doi.org/10.3390/foods13020318

Mejía-Reyes, J. D., García-Cabrera, K. E., Velázquez-Ovalle, G., & Vázquez-Ovando, A. (2022). Capacidad antioxidante: conceptos, métodos de cuantificación y su aplicación en la caracterización de frutos tropicales y productos derivados. *Revista Colombiana de Investigaciones Agroindustriales, 9*(1), 9–33. https://doi.org/10.23850/24220582.4023

Menezes, F. N. D. D., da Cruz Almeida, É. T., da Silva Vieira, A. R., de Souza Aquino, J., dos Santos Lima, M., Magnani, M., & de Souza, E. L. (2021). Impact of cashew (Anacardium occidentale L.) by-product on composition and metabolic activity of human colonic microbiota in vitro indicates prebiotic properties. *Current Microbiology, 78*(6), 2264–2274. https://doi.org/10.1007/s00284-021-02502-z

Milica, G., Bošković, C. M., Čobanović, N., Baltić, M. Ž., Vranešević, J., Samardžić, S., & Maksimović, Z. (2023). Antimicrobial activity of ethanolic extracts from wheat, sunflower and maize crop residues. *Archives of. Veterinary Medicine, 16*(1), 53–67. https://doi.org/10.46784/e-avm.v16i1.315

Morales, P., Barros, L., Ramírez-Moreno, E., Santos-Buelga, C., & Ferreira, I. C. (2014). Exploring xoconostle by-products as sources of bioactive compounds. *Food Research International, 65*, 437–444. https://doi.org/10.1016/j.foodres.2014.05.067

Mudalip, S. K. A., Jemaat, Z., Hashim, N. A., Rajintran, S. K., Khatiman, M. N. B., & Man, R. C. (2025). Separation of natural products from agro-industrial waste: A green and sustainable approach. In *Advances in Separation Sciences* (pp. 395–414). https://doi.org/10.1016/B978-0-323-95292-7.00012-8

Munteanu, I. G., & Apetrei, C. (2021). Analytical methods used in determining antioxidant activity: A review. *International journal of molecular sciences, 22*(7), 3380. https://doi.org/10.3390/ijms22073380

Panaro, M. A., Budriesi, R., Calvello, R., Cianciulli, A., Mattioli, L. B., Corazza, I., Rotondo, N. P., Porro, C., Lamonaca, A., Ferraro, V., Muraglia, M., Corbo, F., Clodoveo, M. L., Monaci, L., Cavalluzzi, M. M., & Lentini, G. (2024). Lentil waste extracts for inflammatory bowel disease (IBD) symptoms control: Anti-inflammatory and spasmolytic effects. *Nutrients, 16*(19), 3327. https://doi.org/10.3390/nu16193327

Parekh, P., Serra, M., Allaw, M., Perra, M., Marongiu, J., Tolle, G., et al. (2022). Characterization of Nasco grape pomace-loaded nutriosomes and their neuroprotective effects in the MPTP mouse model of Parkinson's disease. *Frontiers in Pharmacology, 13*. https://doi.org/10.3389/fphar.2022.935784

Pérez, M. J., Rodriguez, I. F., Zampini, I. C., Cattaneo, F., Mercado, M. I., Ponessa, G., & Isla, M. I. (2020). Prosopis nigra fruits waste characterization, a potential source of functional ingredients for food formulations. *LWT, 132*, 109828. https://doi.org/10.1016/j.lwt.2020.109828

Pérez-Ternero, C., Macià, A., de Sotomayor, M. A., Parrado, J., Motilva, M. J., & Herrera, M. D. (2017). Bioavailability of the ferulic acid-derived phenolic compounds of a rice bran enzymatic extract and their activity against superoxide production. *Food & Function, 8*(6), 2165–2174. https://doi.org/10.1039/c7fo00243b

Pislyagin, E. A., Tarbeeva, D. V., Yurchenko, E. A., Menchinskaya, E. S., Gorpenchenko, T. Y., Pokhilo, N. D., et al. (2025). Neuroprotective activity of oligomeric stilbenes from alpha grape stems in in vitro models of Parkinson's disease. *International Journal of Molecular Sciences, 26*(6), 2411. https://www.mdpi.com/1422-0067/26/6/2411

Quizhpe, J., Ayuso, P., Rosell, M. d. l. Á., Peñalver, R., & Nieto, G. (2024). *Brassica oleracea* var. italica and their by-products as source of bioactive compounds and food applications in bakery products. *Foods, 13*(21). https://doi.org/10.3390/foods13213513

Rahman, M. H., Bajgai, J., Fadriquela, A., Sharma, S., Trinh, T. T., Akter, R.,. .. Lee, K.-J. (2021). Therapeutic potential of natural products in treating neurodegenerative disorders and their future prospects and challenges. *Molecules, 26*(17), 5327. https://doi.org/10.3390/molecules26175327.

Ramírez-Cortina, C. R., Alonso-Gutiérrez, M. S., & Rigal, L. (2012). Valorization of tequila agro-industrial residues for ruminants feeding. *Revista Chapingo serie ciencias forestales y del ambiente, 18*(3), 449–457. https://doi.org/10.5154/r.rchscfa.2011.08.059

Recinella, L., Chiavaroli, A., Veschi, S., Cama, A., Acquaviva, A., Libero, M. L., Leone, S., Di Simone, S. C., Pagano, E., Zengin, G., Menghini, L., Brunetti, L., Izzo, A. A., Orlando, G., & Ferrante, C. (2022). A grape (Vitis vinifera L.) pomace water extract modulates inflammatory and immune response in SW-480 cells and isolated mouse colon. *Phytotherapy Research: PTR, 36*(12), 4620–4630. https://doi.org/10.1002/ptr.7581

Robles-Botero, M. V., Jesús, R. D., Quiroz-Reyes, C. N., & Aguilar-Méndez, M. A. (2020). Caracterización e identificación de compuestos bioactivos con actividad antioxidante de la cáscara, pulpa y semilla del fruto de tejocote (Crataegus mexicana). *TIP. Revista especializada en ciencias químico-biológicas, 23*. https://doi.org/10.22201/fesz.23958723e.2020.0.233

Rojas-García, A., Fernández-Ochoa, Á., Cádiz-Gurrea, M. d. l. L., Arráez-Román, D., & Segura-Carretero, A. (2023). Neuroprotective effects of agri-food by-products rich in phenolic compounds. *Nutrients, 15*(2), 449. https://doi.org/10.3390/nu15020449

Romero-Sáez, M. (2022). Los residuos agroindustriales, una oportunidad para la economía circular. *TecnoLógicas, 25*(54). https://doi.org/10.22430/22565337.2505

Rudra, A., Arvind, I., & Mehra, R. (2021). Polyphenols: Types, sources and therapeutic applications. *International Journal of Home Science, 7*(3), 69–75. https://doi.org/10.22271/23957476.2021.v7.i3a.1182

Sá, A. G. A., Moreno, Y. M. F., & Carciofi, B. A. M. (2020). Plant proteins as high-quality nutritional source for human diet. *Trends in Food Science & Technology, 97*, 170–184. https://doi.org/10.1016/j.tifs.2020.01.011

Sagini, J. P. N., Rossatto, F. C. P., Souza, F., Pilau, E., Quines, C. B., Ávila, D. S., Ligabue-Braun, R., Zimmer, A. R., Pereira, R. I., & Zimmer, K. R. (2024). Inhibition of Staphylococcus epidermidis and Pseudomonas aeruginosa biofilms by grape and rice agroindustrial residues. *Microbial Pathogenesis, 197*, 107019. https://doi.org/10.1016/j.micpath.2024.107019

Sakulrang, S., Razem, M., Mohammadi, N., & Granato, D. (2025). Sustainable, functional food design: Characterizing and utilizing passion fruit by-product extract in anthocyanin-enriched delivery systems. *Future Foods, 11*, 100599. https://doi.org/10.1016/j.fufo.2025.100599

Salazar-López, N. J., Enríquez-Valencia, S. A., Zuñiga Martínez, B. S., & González-Aguilar, G. A. (2023). Residuos agroindustriales como fuente de nutrientes y compuestos fenólicos. *Epistemus (Sonora), 17*(34), 60–69. https://doi.org/10.36790/epistemus.v17i34.265

Saleem, U., Shehzad, A., Shah, S., Raza, Z., Shah, M. A., Bibi, S., et al. (2021). Antiparkinsonian activity of Cucurbita pepo seeds along with possible underlying mechanism. *Metabolic Brain Disease, 36*(6), 1231–1251. https://doi.org/10.1007/s11011-021-00707

Schiebel, C. S., Bueno, L. R., Pargas, R. B., de Mello Braga, L. L. V., da Silva, K. S., Fernandes, A. C. V. U., et al. (2024). Exploring the biological activities and potential therapeutic applications of agro-industrial waste products through non-clinical studies: A systematic review. *Science of The Total Environment, 950*, 175317. https://doi.org/10.1016/j.scitotenv.2024.175317

Seguí, L., & Barrera, C. (2025). Functional ingredients from food waste and by-products: Processing technologies, functional characteristics and value-added applications. *Foods, 14*(5), 847. https://doi.org/10.3390/foods14050847

Senna-Pereira, J., Aparecida-Hister, C., Ubessi, C., & Bosio-Tedesco, S. (2022). Genotoxicity, cytotoxicity and phenolic compounds from aqueous extracts of phyllanthus tenellus Roxb. cultivated under different light conditions. *Pakistan Journal of Biological Sciences: PJBS, 25*(7), 575–585. https://doi.org/10.3923/pjbs.2022.575.585

Sharifi, Z., Jebelli Javan, A., Hesarinejad, M. A., & Parsaeimehr, M. (2023). Application of carrot waste extract and Lactobacillus plantarum in Alyssum homalocarpum seed gum-alginate beads to create a functional synbiotic yogurt. *Chemical and Biological Technologies in Agriculture, 10*(1), 3. https://doi.org/10.1186/s40538-022-00377-1

Sheng, K., Shui, S., Yan, L., Yu, J., Hao, G., Qu, H., et al. (2018). The beneficial effects of dietary grape supplementation on improving cognitive deficits in APP/PS1 double transgenic mice. *Journal of Functional Foods, 49*, 224–234. https://doi.org/10.1016/j.jff.2018.08.030

Singh, G. P., Aayush, K., Chavan, P., Chiu, I., Yan, S., Verma, R., Kuca, K., Kumar, D., & Yang, T. (2024). Valorization of agro-industrial wastes: Converting mango seed into starch nanoparticles-based edible coating for fresh produce preservation. *Innovative Food Science and Emerging Technologies, 95*. https://doi.org/10.1016/j.ifset.2024.103722

Sokolovska, V., & Jankulovska, M. S. (2022). Folin–ciocâlteu method for determination of total polyphenols in onion. *Journal of Agricultural, Food and Environmental Sciences, JAFES, 76*(2), 12–19. https://doi.org/10.55302/JAFES22762012s

Sorrenti, V., Burò, I., Consoli, V., & Vanella, L. (2023, 2019). Recent advances in health benefits of bioactive compounds from food wastes and by-products: Biochemical aspects. *International Journal of Molecular Sciences, 24*(3). https://doi.org/10.3390/ijms24032019

Sun, Y., Zhou, Q. M., Lu, Y. Y., Zhang, H., Chen, Q. L., Zhao, M., & Su, S. B. (2019). Resveratrol inhibits the migration and metastasis of MDA-MB-231 human breast cancer by reversing TGF-β1-induced epithelial-mesenchymal transition. *Molecules, 24*(6), 1131. https://doi.org/10.3390/molecules24061131

Sun, Y., Tao, R., Ju, Y., Duan, J., Xie, Q., Fan, B., & Wang, F. (2024). Natural active products in fruit postharvest preservation: A review. In *Food Frontiers*. John Wiley and Sons Inc. https://doi.org/10.1002/fft2.454

Sur, S., & Panda, C. K. (2017). Molecular aspects of cancer chemopreventive and therapeutic efficacies of tea and tea polyphenols. *Nutrition, 43*, 8–15. https://doi.org/10.1016/j.nut.2017.06.006

Tapia-Blácido, D. R., Garcia, A. L., Beitum, L. R., Zitei-Baptista, L. F., & Aguilar, P. F. (2023). Use of biobased materials from agro-industrial residues in food packaging. In *Advanced applications of biobased materials: food, biomedical, and environmental applications* (pp. 173–229). Elsevier. https://doi.org/10.1016/B978-0-323-91677-6.00020-9

Teniente, S. L., Flores-Gallegos, A. C., Esparza-González, S. C., Campos-Múzquiz, L. G., Nery-Flores, S. D., & Rodríguez-Herrera, R. (2023). Anticancer effect of pomegranate peel polyphenols against cervical cancer. *Antioxidants, 12*(1), 127. https://doi.org/10.3390/antiox12010127

Teshome, E., Forsido, S. F., Rupasinghe, H. P. V., & Olika Keyata, E. (2022). Potentials of natural preservatives to enhance food safety and shelf life: A review. *The Scientific World Journal, 2022*(1), 9901018. https://doi.org/10.1155/2022/9901018

Tesoriere, L., Butera, D., Allegra, M., Fazzari, M., & Livrea, M. A. (2005). Distribution of betalain pigments in red blood cells after consumption of cactus pear fruits and increased resistance of the cells to ex vivo induced oxidative hemolysis in humans. *Journal of agricultural and food chemistry, 53*(4), 1266–1270. https://doi.org/10.1021/jf048134+

Tikoo, G., Singh, B., Sharma, G., & Gupta, C. (2022). Antioxidant bio-constituents from agro-industrial waste and their importance in functional foods. *Mathews Journal of Nutrition & Dietetics, 5*(1), 1–14. https://doi.org/10.30654/MJND.10020

Tiţa, O., Constantinescu, M. A., Tiţa, M. A., Bătuşaru, C., & Mironescu, I. (2024). Sensory, textural, physico-chemical and enzymatic characterization of melted cheese with added potato and carrot peels. *Frontiers in nutrition, 10*. https://doi.org/10.3389/fnut.2023.1260076

Vicente-Zurdo, D., Gómez-Mejía, E., Morante-Zarcero, S., Rosales-Conrado, N., & Sierra, I. (2025). Analytical strategies for green extraction, characterization, and bioactive evaluation of polyphenols, tocopherols, carotenoids, and fatty acids in agri-food bio-residues. *Molecules, 30*(6), 1326. https://doi.org/10.3390/molecules30061326

Vignesh, A., Amal, T. C., Sarvalingam, A., & Vasanth, K. (2024). A review on the influence of nutraceuticals and functional foods on health. *Food Chemistry Advances, 5*, 100749. https://doi.org/10.1016/j.focha.2024.100749

Wu, H., Chen, L., Zhu, F., Han, X., Sun, L., & Chen, K. (2019). The cytotoxicity effect of resveratrol: cell cycle arrest and induced apoptosis of breast cancer 4 T1 cells. *Toxins, 11*(12), 731. https://doi.org/10.3390/toxins11120731

Yadav, S., Pathera, A. K., Islam, R. U., Malik, A. K., & Sharma, D. P. (2018). Effect of wheat bran and dried carrot pomace addition on quality characteristics of chicken sausage. *Asian-Australasian Journal of Animal Sciences, 31*(5), 729–737. https://doi.org/10.5713/ajas.17.0214

Yadav, S., Malik, K., Moore, J. M., Kamboj, B. R., Malik, S., Malik, V. K., et al. (2024). Valorisation of agri-food waste for bioactive compounds: Recent trends and future sustainable challenges. *Molecules, 29*(9), 2055. https://doi.org/10.3390/molecules29092055

Yu, X., Pu, H., & Voss, M. (2024). Overview of anti-inflammatory diets and their promising effects on non-communicable diseases. *British Journal of Nutrition, 132*(7), 898–918. https://doi.org/10.1017/s0007114524001405

Zaini, H. B. M., Sintang, M. D. B., & Pindi, W. (2020). The roles of banana peel powders to alter technological functionality, sensory and nutritional quality of chicken sausage. *Food Science & Nutrition, 8*(10), 5497–5507. https://doi.org/10.1002/fsn3.1847

Zhu, L., Tong, H., Ren, C., Chen, K., Luo, S., Wang, Q., Guo, M., Xu, Y., Hu, M., Fang, J., Xu, J., & Shi, P. (2024). Inflammation unleashed: The role of pyroptosis in chronic liver diseases. *International Immunopharmacology, 141*, 113006. https://doi.org/10.1016/j.intimp.2024.113006

Marine Microbes: A Novel Source of Bioceuticals in Functional Food Development

15

Divya Mori, Mitaliba Gohil, Mansi Mekhiya,
Yashvi Baldhiya, and Nikitasinh Gohil

Abstract

Aquatic microbes, including algae, bacteria, and fungi found in both marine and freshwater ecosystems, are emerging as a novel and sustainable supply of bioactive compounds for food fortification. These microorganisms are rich in essential nutrients, such as vitamins, minerals, polyunsaturated fatty acids, proteins, and antioxidants, which are crucial for human health. Due to their ability to synthesize unique bioactive molecules, aquatic microbes offer a promising alternative to conventional sources for fortifying foods and enhancing their nutritional profile. Furthermore, their rapid growth and adaptability to diverse aquatic environments make them an attractive, scalable option for sustainable production. Incorporating these microbial-derived compounds into food products could address nutritional deficiencies, particularly in populations at risk of malnutrition, while promoting overall health and wellness. This book chapter highlights the potential of aquatic microbes as a sustainable and functional source for food fortification, focusing on their nutritional content, bioactive properties, and the technologies enabling their use in food systems. By integrating aquatic microbes into food fortification strategies, we can enhance the nutritional value of common food items, contributing to healthier diets and improving public health globally.

Keywords

Functional foods · Marine resources · Bioactives · Nutraceuticals · Blue economy

D. Mori · M. Gohil · M. Mekhiya · Y. Baldhiya · N. Gohil (✉)
Faculty of Science, Department of Microbiology, Marwadi University, Rajkot, Gujarat, India
e-mail: nikitasinh.gohil@marwadieducation.edu.in

G. Sanghvi et al. (eds.), *Sustainable Food Fortification*,
https://doi.org/10.1007/978-981-95-1217-1_15

15.1 Introduction

"Fortificare" is a Latin word for "make strong," and "bios" is Greek for "life." In order to address nutrient deficiencies in the general population, food fortification is the process of adding essential elements to food, whether or not they are naturally present. Staple foods are frequently used in this public health approach to give micronutrients that may be deficient in the diet of a population known as "hidden hunger," food fortification is an affordable and scientifically proven way to combat malnutrition. It solves dietary inadequacies, especially in low- and middle-income nations, by adding vital micronutrients to food products, providing a long-term strategy for enhancing public health (Ofori et al., 2022). Lack of vital micronutrients like vitamins and minerals is known as "hidden hunger," which frequently goes unreported because there are no obvious symptoms but can have long-term effects on growth and health (Wakeel et al., 2018).

It is acknowledged that one of the most economical public health initiatives is food fortification. When it comes to treating iron deficiency in pregnant women, it provides a durable substitute for dietary supplements. Fortified foods are an effective way to fight malnutrition because they reach more at-risk groups and help keep nutrient levels steady (Page et al., 2021). If functional foods have extra ingredients that improve health, such as lowering the risk of disease, they are considered fortified products. Despite its benefits, poor food fortification techniques can have detrimental effects on health, particularly in developed countries where vitamin deficits are still present. Deficits in iodine, iron, vitamin A, and zinc were listed as some of the most urgent health hazards in the globe in the 2002 globe Health Report (Cashman, 2018). Innovative and sustainable methods were needed to address these urgent issues, and marine microbes are showing promise as a food fortification solution.

In the varied and nutrient-rich ocean ecosystem, marine microorganisms, such as bacteria, fungi, and microalgae flourish. Oceans are a great source of natural products because of their abundance of marine biodiversity and their unique ability to adapt to harsh environments (Jo et al., 2016). Asian countries are well-known producers of seaweeds as they have been used as a dietary supplement for their bioactive substances, including omega-3 fatty acids and vital vitamins and minerals like zinc, iodine, and selenium, are produced by these microorganisms (Riaz et al., 2020). The capacity of these microorganisms to produce potentially nutritious content in food allows them to serve as active pharmaceutical ingredients for designing nutraceuticals, which are known as "bioceuticals". Marine microorganisms are a sustainable and scalable alternative for nutrient production since they are not limited by freshwater and land requirements like traditional agricultural sources (Senadheera et al., 2023). They even act as renewable resource for new drug development against diabetes, microbial diseases, and inflammation. Therefore, using marine microbes to biofortify food is one of the most promising uses which includes wide ranges of organisms including bacteria, fungi, microalgae, macroalgae, cyanobacteria, etc (Sarker, 2023).

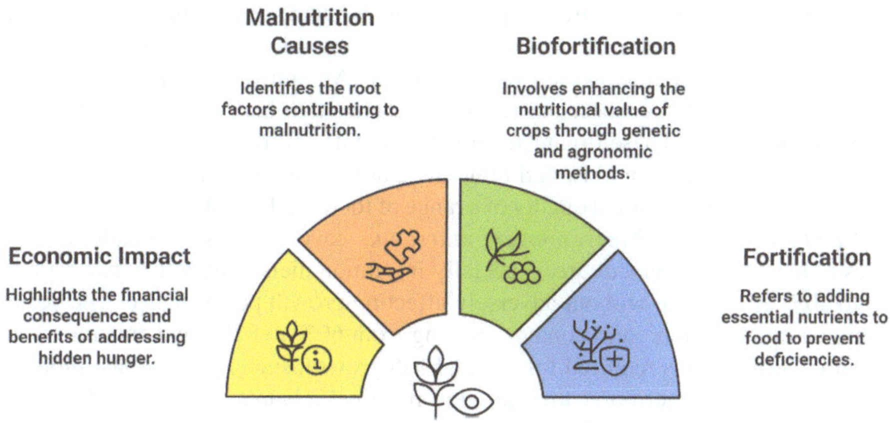

Fig. 15.1 Strategies for combating hidden hunger and malnutrition

Proteins, vitamins (such as B12, D, and E), polysaturated fatty acids (PUFAs), polysaccharides, proteins, peptides, peptide hydrolysates, enzymes, natural pigments, phenolic compounds, and vital minerals are all abundant in marine bacteria like *Vibrio* and *Pseudoalteromonas*, as well as microalgae like *Chlorella* and *spirulina* (Huang et al., 2021; Hosseini et al., 2020) which makes them perfect for addressing malnutrition and micronutrition deficiencies. These substances are perfect for improving food's nutritional profile for health reasons (Shahidi & Ambigaipalan, 2015). Furthermore, certain fungi create enzymes that enhance the nutrients' bioavailability and digestion.

One popular method of fortification is adding micronutrients to staple foods during processing or delivering fortified supplements. Food fortification is essential for enhancing nutritional value, consumer well-being, and reducing health hazards (Fig. 15.1). It reduces disease prevention, improves public health, is affordable, enables targeted intervention, makes fortified foods easily ingested, reaches a diverse population, and reduces stigma, aiding in the fight against malnutrition in lower- and middle-income nations.

15.2 Effectiveness of Fortified Yields in Reducing Malnutrition

The impact of improved yields on tackling malnutrition, especially via bio-based methods, has gained significant acknowledgment as a practical approach to tackle nutritional deficiencies worldwide. The practice of food fortification has been acknowledged as a crucial approach in addressing micronutrient deficiencies and alleviating malnutrition on a global scale. Conventional fortification techniques, including the addition of the mineral to salt or iron to meal, have resulted in notable health advancements. Iodine fortification has played a crucial role in hindering iodine deficiency disorders, and iron fortification has significantly lowered the prevalence of anaemia in different populations. Nonetheless, as the focus on ecological sustainability and nutritious foods intensifies, bio-based methods for food fortification have become increasingly significant. An innovative approach involves the use of aquatic microbes, particularly marine microbial sources, as nutrient-dense supplements. These organisms are rich in vital nutrients including proteins, omega-3 fatty acids, vitamins, and minerals, positioning them as excellent options for improving the nutritional quality of a range of food products (Rohner et al., 2023).

Candida utilis and *Kluyveromyces marxianus* have surfaced as viable alternatives to fishmeal in aquaculture, especially in salmon diets, where they can replace up to 40% of fishmeal without adversely affecting growth performance. The yeasts exhibit a high crude protein content, ranging from 60% to 82%, and encompass all essential amino acids required for optimal fish growth, such as lysine, methionine, and threonine. Furthermore, they provide practical advantages through bioactive compounds such as beta-glucans and mannan-oligosaccharides, which improve immune response and promote gut health. Their excellent digestibility and bioavailability render them an effective protein source, while the presence of naturally occurring B vitamins and nucleotides enhances metabolic functions and aids in cellular repair. The incorporation of such yeasts in aquafeed not only enhances nutritional benefits but also promotes sustainability by decreasing dependence on native fish stocks, thus aiding in environmental conservation efforts (Sy et al., 2015).

Proteins from methanotrophic bacteria, particularly those sourced from *Methylococcus capsulatus*, exhibit significant promise as a substitute protein source in aquaculture, especially for diets of Atlantic salmon. This group of bacteria employs methane as its source of carbon and can be cultivated in a sustainable manner, thereby minimizing the environmental footprint associated with conventional fishmeal-based feeds. Research has shown that the inclusion of protein derived from *Methylococcus capsulatus* in salmon diets can lead to growth rate improvements of up to 36%, exceeding the effectiveness of traditional fishmeal and a plant-based feeds. This enhancement can be linked to the superior protein content, abundant in vital amino acids, alongside bioactive compounds that facilitate effective intake of nutrients and metabolic processes (Zhang et al., 2022).

Although there are many different kinds of biofortification methods, they can be broadly divided into two groups: those that use agronomic strategies and those that use a genetic breeding strategy.

Conventional breeding biofortification uses classic plant breeding methods to improve the nutritional content of crops in a sustainable and natural way. Using this technique, plant types that naturally have larger concentrations of vital elements, such as vitamins, minerals, and amino acids, are chosen and crossed. Scientists and agronomists create superior crop varieties that may aid in addressing the pervasive micronutrient shortages in human diets by carefully selecting parent plants with desired features and breeding them over many generations. In a similar vein, orange maize, often referred to as provitamin A-rich maize, has been produced to assist lower vitamin A deficiency, which is a leading cause of blindness in children. Furthermore, in areas where sweet potatoes and cassava are main crops, high-yield cultivars with higher vitamin A content were developed (Virk et al., 2021).

Transgenic biofortification approach involves the genetic modification of crops to improve their nutritional composition (Jaiswal et al., 2022). This method entails the incorporation of certain genes into a plant's genome to enhance the concentrations of vital vitamins, minerals, or other advantageous chemicals. Transgenic biofortification directly modifies a plant's genetic composition for specific nutritional enhancements, in contrast to traditional biofortification, which depends on specific breeding or agronomic methods. Golden rice: Genetically modified to synthesize beta-carotene, the precursor of vitamin A, to address vitamin A insufficiency. Iron- and zinc-enriched rice and wheat: Enhanced to augment mineral bioavailability, targeting anaemia and malnutrition. High lysine maize: Enhanced necessary amino acid composition. (Garg et al., 2018).

Agronomic biofortification approach involves the genetic modification of crops to improve their nutritional composition. This method entails the incorporation of certain genes into a plant's genome to enhance the concentrations of vital vitamins, minerals, or other advantageous chemicals. (Bhardwaj et al., 2022).

15.3 Factors Affecting Fortification Efficiency

Biofortification is a critical strategy to combat micronutrient deficiencies and malnutrition globally. The efficiency of fortification is influenced by multiple factors spanning biological, technical, economic, and social dimensions (Fig. 15.2). Sustainable approaches, particularly bio-based strategies, aim to enhance the nutritional quality of food while addressing challenges in implementation and scalability.

The effectiveness of food fortification largely depends on the bioavailability of fortified nutrients, which determines how efficiently they are absorbed and utilized by the body. One of the key factors influencing bioavailability is the chemical form of the nutrient. Different chemical forms exhibit varying degrees of absorption; for instance, ferrous sulfate ($FeSO_4$) is more bioavailable than ferric pyrophosphate ($Fe_3(PO_4)_2$) in iron fortification because ferrous iron is more soluble in the digestive tract, making it easier for the body to absorb. Additionally, synergistic and inhibitory interactions within the food matrix play a significant role in nutrient absorption. Certain compounds enhance bioavailability—for example, vitamin C significantly improves iron absorption by reducing ferric iron (Fe^{3+}) to its more absorbable ferrous (Fe^{2+}) form. Conversely, some naturally occurring food components, such as phytates, tannins, and oxalates, can bind to minerals like iron, calcium, and zinc, forming insoluble complexes that hinder their absorption (Mannar & Hurrell, 2018).

Fortification strategies must consider regional food habits, ensuring that staple foods commonly consumed in a population, such as rice in Asia, maize flour in Africa, or wheat flour in Latin America, are chosen as fortification vehicles. Sensory properties significantly influence consumer perception. Certain nutrients may alter food characteristics, such as iron fortification in cereals and dairy, introducing a metallic taste, riboflavin (vitamin B2), imparting a yellowish hue, or omega-3 fatty acids, causing an undesirable fishy odour. To mitigate these effects, food scientists use advanced technologies like microencapsulation, emulsification, and stabilizers

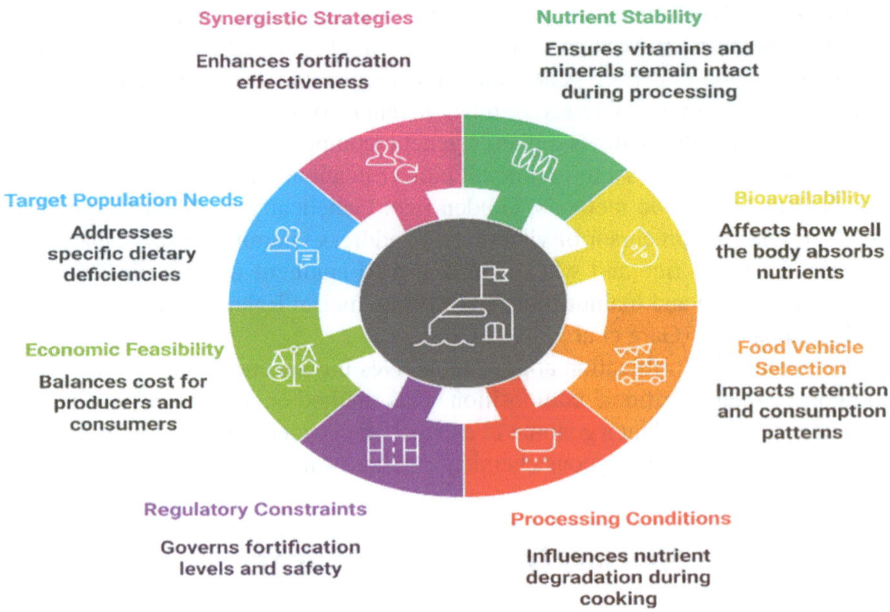

Fig. 15.2 Key factors affecting efficiency of fortification

to mask off-flavours and improve texture (Dahiya et al., 2023). Conducting sensory evaluation tests helps manufacturers refine formulations to maintain consumer acceptability. Moreover, transparent labelling and consumer education about the health benefits of fortification can enhance trust and willingness to adopt fortified foods (Chadare et al., 2019).

The use of marine microbes for biofortification aligns with a sustainable goal by enhancing the nutritional profile of staple foods through conventional breeding, transgenic techniques, or fortified approaches (Gödecke et al., 2018). Unlike conventional food fortification, marine microbes can be cultivated in controlled environments without the need for arable land or freshwater resources. This makes them a scalable and eco-friendly option for addressing global nutrition challenges by adding essential nutrients during food processing.

15.4 Potential Marine Microbes and Their Bioactive Compounds

Ocean microbial life is a dynamic aspect of the marine environment. Marine ecology is a diverse field that has been extensively studied. The deep sea's unique adaptations to severe environments have led to increased interest in it as a possible source of bioactive chemicals.

The ability to thrive in extreme marine environments with pressure, salinity, and temperature, as well as competition for nutrients and symbiosis and highly oxidative environments, has led to the development of unique metabolic and physiological capabilities. Marine bacteria, fungi, and microalgae contribute significantly to natural product production compared to terrestrial microorganisms can be a major prolific producers of valuable primary and secondary metabolites.

Marine microorganisms differ from those found in soil, air, freshwater, industrial effluents, and wastewater due to their unique structural and functional qualities. Bioactive chemicals, such as bryostatins, pentabromo-pseudilin, ectoine, vibriobactin, solonamide, violacein, thiomarinol, and bromoaltero-chromide, have been identified in Gram-positive actinobacteria and filamentous fungi, while Gram-negative bacteria have received less attention.

Stenotrophomonas strains have antimicrobial and antifungal properties which has been isolated from sponges and sea urchins. Heterotrophic bacteria have been isolated from marine ark shell exhibits haemolytic activity (Qeshmi et al., 2019). Marine bacteria *Pseudoalteromonas*, known for producing bioactive substances which inhibits methicillin-resistant *Staphylococcus aureus* (MRSA). *P. luteoviolacea* is known to inhibit the growth of protists, and yellow pigment (tambjamines) producing *P. tunicate* possesses antifungal, antimicrobial, antitumorigenic, immunosuppressive, antiproliferative activities. Lactic acid bacteria (LAB) produce diversified EPSs and have potential applications in foods (Nadzir et al., 2021). *Lactobacilli* are popular for their role in fermented food products as texturing agents and are generally recognized as safe (GRAS) (Zhang et al., 2023). Filamentous gram-positive bacteria are emerging secondary metabolite producers, producing antitumor, antibiotic, anthelmintic, and antifungal agents (Senadheera et al., 2023).

Marine-derived fungi (*Aspergillus sydowii*) contain a variety of constituents, including alkaloids, terpenes, xanthones, anthraquinones, pyrones, sterols, diphenyl ethers, cyclopentenones, and polyketides. They also contain enzymes like α-amylases, lipases, keratinases, xylanases, cellulases, and tannase, which have industrial and biotechnological potential.

Fungi in terrestrial ecosystems form symbiotic relationships with plants and animals, facilitating nutrient cycling. Fungi can grow anywhere in the terrestrial biosphere where moisture and carbon sources exist. Fungi can act as parasites at various levels of the food chain. Fungi in the ocean act as saprotrophs, converting detritus and algae into biomass that zooplankton consume alongside bacterial heterotrophs. This creates a microbial loop that connects refractory polymers produced from primary productivity. Marine fungi played a significant role in microbial evolution due to their diverse chemical composition and biological functions. *Penicillium* sp., known to produce alkaloids, isolated from deep ocean sediment, exhibit antitumor activities. Marine filamentous fungus, namely *Aspergillus glaucus*, produces anthroquinone derivative known as Aspergiolide A have cytotoxic effects. Others such as *Ampelomyces* sp., *Cladosporium* sp., *Pseudallescheria* sp., *Phoma herbarum* were reported for their antimicrobial, antifouling, and antioxidant properties. Marine fungi with diverse genetic origins and extensive metagenomes are a significant source of metagenomic data (Wang et al., 2020).

Actinomycetes are gram-positive bacteria known for producing a range of antibiotics. They form symbiotic relationships with various plants and animals.

Actinomycetes form symbiotic relationships with marine macrofauna, such as cone snails, invertebrates, sponges (e.g., Axinella polypoides, Haliclona), and fora-like insects. These relationships are ecologically significant. Marine-derived actinomycetes, namely, *Nocardiopsis lucentensis* contains essential peptides called Lucentamycins, *Streptomyces* sp. from deep-sea sediments contains Mansouramycin, *Marinispora* contains pyrroles called Lynamicins exhibits antimicrobial (Mahapatra et al., 2019) and showed strong cytotoxic activity against lung and breast cancer (Shin et al., 2018). Marine-derived *Actinomyces* sp. have a great pharmacological importance in pulmonary hypertension (Gupta & Prakash, 2019).

Microalgae, microscopic marine organisms, are found in both littoral and benthic environments and are rich sources of bioactive compounds and biochemicals. These include diatoms, green algae, blue-green algae, dinoflagellates, and yellow-brown flagellates. Diatoms produce essential PUFAs and substances like fucoxanthin, including *Prymnesiophyta*, *Prasinophyta*, *Chlorophyta*, and *Cryptophyta. Nostoc*, *Arthrospira* (Spirulina), edible blue-green algae, and Aphanizomenon are popular species utilized for centuries. However, environmental conditions like abiotic stresses and salinity can affect their pigment and antioxidant components. *Oscillatoria nigroviridis* showed antitrypanosomal and antileishmanial activity, *Ulva lactuca*, *U. intestinalis*, and *Cladophora* sp. showed antiprotozoal activities. Schizochytrium is a key source of omega-3 fatty acids, an alternative to fish oil. Both macro and micro algae are used as functional food for their rich source of vital nutrients, including fibre, polysaccharides, proteins, lipids, minerals, vitamins, beta-carotene, astaxanthin, and phytochemicals (Wang et al., 2021) (Table 15.1). They are commercially used as functional food additives, milk formulations, and dietary supplements. Macroalgae, known as Seaweeds, include numerous nutrients (Rayapu et al., 2017).

Additionally, they include several bioactive chemicals, such as phenolic compounds, which have antibacterial, antiviral, antifungal, anticancer, and antioxidant properties. Seaweeds are often used in East Asian cuisine due to its functional and health benefits. They have high levels of PUFAs, polysaccharides, carotenoids, L-αkainic acid, minerals, and chlorophyll pigments, phlorotannin, phyco-colloids, etc., making them useful in food and pharmaceutical industries. *Dunaliella salina* is a commercially available source of vitamin A and PUFAs supplemented for eye health, skin protection, and immune function (Cheng et al., 2022).

15.5 Bioactive-Derived Organic Compounds: Properties and Applications

Bioactive-derived organic compounds represent a unique category of naturally occurring molecules that can be extracted or synthesized from various biological sources, including microbial compounds, secondary metabolites of plants, and marine bioresources (Fig. 15.3).

Table 15.1 Biological activities of compounds derived by various marine sources

Marine source	Organism	Bioactive compound	Properties/biological activity
Bacteria	*Bacillus velezensis*, *Stenotrophomonas* sp., *Pseudoalteromonas* sp., *P. luteoviolacea*, *P. tunicate*, *S. aureus*, *Lactobacilli*	Lipopeptides (surfactin, iturin, fengycin), Ecteinascidin-743, marinopyrrole	Improve food safety by exhibiting antibacterial activity and enhance food and cosmetic preservation
Fungi	*Aspergillus sydowii*, *Aspergillus glaucus*, *Ampelomuces* sp., *Cladosporium* sp., *Pseudallescheria* sp., *Phoma herbarum*, *Rhodotorula mucilaginosa* *Penicillium brevicompactum*	Alkaloids, phenolic derivatives, carotenoids	Enhance food preservation and safety, used as a natural food colorant due to its carotenoid content anticancer, antioxidant, anti-viral, and antibiotic
Actinomycetes	*Nocardiopsis lucentensis*, *Streptomyces* sp., *Marinispora* sp., *Actinomyces* sp.	Actinomycin D, nystatin, lomaiviticin	Anticancer, antiprotozoal, antiviral, and antibacterial
Algae	Agarophyte, *Gracilaria sjoestedii*, and *G. debilis* (red algae)	Polysaccharides (long chain), phenols, pigments	Used in gelatin manufacturing and as a stabilizing agent in food products, antifungal formulations against *Candida albicans*
	Saccharina japonica, *Sargassum pallidum* (Brown algae)	Polysaccharides (laminarin, alginate, fucoidan), polyphenolic compounds	Food and pharmaceutical industries for cancer treatment, oxidative stress, UV protection

15.5.1 Polysaccharides

Microbially derived polysaccharides are extensively utilized in the food industry as stabilizers, emulsifiers, gelling agents, and water-thickening agents. These polysaccharides can be categorized into cell wall, intercellular, and exocellular polysaccharides. Among them, exocellular polysaccharides (EPSs) are particularly promising due to their purity and ease of isolation. Marine microorganisms produce structurally and functionally diverse EPSs under extreme marine environmental conditions.

Additionally, seaweed serves as a rich source of polysaccharides, particularly sulfated polysaccharides (SPs), which exhibit strong biological and physicochemical characteristics. These polysaccharides are predominantly found in storage and

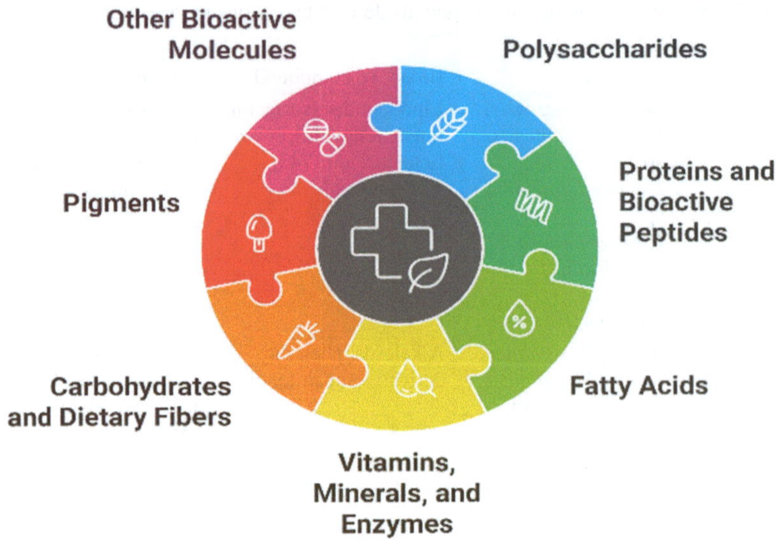

Fig. 15.3 Bioactive-derived organic compounds

cell wall structures. Brown algae contain laminarin, fucoidan, cellulose, sargassum, alginates, and mannitol, whereas green algae possess sulfated galactan, SPs, and xylans.

Red algae are rich in carrageenans, water-soluble sulfated galactan, agars, xylans, and Floridean starch. The polysaccharide content in seaweed species can reach up to 75% on a dry weight basis, with most of these polysaccharides acting as dietary fibres that promote gut health by maintaining intestinal balance. The total fibre content of seaweed varies between 29% and 62%. SPs derived from marine algae exhibit a broad spectrum of bioactivities, including anticoagulant, antioxidant, antiproliferative, antithrombotic, antitumoral, anti-inflammatory, anticomplementary, and antiviral properties. Other storage and structural polysaccharides, such as carrageenan, agar, and alginate, play crucial roles in the food industry by aiding in gel formation, thickening, food product stabilization, and the development of water-soluble films.

The majority of RPS-producing monosaccharides are derived from *Chroococcis, Gloeothece*, and *Synechococcus*. Industrial attention mostly centres on microbial polysaccharides, used as thickening agents to stabilize flow characteristics in aqueous solutions across diverse environments (De Borba Gurpilhares et al., 2018).

Beyond their role in food processing, marine microorganisms produce bioactive compounds with significant health benefits, making them attractive for functional food development. These compounds include antioxidants, antimicrobial agents, antiviral substances, and anti-inflammatory molecules. Antioxidant properties from marine microbes help protect the body from oxidative stress, reducing the risk of chronic diseases such as cardiovascular disorders, diabetes, and neurodegenerative conditions. Marine fungi and bacteria synthesize phenolic compounds that neutralize free radicals and prevent cellular damage.

Furthermore, marine polysaccharides like chitosan and fucoidan (from brown seaweed) possess broad-spectrum antibacterial, antifungal, and antiviral properties, extending the shelf life of food products. Carrageenan, derived from red seaweed, has been explored for its antiviral potential against diseases such as influenza, HIV, and herpes simplex virus. These compounds exhibit tumor-inhibiting potential by inducing apoptosis, suppressing metastasis, and preventing cancer cell proliferation (Ghosh et al., 2022).

15.5.2 Proteins and Bioactive Peptides

Microorganisms such as *Chlorella*, *Dunaliella*, *Micractinium*, *Spirulina*, *Oscillatoria*, *Scenedesmus*, *Chlamydomonas*, and *Euglena* are excellent sources of high-quality protein and serve as sustainable alternatives for food applications due to their favourable amino acid profiles. Marine microorganisms also play a crucial role in enzyme production, offering stability and high activity as they are highly nutritional and have medicinal properties. They are used as a dietary supplement. In addition to enzymes, bioactive peptides extracted from marine organisms exhibit a wide range of therapeutic benefits, including antihypertensive, anticancer, anti-inflammatory, antioxidant, antimicrobial, and immunomodulatory properties, making them valuable in pharmaceutical applications (Chandika et al., 2022).

Bacterial proteins, such as *Methylococcus capsulatus* and yeast-based proteins like *Candida utilis*, offer high-quality, digestible protein with essential amino acids that promote fish growth and enhance environmental sustainability. These microbial protein sources not only lower feed costs but also improve fish health and the overall nutritional value of seafood for human consumption (Cunha & Pintado, 2021). Antitum compounds from marine-derived fungi include verrucarin A, trichodermamides A and B, leptosins, zygosporamide, cephalimysin A, conidiogenone C, fusaranthraquinone, fusarnaphthoquinones A-C, fusarone, cryptosphaerolide, paeciloxocins A and B, and anthraquinone SZ-685C (Hamed et al., 2015).

Apart from their antioxidant properties, marine-derived compounds demonstrate strong antimicrobial and antiviral activities, making them valuable natural food preservatives. Marine bacteria, such as *Bacillus* and *Pseudomonas*, produce antimicrobial peptides (AMPs), including lantibiotics and bacteriocins, which effectively inhibit the growth of foodborne pathogens like *Salmonella*, *E. coli*, and *Listeria*.

Bioactive peptides derived from marine fungi have demonstrated anti-inflammatory effects by modulating immune responses and reducing oxidative stress. These peptides are particularly beneficial for individuals with chronic inflammatory conditions such as arthritis, cardiovascular disorders, and metabolic syndromes.

Table 15.2 Function of fatty acid derived from marine microbes

Fatty acid	Marine source	Function
Gamma-amino-butyric acid (GABA)	*Enterococcus faecium* SH9, *Pseudomonas fluorescens*, *Aeromonas hydrophila*	Neurotransmitter, anti-inflammatory, and antioxidative properties
Arachidonic acid	*Roseobacter sinensis*	Vasoconstrictive and platelet aggregative
Docosahexaenoic acid	*Schizochytrium* sp., *Yarrowia lipolytica*	Brain development and nutritional supplements
Eicosapentaenoic acid	*Shewanella*, *Photobacterium*, *Vibrio*, *Flexibacter*	Antimicrobial and anti-inflammatory nutraceutical
Gamma-linolenic acid	*Spirulina*, *Chlorella* sp.	Tissue integrity and postponement of ageing

15.5.3 Fatty Acids

Marine microorganisms, including bacteria, protists, yeast, and fungi, are known to produce long-chain polyunsaturated fatty acids (PUFAs) in the marine environment. These PUFAs, particularly omega-3 and omega-6 fatty acids, are mainly found in fungi and microalgae. Microalgae such as *Schizochytrium* and *Nannochloropsis* produce eicosapentaenoic acid (EPA) and docosahexaenoic acid (DHA), which are critical for cardiovascular health, brain function, and anti-inflammatory responses (Table 15.2). Unlike conventional fish oil sources, these microalgae-derived omega-3 offer a sustainable solution without depleting marine fish stocks. Marine microalgae, such as Spirulina, contain higher amounts of PUFAs than freshwater microalgal species (Shahidi & Ambigaipalan, 2015). They can also alter PUFA production rates with growth phases. Unlike fish oils, microbial PUFA production is sustainable and independent of climatic or seasonal changes. Microalgae, like *Amphidinium* sp., contain up to 48% of their cell fatty acids as DHA. Various yeast and fungal species are potential sources for producing PUFAs. Innovations in biotechnology, fermentation, and encapsulation techniques are further optimizing the bioavailability and stability of these microbial-based fortification strategies, ensuring their widespread application in functional foods (Lobine et al., 2021).

15.5.4 Carbohydrates and Dietary Fibre

The predominant marine natural carbohydrate is polysaccharides, with a minority being monosaccharides or oligosaccharides. Alginate, the predominant polymer in brown algae, consists of polymannuronic acid, polyguluronic acid, and a linear amalgamation of both polymannuronic and polyguluronic acids. Alginate exists in several forms, including acid and salt, and is regarded as an essential component of the cell wall. Fucoidan has a very complex and extensive structure, in contrast to alginate and laminaran. Chitin and chitosan are natural protein polysaccharide polymers with multimodal properties and unique structural architectures. Over the last decade, both chitin and chitosan derivatives have garnered significant interest in the domains of food, biomedical, culinary, biopharmaceutical, and additional industrial studies, with applications including tissue engineering and wound healing biomaterials (Pati et al., 2021).

15.5.5 Enzymes, Vitamins, and Minerals

Carrageenases are enzymes generated by microorganisms associated with seaweed, mostly from the phyla Proteobacteria or Bacteroidetes. These enzymes are essential for the breakdown of carrageenans, the sulphur-containing polysaccharides present in red seaweeds. Carrageenases hydrolyse internal $(1 \rightarrow 4)$-β-D-glycosidic connections, deconstructing large carrageenan structures into smaller oligosaccharides. This process is crucial for maritime carbon cycling and has notable biotechnological uses, such as the synthesis of bioactive chemicals, prebiotics, and biofuels (Ghanbarzadeh et al., 2018). Galactosidases are enzymes synthesized by yeasts that are essential for the hydrolysis of oligosaccharides present in soybean products. These enzymes decompose complex galactooligosaccharides (GOS), including raffinose and stachyose, into simpler sugars, enhancing digestibility and diminishing anti-nutritional factors in soy-based meals. The enzymatic activity of galactosidases is advantageous in food processing, since it improves nutritional bioavailability and alleviates gastrointestinal discomfort from undigested oligosaccharides. Galactosidases generated from yeast are extensively used in biotechnology, especially in the fermentation process of soy products and the manufacture of lactose-free dairy substitutes (De Borba Gurpilhares et al., 2018).

15.5.6 Pigments

Light-absorbing pigments called phycobilin are mostly found in glaucophytes, red algae, cyanobacteria, and some cryptomonads. The fact that these pigments are attached to phycobiliproteins—water-soluble proteins—makes them special. Because of their effectiveness in absorbing light, organisms may flourish in a variety of watery environments. Phycobilin's are useful for therapeutic applications since they also have anticancer and antioxidant qualities (Muñoz-Miranda & Iñiguez-Moreno, 2023; Patel et al., 2020).

Chromobacterium violaceum, along with other bacteria, including *Janthinobacterium* and *Pseudoalteromonas*, create the violet pigment known as vitexin. Numerous biological actions, such as antibacterial, anticancer, antioxidant, antifungal, and antiparasitic properties, are shown by vitexin. In a variety of tumour cell lines, violacein exhibits cytotoxicity, causing apoptosis and blocking signalling pathways (Patel et al., 2024). Its antioxidant properties further emphasize its potential as a treatment for illnesses linked to oxidative stress (Durán et al., 2016).

Algae such as *Dunaliella salina Heterochlorella luteoviridis, Chlorella ellipsoidea* are naturally producers of carotenoids, composed of yellow to orange pigments that are vital to photosynthesis and provide many health advantages. During photosynthesis, fucoxanthin is a strong antioxidant with a high energy transfer efficiency, while β-carotene is a precursor to vitamin A (Imchen & Singh, 2022) carotenoids have photoprotective qualities that protect skin from UV rays and anticancer effects by preventing tumour formation. Algal-derived carotenoids find extensive use as natural food colouring agents, feed additives, and therapeutic ingredients in medications and cosmetics (Sathasivam & Ki, 2018).

Phycocyanin is a blue-coloured protein, mostly found in cyanobacteria like *Aphanizomenon flos-aquae* and *Spirulina platensis*. It is an extra pigment that works with chlorophyll to help photosynthesis by taking in red and orange light. Because of its bright blue colour, it's a famous natural food and drink dye that goes by names like "Lina Blue" or "EXBERRY Shade Blue." It's often used in candy and ice cream (Morya et al., 2023).

Pseudomonas otitidis DDB2 generates the black pigment known as nano-melanin. Its capacity to harm bacterial membranes and impede development is responsible for these effects. Due to its special qualities, which include photoprotection, hydrophobicity, antioxidant activity, and potential uses as a natural dye, antioxidant additive, as well as imaging contrast agent, nano-melanin is very adaptable for both industrial and medicinal settings (Deepthi et al., 2021).

15.5.7 Other Marine Biomolecules

Marine algae are a rich source of phenolic compounds, making them highly potent bioactive resources in marine ecosystems. Algae are rich in pigments, phenolic compounds, and vitamins, which are essential for autotrophic organisms to capture solar energy during photosynthesis. The primary types of photosynthetic pigments in marine life include phycobiliproteins, carotenoids, and chlorophylls. Carotenoids act as potent antioxidants, neutralizing reactive oxygen species (ROS). B-carotene, with provitamin A potential, has demonstrated cancer prevention abilities and a reduced risk of cardiovascular and ophthalmological ailments. Tocopherol, a potent antioxidant, is found in α-, β-, and δ-tocopherol forms across microalgae species and can aid in preventing cardiovascular diseases (Cheng et al., 2022).

Seaweed species like *Undaria* sp., *Laminaria* sp., *Porphyra* sp., and *Fucus* sp. Phlorotannins contribute to numerous biological activities, including antioxidant, anti-tumour, anti-inflammatory, antidiabetic, antiallergic, and antihypertensive effects. Marine fungal strains are known for producing polyketide-derived alkaloids, terpenes, peptides, and mixed biosynthesis compounds (Catarino et al., 2021). These secondary metabolites include compounds such as (E)-4-methoxy-5-(3-methoxybut-1-enyl)-6-methyl-2H-pyran-2-one, norliquexanthone, costaclavine, fumgaclavine A, C, and xanthocillin. Other compounds include chromone derivatives, antibacterial and cytotoxic activities, and anticancer activity (Circuncisão et al., 2018). Paeciloxocins A and B are new aromatic metabolites isolated from the mangrove fungus *Paecilomyces* sp. collected from the Taiwan Strait. They exhibited cytotoxicity against the HepG2 cell line. The monomeric C-methylated hexaketide (2E,4E)-1-(2,6- dihydroxy-3,5-dimethyl-phenyl) hexa-2,4-dien-1-one has strong activity against HeLa (cervical cancer) and SW-620 cells.

15.6 Exploring the Combinatorial Effects of Marine Microbes and Bioactive Compounds for Health Benefits

15.6.1 Microalgae + Probiotics

The amalgamation of marine microalgae, including *Spirulina, Dunaliella*, and *Haematococcus*, with probiotics such as *Bifidobacterium* and *Lactobacillus* provides substantial advantages in human nutrition and functional foods. Microalgae are abundant in nutritious proteins, antioxidants, and critical omega-3 fatty acids. When combined with probiotics, these bioactive substances enhance bioavailability, hence enhancing the absorption of vital vitamins, such as B12, A, and E, along with minerals required for metabolic activities. The synergistic impact of microalgae and probiotics offers natural anti-inflammatory as well as detoxifying properties, aiding in the reduction of oxidative stress and promoting gut and liver health (Perković et al., 2022).

15.6.2 Marine Microorganism + Adaptogenic Herb

The adaptogenic herbs *Schisandra chinensis* (Panossian et al., 2020) and probiotic-enhanced marine microorganisms, including *Bacillus subtilis,* work in unison to promote gut health, immunological function, and endurance. The well-researched probiotic *Bacillus subtilis* stimulates the immune system by regulating beneficial bacteria populations, improves digestion, balances the gut microbiota, and increases nutrient absorption. The adaptogenic and antioxidant qualities of *Schisandra chinensis* assist the body in fighting stress, enhancing liver function, and increasing stamina and mental clarity. Together, *Bacillus subtilis* increases the bioavailability of lignans and polyphenols, two of *Schisandra's* active ingredients, so optimizing their effectiveness (Patel et al., 2020).

The adaptogenic herb Ashwagandha (*Withania somnifera*) and omega-3-rich microalgae (*Isochrysis galbana*) form a potent synergy that enhances cognitive function, resilience to stress, and inflammation reduction. Docosahexaenoic acid (DHA) and eicosapentaenoic acid (EPA) are two essential omega-3 fatty acids that are indispensable for the maintenance of brain health, synaptic function, and neuroprotection. *Isochrysis galbana* is an exceptional source of these fatty acids. These fatty acids are essential for the preservation of the integrity of cell membranes and the reduction of neuroinflammation (Wróbel-Biedrawa & Podolak, 2024).

15.6.3 Marine Bacteria + Antioxidants

Bacillus indicus HU36 and carotenoids β-carotene and lycopene may effectively reduce oxidative stress by interfering with Fe(II)-induced lipid peroxidation via synergistic antioxidant activity. *Bacillus indicus* HU36, a probiotic strain that produces high quantities of carotenoids, neutralizes ROS and prevents lipid peroxidation to protect cells from oxidative damage. Beta-carotene and lycopene, as potent lipid-phase antioxidants, enhance this action by scavenging free radicals and shielding membranes in cells from iron-catalysed oxidative destruction. Endogenous bacterial carotenoid synthesis and exogenous carotenoids boost antioxidant capacity, lowering oxidative stress and maintaining cellular health.

Pseudoalteromonas spp. and vitamin E (α-tocopherol) form a potent synergy that enhances antioxidant defences and improves the stability of biological systems against oxidative stress. Biologically active substances, exopolysaccharides, and antioxidant enzymes are among the products of *Pseudoalteromonas* spp., a marine bacterium. Through synergistic scavenging of free radicals and reinforcement of cellular defence mechanisms, the efficacy of α-tocopherol is further enhanced by *Pseudoalteromonas*-derived antioxidants. This combination has substantial implications; it provides defence from oxidative stress-related diseases, ageing, and inflammation-driven disorders (Sy et al., 2015).

15.6.4 Microalgae + Polyphenols

Nannochloropsis gaditana are recognized for their ability to improve skin health, reduce inflammation, and enhance antioxidant activity by modulating genes associated with DNA damage repair, hydration, and elasticity. Upon combining with flavonoids such as quercetin as well as apigenin-*O*-rutinoside, which act as potent antioxidants that fight free radicals, the synergy enhances anti-ageing effects by promoting metabolic health, reducing oxidative stress, and enhancing cellular repair mechanisms. This combination is especially advantageous in cosmeceutical formulations that aim to protect the skin and improve vitality (Letsiou et al., 2017).

Rich in polyphenols, flavonoids, and fatty acids like γ-linolenic acid (GLA), *Spirulina platensis* has shown strong antioxidant activity, especially in ethanolic preparations that include phenolic components like apigenin and quercetin. Red seaweed is a rich source of quercetin, a flavonoid that improves this synergy by lowering oxidative stress, decreasing inflammation, and offering extra free radical scavenging capabilities. When combined, these substances support stronger immune responses, better cellular repair systems, and defence against long-term illnesses, including diabetes and cardiovascular conditions (Bellahcen et al., 2020).

15.7 Future Prospective

The global food fortification market is expected to expand at a compound annual growth rate (CAGR) of 6.4–10.4%. The future of aquatic microbes as a source for nutrition supplements is promising, driven by advancements in biotechnology, sustainability concerns, and the increasing demand for functional foods and nutraceuticals. Growth encompasses traditional staples such as cereals and dairy, as well as functional foods, plant-based alternatives, and performance beverages designed for specific health objectives, including immunity, cognitive wellness, and metabolic health. In the food sector, the procedure, cost, and consumer preference for natural ingredients must be considered. Bio-fortification strategies are needed to combat malnutrition, but conventional breeding, transgenic approaches, and agronomic and microbe-mediated approaches are limited. Transgenic crops' global adaptability is under question due to social, ethical, political, and environmental issues. A sustainable, social, economic, and environmentally friendly approach, such as microbial bio-fortification, enriches nutrient contents in crops in a sustainable manner. Identifying, isolating, and applying novel microorganisms capable of facilitating mineral availability and enhancing crop uptake is crucial. Additionally, incorporating microbial populations as partners for mineral bio-fortification of crops is essential. Future challenges include a synergism of technologies to provide nutritious food for growing populations using sustainable and environmentally friendly technologies.

References

Bellahcen, T. O., AAmiri, A., Touam, I., Hmimid, F., Amrani, A. E., Cherif, A., & Cherki, M. (2020). Evaluation of Moroccan microalgae: Spirulina platensis as a potential source of natural antioxidants. *Journal of Complementary and Integrative Medicine, 17*(3). https://doi.org/10.1515/jcim-2019-0036

Bhardwaj, A. K., Chejara, S., Malik, K., Kumar, R., Kumar, A., & Yadav, R. K. (2022). Agronomic biofortification of food crops: An emerging opportunity for global food and nutritional security. *Frontiers in Plant Science, 13*. https://doi.org/10.3389/fpls.2022.1055278

Cashman, K. D. (2018). Vitamin D deficiency: A public health issue in high- and low-income countries or just hype? *World Review of Nutrition and Dietetics*, 206–214. https://doi.org/10.1159/000484391

Catarino, M. D., Amarante, S. J., Mateus, N., Silva, A. M. S., & Cardoso, S. M. (2021). Brown algae phlorotannins: A marine alternative to break the oxidative stress, inflammation and cancer network. *Foods, 10*(7), 1478. https://doi.org/10.3390/foods10071478

Chadare, F. J., Idohou, R., Nago, E., Affonfere, M., Agossadou, J., Fassinou, T. K., Kénou, C., Honfo, S., Azokpota, P., Linnemann, A. R., & Hounhouigan, D. J. (2019). Conventional and food-to-food fortification: An appraisal of past practices and lessons learned. *Food Science & Nutrition, 7*(9), 2781–2795. https://doi.org/10.1002/fsn3.1133

Chandika, P., Tennakoon, P., Kim, T., Kim, S., Je, J., Kim, J., Lee, B., Ryu, B., Kang, H., Kim, H., Kim, Y., Kim, C., Choi, I., Park, W., Yi, M., & Jung, W. (2022). Marine biological macromolecules and chemically modified macromolecules; potential anticoagulants. *Marine Drugs, 20*(10), 654. https://doi.org/10.3390/md20100654

Cheng, P., Li, Y., Wang, C., Guo, J., Zhou, C., Zhang, R., Ma, Y., Ma, X., Wang, L., Cheng, Y., Yan, X., & Ruan, R. (2022). Integrated marine microalgae biorefineries for improved bioactive compounds: A review. *The Science of the Total Environment, 817*, 152895. https://doi.org/10.1016/j.scitotenv.2021.152895

Circuncisão, A. R., Catarino, M. D., Cardoso, S. M., & Silva, A. M. S. (2018). Minerals from macroalgae origin: Health benefits and risks for consumers. *Marine Drugs, 16*(11), 400. https://doi.org/10.3390/md16110400

Cunha, S. A., & Pintado, M. E. (2021). Bioactive peptides derived from marine sources: Biological and functional properties. *Trends in Food Science & Technology, 119*, 348–370. https://doi.org/10.1016/j.tifs.2021.08.017

Dahiya, D., Terpou, A., Dasenaki, M., & Nigam, P. S. (2023). Current status and future prospects of bioactive molecules delivered through sustainable encapsulation techniques for food fortification. *Sustainable Food Technology, 1*(4), 500–510. https://doi.org/10.1039/d3fb00015j

De Borba Gurpilhares, D., Cinelli, L. P., Simas, N. K., Pessoa, A., Jr., & Sette, L. D. (2018). Marine prebiotics: Polysaccharides and oligosaccharides obtained by using microbial enzymes. *Food Chemistry, 280*, 175–186. https://doi.org/10.1016/j.foodchem.2018.12.023

Deepthi, S., Mohan, K., Mishra, N., & Agsar, D. (2021). Melanin production by Pseudomonas sp. and in silico comparative analysis of tyrosinase gene sequences. *Biotechnologia, 102*(4), 411–424. https://doi.org/10.5114/bta.2021.111106

Durán, N., Justo, G. Z., Durán, M., Brocchi, M., Cordi, L., Tasic, L., Castro, G. R., & Nakazato, G. (2016). Advances in Chromobacterium violaceum and properties of violacein-its main secondary metabolite: A review. *Biotechnology Advances, 34*(5), 1030–1045. https://doi.org/10.1016/j.biotechadv.2016.06.003

Garg, M., Sharma, N., Sharma, S., Kapoor, P., Kumar, A., Chunduri, V., & Arora, P. (2018). Biofortified crops generated by breeding, agronomy, and transgenic approaches are improving lives of millions of people around the world. *Frontiers in Nutrition, 5*. https://doi.org/10.3389/fnut.2018.00012

Ghanbarzadeh, M., Golmoradizadeh, A., & Homaei, A. (2018). Carrageenans and carrageenases: versatile polysaccharides and promising marine enzymes. *Phytochemistry Reviews, 17*(3), 535–571. https://doi.org/10.1007/s11101-018-9548-2

Ghosh, S., Sarkar, T., Pati, S., Kari, Z. A., Edinur, H. A., & Chakraborty, R. (2022). Novel bioactive compounds from marine sources as a tool for functional food development. *Frontiers in Marine Science, 9*. https://doi.org/10.3389/fmars.2022.832957

Gödecke, T., Stein, A. J., & Qaim, M. (2018). The global burden of chronic and hidden hunger: Trends and determinants. *Global Food Security, 17*, 21–29. https://doi.org/10.1016/j.gfs.2018.03.004

Gupta, C., & Prakash, D. (2019). Nutraceuticals from microbes of marine sources. In *IntechOpen eBooks*. https://doi.org/10.5772/intechopen.82369

Hamed, I., Özogul, F., Özogul, Y., & Regenstein, J. M. (2015). Marine bioactive compounds and their health benefits: A review. *Comprehensive Reviews in Food Science and Food Safety, 14*(4), 446–465. https://doi.org/10.1111/1541-4337.12136

Hosseini, S. F., Rezaei, M., & McClements, D. J. (2020). Bioactive functional ingredients from aquatic origin: A review of recent progress in marine-derived nutraceuticals. *Critical Reviews in Food Science and Nutrition, 62*(5), 1242–1269. https://doi.org/10.1080/10408398.2020.1839855

Huang, I., Pietrasiak, N., Gobler, C. J., Johansen, J. R., Burkholder, J. M., D'Antonio, S., & Zimba, P. V. (2021). Diversity of bioactive compound content across 71 genera of marine, freshwater, and terrestrial cyanobacteria. *Harmful Algae, 109*, 102116. https://doi.org/10.1016/j.hal.2021.102116

Imchen, T., & Singh, K. S. (2022). Marine algae colorants: Antioxidant, anti-diabetic properties and applications in food industry. *Algal Research, 69*, 102898. https://doi.org/10.1016/j.algal.2022.102898

Jaiswal, D. K., Krishna, R., Chouhan, G. K., De Araujo Pereira, A. P., Ade, A. B., Prakash, S., Verma, S. K., Prasad, R., Yadav, J., & Verma, J. P. (2022). Bio-fortification of minerals in crops: Current scenario and future prospects for sustainable agriculture and human health. *Plant Growth Regulation, 98*(1), 5–22. https://doi.org/10.1007/s10725-022-00847-4

Jo, C., Khan, F. F., Khan, M. I., & Iqbal, J. (2016). Marine bioactive peptides: Types, structures, and physiological functions. *Food Reviews International, 33*(1), 44–61. https://doi.org/10.1080/87559129.2015.1137311

Letsiou, S., Kalliampakou, K., Gardikis, K., Mantecon, L., Infante, C., Chatzikonstantinou, M., Labrou, N. E., & Flemetakis, E. (2017). Skin protective effects of Nannochloropsis gaditana extract on H2O2-stressed human dermal fibroblasts. *Frontiers in Marine Science, 4*. https://doi.org/10.3389/fmars.2017.00221

Lobine, D., Rengasamy, K. R. R., & Mahomoodally, M. F. (2021). Functional foods and bioactive ingredients harnessed from the ocean: Current status and future perspectives. *Critical Reviews in Food Science and Nutrition, 62*(21), 5794–5823. https://doi.org/10.1080/10408398.2021.1893643

Mahapatra, G. P., Raman, S., Nayak, S., Gouda, S., Das, G., & Patra, J. K. (2019). Metagenomics approaches in discovery and development of new bioactive compounds from marine actinomycetes. *Current Microbiology, 77*(4), 645–656. https://doi.org/10.1007/s00284-019-01698-5

Mannar, M. V., & Hurrell, R. F. (2018). Food fortification. In *Elsevier eBooks* (pp. 3–11). https://doi.org/10.1016/b978-0-12-802861-2.00001-8

Morya, S., Chattu, V. K., Khalid, W., Khalid, M. Z., & Siddeeg, A. (2023). Potential protein phycocyanin: An overview on its properties, extraction, and utilization. *International Journal of Food Properties, 26*(2), 3160–3176. https://doi.org/10.1080/10942912.2023.2271686

Muñoz-Miranda, L. A., & Iñiguez-Moreno, M. (2023). An extensive review of marine pigments: Sources, biotechnological applications, and sustainability. *Aquatic Sciences, 85*(3). https://doi.org/10.1007/s00027-023-00966-8

Nadzir, M. M., Nurhayati, R. W., Idris, F. N., & Nguyen, M. H. (2021). Biomedical applications of bacterial exopolysaccharides: A review. *Polymers, 13*(4), 530. https://doi.org/10.3390/polym13040530

Ofori, K. F., Antoniello, S., English, M. M., & Aryee, A. N. A. (2022). Improving nutrition through biofortification–A systematic review. *Frontiers in Nutrition, 9*. https://doi.org/10.3389/fnut.2022.1043655

Page, M. J., McKenzie, J. E., Bossuyt, P. M., Boutron, I., Hoffmann, T. C., Mulrow, C. D., Shamseer, L., Tetzlaff, J. M., Akl, E. A., Brennan, S. E., Chou, R., Glanville, J., Grimshaw, J. M., Hróbjartsson, A., Lalu, M. M., Li, T., Loder, E. W., Mayo-Wilson, E., McDonald, S., et al. (2021). The PRISMA 2020 statement: An updated guideline for reporting systematic reviews. *BMJ*, n71. https://doi.org/10.1136/bmj.n71

Panossian, A. G., Efferth, T., Shikov, A. N., Pozharitskaya, O. N., Kuchta, K., Mukherjee, P. K., Banerjee, S., Heinrich, M., Wu, W., Guo, D., & Wagner, H. (2020). Evolution of the adaptogenic concept from traditional use to medical systems: Pharmacology of stress- and aging-related diseases. *Medicinal Research Reviews, 41*(1), 630–703. https://doi.org/10.1002/med.21743

Patel, N., Dwivedi, M., Jadeja, S., & Begum, R. (2020). Antibacterial activity of marine bacterial pigments obtained from Arabian Sea water samples. *Journal of Pure and Applied Microbiology, 14*(1), 517–526. https://doi.org/10.22207/jpam.14.1.54

Patel, N., Maharshi, A., & Gohil, N. (2024). Microbial pigments: A potential substitute of synthetic colorants in the food and healthcare sectors. In *Microbial products for health and nutrition* (pp. 191–220). https://doi.org/10.1007/978-981-97-4235-6_8

Pati, S., Sarkar, T., Sheikh, H. I., Bharadwaj, K. K., Mohapatra, P. K., Chatterji, A., Dash, B. P., Edinur, H. A., & Nelson, B. R. (2021). γ-Irradiated chitosan from Carcinoscorpius rotundicauda (Latreille, 1802) improves the shelf life of refrigerated aquatic products. *Frontiers in Marine Science, 8*. https://doi.org/10.3389/fmars.2021.664961

Perković, L., Djedović, E., Vujović, T., Baković, M., Paradžik, T., & Čož-Rakovac, R. (2022). Biotechnological enhancement of probiotics through co-cultivation with algae: Future or a trend? *Marine Drugs, 20*(2), 142. https://doi.org/10.3390/md20020142

Qeshmi, F. I., Homaei, A., Fernandes, P., Hemmati, R., Dijkstra, B. W., & Khajeh, K. (2019). Xylanases from marine microorganisms: A brief overview on scope, sources, features and potential applications. *Biochimica Et Biophysica Acta (BBA) – Proteins and Proteomics, 1868*(2), 140312. https://doi.org/10.1016/j.bbapap.2019.140312

Rayapu, L., Makkar, F., Chakraborty, K., & Valluru, L. (2017). Phytochemical evaluation and antimicrobial activity of gracilaria opuntia: An important anti-diabetic red marine macroalgae. *International Journal of Current Pharmaceutical Research, 9*(6), 37. https://doi.org/10.22159/ijcpr.2017v9i6.23426

Riaz, U., Aziz, H., Anum, W., Mehdi, S. M., Murtaza, G., & Jamil, M. (2020). Biofortification technologies used in agriculture in relation to micronutrients. In *Springer eBooks* (pp. 225–239). https://doi.org/10.1007/978-3-030-49856-6_9

Rohner, F., Wirth, J. P., Zeng, W., Petry, N., Donkor, W. E., Neufeld, L. M., Mkambula, P., Groll, S., Mbuya, M. N., & Friesen, V. M. (2023). Global coverage of mandatory large-scale food fortification programs: A systematic review and meta-analysis. *Advances in Nutrition, 14*(5), 1197–1210. https://doi.org/10.1016/j.advnut.2023.07.004

Sarker, P. K. (2023). Microorganisms in fish feeds, technological innovations, and key strategies for sustainable aquaculture. *Microorganisms, 11*(2), 439. https://doi.org/10.3390/microorganisms11020439

Sathasivam, R., & Ki, J. (2018). A review of the biological activities of microalgal carotenoids and their potential use in healthcare and cosmetic industries. *Marine Drugs, 16*(1), 26. https://doi.org/10.3390/md16010026

Senadheera, T. R. L., Hossain, A., & Shahidi, F. (2023). Marine bioactives and their application in the food industry: A review. *Applied Sciences, 13*(21), 12088. https://doi.org/10.3390/app132112088

Shahidi, F., & Ambigaipalan, P. (2015). Novel functional food ingredients from marine sources. *Current Opinion in Food Science, 2*, 123–129. https://doi.org/10.1016/j.cofs.2014.12.009

Shin, D., Byun, W. S., Moon, K., Kwon, Y., Bae, M., Um, S., & Oh, D. C. (2018). Coculture of marine Streptomyces sp. with Bacillus sp. produces a new piperazic acid-bearing cyclic peptide. *Frontiers in chemistry, 6*, 498. https://doi.org/10.3389/fchem.2018.00498

Sy, C., Dangles, O., Borel, P., & Caris-Veyrat, C. (2015). Interactions between carotenoids from marine bacteria and other micronutrients: Impact on stability and antioxidant activity. *Marine Drugs, 13*(11), 7020–7039. https://doi.org/10.3390/md13117020

Virk, P. S., Andersson, M. S., Arcos, J., Govindaraj, M., & Pfeiffer, W. H. (2021). Transition from targeted breeding to mainstreaming of biofortification traits in crop improvement programs. *Frontiers in Plant Science, 12*. https://doi.org/10.3389/fpls.2021.703990

Wakeel, A., Farooq, M., Bashir, K., & Ozturk, L. (2018). Micronutrient malnutrition and biofortification: Recent advances and future perspectives. In *Elsevier eBooks* (pp. 225–243). https://doi.org/10.1016/b978-0-12-812104-7.00017-4

Wang, C., Tang, S., & Cao, S. (2020). Antimicrobial compounds from marine fungi. *Phytochemistry Reviews, 20*(1), 85–117. https://doi.org/10.1007/s11101-020-09705-5

Wang, X., Zhang, Z., Zhang, S., Yang, F., Yang, M., Zhou, J., Hu, Z., Xu, X., Mao, G., Chen, G., Xiang, W., Sun, X., & Xu, N. (2021). Antiaging compounds from marine organisms. *Food Research International, 143*, 110313. https://doi.org/10.1016/j.foodres.2021.110313

Wróbel-Biedrawa, D., & Podolak, I. (2024). Anti-neuroinflammatory effects of adaptogens: A mini-review. *Molecules, 29*(4), 866. https://doi.org/10.3390/molecules29040866

Zhang, Q., Liang, H., Longshaw, M., Wang, J., Ge, X., Zhu, J., Li, S., & Ren, M. (2022). Effects of replacing fishmeal with methanotroph (Methylococcus capsulatus, Bath) bacteria meal (FeedKind®) on growth and intestinal health status of juvenile largemouth bass (Micropterus salmoides). *Fish & Shellfish Immunology, 122*, 298–305. https://doi.org/10.1016/j.fsi.2022.02.008

Zhang, J., Xiao, Y., Wang, H., Zhang, H., Chen, W., & Lu, W. (2023). Lactic acid bacteria-derived exopolysaccharide: Formation, immunomodulatory ability, health effects, and structure-function relationship. *Microbiological Research, 274*, 127432. https://doi.org/10.1016/j.micres.2023.127432

MIX
Papier aus verantwortungsvollen Quellen
Paper from responsible sources
FSC® C105338

If you have any concerns about our products,
you can contact us on
ProductSafety@springernature.com

In case Publisher is established outside the EU,
the EU authorized representative is:
Springer Nature Customer Service Center GmbH
Europaplatz 3, 69115 Heidelberg, Germany

Printed by Libri Plureos GmbH
in Hamburg, Germany